AF527282

THEORY AND DESIGN OF PRESSURE VESSELS

THEORY AND DESIGN OF PRESSURE VESSELS

John F. Harvey, P.E.

CBS Publishers & Distributors Pvt. Ltd.

New Delhi • Bengaluru • Chennai • Kochi • Kolkata • Mumbai
Hyderabad • Uttarakhand • Nagpur • Patna • Pune • Jharkhand

ISBN: 81-239-1041-X

First Indian Edition: 1987
Reprint: 2001

Copyright © 1985 by Van Nostrand Reinhold Company, Inc.

Reprinted in India by CBS Publishers & Distributors in arrangement with Van Nostrand Reinhold Company, Inc.

All rights reserved. No part of this book may be reproduced or transmitted in any form or by any means, electronic or mechanical, including photocopying, recording, or any information storage and retrieval system without permission, in writing, from the publisher.

Published by **Satish Kumar Jain** and produced by **Varun Jain** for
CBS Publishers & Distributors Pvt. Ltd.,
4819/XI Prahlad Street, 24 Ansari Road, Daryaganj, New Delhi - 110002
delhi@cbspd.com, cbspubs@airtelmail.in • www.cbspd.com
Ph.: 23289259, 23266861, 23266867 • Fax: 011-23243014

Corporate Office: 204 FIE, Industrial Area, Patparganj, Delhi - 110 092
Ph: 49344934 • Fax: 011-49344935
E-mail: publishing@cbspd.com • publicity@cbspd.com

Branches:

- ***Bengaluru:*** 2975, 17th Cross, K.R. Road, Bansankari 2nd Stage, Bengaluru - 70 • Ph: +91-80-26771678/79 • Fax: +91-80-26771680 E-mail: cbsbng@gmail.com, bangalore@cbspd.com
- ***Chennai:*** No. 7, Subbaraya Street, Shenoy Nagar, Chennai - 600030 Ph: +91-44-26681266, 26680620 • Fax: +91-44-42032115 E-mail: chennai@cbspd.com
- ***Kochi:*** Ashana House, 39/1904, A.M. Thomas Road, Valanjambalam, Ernakulum, Kochi • Ph: +91-484-4059061-65 Fax: +91-484-4059065 • E-mail: cochin@cbspd.com
- ***Kolkata:*** 6-B, Ground Floor, Rameshwar Shaw Road, Kolkata - 700014 Ph: +91-33-22891126/7/8 • E-mail: kolkata@cbspd.com
- ***Mumbai:*** 83-C, Dr. E. Moses Road, Worli, Mumbai - 400018 Ph: +91-9833017933, 022-24902340/41 • E-mail: mumbai@cbspd.com

Representatives:

- Bhubaneswar 0-9911037372 • Hyderabad 0-9885175004 • Jharkhand 0-9811541605
- Nagpur 0-9021734563 • Patna 0-9334159340 • Pune 0-9623451994
- Uttarakhand 0-9716462459 • Dhaka (Bangladesh) 01912-003485

Printed at:
Neekunj Print Process, Delhi (India)

PREFACE

This book is intended to serve as an introduction to pressure vessel design for the student by presenting the fundamental theory, and as a design basis for the practicing engineer by demonstrating the application of this theory with illustrations and examples of the solution of practical problems. In each chapter, design methods are illustrated by numerical problems in which the arithmetical work has been made simple in order to focus attention upon the theory involved, thereby avoiding the danger of it becoming merely a mechanical operation with an incomplete understanding of the theory and its significance. A considerable amount of new material and references have been added in a manner so as to preserve an orderly presentation. The result has been a complete revision and substantial increase in the size of original text, including a new chapter on buckling collapse. The latter has been prompted by the rapid growth in undersea exploration and mineral mining potential. Deep diving submersibles, offshore platforms, and deep drilling casings have made a priority of light weight, high strength materials in order to save weight and costs. The result is that elastic and plastic buckling instability have become increasingly important as failure modes.

Chapter 1 introduces the basic design considerations, analytical and experimental stress analysis methods, and the interaction of material properties and fabrication methods. The significance and consequences of applied and residual stresses under service loading and operating environment are emphasized. Pressure vessel terminology is presented, and basic membrane ligament theory developed.

Chapter 2 covers the basic theory of membrane stress and deflection analysis of axisymmetric vessels and its application to commonly encountered cylinders, spheres, ellipsoids, cones, and tori. An analysis of intersecting spheres and diaphragm vessels, upon which many deep diving oceanographic submersibles are based, is presented. Correspondingly, instability in the knuckle region of large, shallow, dished heads is investigated because of their extensive use. Also, an analysis of vessels for ultra-high pressure, employing thick-wall, multilayer, cascade, segmented, yoke, and wedge principles of design is presented. An introduction to steady-state and transient thermal stresses and their evaluation in terms of the full restraint thermal stress by analytical and graphical methods is given.

Chapter 3 covers the bending of flat, solid, circular plates under uniform and concentrated loading, and the effect of local flexibility at clamped-edge supports. Likewise, the design of stacked and built-up plates is developed and the design of orthogonal and concentric reinforced and perforated plates is analyzed together with expanded and welded tube-to-tubesheet joints. Expanded tube joints, wherein union is achieved through residual stress, are treated in depth. Not only is this the oldest and most widely used joining method, but it is the only one available for many pressure components employing unweldable dissimilar materials.

Chapter 4 presents the determination by the elastic foundation method of secondary bending and direct stresses encountered in vessels as a result of differential dilation of their parts. An analysis of bimetallic joints, frequently used in the chemical, petroleum and power industry, and their optimum location is developed. Likewise, an introduction to flange design is presented, together with a discussion of elastic foundation attenuation factors and their part in appraising the extent of secondary stresses and opening reinforcement limits in vessels.

Chapter 5 discusses the failure analysis and failure prevention of materials in their environment, for example, irradiation damage, hydrogen embrittlement, elevated and cryogenic temperatures, the multitudinous effects on fatigue life, etc. The economic trend to high strength materials, with their associated susceptibility to brittle fracture, has focused considerable attention to this subject. Applicable theories of failure and fracture mechanics analysis methods are presented to predict their behavior. This involves a basic understanding of the structure of metals and their elastic and plastic behavior. Fatigue is a prime cause of vessel failure, but it is completely amenable to prevention; accordingly, low and high cycle fatigue behavior, life prediction and damage accumulation are given together with a fatigue theory to account for multidirectional applied and residual stress conditions. High temperatures in the petroleum, chemical, and power industries have accentuated the problem of creep and rupture. Hence, life and strain fraction concepts for predicting service life, methods of estimating the relaxation of bolted joints, and the effect of the thermal stress relief of residual stresses in welded vessels to enhance safety are developed. Ways of establishing optimum metal forming temperatures for fabrication are presented, as are methods of evaluating material degradation at elevated temperatures. High temperatures can also produce opposite environmental effects; for instance, they increase the hydrogen embrittlement of petroleum processing vessels, while mitigating the neutron embrittlement of nuclear reactor vessels. All products have flaws which originate with the parent material, result from the fabrication method, or are designed-in by construction details. Hence it is important to appreciate this reality and appraise its significance for the intended service. In this respect, crack acceptance criteria, defect size evaluation, and design methods for attaining the required vessel life are discussed.

Chapter 6 covers design-construction features and their effect on geometric and thermally induced stress concentrations. Particular attention is given to those encountered under static and dynamic conditions at vessel openings, nozzles and structural supports and the means of coping with them to insure maximum integrity. The theory and practice of reinforced openings for radial, nonradial, and multiple nozzle arrangements are covered, as is designing their thermal sleeves which are vital to vessel thermal-shock protection. Stresses in bolts and the design of bolted and nonbolted closures receive extensive treatment as do the use of crack arrest features to negate brittle fracture potential. Vessel support skirts, saddles, and attachments impose both structural and thermal loadings that must be reconciled with the vessel pressure stresses. Gaskets are essential to removable closures and their type, design, and control parameters are presented. Thin claddings are widely used to prevent corrosion of the base metal or contamination of the media, and their life is appraised.

Chapter 7 introduces fabrication-construction methods and their economic potential. This includes novel innovations to satisfy a unique requirement, such as filament-wound, multilayer, wire-wrapped, link-belt, coiled, prestressed steel and concrete, and other vessels. Material selection is by basic cost per unit of stress and optimum safety factors utilizing advanced composites, high strength materials, or those with enhanced properties such as that obtained from directional solidification for use in a creep rupture environment. Fabrication methods like modular construction, cryogenic and high energy forming, adhesive joints, metallurgical bonding and healing of internal defects by hot isostatic pressing, etc.—all to achieve the economic goal of low cost and long service life—are presented.

Chapter 8 introduces the stability theory of plain and stiffened vessels, and its application to the design of thin, intermediate and thick walled ones. The overall stabilizing effect of structural stiffeners, as well as their effect in regions of high local compressive stress, is covered from initial design, failure analysis, and repair viewpoints. This forms the basis for the development of regulatory codes and compliance standards. The fabrication tolerances and construction details of noncircularity, local thinning, unreinforced openings, and stress concentrators are not self-limiting or self-compensating in a buckling phenomenon, and these critical effects are evaluated.

In the preparation of this book, the author has drawn from his previous publications, lecture notes used in teaching at the University of Akron, engineering experience with The Babcock & Wilcox Company, and as a member of numerous committees of the American Society of Mechanical Engineers, and as chairman of the Pressure Vessel Research Committee.

JOHN F. HARVEY

NOTATION

A	Cross-sectional area
B	Distance
a,b,c,d	Distances
D	Flexural rigidity
d	Diameter
E	Modulus of elasticity
e	Unit strain
e_x,e_y,e_z	Unit strains in x, y and z directions
$e_{\text{Y.P.}}$	Yield point strain
F	Force
H	Height, thickness
h	Thickness
I	Moment of inertia
J	Polar moment of inertia
K_t	Stress concentration factor
K_f	Fatigue strength reduction factor
k	Modulus of foundation, stress
L	Distance, span
l	Length, span
M	Bending moment
M_R	Resisting bending moment
M_{ult}	Ultimate bending moment
$M_{\text{Y.P.}}$	Bending moment at which yielding begins
M_t	Twisting moment
P,Q	Concentrated forces
p	Pressure
q	Load per unit length, sensitivity factor
R	Radius
r	Radius, radius of curvature
s	Length
T	Twisting moment, temperature
u	Rate of strain, displacement in x direction

V	Shearing force, velocity
W	Weight
w	Displacement in z direction
x,y,z	Rectangular coordinates
Z	Section modulus
ksi	Kips per square inch (1 kip = 1,000 pounds)
psi	Pounds per square inch
α	Angle, coefficient of thermal expansion, numerical coefficient
β	Angle, numerical coefficient
γ	Shearing strain
Δ	Distance, deflection, difference
δ	Total elongation, total deflection, distance
θ	Angle
μ	Poisson's ratio
ρ	Distance, radius
σ	Unit normal stress
$\sigma_1, \sigma_2, \sigma_3$	Principal stresses
$\sigma_x, \sigma_y, \sigma_z$	Unit normal stresses on planes perpendicular to the x, y, and z axes
σ_E	Unit stress at endurance limit
σ_{ult}	Ultimate stress
$\sigma_{Y.P.}$	Yield point stress
τ	Unit shear stress
$\tau_{xy}, \tau_{yz}, \tau_{zx}$	Unit shear stresses on planes perpendicular to the x, y, and z axes and parallel to the y, z, and x axes
$\tau_{Y.P.}$	Yield point stress in shear
ϕ	Angle

CONTENTS

THEORY AND DESIGN OF PRESSURE VESSELS

1

Pressure Vessels

1.1. Introduction

Pressure vessels are leakproof containers. They may be of any shape and range from beverage bottles to the sophisticated ones encountered in engineering construction, Fig. 1.1. In the latter, high pressures, extremes of temperature, and severity of functional performance requirements pose exacting design problems. The word "design," as used here, does not mean only the calulation of the detail dimensions of a member, but rather is an all-inclusive term incorporating: (1) the reasoning that established the most likely mode of damage or failure, (2) the method of stress analysis employed and significance of results, and (3) the selection of material type and its environmental behavior. New concepts in design and new materials available for construction are challenging the ingenuity of engineers, and the resulting complexity of problems arising from every area affect both safety and cost effectiveness.

The ever-increasing use of vessels for storage, industrial processing, and power generation under unusual conditions of pressure, temperature, and environment has given special emphasis to analytical and experimental methods for determining their operating stresses. Of equal importance is appraising the meaning or significance of these stresses. This appraisal entails means of determining the value and extent of the stresses and strains, establishing the behavior of the material involved, and evaluating the compatibility of these two factors in the media or environment to which they are subjected. A knowledge of material behavior is required not only to avoid failures, but equally to permit maximum economy of material choice and amount used. For instance, if the stresses or strains in a structure are unduly low, its size becomes larger than necessary and the economic potential of the material is not reached. Developments in the space, nuclear, and chemical industries have placed new demands on materials suitable for extremes in temperature, impact, and fatigue. Sometimes these appli-

(a)

Fig. 1.1. (*a*) Oil Refinery (Courtesy Standard Oil of Indiana. (*b*) Steam Generator for a Nuclear Power Plant. (*c*) Boiler Drum in Fabrication. (*d*) Extended Surface U-Bend Heater Section (*Courtesy The Babcock & Wilcox Company*)

cations also require consideration of other environmental effects, such as corrosion, neutron bombardment, hydrogen embrittlement, etc. The characteristics of materials when subjected to the action of stresses and strains are called mechanical properties.

The stresses, i.e., forces acting on vessels, produce changes in their dimensions known as strains. The determination of the relationship between the external forces applied to a vessel and the stresses and strains within the vessel form the basis of this field of stress analysis. The basic interaction of stresses and strains is well illustrated by the conventional tensile test specimen, Fig. 1.2, from which come the basic data required to pursue this stress analysis. In Fig. 1.2*a* the axial external force P is resisted by uniformly distributed internal forces over an area A_0, called stress, of magnitude

$$\sigma = \frac{P}{A_0} \tag{1.1.1}$$

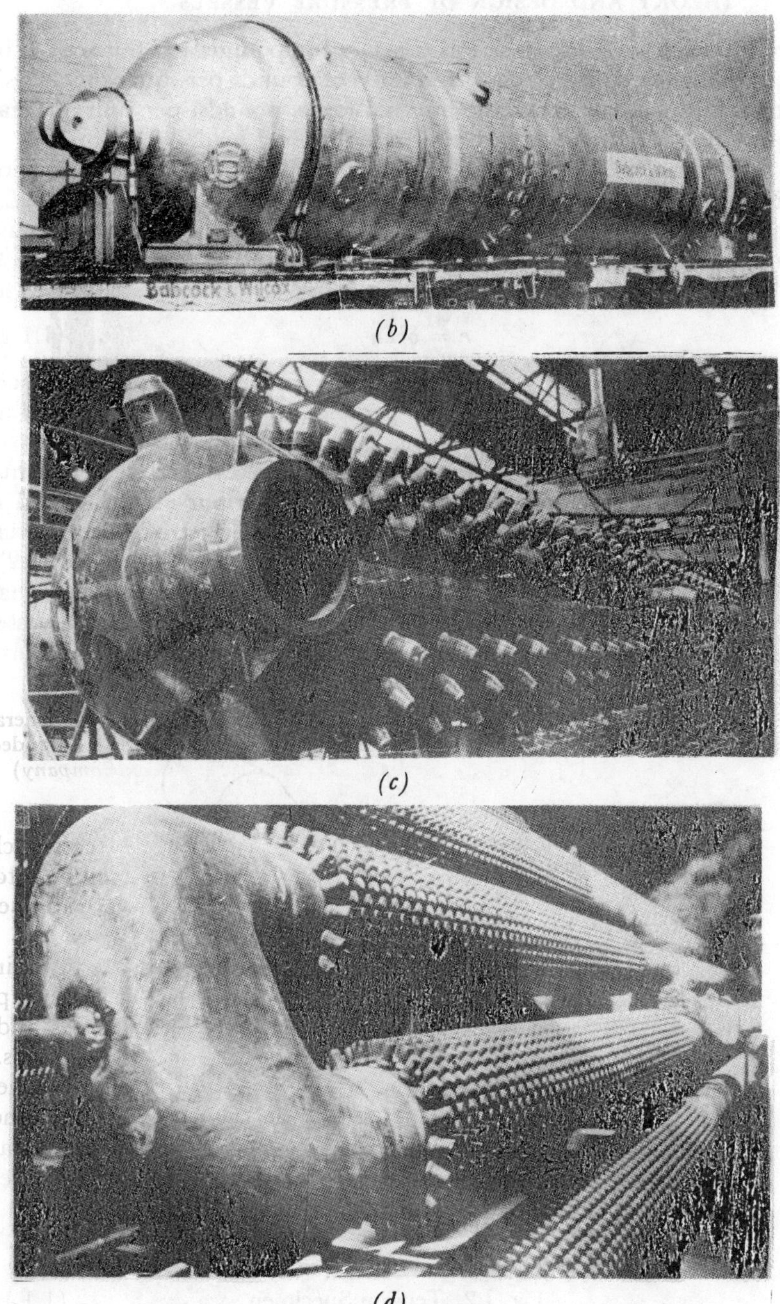

(b)

(c)

(d)

Fig. 1.1. (*Continued*)

In Equation 1.1.1 the force P is measured in pounds and the area A_0 in square inches, so the stress σ is in units of pounds per square inch, psi. Stress may be defined as the internal force, pounds, per unit of area, square inch, induced by an externally applied load.

The force P also produces a stretching or elongation of the specimen, Fig. 1.2*b*. Since the stress is uniform throughout the gage length L_0, and on the basis that the material is perfectly homogeneous, it is assumed that this uniform stress will produce uniform total elongation δ. The elongation per unit of length e is called strain and is expressed as

$$e = \delta / L_0 \tag{1.1.2}$$

The elongation δ and gage length L_0 are measured in inches; hence, the strain e is measured in inches per inch. Strain may be defined as the change in unit length resulting from stress.

Basic analytical stress solutions involve the stress-strain relationship of the material. The relationship differs for various materials and is determined by a simple tension or compression test of the material from which a stress-strain diagram is plotted (see paragraph 5.2). Figure 1.3 shows several typical types of diagrams. Figure 1.3*a* is that for a ductile steel as commonly used in structural and pressure vessel design in which a large deformation is produced prior to rupture.

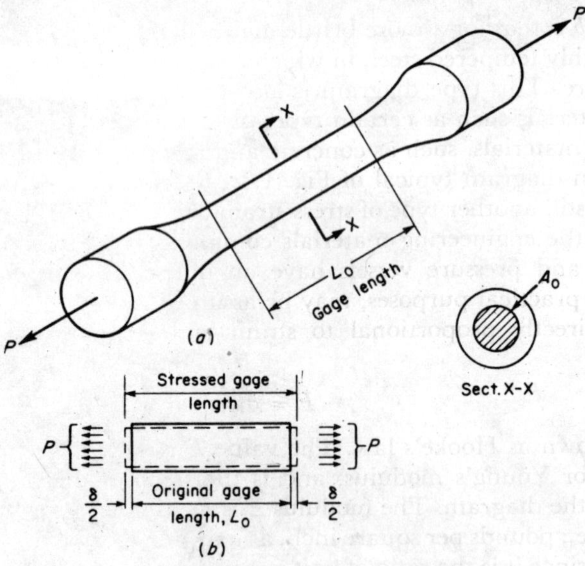

Fig. 1.2. Tension Specimen

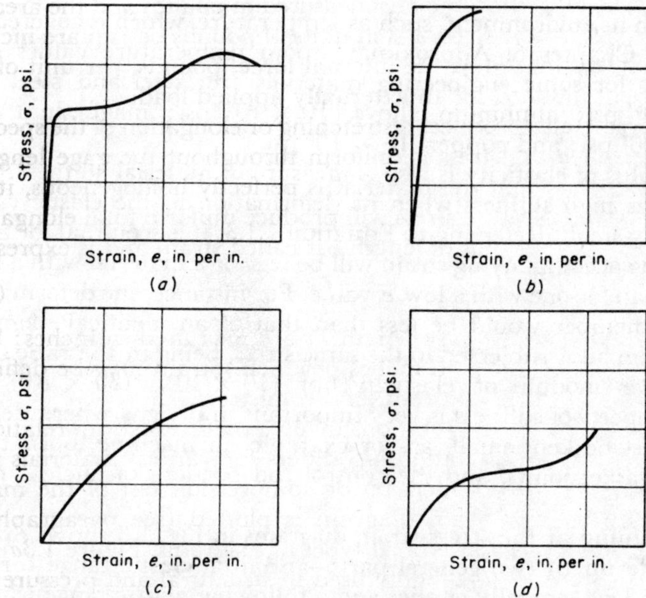

Fig. 1.3 Some Typical Types of Stress-Strain Diagrams

Figure 1.3*b* is that for a more brittle material as would be characteristic of a highly tempered steel, in which a small deformation occurs before rupture. This type diagram is also typical for some nonmetallic brittle materials such as certain types of plastics and plaster of Paris. Still other materials, such as concrete and malleable cast iron, have a stress-strain diagram typical of Fig. 1.3*c*. Rubber, a very elastic material, has still another type of stress-strain curve as shown in Fig. 1.3*d*. However, the engineering materials commonly used in the design of structures and pressure vessels have an initial stress-strain relation which, for practical purposes, may be assumed linear, indicating that stress is directly proportional to strain and is represented by the equation

$$E = \sigma/e \tag{1.1.3}$$

This is known as Hooke's law. The value E is called the modulus of elasticity, or Young's modulus, and is the slope of this straight line portion of the diagram. The modulus E is measured in the same units as stress; i.e., pounds per square inch, as may be concluded from Equation 1.1.3, since it is the ratio of unit stress, psi, and unit strain, in./in., which is a pure number. This value varies with the type of material, and with its environment, such as temperature, which is covered more fully in Chapter 5. Approximate room temperature values of this modulus for some engineering materials are: steel and steel alloys, 30×10^6 psi; aluminum alloys, 10×10^6 psi; magnesium alloys, $6 \cdot 5 \times 10^6$ psi; and copper, 16×10^6 psi.

Modulus of elasticity is also a measure of material stiffness. A material has high stiffness when its deformation in the elastic range is relatively small. Referring to Equation 1.1.3, it is seen that for a given stress, the accompanying strain will be less for a material with a high E value than for one with a low E value. For instance, the deformation of a steel member would be less than that of an identical member of aluminum alloy subjected to the same stress, being in the ratio of their respective moduli of elasticity; or $(10 \times 10^6) : (30 \times 10^6) = 1{:}3$. This property of stiffness is very important in designs where deformation must be kept small, as, for example, in machine tools, turbine rotors, gasket joints, and the control rod portions of nuclear reactor vessels.

Examining of the stress-strain diagrams in Fig. 1.3 shows that they are made up of two general parts—an initial elastic range for which Hooke's law generally applies and a following plastic range where the strains become large and this law no longer applies. The elastic and plastic ranges for a ductile steel commonly used in engineering constructions are given in Fig. 1.4.

Elasticity is the property of a material to return to its original shape after removal of the load. The elastic range is the first stage of loading wherein the material returns to its original shape after unloading. Most engineering designs require that permanent deformations be avoided in order to assure proper functional performance and continuous reliable service; hence, it is desirable to define the more important mechanical properties covering the elastic range.

Figure 1.5*a* is an enlargement of the elastic range of the stress-strain diagram of Fig. 1.4. The following mechanical properties are descriptive of this range.

Proportional limit is the greatest stress, as represented by point a in Fig. 1.5*a*, that a material can withstand without deviating from the direct proportionality of stress to strain.

Elastic limit is the maximum stress, as represented by point b in Fig. 1.5*a*, which a material is capable of withstanding without permanent deformation upon complete release of the stress. Determining the elastic limit is very difficult and requires the use of sensitive strain measuring instruments. However, it closely approximates the more readily determined value of the proportional limit.

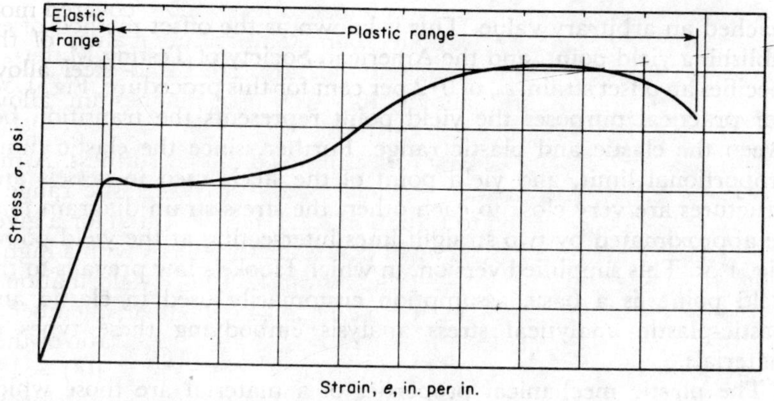

Fig. 1.4. Elastic and Plastic Strain Range

Yield point is the stress, represented by point *c* in Fig. 1.5*a*, at which there occurs a marked increase in strain without an increase in stress. This phenomenon of yielding is due to a sudden plastic flow of the material and is associated with the slippage along planes of weakness of unfavorably located individual crystals of the material, discussed subsequently in Chapter 5. This value is determined from the stress-strain diagram. Some materials have a diagram as shown in Fig. 1.5*a* with an upper yield point, *c*, and a lower yield point, *d*. It is readily evidenced during testing by a drop in the beam or halt in the dial of the testing machine. Other materials do not have such definite properties, and for these it is the customary practice to define the yield strength as that stress where a permanent set or deformation has

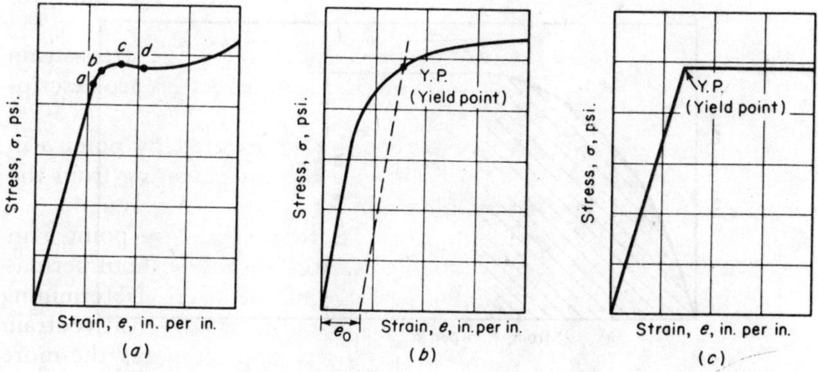

Fig. 1.5. Determination of Elastic Strength

reached an arbitrary value. This is known as the offset method of establishing yield point, and the American Society of Testing Materials specifies an offset strain, e_0, of 0·2 per cent for this procedure, Fig. 1.5*b*. For practical purposes the yield point represents the transition between the elastic and plastic range. Further, since the elastic limit, proportional limit, and yield point of the steels used in vessels and structures are very close to each other, the stress-strain diagram may be approximated by two straight lines intersecting at the yield point, Fig. 1.5*c*. This simplified version, in which Hooke's law prevails to the yield point, is a basic assumption customarily used in elastic and elastic-plastic analytical stress analysis embodying these types of materials.

The plastic mechanical properties of a material are those which measure its ability to resist rupture, undergo deformation, and absorb

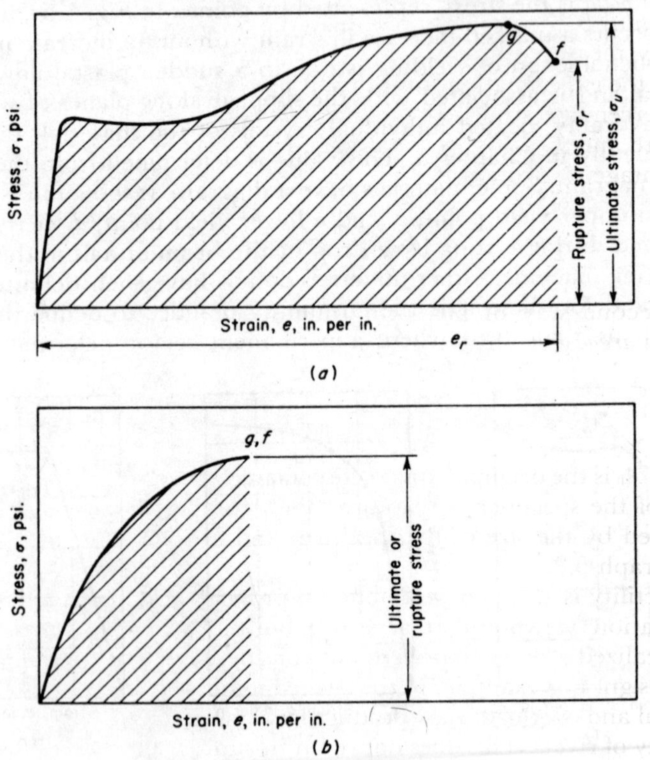

Fig. 1.6. Determination of Plastic Strength

energy. These are obtained from the same stress-strain diagram of the material as are the elastic mechanical properties. The three prime properties that describe the plastic range of a material are defined and described in the following.

Ultimate strength is the maximum stress, as represented by point g, Fig. 1.6*a*, that the material will withstand. This stress equals the maximum load divided by the original cross-sectional area of the specimen. Occasionally, the rupture stress, as represented by point f, Fig. 1.6*a*, is used to measure plastic strength. The rupture stress is equal to the rupture load divided by the original cross-sectional area of the specimen. For many ductile materials these two values are close; for brittle materials they are identical, Fig. 1.6*b*. Ultimate strength is an important property in all engineering designs. This is particularly so in pressure vessels where the material is under tension, much as it is in a tensile test specimen, and under conditions of static loading it establishes the bursting pressure of the vessel.

Ductility is the property of a material to undergo deformation. It is usually measured in two ways. One way is by the *percentage elongation* of the gage length at time of rupture, which is also the corresponding percentage strain. If, in Fig. 1.2, L_r is the final gage length at rupture and L_0 is the original gage length of the specimen, then the percentage elongation is

$$d_e = \frac{L_r - L_0}{L_0} 100 \tag{1.1.4}$$

The second way of measuring ductility is by the *percentage reduction in area*, given by

$$d_a = \frac{A_0 - A_r}{A_0} 100 \tag{1.1.5}$$

where A_0 is the original cross-sectional area and A_r is the cross-sectional area of the specimen at rupture. These properties are somewhat influenced by the size of the specimen and are discussed more fully in paragraph 5.2.

Ductility is an important material property from both a design and fabrication viewpoint. It acts as a built-in excessive stress adjuster for localized stresses that were not considered or not contemplated in the design. For example, in the riveted joints of bridges and buildings, normal and overloads may produce local yielding of the member in the vicinity of rivets. This does not result in rupture, however, since structural steel has a high ductility which results in a redistribution and reduction in the local stresses occurring at points of stress concentration (Chapter 6). Similarly, in pressure vessels at joints, nozzles, openings, etc., high stresses develop locally which are subsequently reduced by plastic flow of the material. Ductility is also an important material property in fabrication and processing, such as rolling, forging, drawing, and extruding. If the ductility is not adequate, the large deformations produced in these operations result in rupture of the material. For instance, the steel plates used to fabricate the shells and heads of the vessels shown in Figs. 1.1 and 1.16 must have sufficient ductility to permit them to be bent from a flat plate to their final curvatures without cracking. Frequently it is necessary to heat the material to high temperatures to increase its ductility during the forming operation, paragraph 5.20.

Toughness is the ability of a material to absorb energy during plastic deformation. It is often measured by the energy absorbed per unit of volume in stressing to rupture and is called the *modulus of toughness.* Since total energy is equal to force times distance, the energy per unit of volume is equal to stress times strain and is in units of inch-pounds per cubic inch. Accordingly, the modulus of toughness, or total energy to rupture, is

$$T_0 = \int_0^{e_r} \sigma \, de \tag{1.1.6}$$

which is the total area under the stress-strain diagram, Fig. 1.6. This area can be determined by a planimeter, or by approximate methods. One convenient method that is employed for many ductile materials is to use the product of the ultimate stress times the strain at rupture, $\sigma_u \, e_r$ as an approximate measure of the area under the stress-strain diagram. Materials of high toughness have high strength, as well as large ductility. Brittle materials have low toughness, since they have only small plastic deformations before rupture. Toughness is a most desirable property in parts, structures, or vessels subject to mechanical or thermal shock.

A particular type of toughness is that measured in the presence of a notch, and is called "notch toughness." When the notch is a very sharp one, a crack, this material property becomes the basis for determining the stresses in the region of a crack called Fracture Mechanics, Par. 5.22.

1.2 Methods for Determining Stresses

Stress analyses can be performed by analytical or experimental means. The analytical method involves a rigorous mathematical solution based on an applicable theory of elasticity, plasticity, creep, etc., and it is the most direct and inexpensive approach when the problem adapts itself to such a solution. It requires a general solution to the equilibrium equation equating internal resisting stresses of the material and that imposed by the gradually applied static external forces. When the member is a simple one such that the mathematical equation can be readily written to describe the continuous strain behavior of the material throughout, the ordinary equations for direct stress, P/A, bending stress, Mc/I, and torsional shear stress, Tc/J, can be developed (where P is the load applied at the centroid of the cross section of area A which is under consideration; M is the bending moment applied to the cross section which has a moment of inertia I about its neutral axis and c is the distance from this axis, and T is the twisting moment acting on the cross section which has a polar moment of inertia J). However, when the member has geometric shape discontinuities, such as internal holes or external abrupt cross-section changes, it becomes difficult to express the continuous internal strain distribution mathematically and obtain a particular solution to the equations. When the problem is too complex and beyond analytical solution, or when a check or evaluation of an analytical solution is desired, recourse must be made to experimental means.[1,20] Three of the more commonly used methods follow.

1. *Strain Gage*

This method consists of measuring the surface strains on an actual vessel or structure, or scale model, with mechanical or electrical resistance strain gages. Two of the more commonly used mechanical gages are the *Berry* and *Huggenberger* strain gages, Fig. 1.7. Each works on the principle of multiplying strains by mechanical leverage. The Berry strain gage, Fig. 1.7*a*, consists of a frame to which is attached a stationary contact point, and a second movable contact point on the short leg of a bell-crank lever. The long leg of the lever is in contact with a dial indicator which is used to measure the strain. The lever ratio is usually one to five; and the gage length is 2 or 8 in. A more sensitive mechanical strain gage is the Huggenberger, Fig. 1.7*b*, which embodies a more elaborate multiplying lever system to give scale division readings[2] to 0.0001 in. The gages are available in ½-in. and 1-in. gage lengths. In recent years mechanical gages have been largely supplanted by electric resistance gages. These have the advantage of ease of application and adaptability to difficult locations, such as sharp corners or the inside wall of a pressure vessel. The SR–4 gage, Fig. 1.8, is the most commonly used electric strain gage. It consists of a short

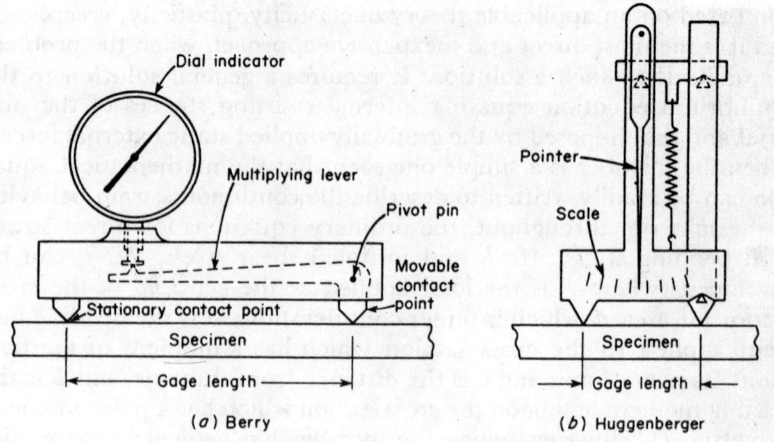

Fig. 1.7. Mechanical Type Strain Gages

length of fine wire held to the material by an adhesive so that any strain on the material is transmitted to the wire strain gage. The change in length and cross-sectional dimensions of the wire produced by the stress cause a change in the electrical resistance. This change in electrical resistance is calibrated and used as a measure of the strain. These gages are available in gage lengths of $\frac{1}{32}$ in. to 1 in. Figure 1.9 shows a group of electrical strain gages mounted on the bolted flanged head of a large reactor vessel.

If strain gages are placed in pairs opposite each other across the thickness of a member, the magnitude of the direct and bending stress can be determined. This is useful, for instance, in evaluating the primary membrane stress and the secondary bending stress in vessels. The material of test models can be the same as that of the prototype,or it

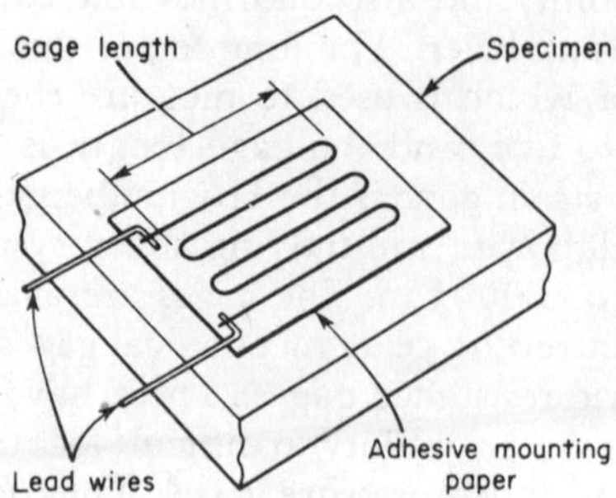

Fig. 1.8. Schematic Arrangement of SR-4 Type Electric Strain Gage

Fig. 1.9. Electrical Strain Gage Test of the Bolted Flanged Head of a Nuclear Reactor Vessel (*Courtesy The Babcock & Wilcox Company*)

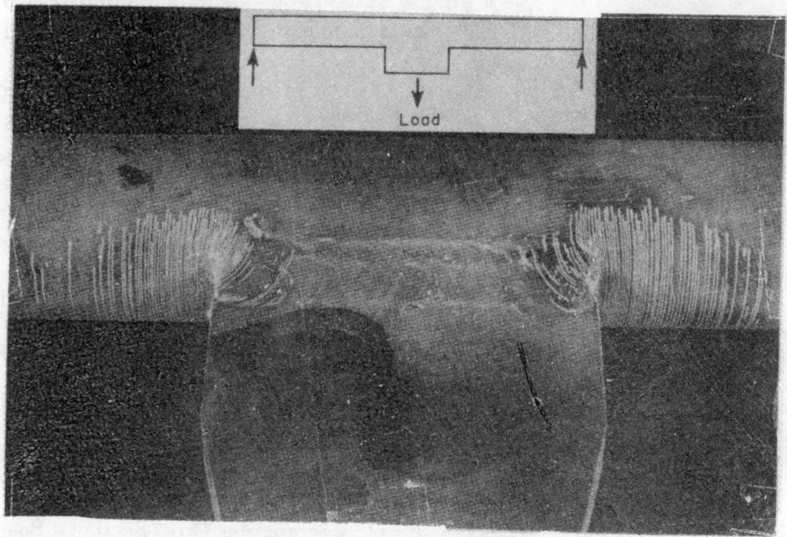

Fig. 1.10 Brittle Coating Stress Indications Near a Beam Lug (*Courtesy The Babcock & Wilcox Company*)

can be any other following the same type elastic behavior as the contemplated material. Of course, a full-scale model constructed with the actual material and by the contemplated fabrication process eliminates all reservations of small-scale model similitude, material behavior, and fabrication tolerances. This method, which is admittedly expensive and time consuming, can be aided by qualitative methods, such as brittle or other applied surface coatings that crack in the region of high stress, thereby allowing strain gages to be more strategically located.[3,12] Figure 1.10 shows such a coating that has the characteristics of cracking perpendicular to the direction of principal stress applied to a heavy lug welded to a beam. The coating cracks first, starting at the edge of the lug, showing that the lug strengthened the beam within its width, but moved the point of maximum stress to the edge of the lug.

2. *Photoelastic*

This consists of optically measuring the principal stress differences in isotropic transparent material models which become doubly refractive when polarized light is passed through the model.[4,23] In two-dimensional models this method gives the average stress throughout the

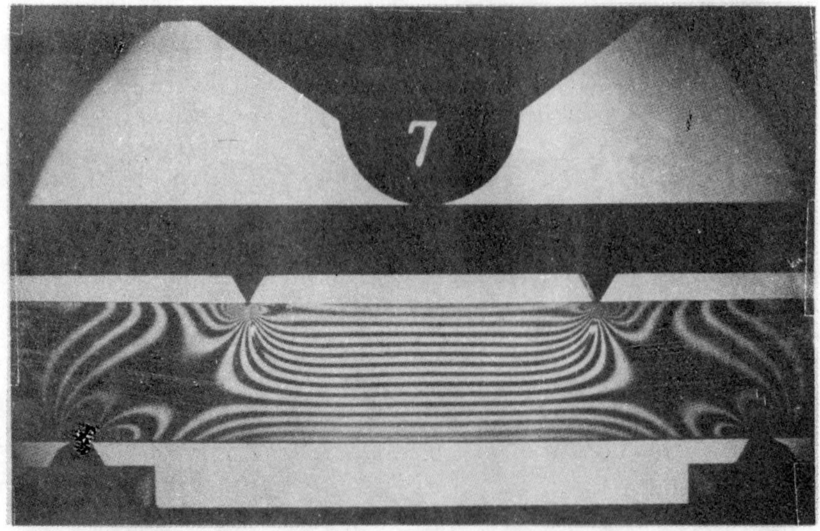

Fig. 1.11. Photoelastic Model of a Simple Rectangular Cross-Sectional Beam (*Courtesy The Babcock & Wilcox Company*)

thickness, whereas in three-dimensional models, using newly developed freeze techniques, the stress throughout the thickness can be determined. Normally, these models are relatively small and inexpensive. Recent extensions of this method have also been made to permit analyses of actual structures. It consists of bonding thin sheets of photoelastic plastic to the part to be analyzed. The surface strains of the part are transferred to the plastic coating, and measurements made by reflected light.[5,6] Figure 1.11 shows a photoelastic model of a rectangular beam loaded by equal loads placed equal distances from each end support so as to place the center portion of the beam in pure bending. The parallel dark lines or fringes indicate that at a short distance from the points of application of the loads the stress distribution is uniform and the same in all vertical cross sections. This is a region in which the basic bending stress formula, Mc/I, holds true. In the region of the loads and supports it is seen that these fringes are not uniform, and hence the simple bending stress formula would not be completely applicable. The local disturbances are called *stress concentrations*. They occur in regions of concentrated loads or abrupt changes in geometry

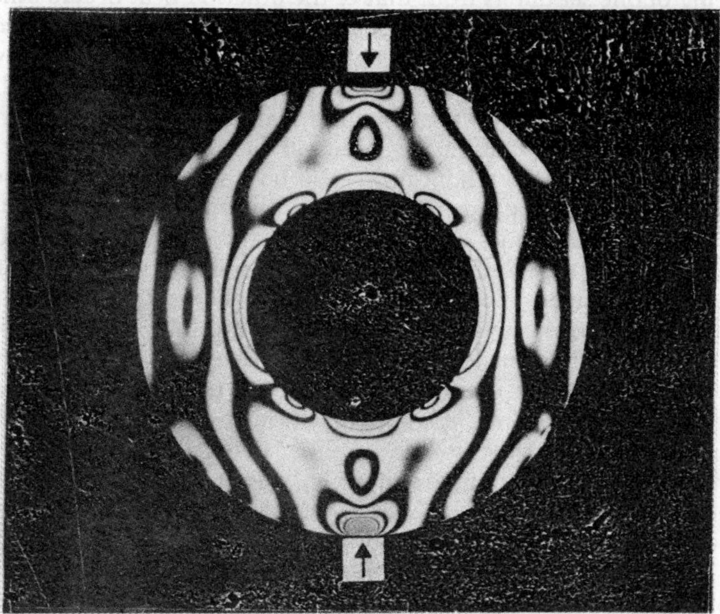

Fig. 1.12. Photoelastic Model of a Thick Ring Diametrically Loaded (*Courtesy The Babcock & Wilcox Company*)

and are discussed in Chapter 6. The photoelastic method is particularly useful in analyzing complicated shapes, such as the thick ring shown in Fig. 1.12. It also gives an overall picture of the stress distribution and indicates regions of both high stress, where changes in contour can be made to reduce the concentration, and regions of low stress, where material can be removed without detriment to the general strength of the structure.[25, 26]

3. *Moiré Method*

The moiré fringe technique*[,7,8,9,13] is one of the most adaptable experimental stress analysis methods for evaluating thermal, as well as pressure, stresses. It can be applied directly to metals in models or actual structures; hence, restrictions imposed by similitude conditions when materials other than metals are used to simulate the behavior of metals in thermal stress analysis can be eliminated. When grids with[27] periodic rulings are made to overlap, interference patterns called moiré fringes are produced. The simplest form of moiré pattern arises from the parallel superposition of two sets of parallel lines when the spacing of one set differs from that of the other. A beat occurs when a line of one figure falls exactly between two lines of the other figure, or when the lines are not wide enough to fill a space completely by an apparent broadening of the lines as the two figures move out of phase. The more closely the two sets of rulings match each other, the farther apart the beats are. Thus, if the rulings are a millimeter apart but one set is in error by 0.001 millimeter, the beat will occur every meter. Hence, the moiré pattern represents an enormous magnification (in this case a million times = 1000/0.001) in the difference in length of the spacings. This is similar to the principle of the vernier used on the machinist's caliper and surveyor's transit. The application of this principle to stress analysis consists of placing fine, regularly spaced lines on the undeformed test specimen and also on a transparent screen and then deforming the test specimen. Moiré fringes are formed when the transparent master grid is superpositioned upon the deformed grid from which displacements and strains can be determined. For instance, a specimen strained uniformly in tension in a direction perpendicular to its screen lines, with the master screen superpositioned onto it without rotation is shown in Fig. 1.13. Interference fringes will

*Moiré is the French word for "watered." In English it is most frequently heard in the term "moiré silk," a fabric that has a shimmering appearance resembling the reflections on the surface of a pool of water.

occur every nth line of the undeformed master screen where there is the maximum mismatch with the $(n - 1)$th line of the elongated specimen grid or at every $(n + 1)$th line of the compressed specimen grid. If "a" is the distance between two neighboring screen lines, and "d" the distance between two neighboring interference fringes, Fig. 1.13, the strain "e" is then expressed by

$$d = na \tag{1.2.1}$$

$$d = (n - 1)a(1 + e) \tag{1.2.2}$$

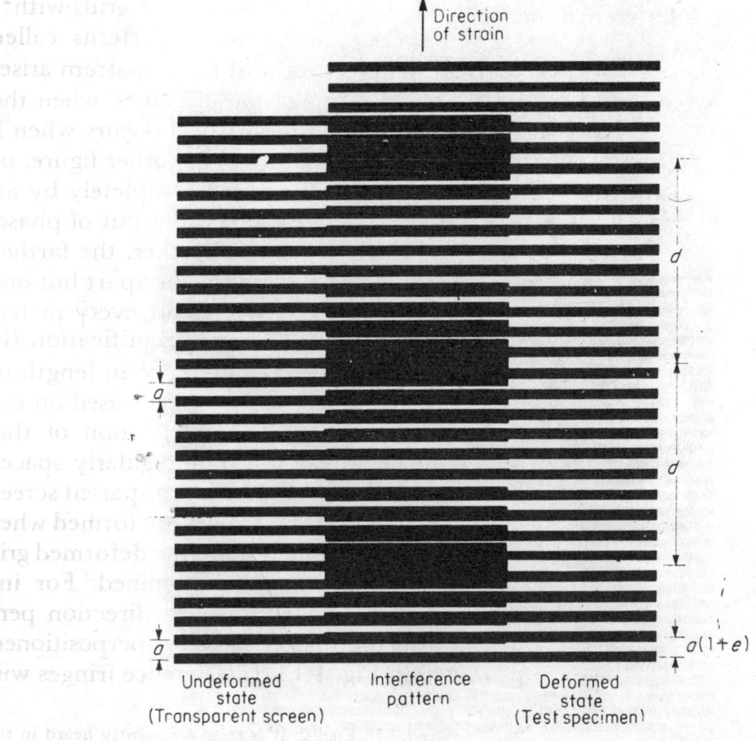

Fig. 1.13. Interference Fringes Due to Elongation Perpendicular to Screen Lines[7]

from which

$$e = \frac{a}{d - a} \tag{1.2.3}$$

$$e = -\frac{a}{d + a} \tag{1.2.4}$$

For small strians, "a" is extremely small compared to "d", so Eqs. 1.2.3 and 1.2.4 can be written

$$e = \pm\frac{a}{d} \tag{1.2.5}$$

If the master screen is slightly rotated by an angle α with respect to the undeformed screen on the specimen, interference lines appear as shown in Fig. 1.14. Referring to Fig. 1.15, the distance "h" between neighboring interference lines is

$$h = x_1 + x_2 \tag{1.2.6}$$

where

$$x_1 = \frac{a}{\tan \alpha} \tag{1.2.7}$$

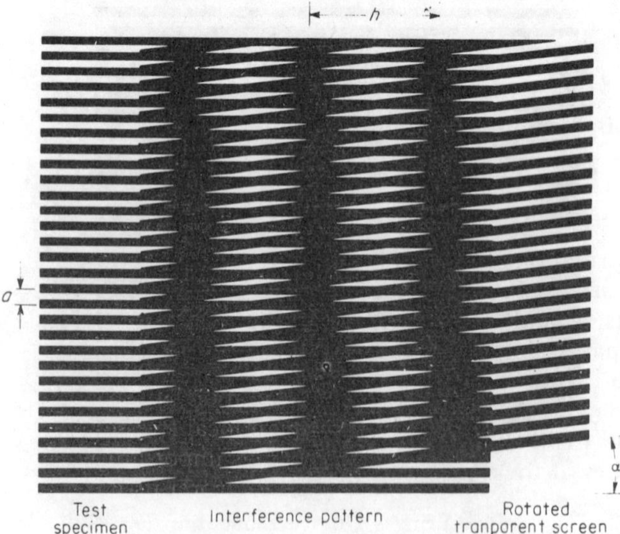

Fig. 1.14. Interference Fringes Due to Rigid-Body Rotation of One Screen

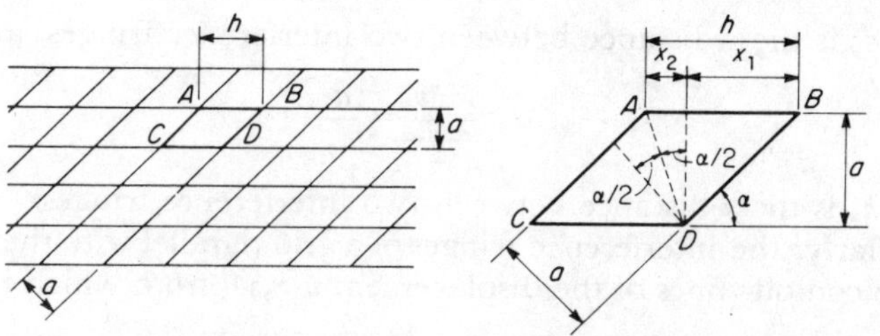

Fig. 1.15. Rotation Effect on Interference Fringes

and

$$x_2 = a \tan \frac{\alpha}{2} \tag{1.2.8}$$

Substituting Eqs. 1.2.7 and 1.2.8 in Eq. 1.2.6 gives

$$h = \frac{a}{\tan \alpha} + a \tan \frac{\alpha}{2} \tag{1.2.9}$$

and introducing the trigonometric identities

$$\tan \frac{\alpha}{2} = \frac{1 - \cos \alpha}{\sin \alpha}, \text{ and } \tan \alpha = \frac{\sin \alpha}{\cos \alpha} \tag{1.2.10}$$

gives

$$\sin \alpha = \frac{a}{h} \tag{1.2.11}$$

then for small rotations (small angles)

$$\alpha = \frac{a}{h} \tag{1.2.12}$$

When deformation of the specimen causes the fringe rotation, α is then the shear strain. From the basic moiré technique for measuring displacements, u and v in the OX and OY axis direction, respectively, the principal strains can be determined.[10,11,14,21] This is done by providing two orthogonal sets of screen lines on the surface of the specimen and viewing the loaded specimen first with the screen lines on the master grid parallel to the OX axis. The interference fringes are the contour lines of the displacement $v(x,y)$, so that for small strains

$$e_y = \frac{\partial v}{\partial y} = \frac{a}{d_y} \tag{1.2.13}$$

where d_y is the y-distance between two interference fringes, and

$$\alpha_x = \frac{\partial v}{\partial x} = \frac{a}{h_x} \tag{1.2.14}$$

where h_x is the x-distance between two interference fringes.

Similarly, the interference fringes of a grid parallel with the OY axis are the contour lines of the displacement $u(x,y)$, from which it follows that

$$e_x = \frac{\partial u}{\partial x} = \frac{a}{d_x} \tag{1.2.15}$$

where d_x is the x-distance between two interference fringes, and

$$\alpha_y = \frac{\partial u}{\partial y} = \frac{a}{h_y} \tag{1.2.16}$$

with h_y being the y-distance between two interference fringes. From the three strain components, e_x, e_y and $\gamma = \alpha_x - \alpha_y$, the principal strains and their direction can be determined. Figure 1.16 shows the moiré

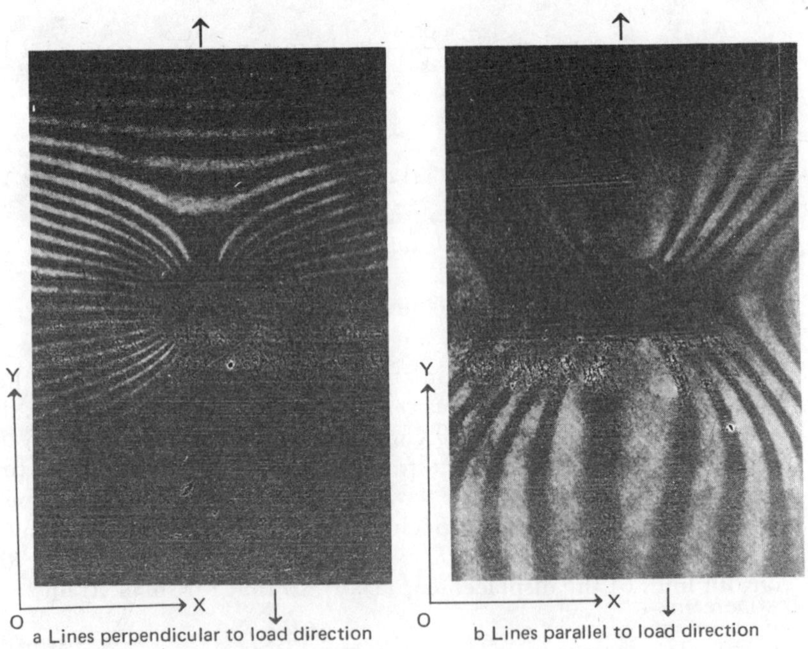

a Lines perpendicular to load direction   b Lines parallel to load direction

Fig. 1.16. Moiré Pattern in the Region of a Hole in a Tensile Specimen

pattern on a plate with a small, central circular hole subject to tension in the vertical, or OY direction. The regular spacing of horizontal fringes in Figure 1.16*a*, and vertical fringes in Figure 1.16*b* at relatively short distances away from the hole indicates that the effect of the hole is very local. In contrast, the closeness of these fringes adjacent to the hole edge of its horizontal axis indicates high strains in this local area. The steep angle of these fringes in a cross-shaped region 45° to the axis denotes an area of high shear, Art. 5.5.

This method of experimental stress analysis is adaptable to models as well as prototypes, to mechanically and thermally produced stresses, and is particularly valuable in high temperature stress analysis in which case the grid pattern is etched directly on the metal structure.[28, 29]

Although these are the three most widely used experimental methods, others which employ optical and acoustical holography, ultrasonics, X-ray diffraction, brittle models, plastic models, soap film, electric analogy, fluid flow analogy, repeated stress, plastic flow, etc., are frequently used for particular applications.

1.3 Stress Significance

Analytical formulas for the evaluation of stresses are usually based on elastic theory and elastic behavior of the material, i.e., material which conforms to Hooke's law, and it may at first be thought that materials which follow this behavior right up to the breaking point would be the most desirable for use. This is not the case, however; for instance, plaster of Paris has a perfectly straight stress-strain curve up to the breaking point but, of course, is not a suitable material for a structural member or pressure vessel, strictly because it is totally elastic and not partially plastic in its behavior. It is this plastic property of the material, with its ability to give or yield under high peak or local stress and so accommodate the applied loading by a more favorable distribution of internal stress, that is the most important property of a pressure vessel material. The elastically computed or actual strength of most members, considering the structure as a whole, would be considerably reduced if it were not accompanied by plastic deformation at various relatively small portions of the member where high local stresses occur.

It is important not only to determine the value of a stress, but also to interpret its meaning or significance—the two go "hand-in-glove." Determination of stress significance requires a knowledge of:

1. The type and nature of the applied loading and the resulting stress distribution or pattern within the member. For instance, is the

applied loading mechanical or thermal, of a steady (static) or unsteady (variable or cyclic) nature, and is the resulting stress pattern uniform, or does it have high peak values?

2. The ductile and plastic properties of the material. For instance, are the properties of the material such that internal yielding or readjustment of strain can reduce the effects of local stress concentrations?

3. The toughness or adaptability of the material under adverse working conditions or environments.[15] For instance, are the properties of the material sufficient to absorb applied impact or shock loadings?

The strength of a member does not depend only on the value of the maximum stress or strain in the member, but also on the external shape readjustment that the member itself can make to one more favorable than that assumed in the design, and on the plastic property of the material to permit internal stress adjustment.

1. *Types of Loading and The Stress Pattern*

Structures are subjected to two basic types of loading: namely, steady or static and unsteady (variable, cyclic or impact). Although practically all structures encounter variable or cyclic loading, those on most building structures, many machine members, and an appreciable number of pressure vessels, such as boiler steam drums, may be assumed to be statically loaded without introducing serious error. Such structures made of ductile materials and subject to static loads fail by gross yielding. The ductility of the material allows a redistribution of stresses by plastic flow to attenuate points of high local values toward a pattern more favorable to maximum resistance. Hence, the stresses in a large portion or volume are involved in the behavior of the member, and the basic primary stress analysis equations are significant in determining their strength and stiffness.

However, when the loading is such that the member is subjected to a considerable number of stress cycles, even though the material is ductile, appreciable error can be introduced by considering a static loading condition to exist in appraising integrity on the basis of simple elastic formulas. Under such conditions failure occurs due to a condition known as fatigue. This failure does not involve a sufficient amount of metal to make these formulas representative of the action prevailing; since in fatigue, failure is due primarily to a highly localized stress which causes a minute fracture that gradually spreads until the member is ruptured. This type of failure is of particular importance in pressure vessels for hydraulic or pneumatic accumulators.

Impact or shock loading can be imposed on structures, including nuclear vessels, by earthquake, explosions, or collision of mobile equipment. This requires design considerations to accomplish transfer of the kinetic energy throughout the vessel, absorption of this energy within the vessel and associated structure, and use of materials of adequate toughness.

The stress distribution near the point of load applications, such as the point of contact of the beam load in Fig. 1.11, or the support bracket on a vessel, may vary greatly from the assumed pattern on which the ordinary equations are based, and these local stress values may be relatively high. Even though the material is ductile and a measure of stress redistribution can occur, these local stresses can be significant ones and are frequently responsible for failures. In relatively brittle materials, or in ductile material subjected to cyclic loading, stresses at the points of load application may control the strength of the member rather than the stresses given by the ordinary equations. This is particularly important in vessels which are designed as membrane or tension members and cannot resist large bending moments perpendicular to their surface and yet for practical purposes must have support brackets, lifting lugs, nozzles, etc., attached to them.

2. *Initial or Residual Stress*

The basic equations for determining stresses are based on the assumption that the stresses in a member are caused only by external loads, and residual stresses set up in the fabrication or construction processes, such as weld shrinkage, casting cooling, metal heat treatment, etc., are not considered. Although these stresses are secondary, since their value is self-limiting (they are not produced by unrelenting external loads), they may be of great importance in brittle materials, and even in ductile material when the material is subject to fatigue loading. Equally important is the danger of creating, in conjunction with the applied loading stresses, a three-dimensional stress pattern in thick sections that is restrictive to the redistribution of high localized peak stresses through yielding. It is for this reason that stress relieving of thick vessels, usually required by construction codes, is much more important than thin ones in which the state of stress is essentially two dimensional.

3. *Shape of Member*

The basic assumption for continuity of action in a member on which ordinary formulas for direct stress and bending stress are based, requiring that a plane section remain plane after bending, cannot hold near points of abrupt changes in section due to the restraining influence of this stiffer portion on adjacent sections. Figure 1.10 illustrates this condition. The stresses in the region influenced by these geometrical shape discontinuities are higher than predicted by the assumed mathematical law of distribution on which the ordinary stress formulas are based, and are known as localized or concentration stresses. The errors introduced by the use of the ordinary formulas for the design of members with abrupt changes of section are generally not serious if the load is static and the material ductile so as to permit a slight measure of plastic flow; hence, the member acts more nearly as assumed.

The practicality of this is attested by the satisfactory behavior of most buildings, vessels, and machines. These localized stresses are, however, most important in brittle materials, even under static loads, since under such conditions a redistribution or transfer of stress from the highly overstressed material to adjacent lower stressed material does not take place and rupture of the member results. They are equally significant when cyclic loading is involved, even when the material is ductile, since the region of high stress acts as a focal point from which fatigue failure can stem.

The problem of evaluating localized stresses in vessels has assumed major importance in the last decade as engineering advancements have placed unusual pressure, temperature, and environment demands on pressure vessels. The petroleum and chemical processes require operating pressure in the 5,000 to 200,000 psi range. The rapidly expanding cryogenics industry has introduced low temperature conditions to minus 425°F. The nuclear power plant has given rise to high pressure, high temperature, and special cyclic and material irradiation operating conditions. All these requirements have focused considerable attention on the stress analysis, materials of construction, and economics of design of vessels for these services.

1.4 Design Approach

The design of most structures is based on formulas that are known to be approximate. The unknown items, such as extent of yielding and the omitted factors in design and material behavior, are considered to be provided for by the use of working stresses that are admittedly below those at which the member will fail. This "factor of safety" or "factor of ignorance" approach,[16] although possessing virtue of having worked well in the past for ductile materials under static loading and

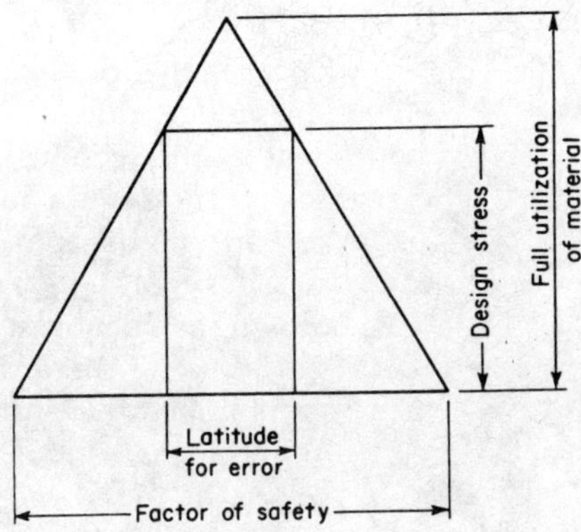

Fig. 1.17. Triangle of Knowledge

providing the designer with preliminary sizing data, is yielding to more refined analytical and experimental methods. This improvement will continue as knowledge and cognizance of influencing design and material parameters increase and are put to engineering and economic use. This might be illustrated by the triangle of knowledge, Fig. 1.17, which indicates that as our ignorance decreases with discovery or recognition of more of the factors affecting behavior and proper account is taken in design analysis, the latitude for error decreases; accordingly, the potential properties of the material can be more fully utilized with confidence.

The safety demands of nuclear reactors, deep-diving submersibles, space vehicles and chemical retorts have accelerated pressure vessel material behavior and stress analysis knowledge. For instance, the nuclear reactor, Fig. 1.18, with its extremely large, heavy section cover flanges and nozzle reinforcement, operating under severe thermal transients in a neutron irradiation environment, has focused considerable attention on research in this area which has been directly responsible for improved materials, knowledge of their behavior in specific environments, and new stress analysis methods.

High strength materials, created by alloying elements, manufacturing processes or heat treatments, are developed to satisfy economic or engineering demands, such as reduced vessel thickness. They are continually being tested to establish design limits consistent with their higher strength, and adapted to vessel design as experimental and fabrication knowledge justifies their use. There is no one perfect pressure vessel material suitable for all environments, but material selec-

(*a*)

(*b*)

Fig. 1.18. Vessel and Closure Head During Fabrication (*Courtesy The Babcock & Wilcox Company*)

tion must match application and environment. This has become especially important in chemical reactors because of the embrittlement effects of gaseous absorption, and in nuclear reactors because of the irradiation damage from neutron bombardment.

Major improvements, extensions and developments in analytical and experimental stress analysis are permitting fuller utilization of material properties with confidence and justification. Many formerly insoluble equations of elasticity are now yielding to computer adaptable solutions, such as the finite element method,[24] Fig. 1.19; and these together with new experimental techniques[19,22] have made possible the stress analysis of structural discontinuities at nozzle openings, attachments, etc. This is significant because 80 per cent of all pressure vessel failures are caused by the high localized stresses associated with these "weak link" construction details. Hence, stress concentrations at vessel nozzle openings, attachments and weldments are of prime importance, and methods for minimizing them through better designs and analyses are the keys to long pressure vessel life. Control of proper construction details assures a vessel of balanced design and maximum integrity.

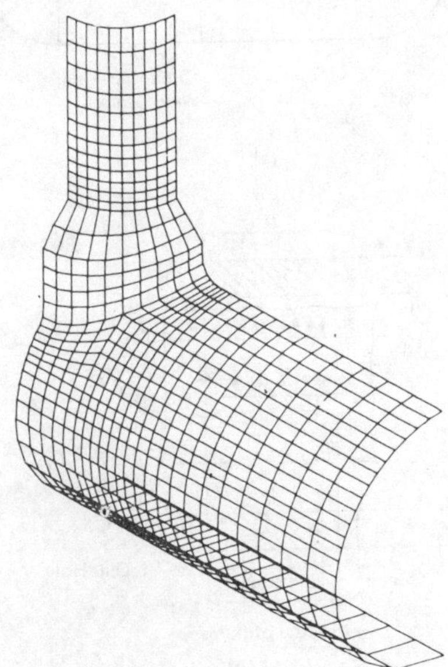

Fig. 1.19. Mesh for Finite Element Stress Analysis at a Nozzle-Shell Juncture[17,18]

1.5. Terminology and Ligament Efficiency

Pressure components vary widely in shape and complexity according to the functions that they must perform. However, they generally consist of a few basic parts, such as cylinders, rings, and various shaped closure heads. Metal pressure components are fabricated by a welding process, concrete components by onsite placement in forms, and composite material components by winding on mandrels and the use of adhesives. Figures 1.20*a* shows the construction features and terminology of a simple metal vessel fabricated by a welding process. When the vessel diameter is in the size range of procurable tubular products, the cylindrical part is normally so selected; however, when diameters

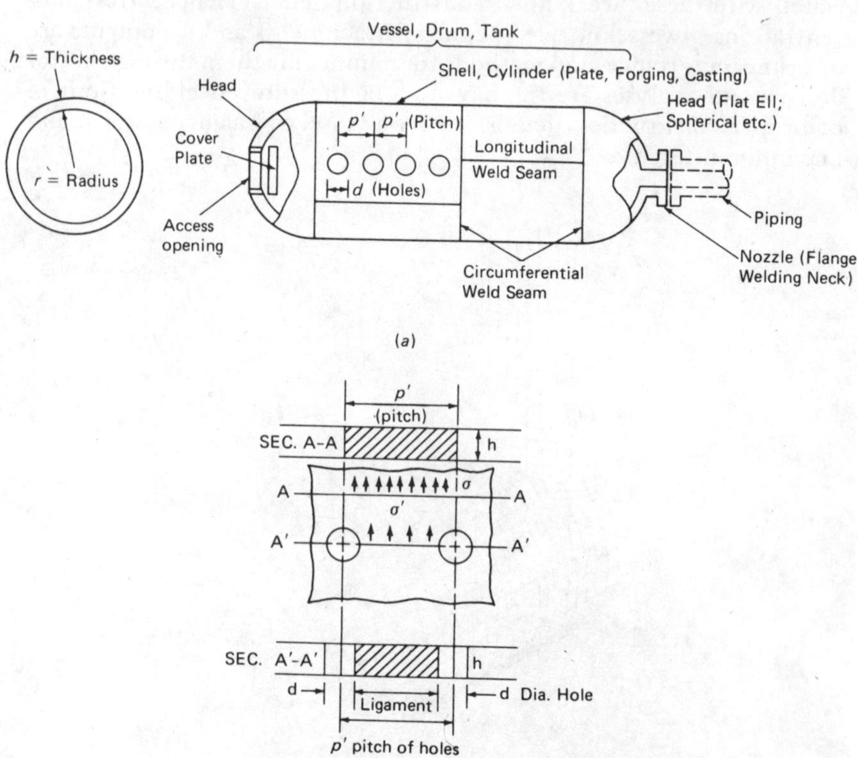

Fig. 1.20. (*a*) Vessel Terminology. (*b*) Vessel Ligament.

(*a*)

Fig. 1.21. Ligaments in (*a*) Cylindrical Vessel, (*b*) Spherical Vessel (*Courtesy The Babcock & Wilcox Company*)

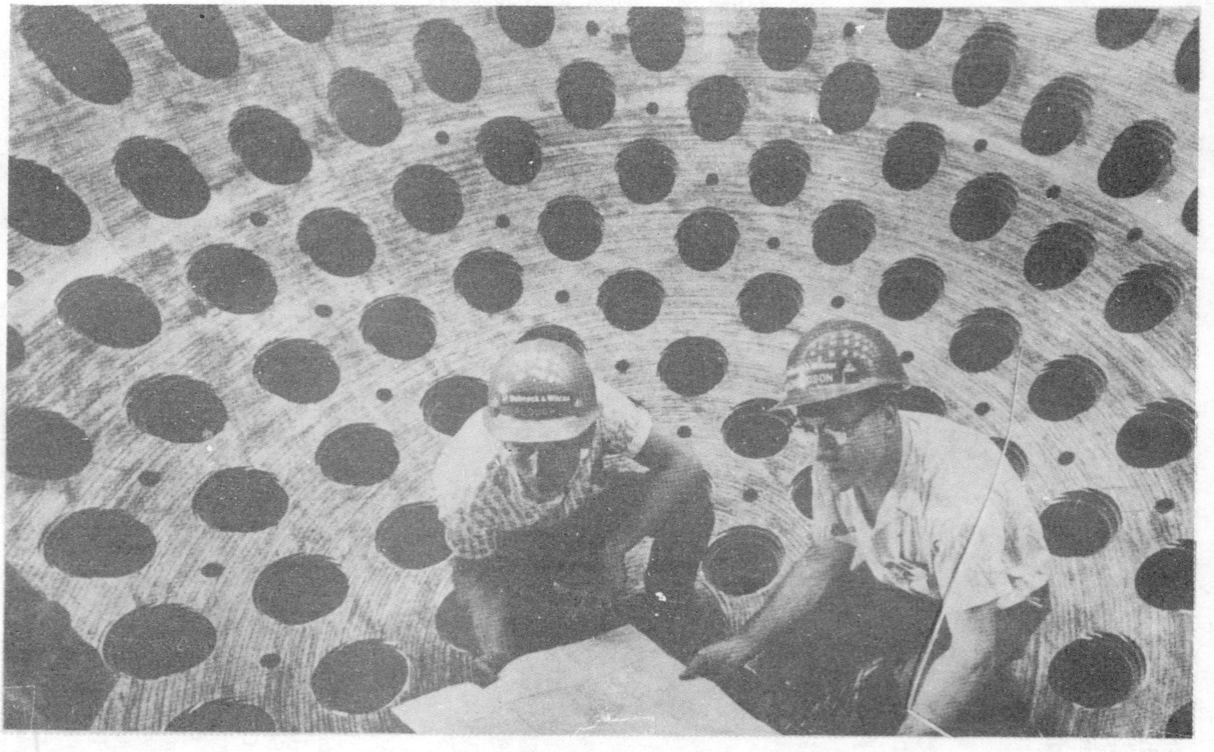

(b)

Fig. 1.21. *(Continued)*

exceed these, rolled plate, partial forgings, or castings welded into cylinders are employed.

Vessels must have openings for functional purposes. Boiler drums and heat exchangers are examples of these which employ rows of holes, diameter d, spaced at a pitch p', Fig. 1.20(b). The material remaining between these holes are called ligaments, and the cross-sectional area of the ligament compared to the area in a normal unpierced cross section of width p' is called the ligament efficiency:

$$\text{Ligament efficiency} = \frac{\text{Area in net section}}{\text{Area in normal section}} = \frac{p' - d}{p'} \quad (1.5.1)$$

Ligament efficiency is a convenient way to compare the membrane or average stress in various ligaments, σ', throughout a vessel with the normal stress occuring at an unpierced location, σ. Referring to Fig. 1.20*b*, from the continuity of force equilibrum of that in a normal section of width equal to pitch, p', and that on a ligament, $p' - d$,

$$p'h\sigma = (p' - d)h\sigma' \quad (1.5.2)$$

$$\sigma' = \sigma \frac{p'}{p' - d} \quad (1.5.3)$$

$$\sigma' = \frac{\sigma}{\text{Lig. eff.}} \quad (1.5.4)$$

Figure 1.21*a* shows longitudinal and circumferential ligaments in a cylindrical vessel, while Fig. 1.21*b* shows ligaments in a spherical vessel during fabrication.

REFERENCES

1. W. M. Murry, "A Comparison of Stress Analysis Methods," *Strain Gage Readings*, Vol. III, No. 6, February, 1961.

2. Joseph Marin, *Mechanical Behavior of Engineering Materials*, Prentice-Hall, Inc., Englewood Cliffs, N.J., 1962.

3. C. A. Crites, "Brittle Coating Methods for Stress Analysis," *Product Engineering*, November 27, 1961.

4. Tuppeny and Kobayashi, "Manual on Experimental Stress Analysis," Society for Experimental Mechanics, Westport, Conn., 1966.

5. Zander and Maier, "Six New Techniques for Photoelastic Coatings," *Product Engineering*, July, 24, 1961.

6. Hovanesian, Eggenberger and Hung, "Full-Field Generation in Photoelastic Coating," *Experimental Mechanics*, Vol. 12, No. 4, April, 1972.

7. A. Vinckier and R. Dechaene, "Use of Moiré Effect to Measure Plastic Strains," *ASMA Transactions*, Paper No. 59-Met. 7. 1959.

8. P. Dantu, "Extension of the Moiré Method to Thermal Problems," *Experimental Mechanics*, March, 1964.

9. C. A. Sciammarella and B. E. Ross, "Thermal Stresses in Cylinders by the Moiré Method," *Experimental Mechanics*, October, 1964, Vol. 4, No. 10.

10. V. J. Parks and A. J. Durelli, "Various Forms of the Strain-Displacement Relations Applied to Experimental Strain Analysis," *Experimental Mechanics*, February, 1964.

11. L. P. Martin and F. D. Ju, "The Moiré Method for Measuring Large Plane Deformations," ASME Paper 69-APMW-21, 1969.

12. A. J. Durelli, V. J. Parks and V. Pavlin, "Brittle-Coating-Patterns Evaluation," *Experimental Mechanics*, August 1974.

13. L. Johnson, "Moiré Technique for Measuring Strains During Welding," *Experimental Mechanics*, April, 1974.

14. R. E. Rowlands, J. A. Jensen and K. D. Winters, "Differentiation Along Arbitrary Orientations," *Experimental Mechanics*, March 1978.

15. A. P. Gelman, "Determining Interaction Effects in Seismic Analysis of Conponents," ASME Publication Paper 75-PVP-50, 1975.

16. My Dao-Thien and M. Massoud, "On the Relation Between the Factor of Safety and Reliability," ASME Publication Paper 73-WA/DE-1, 1973.

17. S. E. Moore and J. W. Bryson, "Design Criteria for Piping and Nozzles Program Quarterly Progress Report for April-June 1976," Oak Ridge National Laboratory, ORNL/NUREG/TM-91, 1977.

18. J. W. Bryson, W. G. Johnson and B. R. Bass, "Stresses in Reinforced Nozzle–Cylinder Attachments Under Internal Pressure Loading Analyzed by the Finite-Element Method–A Parameter Study," Oak Ridge National Laboratory Report ORNL/NUREG-4, October 1977.

19. R. H. Bryan, T. M. Cate, P. P. Holz, T. A. King, J. G. Merkle, G. C. Robinson, G. C. Smith, J. E. Smith and G. D. Whitman, "Test of six-Inch Thick Pressure Vessels Series 3, Intermediate Test Vessel V-7A Under Sustained Loading," Oak Ridge National Laboratory Report ORNL/NVREG-9, February, 1978.

20. A. W. Hendry, *Elements of Experimental Stress Analysis*, Pergamon Press Ltd., London, 1977.

21. J. Buitrago and A. J. Durelli, "On the Interpretation of Shadow-moiré fringes," *Experimental Mechanics*, June 1978.

22. J. E. Sollid and K. A. Stetson, "Strains for Holographic Data," *Experimental Mechanics*, June, 1978.

23. J. F. Doyle and H. T. Danyluk, "Integrated Photoelasticity for Axisymmetric Problems," *Experimental Mechanics*, June, 1978.

24. R. K. Mueller, "Mathematical and Physical Models," *Experimental Mechanics*, July, 1978.

25. A. S. Redner, "Photoelastic Coatings," *Experimental Mechanics*, November, 1980.

26. R. J. Sanford, "Photoelastic Holography–A Modern Tool for Stress Analysis," *Experimental Mechanics*, December 1980.

27. B. W. Cotterman and P. D. Heimdahl, "Classroom Moiré Demonstrator," *Experimental Techniques*, January 1983.

28. C. A. Sciammarella, "The Moiré Method–A Review," *Experimental Mechanics*, November, 1982.

29. G. Cloud, R. Radke, and J. Peiffer, "Moiré Gratings for High Temperatures and Long Times," *Experimental Mechanics*, Vol. 19, No. 10, October 1979.

2

Stresses in Pressure Vessels

2.1 Introduction

Pressure vessels commonly have the form of spheres, cylinders, ellipsoids, or some composite of these. Such composites are illustrated in the vessel shapes of Figs. 2.1 and 2.2. In practice, vessels are usually composed of a complete pressure-containing shell together with flange rings and fastening devices for connecting and securing mating parts. As the name implies, their main purpose is to contain a media under pressure and temperature; however, in doing so they are also subjected to the action of steady and dynamic support loadings, piping reactions, and thermal shocks which require an overall knowledge of the stresses imposed by these conditions on various vessel shapes and appropriate design means to ensure safe and long life.

When vessels or shells are considered to be formed of plate in which the thickness is small in comparison with the other dimensions, and as such offer little resistence to bending perpendicular to their surface, they are called "membranes," and the stresses calculated by neglecting bending are called "membrane stresses." A piece of writing paper is very resistant to forces in its plane, but can offer little resistance to bending perpendicular to its plane. In one sense, this is a desirable condition for it permits the vessel to deform readily without incurring large bending stresses at points of discontinuity, Chapter 4. Membrane stresses are average tension or compression stresses over the thickness of the vessel wall and are considered to act tangent to its surface. Most vessels for boiler drums, accumulators, or chemical and nuclear vessels fall in this category.

2.2 Stresses in a Circular Ring, Cylinder, and Sphere

If a thin circular ring is subjected to the action of radial forces uniformly distributed along its circumference, hoop forces will be pro-

Fig. 2.1. Nuclear Power Plant Containment Vessel (*Courtesy Chicago Bridge & Iron Co.*)

duced throughout its thickness which act in a tangential direction. A uniform enlargement of the ring will take place if the acting forces are radial outward, or contraction will occur if the acting forces are radial inward. The magnitude of the force F in the ring can be found by cutting the ring at a horizontal diametrical section giving the free body shown in Fig. 2.3. If the force per unit length of cirumference is q, and r is the radius of the ring, the force acting on an element of the ring is $qrd\phi$. Taking the sum of the vertical components of all the forces

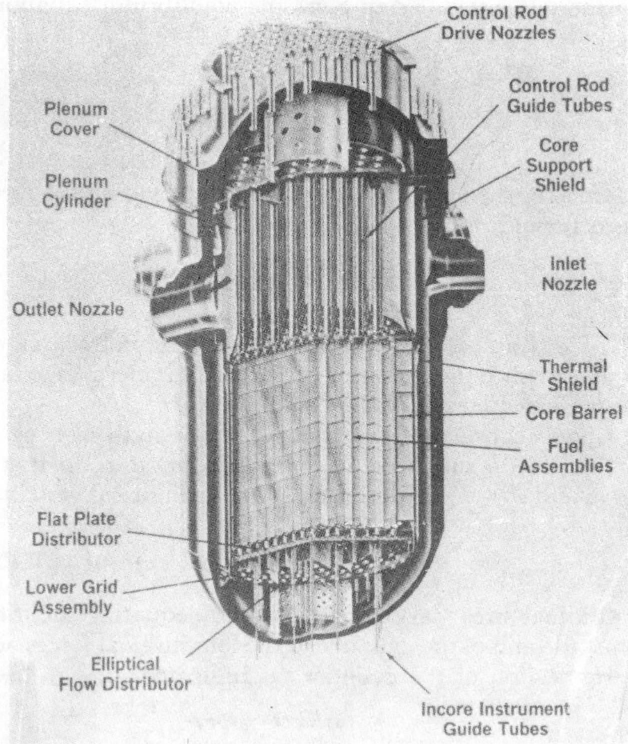

Fig. 2.2. Nuclear Reactor Vessel (*Courtesy The Babcock & Wilcox Company*)

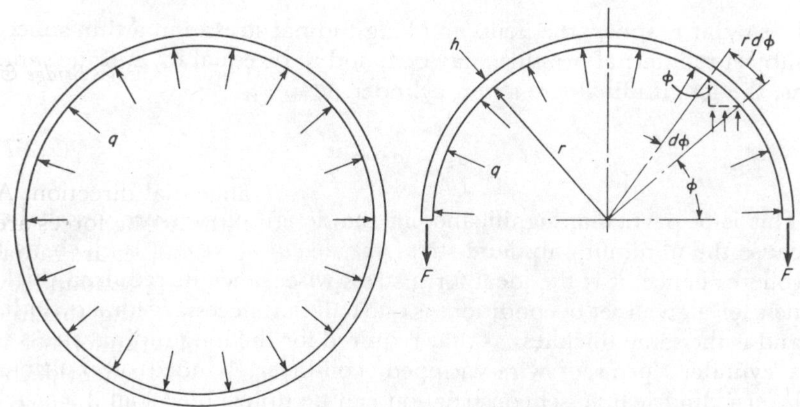

Fig. 2.3. Radial and Hoop Stresses in a Thin Ring

acting on the semicircular ring gives the equilibrium equation:

$$2F = 2\int_0^{\pi/2} qr \sin \phi d\phi = 2qr \tag{2.2.1}$$

$$F = qr \tag{2.2.2}$$

The unit stress in the ring can be obtained by dividing the force F by the cross-sectional area A of the ring.

$$\sigma_2 = \frac{qr}{A} \tag{2.2.3}$$

In Eq. 2.2.1, $r \sin \phi d\phi$ is the projection of a circumferential element on a diameter; hence the right side of Eq. 2.2.1. is merely the unit force times the projected length of the contact surface.

If the ring is considered a section of unit length of a cylindrical vessel of thickness h subjected to internal pressure p, so that in Eq. 2.2.3, $q = p$ and $A = h$, the hoop stress in a cylindrical vessel becomes

$$\sigma_2 = \frac{pr}{h} \tag{2.2.4}$$

The longitudinal stress can be calculated by equating the total pressure against the end of the cylinder to the longitudinal forces acting on a transverse section of the cylinder, as indicated in Fig. 2.4, giving

$$\sigma_1 h 2\pi r = p\pi r^2 \tag{2.2.5}$$

$$\sigma_1 = \frac{pr}{2h} \tag{2.2.6}$$

In similar manner, the hoop and longitudinal stresses in a thin sphere subject to internal pressure may be found to be equal to, and the same as, the longitudinal stress in a cylinder.

$$\sigma_1 = \sigma_2 = \frac{pr}{2h} \tag{2.2.7}$$

This is of particular significance in the design of pressure vessels because the minimum absolute stress value $\sigma_1 = \sigma_2 = \sigma_{min.}$ is given by a sphere; hence, it is the ideal form stress-wise. Since its required thickness for a given set of conditions is one half that necessary for a cylinder and is the same thickness as that required for the longitudinal stress in a cylinder, forms of wire-wrapped, coil-layer, banded or multiple-layer cylindrical vessel construction can be utilized,[1,2,3] Fig. 2.5. Such constructions have been widely used in the chemical and petroleum industry where they permit a material selection compatible with the

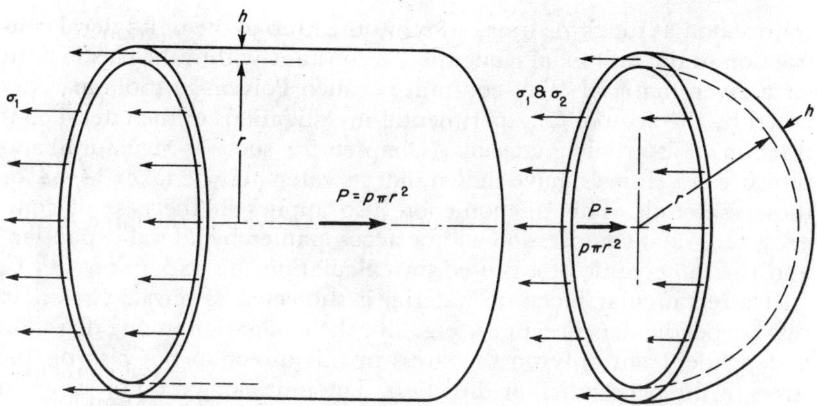

Fig. 2.4. Longitudinal Stress in a Cylinder and Sphere

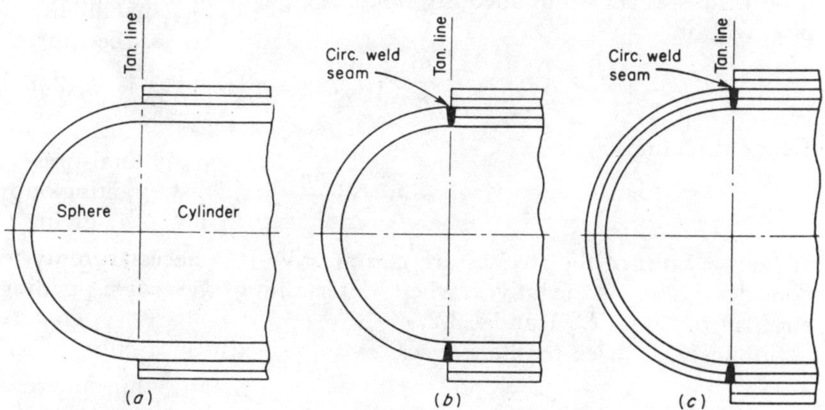

Fig. 2.5. Vessel Construction Employing Banded or Multiple-Layer Cylinders and Hemispheres

contained media for the inner layer, and economical strength material for the media non-contact portion (see Fig. 6.36). They also provide a means of instituting a prestress within the vessel wall, and act as arrestors to fast running cracks in environments conducive to brittle fracture.[4]

2.3 Poisson's Ratio

If a bar is subjected to axial tension, it is elongated not only in the axial direction, but experiments have shown that it undergoes lateral contraction at the same time, and that the ratio of the unit lateral contraction to the unit axial elongation is constant within the elastic limit for a given material. This constant is called Poisson's ratio and is denoted by the symbol μ. Experimental investigations of the lateral contraction of isotropic materials (Chapter 5), such as structural and pressure vessel steels, have shown that its value may be taken as 0.3 for these materials. This phenomenon also applies in the case of compression. Axial compression will be accompanied by lateral expansion, and the same value of μ is used for calculating this expansion.

If a rectangular block of material is subjected to tensile stresses in two perpendicular directions, Fig. 2.6, the elongation in one direction is dependent not only on the stress in this direction but also on the stress in the perpendicular direction. The unit elongation or strain in the direction of the tensile stress σ_1 is σ_1/E. The tensile stress σ_2 will produce lateral contraction in the direction of σ_1 equal to $\mu\sigma_2/E$, so that if both stresses act simultaneously the unit elongation in the direction of σ_1 will be

$$e_1 = \frac{\sigma_1}{E} - \frac{\mu\sigma_2}{E} \tag{2.3.1}$$

In the direction of σ_2,

$$e_2 = \frac{\sigma_2}{E} - \frac{\mu\sigma_1}{E} \tag{2.3.2}$$

If one or both of the stresses are compressive it is necessary only to consider these as negatives when determining the corresponding strains from Eqs. 2.3.1 and 2.3.2.

Similarly, if three tensile stresses, σ_1, σ_2, σ_3, exist on a cube of iso-

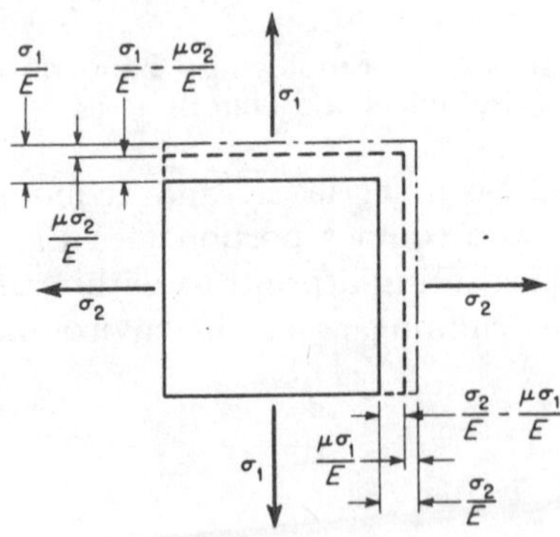

Fig. 2.6. Strain Due to Two Principal Stresses

tropic material, the strain in the direction of σ_1 is

$$e_1 = \frac{\sigma_1}{E} - \frac{\mu\sigma_2}{E} - \frac{\mu\sigma_3}{E} \tag{2.3.3}$$

The stresses existing in a vessel may be determined experimentally from actual strain measurements made on the vessel by employing Eqs. 2.3.1 and 2.3.2, which gives the stresses σ_1 and σ_2 as functions of the strains e_1 and e_2 as

$$\sigma_1 = \frac{(e_1 + \mu e_2)E}{1 - \mu^2} \tag{2.3.4}$$

$$\sigma_2 = \frac{(e_2 + \mu e_1)E}{1 - \mu^2} \tag{2.3.5}$$

2.4 Dilation of Pressure Vessels

Dilation, or radial growth, of a pressure vessel can be obtained by integrating the hoop strain in the vessel wall from an axis through the center of rotation and parallel to a radius. Thus in Fig. 2.7 the dilation is

$$\delta = \int_0^{\pi/2} e_2 r \cos\phi d\phi = e_2 r \tag{2.4.1}$$

and substituting the value of e_2 from Eq. 2.3.2 gives

$$\delta = r\left(\frac{\sigma_2}{E} - \frac{\mu\sigma_1}{E}\right) \tag{2.4.2}$$

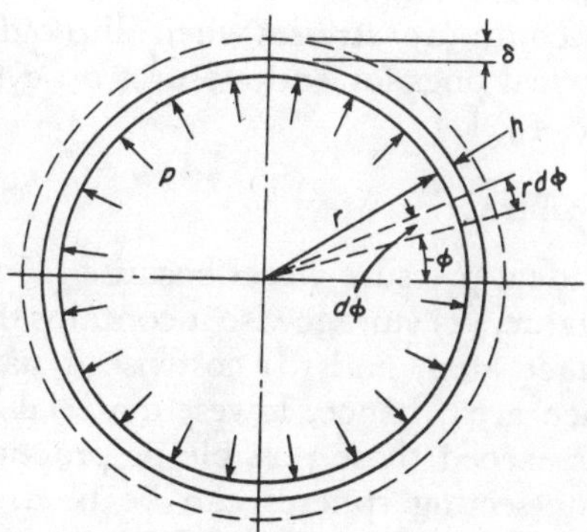

Fig. 2.7. Dilation of Vessel Due to Internal Pressure

The dilation of a cylindrical vessel is found from Eq. 2.4.2, by substituting the value σ_1 from Eq. 2.2.6 and σ_2 from Eq. 2.2.4, to be

$$\delta = \frac{pr^2}{2hE}(2-\mu) \tag{2.4.3}$$

The dilation of a spherical vessel is also found from Eq. 2.4.2, by substituting the value of σ_1 and σ_2 from Eq. 2.2.7, to be

$$\delta = \frac{pr^2}{2hE}(1-\mu) \tag{2.4.4}$$

Likewise the growth of a conical vessel can be found to be (see paragraph 2.6.3)

$$\delta = \frac{pr^2(2-\mu)}{2hE\cos\alpha} \tag{2.4.5}$$

The equatorial dilation of an ellipsoidal vessel is dependent upon the major-to-minor axis ratio a/b, and is found by substituting the value of σ_1 from Eq. 2.6.28 and σ_2 from Eq. 2.6.29 in Eq. 2.4.2, noting $a = r$, to give

$$\delta = \frac{r}{E}\left[\frac{pa}{h}\left(1 - \frac{a^2}{2b^2}\right) - \frac{\mu pa}{2h}\right] \tag{2.4.6}$$

$$\delta = \frac{pr^2}{hE}\left(1 - \frac{a^2}{2b^2} - \frac{\mu}{2}\right) \tag{2.4.7}$$

The equatorial dilation is not always positive or outward from the center, as with a cylinder or sphere, but may be inward depending upon the a/b ratio. For instance, if the vessel material is steel which has a Poisson's Ratio $\mu = 0.3$, Eq. 2.4.7 shows that the equatorial dilation will be negative, or inward, for $a/b > 1.3$. It is this behavior that causes an increase in the discontinuity stresses when ellipsoidal heads are used instead of hemispherical ones for end closures on cylindrical shells of equal thickness, Par. 4.7.1.

2.5 Intersecting Spheres

The sphere is an ideal pressure vessel because (1) stresswise it gives the lowest possible value, (2) storagewise it contains the largest volume with minimum surface area, and (3) costwise it has both minimum thickness and surface area, hence, lowest material weight and cost. When requirements exceed those possible or practicable for a single sphere, multiple intersecting spheres can be used. The basic membrane stresses can be found from Eq. 2.2.7, and the stresses on the intersection reinforcement member for equal size spheres, Fig. 2.8, by noting that the unit pull of the sphere wall is $pr/2$ which gives an out-

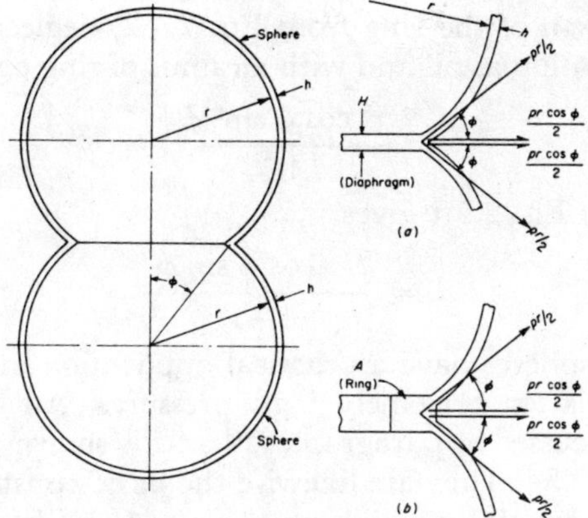

Fig. 2.8. Forces acting at the Juncture of Intersecting Spheres

ward radial component at the intersection of $(pr \cos \phi)/2$ for each sphere, or a total of $pr \cos \phi$. The stress in a sphere is $pr/2h$, and corresponding unit elongation is $pr(1 - \mu)/2Eh$. A reinforcement diaphragm, Fig. 2.8*a*, must dilate the same as the spheres at the intersection in order to eliminate bending in the shell. This is accomplished by designing the diaphragm so that under a radial pull of $pr \cos \phi$ per unit of intersection circumference it will grow the same amount as the natural radial growth of the spherical vessel at the radius of the intersecting circle. The radial growth of the intersecting circle, with radius $r \sin \phi$, is

$$\delta_1 = \frac{pr^2(1-\mu) \sin \phi}{2Eh} \tag{2.5.1}$$

The radial growth of a solid continuous diaphragm of thickness H, neglecting the effect of pressure on the diaphragm, is

$$\delta_2 = \frac{pr \cos \phi(1-\mu) r \sin \phi}{EH} \tag{2.5.2}$$

and equating Eqs. 2.5.1 and 2.5.2 gives a diaphragm thickness of

$$H = 2h \cos \phi$$

If a reinforcing ring is used, Fig. 2.8*b*, the same procedure can be followed by letting A equal the cross-sectional area of the ring; the hoop stress on the ring of Eq. 2.2.3 then becomes

$$\sigma_2 = \frac{pr^2 \cos \phi \sin \phi}{A} \tag{2.5.3}$$

The radial growth of the ring from Eq. 2.4.2, neglecting the second term of this equation as nil and with a radius of ring equal to $r \sin \phi$, is

$$\delta_2 = \frac{pr^3 \cos \phi \sin^2 \phi}{AE} \tag{2.5.4}$$

Equating this to Eq. 2.5.1 gives

$$A = \frac{2hr \cos \phi \sin \phi}{1-\mu} \tag{2.5.5}$$

Intersecting spheres have a practical application in the economical design of vessels for extremely high pressures. An example of this approach is discussed in paragraph 7.9 and is shown subsequently in Figs. 7.33 and 7.34. They are likewise the basic construction used for deep-diving submersibles which must embody both minimum weight for buoyancy and maximum strength for pressure.[5]

The effect of non-circularity and residual stress on the collapse of spheres is discussed in paragraph 8.6.

2.6 General Theory of Membrane Stresses in Vessels Under Internal Pressure

The membrane stresses in vessels of revolution, including those of complicated geometry, can be evaluated from the equations of statics provided they are loaded in a rotationally symmetrical manner—the pressure loading need not be the same throughout the entire vessel but only on any plane perpendicular to the axis of rotation 0–0, Fig. 2.9.

In the vessel of Fig. 2.9*a* if an element *abef* is cut by two meridional sections, *ab* and *ef*, and by two sections *ae* and *bf* normal to these meridians, it is seen that a condition of symmetry exists and only normal stresses act on the sides of this element. Let:

- σ_1 = longitudinal or meridional stress (stress in the meridional direction)
- σ_2 = hoop stress (stress along a parallel circle)
- h = thickness of vessel
- ds_1 = element dimension in the meridional direction (face *ab* and *ef*)
- ds_2 = element dimension in the hoop direction (face *ae* and *bf*)
- r_1 = longitudinal or meridional radius of curvature
- r_2 = radius of curvature of the element in the hoop direction (perpendicular to the meridian)
- p = pressure

Referring to Fig. 2.9*a*, the total forces acting on the sides of the element are $\sigma_1 h ds_2$ and $\sigma_2 h ds_1$. The force $\sigma_2 h ds_1$ has a component in a direction

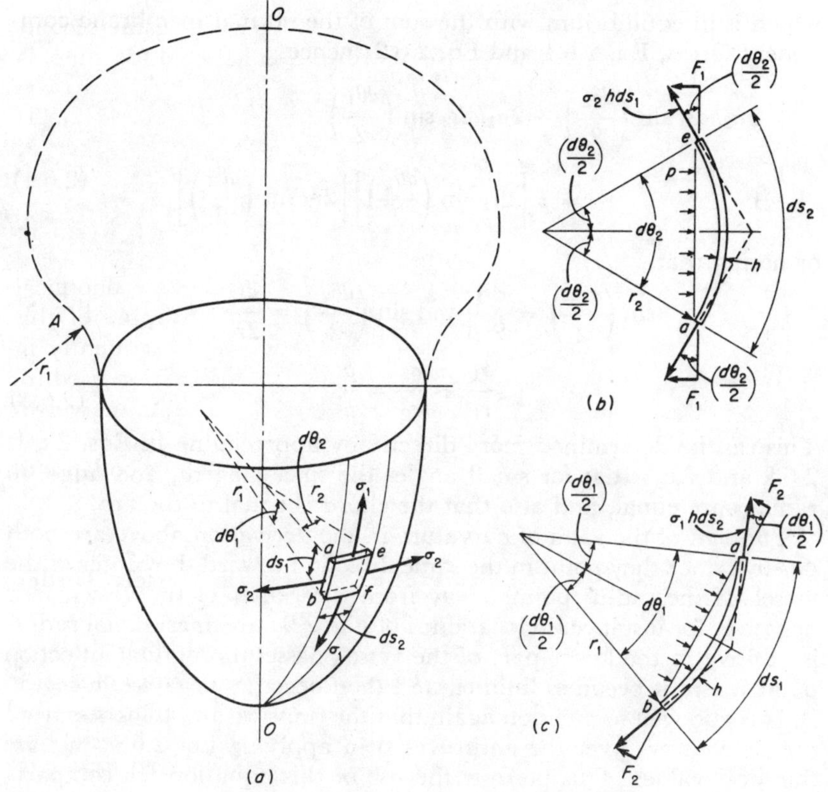

Fig. 2.9. Membrane Stresses in Vessels

normal to the element, Fig. 2.9*b*, of

$$2F_1 = 2\sigma_2 h ds_1 \sin\left(\frac{d\theta_2}{2}\right) \tag{2.6.1}$$

and similarly the force $\sigma_1 dhs_2$ has a component in a direction normal to the element, Fig. 2.9*c*, of

$$2F_2 = 2\sigma_1 h ds_2 \sin\left(\frac{d\theta_1}{2}\right) \tag{2.6.2}$$

The normal pressure force on the element is

$$P = p\left[2r_1 \sin\left(\frac{d\theta_1}{2}\right)\right]\left[2r_2 \sin\left(\frac{d\theta_2}{2}\right)\right] \tag{2.6.3}$$

which is in equilibrium with the sum of the normal membrane component forces, Eq. 2.6.1 and Eq. 2.6.2; hence,

$$2\sigma_2 h ds_1 \sin\left(\frac{d\theta_2}{2}\right) + 2\sigma_1 h ds_2 \sin\left(\frac{d\theta_1}{2}\right) = p\left[2r_1 \sin\left(\frac{d\theta_1}{2}\right)\right]\left[2r_2 \sin\left(\frac{d\theta_2}{2}\right)\right] \tag{2.6.4}$$

or noting that

$$\sin\left(\frac{d\theta_1}{2}\right) = \frac{ds_1}{2r_1} \text{ and } \sin\left(\frac{d\theta_2}{2}\right) = \frac{ds_2}{2r_2};$$

$$\frac{\sigma_1}{r_1} + \frac{\sigma_2}{r_2} = \frac{p}{h} \tag{2.6.5}$$

This can be determined more directly by appreciating in Eqs. 2.6.1, 2.6.2 and 2.6.3 that for small angles the sine, tangent, and angle in radians are equal, and also that the chord is equal to the arc.

The sign of the radii of curvature in the derivation above are both positive since they point in the same direction toward the center of the vessel. If the radius points away from the center of the vessel, it is negative; for instance, in location *A* of Fig. 2.9*a*, the meridional radius is positive in the lower part of the vessel, passes through an inflection point where it becomes infinite, and then becomes negative in region *A*. It is also well to mention again that the pressure and thickness need not be constant over the entire vessel in applying Eq. 2.6.5, but are the local values. This permits the use of this equation for compartmented vessels which operate at different pressures, or those subject to varying pressures due to a head of fluid, and likewise embody corresponding thickness adjustments.

Some applications of this equation for commonly used geometric shapes are discussed below.[6, 7, 8, 9, 52, 58]

1. *Cylindrical Vessel Under Internal Pressure*

In the case of the cylinder portion of a vessel under internal pressure p, the hoop radius $r_2 = r$, the longitudinal radius $r_1 = \infty$, and each is constant throughout the entire cylinder. Substituting these values into Eq. 2.6.5 gives

$$\frac{\sigma_1}{\infty} + \frac{\sigma_2}{r} = \frac{p}{h} \tag{2.6.6}$$

$$\sigma_2 = \frac{pr}{h} \text{ (hoop stress)} \tag{2.6.7}$$

The longitudinal stress can be found, as in paragraph 2.2, by equating the longitudinal forces producing extension to the total pressure force on this cross section of the vessel

$$\sigma_1 2\pi rh = p\pi r^2 \tag{2.6.8}$$

$$\sigma_1 = \frac{pr}{2h} \text{ (longitudinal stress)} \tag{2.6.9}$$

2. *Spherical Vessel Under Internal Pressure*

In the case of a sphere, the longitudinal and hoop radii are equal, $r_1 = r_2 = r$, and from symmetry it follows that $\sigma_1 = \sigma_2 = \sigma$. Thus, Eq. 2.6.5 becomes

$$\sigma = \frac{pr}{2h} \tag{2.6.10}$$

3. *Conical Vessel Under Internal Pressure*

In this case, Fig. 2.10, it is seen that $r_1 = \infty$, just as in the case of a cylinder, since its generatrix is a straight line, and $r_2 = r/\cos\alpha$. Thus from Eq. 2.6.5,

$$\sigma_2 = \frac{pr}{h\cos\alpha} \tag{2.6.11}$$

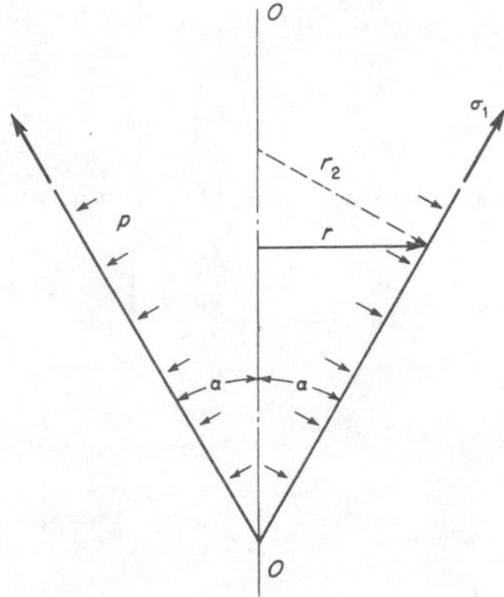

Fig. 2.10 Stresses in a Conical Vessel

from which it is seen that (1) the hoop stress approaches that in a cylinder as α approaches zero, and (2) the stress becomes infinitely large as α approaches 90° and the cone flattens out into a plate. The latter merely verifies the assumption that a flat membrane cannot take loads perpendicular to its plane.

The longitudinal stress can be found by equating the axial component of this force in the vessel wall to the total pressure force on a plane perpendicular to the axis of revolution:

$$\sigma_1 h 2\pi r \cos\alpha = p\pi r^2 \tag{2.6.12}$$

$$\sigma_1 = \frac{pr}{2h\cos\alpha} \tag{2.6.13}$$

and comparable deductions to those above from the hoop stress can be made.

4. *Ellipsoidal Vessel Under Internal Pressure*

Ellipsoidal shaped heads are frequently used for the end closure of cylindrical shells for steam boilers, reactors, and storage vessels in order to accommodate special space or volume requirements.[10,11] In such constructions a half of an ellipsoid is used, Fig. 2.11. Since the radius of curvature varies from point to point, the solution of Eq. 2.6.5 becomes somewhat more complicated than for those geometric shapes of constant radii. An ellipse of semi-major axis a and semi-minor axis b is described in orthogonal coordinates x and y by the equation

$$b^2x^2 + a^2y^2 = a^2b^2 \tag{2.6.14}$$

$$y = \pm\frac{b}{a}\sqrt{a^2 - x^2} \tag{2.6.15}$$

The radius of curvature at any point* is given by

$$\rho = \frac{\left[1 + \left(\frac{dy}{dx}\right)^2\right]^{3/2}}{\frac{d^2y}{dx^2}} \tag{2.6.16}$$

Differentiating Eq. 2.6.15 gives

$$\frac{dy}{dx} = \frac{-bx}{a\sqrt{a^2 - x^2}} = \frac{-b^2x}{a^2y}, \text{ and } \frac{d^2y}{dx^2} = \frac{-ba^2}{a\sqrt{(a^2 - x^2)^3}} = \frac{-b^4}{a^2y^3} \tag{2.6.17}$$

*See any text on calculus. This equation gives the radius of curvature of any point on a continuous curve with the sign merely mathematically indicating the direction of concavity.

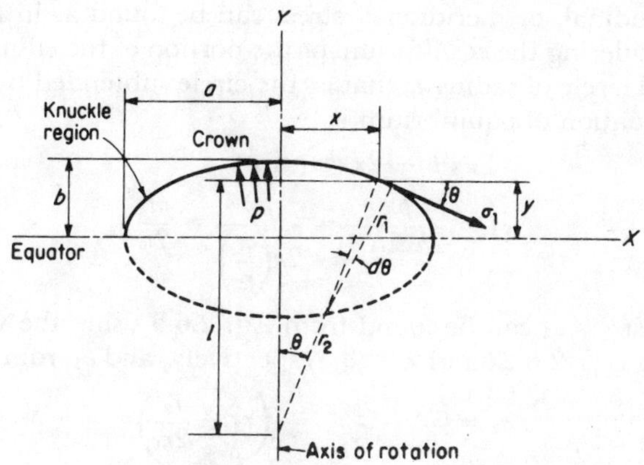

Fig. 2.11. Stress in an Ellipsoid

Substituting these values in Eq. 2.6.16 gives

$$r_1 = |\rho| = \frac{\left[1 + \left(\dfrac{b^2x}{a^2y}\right)^2\right]^{\frac{3}{2}}}{\dfrac{b^4}{a^2y^3}} = \frac{[a^4y^2 + b^4x^2]^{\frac{3}{2}}}{a^4b^4} \tag{2.6.18}$$

The radius of curvature r_2 in the hoop direction is the length of the normal from the ellipse to the axis of rotation. The slope of the tangent to the ellipse at coordinates x and y is given by

$$\tan\theta = \frac{dy}{dx} = \frac{bx}{a\sqrt{a^2 - x^2}} \tag{2.6.19}$$

Noting $\tan\theta = x/l$ and substituting the value of $\tan\theta$ from Eq. 2.6.19 gives

$$l = \frac{a}{b}\sqrt{a^2 - x^2} \tag{2.6.20}$$

Also

$$r_2 = \sqrt{l^2 + x^2} \tag{2.6.21}$$

and substituting the value of l from Eq. 2.6.20 gives

$$r_2 = \sqrt{\frac{a^2(a^2 - x^2)}{b^2} + x^2} = \frac{(a^4y^2 + b^4x^2)^{\frac{1}{2}}}{b^2} \tag{2.6.22}$$

or

$$r_1 = r_2^3\frac{b^2}{a^4} \tag{2.6.23}$$

The longitudinal, or meridional, stress can be found as in Paragraph 2.2 by considering the equilibrium of the portion of the ellipsoid above the parallel circle of radius x; that is, the circle subtended by the angle 2θ. The equation of equilibrium is

$$\pi x^2 p - 2\pi x h \sigma_1 \sin\theta = 0 \tag{2.6.24}$$

$$\sigma_1 = \frac{px}{2h\sin\theta} = \frac{px}{2h\left(\dfrac{x}{r_2}\right)} = \frac{pr_2}{2h} \tag{2.6.25}$$

The hoop stress σ_2 can be found from Eq. 2.6.5 using the values of r_1 and r_2 from Eqs. 2.6.22 and 2.6.23, respectively, and σ_1 from Eq. 2.6.25

$$\sigma_2 = \frac{pr_2}{h} - \frac{r_2\sigma_1}{r_1} = \frac{p}{h}\left(r_2 - \frac{r_2^2}{2r_1}\right) \tag{2.6.26}$$

At the crown $r_1 = r_2 = a^2/b$ and from Eqs. 2.6.25 and 2.6.26

$$\sigma_1 = \sigma_2 = \frac{pa^2}{2bh} \tag{2.6.27}$$

At the equator $r_1 = b^2/a$ and $r_2 = a$, so from Eq. 2.6.25

$$\sigma_1 = \frac{pa}{2h} \tag{2.6.28}$$

which is the same as the longitudinal stress in a cylinder, while from Eq. 2.6.26

$$\sigma_2 = \frac{pa}{h}\left(1 - \frac{a^2}{2b^2}\right) \tag{2.6.29}$$

and it is seen that the hoop stress becomes compressive if $a/b > 1.42$.

As the a/b ratio increases above 1.42, the location of the maximum shearing stress, to which the failure of ductile materials subscribe, Par. 5.15, shifts from the center of the crown where the maximum shearing stress is, noting the average radial stress σ_r through the thickness is $p/2$,

$$\tau_{cmax} = \frac{\sigma_1 - \sigma_r}{2} = \frac{\dfrac{pa^2}{2bh} - \left(-\dfrac{p}{2}\right)}{2} = \frac{p}{4}\left(\frac{a^2}{bh} + 1\right) \tag{2.6.30}$$

to a maximum at the equator where

$$\tau_{emax} = \frac{\sigma_1 - \sigma_2}{2} = \frac{\dfrac{pa}{2h} - \dfrac{pa}{h}\left(1 - \dfrac{a^2}{2b^2}\right)}{2} \tag{2.6.31}$$

$$= \frac{pa}{4h}\left(\frac{a^2}{b^2} - 1\right) \tag{2.6.32}$$

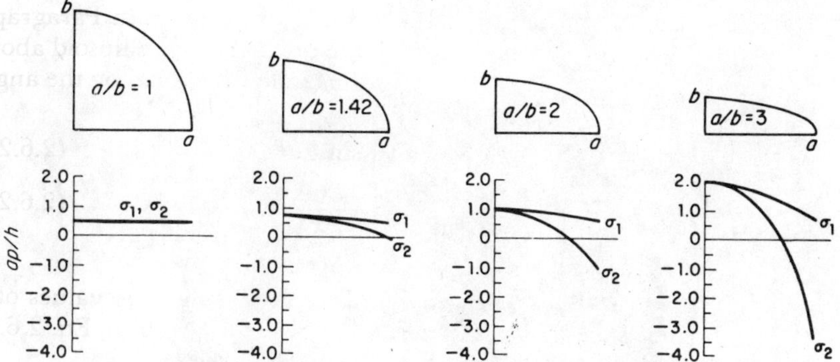

Fig. 2.12. Ratio of Stress in an Ellipsoid to Stress in a Cylinder with Variation in Ratio of Major-to-Minor Axis

The variation in stress throughout an ellipsoid for increasing a/b ratios is shown in Fig. 2.12. The meridional stress remains tensile throughout the ellipsoid for all a/b ratios, being a maximum at the crown and diminishing in value to a minimum at the equator. The hoop stress is also tensile in the crown region but this decreases as the equator is approached where it becomes compressive for a/b ratios greater than 1.42. A ratio $a/b = 1$, a sphere, gives the lowest stress. If $a/b = 2$ a maximum tensile stress of pa/h, which is the same as the hoop stress in a cylinder, occurs at the center of the crown and a hoop compressive stress of equal magnitude occurs at the equator. Many construction codes and specifications restrict the use of elliptical heads to those with a maximum major-to-minor axis ratio of 2.0. As the ratio a/b is further increased the greatest stress in the crown is still tension and lies at the center, but is far exceeded in magnitude by the compressive hoop stress in the knuckle region and at the equator. It is this compressive stress that can cause:

1. Local buckling of thin heads due to the high hoop compressive stress,[12, 13, 14, 15, 38, 39, 54] Fig. 2.13.
2. Local failure due to the high shear stress developed,[16, 40] Fig. 2.14.

Likewise, torospherical or dished heads,[53] which simulate ellipsoidal ones by a compound curve composed of a crown radius and a knuckle radius, Fig. 2.15, should have a large knuckle radius in order to minimize the hoop stresses in this region.[17, 18, 19] Many pressure vessel construction codes recognize this fact and specify a minimum permissible

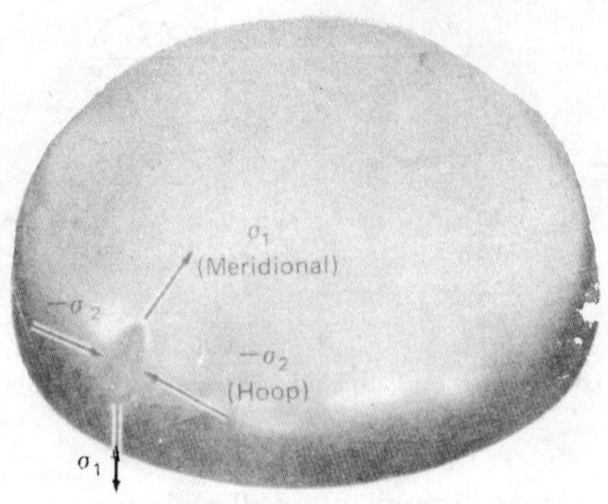

Fig. 2.13. Initial Buckle in the Knuckle Region of an Ellipsoidal Model Test Head Subject to Internal Pressure[15]

knuckle radius. For instance, the ASME unfired Pressure Vessel Code specifies the minimum value of the knuckle radius as 6 percent of the crown radius.

The local buckling of thin elliptical or torospherical heads in the knuckle[41,42,43] region is akin to that in the thin webs of beams and girders at the high shear stress region of their supports. In these thin vessel heads, this phenomenon can be prevented by providing:

1. An entire head of adequate thickness, or
2. An annular knuckle region of adequate thickness,[44] or
3. Annular structural stiffeners in the knuckle region.

The first two methods are generally more economical for initial construction; whereas, the third method is more economical and more readily adaptable to repairs and alterations.

The buckling pressure is not sensitive to geometric imperfections, and initial wrinkles less than the head thickness do not increase during pressurization.[59] This is attributed to the stable post buckling behavior under internal pressure, and is in great contrast to the situation where the head is under external pressure and an unstable condition

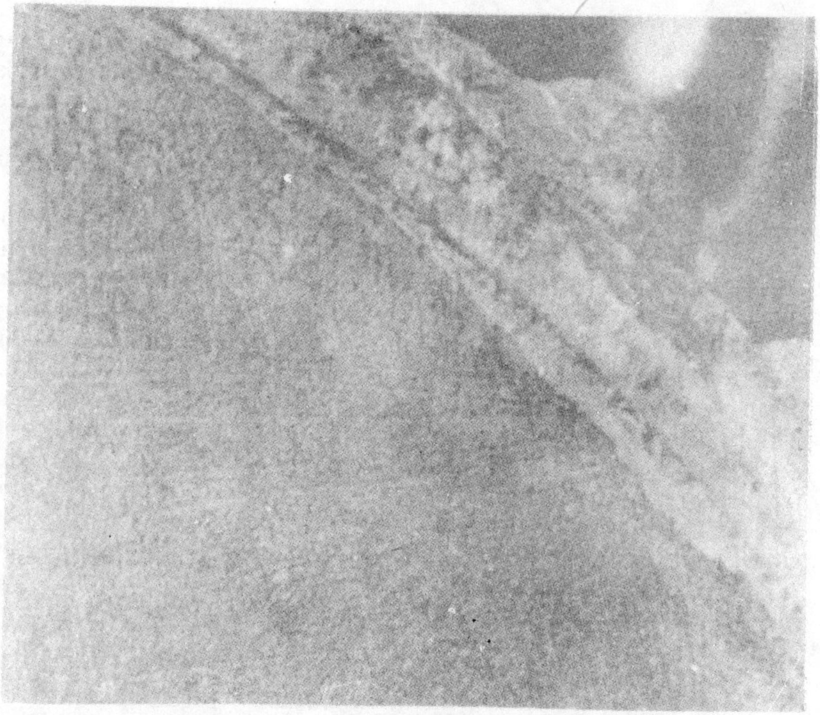

Fig. 2.14. Lueders' Lines on the Outside Surface of the Knuckle of a Shallow Ellipsoidal Head Subject to Internal Pressure Indicating Yielding of the Metal in This Highly Stressed Region[16]

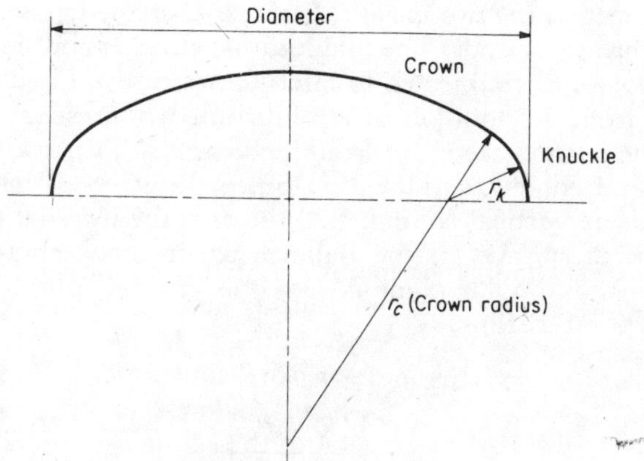

Fig. 2.15. Torospherical Head

prevails wherein imperfections are of paramount importance, Par. 8.8. As with most buckling phenomena, predictions based on the tangent or reduced modulus are conservative while those employing Young's modulus are unconservative, and combination of the two seem to be in closer agreement with experiment, Par. 8.2.4 and 8.6.2.

This local instability under internal pressure does not represent total failure as compared to buckling of vessels under external pressure which leads to total collapse, Chapter 8. While the design approaches described above to prevent this local buckling require increasing the bending stiffness by local reinforcement (increase in thickness or stiffeners), another approach is to "correct" the shape of the pressure vessel in the region of impending instability by minimizing or eliminating the initating harmful compressive stresses; i.e., create a "buckle-free" shape in which only a tension stress state exists, whereby the possibility of buckling is excluded. This approach is well adapted to very large thin vessels, such as fluid storage tanks. In the frequently used cylindrical pressure vessel with dished end closures subjected to internal constant pressure, the "buckle-free" shape is in close agreement with the 2:1 ellipsoidal closure.[60]

The buckling of ellipsoidal and torospherical heads under external pressure, and the stabilizing effect of ring stiffeners is given in Chapter 8.

2.7 Torus Under Internal Pressure

A torus or doughnut shape, or part thereof, is one of the most useful and widely used shapes in vessel construction. Figure 2.16 is a photograph of a steam generator during fabrication showing the 180° torus section connecting the two parallel legs of a U-shaped vessel, as well as the numerous tube bends. The longitudinal stress σ_1 and hoop stress σ_2 in the wall of a torus subjected to internal pressure p, Fig. 2.17, can be calculated from a condition of equilibrium with respect to vertical forces on the portion aba_1b_1 cut from the vessel by a vertical cylinder of radius R_0 and conical surface aoa_1. Since the stress σ_2 acting on the circle R_0 has no vertical component, nor does the internal pressure on the cylindrical surface bc, the balance occurs between the internal pressure on the annular plane ac and the vertical component of the stress σ_2 at point a; thus,

$$\pi p(R^2 - R_0^2) - \sigma_2 h 2\pi R \sin\theta = 0 \tag{2.7.1}$$

$$\sigma_2 = \frac{p(R^2 - R_0^2)}{2hR \sin\theta} \tag{2.7.2}$$

Fig. 2.16. Steam Generator During Fabrication Showing the Use of Torus Sections (*Courtesy The Babcock & Wilcox Company*)

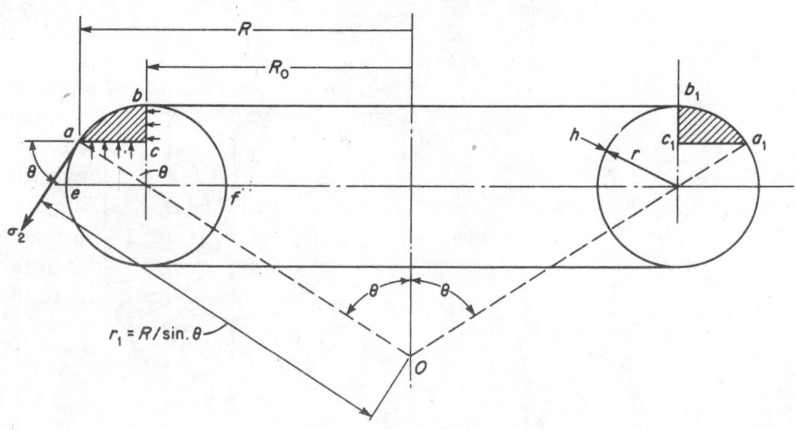

Fig. 2.17. Stresses in a Torus

where R is the radius at any point, and R_0 is the radius of the centerline. Since $R = R_0 + r \sin\theta$,

$$\sigma_2 = \frac{p[(R_0 + r\sin\theta)^2 - R_0^2]}{2h(R_0 + r\sin\theta)\sin\theta} \tag{2.7.3}$$

$$\sigma_2 = \frac{pr}{2h}\,\frac{2R_0 + r\sin\theta}{R_0 + r\sin\theta} \tag{2.7.4}$$

On the centerline of the torus, $\theta = 0$ and Eq. 2.7.4 reduces to

$$\sigma_2 = \frac{pr}{h} \tag{2.7.5}$$

which is the hoop stress in a straight cylinder of radius r. The minimum hoop stress occurs at the outside of the torus on an axial plane of symmetry through the point e, $\theta = \pi/2$, Fig. 2.18, of

$$\sigma_2 = \frac{pr}{2h}\,\frac{2R_0 + r}{R_0 + r} \tag{2.7.6}$$

and is a maximum at the crotch point f, $\theta = 3\pi/2$, of

$$\sigma_2 = \frac{pr}{2h}\,\frac{2R_0 - r}{R_0 - r} \tag{2.7.7}$$

The longitudinal stress σ_1 can be calculated from Eq. 2.6.5, noting that $r_1 = R/\sin\theta$, so that

$$\frac{\sigma_1 \sin\theta}{R} + \frac{\sigma_2}{r} = \frac{p}{h} \tag{2.7.8}$$

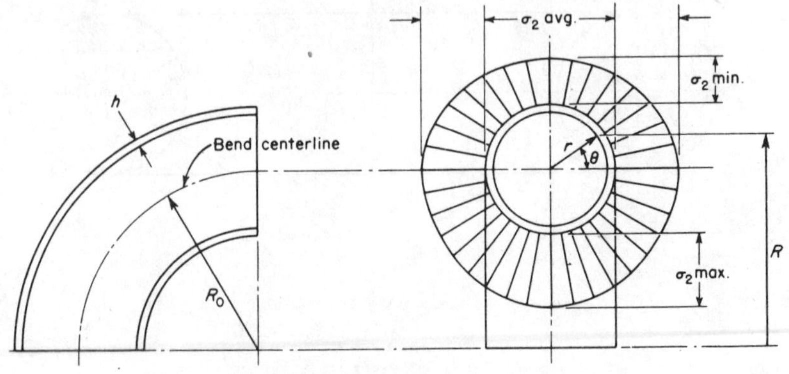

Fig. 2.18. Variation in Hoop Stress in a Bend

which gives, when substituting $R = R_0 + r \sin \theta$, a value of

$$\sigma_1 = \frac{pr}{2h} \tag{2.7.9}$$

and is the same as for a straight cylinder. Its value is independent of location on the torus.

Hence, in a torus of uniform thickness, both the principal stresses are tensile, with the hoop stress the greater and reaching a maximum at the crotch where failure would be expected to occur first. Figure 2.18 shows the variation in this stress around a cross section through the torus, whereas Fig. 2.19 shows the variation in this stress at the crotch point of maximum intensity with the radius of bend centerline, from which it is seen that this stress becomes large for small bend radii (doughnut with a small hole). Conventional pipe or tube bends[45] are made by pushing or pulling them around a former, of the required radius. The operation is usually performed cold when the size is small and/or the bend radius generous; when the size is large and/or the bend radius sharp, hot forming is done. The natural redistribution of

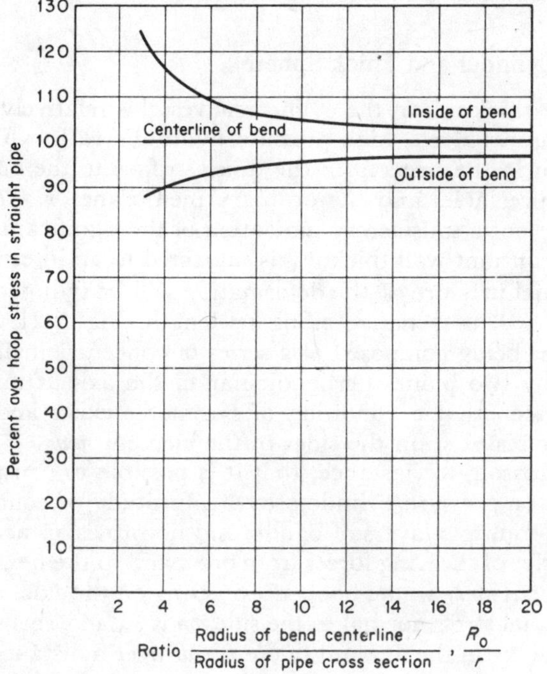

Fig. 2.19. Variation in Hoop Stress with Bend Radii

metal which occurs during bending, thinning at the outside and thickening at the inside, is a compensating factor of the same order as the acting stress; hence, the requirement that conventional pipe bends be made of thicker material to adjust for thinning during bending is seldom warranted for the ratios of R_0/r customarily used. In fact, this and other associated factors, such as strain hardening, usually result in pressure failures occurring in the straight portion of pipes or tubes. This is illustrated in the internal and external pressure tests of the tube bend of Fig. 2.20, showing the rupture to occur in the straight cylindrical portion under internal pressure. When failure was forced to occur in the torus portion, Fig. 2.20*d*, the rupture took place on the centerline of the bend where the stress is the same as that in a straight cylinder and, incidentally, where the material tensile strength had been increased the least by strain hardening (paragraph 5.7) from the fabricating process, but which nevertheless accounted for a 20 per cent increase in bursting pressure. Collapse under external pressure also occurred first in the straight portion, Fig. 2.20*c*, due to the stabilizing effect of the double curvature in the torus region. For this particular size torus, the collapse pressure was 93 per cent higher, Fig. 2.20*e*, than that for a cylinder of the same size and thickness.

2.8 Thick Cylinder and Thick Sphere

When the thickness of the cylindrical vessel is relatively large, as in the case of gun barrels, high-pressure hydraulic ram cylinders, etc., the variation in the stress from the inner surface to the outer surface becomes appreciable, and the ordinary membrane or average stress formulas are not a satisfactory indication of the significant stress. If a cylinder of constant wall thickness is subjected to an internal pressure p_i and external pressure p_0, the deformation will be symmetrical about its axis and will not change along its length, Fig. 2.21. It may be thought of as being composed of a series of concentric cylinders. If a ring is cut by two planes perpendicular to the axis at unit distance apart, it is seen that a condition of symmetry exists and hence no shearing stresses exist on the sides of the element mnm_1n_1, Fig. 2.21*a*. This is the reason, for instance, that it is possible to construct heavy gun barrels or pressure cylinders by a multitude of concentric thin cylinders or multiple layers of bands, and no provision need be made for the transfer of shearing forces from one band to the next. Considering the element mnm_1n_1, the hoop stress acting on the sides mm_1 and nn_1 is σ_t. The radial stress normal to the side mn is σ_r, and this stress varies with the radius r in the amount of $(d\sigma_r/dr)dr$ over a distance dr. There-

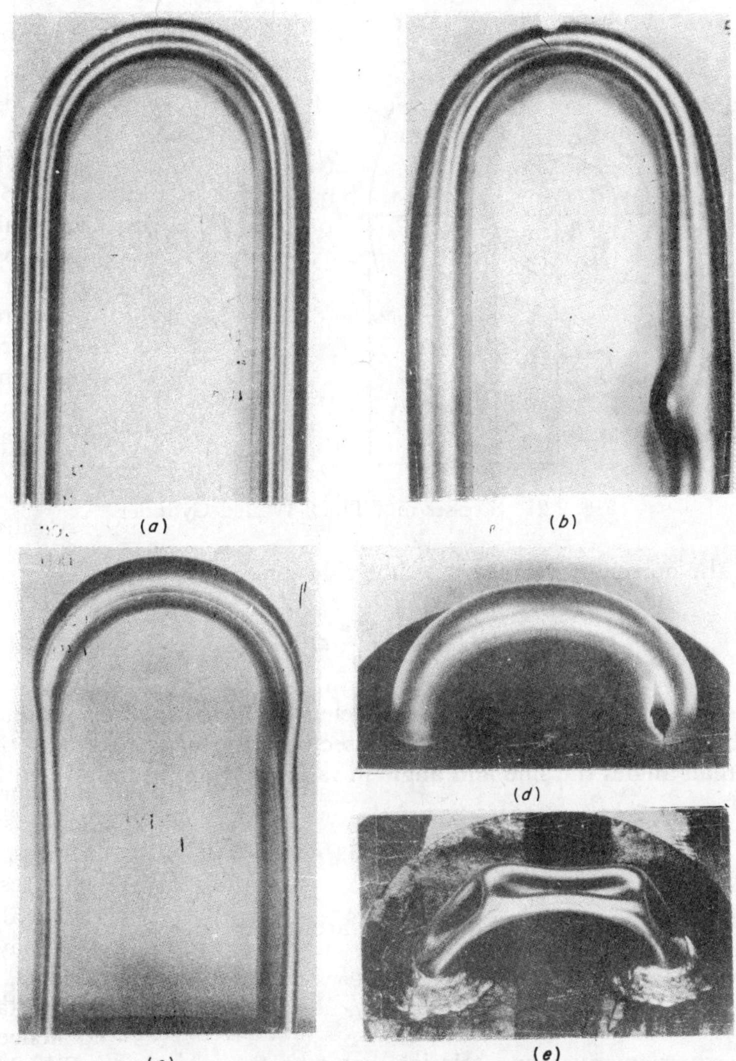

Fig. 2.20. Failure of Cylindrical and Torus Sections.
(R_0 = 1.3125 in., r = 0.311 in., h = 0.058 in., T.S. = 87,000 psi)
(*a*) Unpressurized Vessel, (*b*) Failure at Internal Pressure, P_i = 17,500 psi, Showing Gross Deformation and Rupture in the Cylindrical Section, (*c*) Failure at External Pressure, p_0 = 6,800 psi, Showing Collapse of the Cylindrical Section, (*d*) Rupture of Torus Section at Internal Pressure p_i = 21,000 psi, (*e*) Collapse of Torus Section at External Pressure p_0 = 12,200 psi

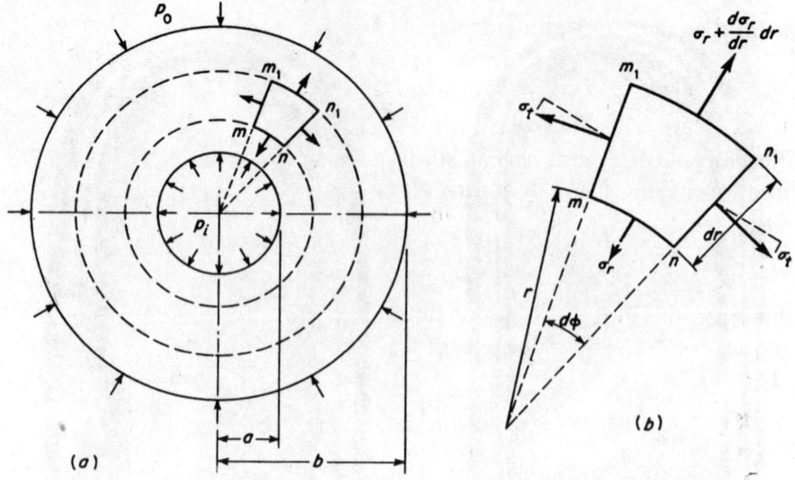

Fig. 2.21. Stresses in a Thick-Walled Cylinder

fore, the normal radial stress on the side m_1n_1 is

$$\sigma_r + \frac{d\sigma_r}{dr} dr \tag{2.8.1}$$

The equation of equilibrium for the element is obtained by summing up the forces in the direction of the bisector of the angle $d\phi$, noting that for small angles the sine and angle in radians are substantially equal. Then

$$\sigma_r r d\phi + \sigma_t dr d\phi - \left(\sigma_r + \frac{d\sigma_r}{dr} dr\right)(r + dr) d\phi = 0 \tag{2.8.2}$$

and if small quantities of high order are neglected,

$$\sigma_t - \sigma_r - r\frac{d\sigma_r}{dr} = 0 \tag{2.8.3}$$

The equation gives one relation between the stresses σ_t and σ_r. A second relation can be obtained from the deformation of the cylinder and from the assumption that the longitudinal strain of all fibers is equal. The deformation of the cylinder is then symmetrical with respect to the axis and consists of a radial displacement of all points in the wall of the cylinder. Hence, this displacement is constant in the circumferential direction but varies with distance along a radius. If u denotes the radial displacement of a cylindrical surface of radius

radial displacement of a surface of radius $r + dr$ is

$$u + \frac{du}{dr}\,dr \tag{2.8.4}$$

Therefore, an element mnm_1n_1 undergoes a total elongation in a radial direction of $(du/dr)dr$, or a unit elongation of

$$e_r = \left(\frac{du}{dr}\right)\frac{dr}{dr} = \frac{du}{dr} \tag{2.8.5}$$

In a circumferential direction the unit elongation of the same element is equal to the unit elongation of the corresponding radius, paragraph 2.4, or

$$e_t = \frac{u}{r} \tag{2.8.6}$$

Then from Eqs. 2.3.4 and 2.3.5 a second set of expressions for the stresses in terms of the strains becomes

$$\sigma_r = \frac{E}{1-\mu^2}\left(\frac{du}{dr} + \mu\,\frac{u}{r}\right) \tag{2.8.7}$$

$$\sigma_t = \frac{E}{1-\mu^2}\left(\frac{u}{r} + \mu\,\frac{du}{dr}\right) \tag{2.8.8}$$

These stresses are interdependent since they are expressed in terms of one function u. By substituting the values for σ_r and σ_t from Eqs. 2.8.7 and 2.8.8 into Eq. 2.8.3, the following equation for determining u is obtained:

$$\frac{d^2u}{dr^2} + \frac{1}{r}\frac{du}{dr} - \frac{u}{r^2} = 0 \tag{2.8.9}$$

The general solution of this equation is

$$u = C_1 r + \frac{C_2}{r} \tag{2.8.10}$$

Substituting from Eq. 2.8.10, noting $du/dr = C_1 - C_2/r^2$, into Eqs. 2.8.7 and 2.8.8 gives

$$\sigma_r = \frac{E}{1-\mu^2}\left[C_1(1+\mu) - C_2\,\frac{1-\mu}{r^2}\right] \tag{2.8.11}$$

$$\sigma_t = \frac{E}{1-\mu^2}\left[C_1(1+\mu) + C_2\,\frac{1-\mu}{r^2}\right] \tag{2.8.12}$$

The constants C_1 and C_2 can be determined from the conditions at the inner and outer surfaces of the cylinder where the pressure, i.e., the normal stresses σ_r, are known. For instance, if p_i denotes the internal pressure and p_o denotes the external pressure, the conditions at the inner and outer surfaces of the cylinder are

$$\sigma_{r_a} = -p_i \quad \text{and} \quad \sigma_{r_b} = -p_o \tag{2.8.13}$$

The negative sign on the right-hand side of these equations indicates that the stress is compressive, because the normal stress is considered positive for tension. Substituting the expressions for σ_r from Eq. 2.8.11 into Eq. 2.8.13 gives equations for determining the constants C_1 and C_2 from which

$$C_1 = \frac{1-\mu}{E}\,\frac{a^2p_i - b^2p_o}{b^2 - a^2}; \qquad C_2 = \frac{1+\mu}{E}\,\frac{a^2b^2(p_i - p_o)}{b^2 - a^2} \tag{2.8.14}$$

Placing the values of these into Eqs. 2.8.11 and 2.8.12 gives the general expressions for the normal stresses:

$$\sigma_r = \frac{a^2p_i - b^2p_o}{b^2 - a^2} - \frac{(p_i - p_o)a^2b^2}{r^2(b^2 - a^2)} \tag{2.8.15}$$

$$\sigma_t = \frac{a^2p_i - b^2p_o}{b^2 - a^2} + \frac{(p_i - p_o)a^2b^2}{r^2(b^2 - a^2)} \tag{2.8.16}$$

Inspection of Eqs. 2.8.15 and 2.8.16 indicates that the maximum value of σ_t occurs at the inner surface, and maximum σ_r will always be the larger of the two pressures, p_i and p_o. These equations are known as the Lame solution, or thick-cylinder formulas. It is noted that the sum of these two stresses remains constant; hence the deformation of all elements in the axial direction is the same, and cross sections of the cylinder remain plane after deformation, thereby fulfilling the original assumption.

The maximum shearing stress at any point in the cylinder is equal to one half the algebraic difference of the maximum and minimum principal stresses at that point. Since the longitudinal (axial) stress is an intermediate value between σ_r and σ_t,

$$\tau = \frac{\sigma_t - \sigma_r}{2} = \frac{(p_i - p_o)}{b^2 - a^2}\,\frac{a^2b^2}{r^2} \tag{2.8.17}$$

1. *Cylinder Under Internal Pressure Only*

In this particular case which covers most of the practical vessel applications $p_o = 0$, Eqs. 2.8.15 and 2.8.16 reduce to:

$$\sigma_r = \frac{a^2 p_i}{b^2 - a^2}\left(1 - \frac{b^2}{r^2}\right) \tag{2.8.18}$$

$$\sigma_t = \frac{a^2 p_i}{b^2 - a^2}\left(1 + \frac{b^2}{r^2}\right) \tag{2.8.19}$$

These equations show that both stresses are maximum at the inner surface where r has the minimum value; σ_r is always a compressive stress, and smaller than σ_t; and σ_t a tensile stress which is maximum at the inner surface of the cylinder equal to

$$\sigma_{t_{\text{max.}}} = \frac{p_i(a^2 + b^2)}{b^2 - a^2} \tag{2.8.20}$$

From Eq. 2.8.20 it is seen that $\sigma_{t_{\text{max.}}}$ is always numerically greater than the internal pressure, but approaches this value as b increases. The minimum value of σ_t is at the outer surface of the cylinder and is always less than that at the inner surface by the value of the internal pressure p_i. Figure 2.22 illustrates this variation through the wall of a thick cylinder of ratio $K = \dfrac{\text{outside radius}}{\text{inside radius}} = 2.0$. In designing for very high pressure, these observations point out the necessity of using comparably high yield point materials, or using design-construction features that will create an initial residual compressive stress on the inner surface to help counterbalance the high applied stress at this location, such as hoops shrunk on the barrels of guns, or cylinders strained be-

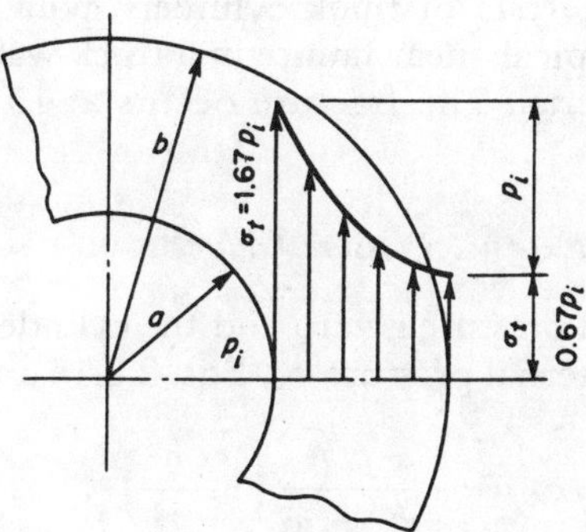

Fig. 2.22. Variation in Tangential Stress Through the Wall of a Thick Cylinder, $K = b/a = 2$

TABLE 2.1. RATIO OF MAXIMUM TO AVERAGE STRESS ON A CYLINDER FOR VARIOUS VALUES OF K (RATIO OF OUTSIDE TO INSIDE RADIUS)

	$K = b/a = 1 + h/a$					
	1.1	1.2	1.4	1.6	1.8	2.0
$\sigma_{max}/\sigma_{avg.}$	1.05	1.10	1.23	1.37	1.51	1.67

yond the yield point by hydraulic pressure so that upon release of the pressure the metal at the bore remains in a state of residual compression and the outer layers in moderate tension.

A comparison between the maximum stress obtained by the thick-cylinder formula, Eq. 2.8.20, and that obtained by the thin cylinder or average stress formula, Eq. 2.2.4, is shown in Table 2.1 for various values of K, and indicates that for small wall thicknesses there is little difference; for instance, with a wall thickness of 20 per cent of the inside radius, the maximum stress is only 10 per cent higher than the average stress.

The shearing stress is a maximum on the inner surface and from Eq. 2.8.17 for $r = a$ gives

$$\tau = \frac{p_i b^2}{b^2 - a^2} \tag{2.8.21}$$

This equation for the shearing stress[46] is particularly significant from a design viewpoint as a criteria of failure, since it correlates very well with actual rupture tests of thick cylinders, paragraph 5.15. Figure 7.16*b* shows the typical shear failure in a thick-walled cylinder made from a ductile material. The fracture occurs at 45° to the maximum principal stress.

2. *Cylinder Under External Pressure Only*

When the internal pressure is zero and the cylinder is subject to only the action of the external pressure p_o, Eqs. 2.8.15 and 2.8.16 reduce to

$$\sigma_r = -\frac{p_o b^2}{b^2 - a^2}\left(1 - \frac{a^2}{r^2}\right) \tag{2.8.22}$$

$$\sigma_t = -\frac{p_o b^2}{b^2 - a^2}\left(1 + \frac{a^2}{r^2}\right) \tag{2.8.23}$$

These equations show that both σ_r and σ_t are compressive stresses with σ_t always being numerically greater than σ_r just as it was in the case of internal pressure only. The maximum tangential compressive stress σ_t occurs on the inside surface of value

$$\sigma_{t_a} = -\frac{2p_o b^2}{b^2 - a^2} \tag{2.8.24}$$

whereas the maximum radial stress σ_r is at the outer surface and equal to p_0; i.e., the maximum values of σ_t and σ_r do not occur at the same point in the cylinder in this case.[47] When the external pressure is reversed in direction, as could result from a vacuum surrounding the cylinder or a series of outwardly directed uniformly applied loads, p_0 is replaced by $-p_0$. When the ratio b/a becomes very large, it is noted that the maximum stress approaches twice the value of the external pressure which agrees with that found in paragraph 6.3 for a small hole in a large plate subject to uniformly distributed radial forces. The buckling collapse of thick-walled cylinders is given in Par. 8.4.

3. *Deformation of a Thick Cylinder*

The radial displacement of any point in the wall of the cylinder can be found from Eq. 2.8.10 by substituting the values of the constants C_1 and C_2 from Eq. 2.8.14, which gives

$$u = \frac{1-\mu}{E}\,\frac{a^2p_i - b^2p_o}{b^2 - a^2}\,r + \frac{1+\mu}{E}\,\frac{a^2b^2(p_i - p_o)}{(b^2 - a^2)r} \tag{2.8.25}$$

In the case of a cylinder subjected to internal pressure p_i only, the radial displacement at the inner surface $r = a$ from Eq. 2.8.25 is:

$$u_a = \frac{p_i a}{E}\left(\frac{a^2 + b^2}{b^2 - a^2} + \mu\right) \tag{2.8.26}$$

and at the outer surface is:

$$u_b = \frac{2p_i\, a^2 b}{E(b^2 - a^2)} \tag{2.8.27}$$

When the cylinder is subjected to external pressure p_0 only, the displacement of the inner surface, $r = a$, is:

$$u = -\frac{2p_o a b^2}{E(b^2 - a^2)} \tag{2.8.28}$$

and at the outer surface is:

$$u = -\frac{bp_o}{E}\left(\frac{a^2+b^2}{b^2-a^2}-\mu\right) \tag{2.8.29}$$

The minus sign indicates that the displacement is toward the axis of the cylinder.[48,49]

4. *Stresses in a Thick Sphere*

In a manner similar to that used to establish the stresses in a thick walled cylinder, those in a thick sphere can be found to be

$$\sigma_r = \frac{p_i\left(1-\dfrac{b^3}{r^3}\right)}{\left(\dfrac{b^3}{a^3}-1\right)} - \frac{p_o\dfrac{b^3}{a^3}\left(1-\dfrac{a^3}{r^3}\right)}{\left(\dfrac{b^3}{a^3}-1\right)} \tag{2.8.30}$$

$$\sigma_t = \frac{p_i\left(1+\dfrac{1}{2}\dfrac{b^3}{r^3}\right)}{\left(\dfrac{b^3}{a^3}-1\right)} - \frac{p_o\dfrac{b^3}{a^3}\left(1+\dfrac{1}{2}\dfrac{a^3}{r^3}\right)}{\left(\dfrac{b^3}{a^3}-1\right)} \tag{2.8.31}$$

The maximum stress is always the tangential stress, σ_t, and occurs at the inside surface, $r = a$. In the case of a sphere subject to internal pressure only $p_o = 0$; this is, per Eq. 2.8.31,

$$\sigma_{t\max} = \frac{p_i}{2}\,\frac{2a^3+b^3}{b^3-a^3} \tag{2.8.32}$$

2.9 Shrink-Fit Stresses in Builtup Cylinders

Cylindrical vessels can be reinforced by shrinking on an outer cylindrical liner so that a contact pressure is produced between the two. This is usually done by making the inside radius of the outer cylinder smaller than the outside radius of the inner one and assembling the two after first heating the outer one. (The reverse procedure of cooling the inner cylinder with Dry Ice or liquid gases has also been used.) A contact pressure is developed after cooling dependent upon the initial interference of the two cylinders. Its magnitude and the stresses it produces are calculable by the equations of paragraph 2.8.3. As an example, prior to assembly the outside radius b of the inner cylinder in Fig. 2.23 was larger than the inside radius of the outer cylinder by an amount δ, creating a pressure p between the cylinders after assembly.

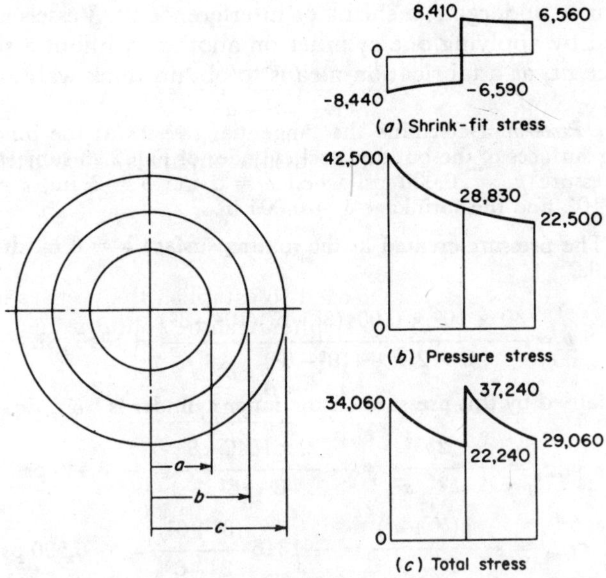

Fig. 2.23 Shrink-Fit Tangential Stresses in a Cylinder

Its value can be determined from the condition that the increase in the inner radius of the outer cylinder plus the decrease in the outer radius of the inner cylinder must equal δ. Thus, from Eqs. 2.8.26 and 2.8.29,

$$\frac{bp}{E}\left(\frac{b^2+c^2}{c^2-b^2}+\mu\right)+\frac{bp}{E}\left(\frac{a^2+b^2}{b^2-a^2}-\mu\right)=\delta \tag{2.9.1}$$

or

$$p=\frac{E\delta}{b}\,\frac{(b^2-a^2)(c^2-b^2)}{2b^2(c^2-a^2)} \tag{2.9.2}$$

If such a builtup cylinder is now subjected to internal pressure, the stresses produced by this pressure are the same as those in a solid-wall cylinder of thickness equal to the sum of those of the individual cylinders c–a. These stresses are superposed on the shrink-fit stresses discussed previously. The latter are compressive at the inner surface of the cylinder which reduces the maximum tangential tensile stress due to the internal pressure at this point, thereby creating a more favorable stress distribution than in the case of a solid-wall cylinder (see the illustrative problem below). This procedure of building up cylinders is used in the construction of gun barrels and of vessels for extremely high internal pressures. This favorable stress pattern does not exist unless the parts have undergone a shrink or interference fit. Vessels are often constructed by applying one cylinder on another, without a shrink or interference fit, as a fabrication means to obtain thick walls.

Illustrative Problem. Determine the tangential stresses at the inner, outer and mating surfaces of the builtup steel cylinder of Fig. 2.23 subjected to an internal pressure $p_i = 20{,}000$ psi when $a = 6$ in., $b = 8$ in., $c = 10$ in., $E = 30 \times 10^6$, and the shrinkage $\delta = 0.004$ in.

Solution: The pressure created at the mating surface $b = 8$ in. due to δ is from Eq. 2.9.2:

$$p = \frac{30 \times 10^6 \times 0.004(8^2-6^2)(10^2-8^2)}{8 \times 2 \times 8^2(10^2-6^2)} = 1846 \text{ psi.}$$

The stress caused by this pressure on the inner cylinder is from Eq. 2.8.23:

$$\sigma_{t_{r=6}} = -\frac{2pb^2}{b^2-a^2} = -\frac{2 \times 1846 \times 8^2}{8^2-6^2} = -8{,}440 \text{ psi}$$

$$\sigma_{t_{r=8}} = -\frac{p(b^2+a^2)}{b^2-a^2} = -1846\frac{(8^2+6^2)}{8^2-6^2} = -6{,}590 \text{ psi}$$

The stress caused by this pressure on the outer cylinder is from Eq. 2.8.19

$$\sigma_{t_{r=8}} = p\frac{(b^2+c^2)}{c^2-b^2} = 1846\frac{(8^2+10^2)}{10^2-8^2} = 8{,}410 \text{ psi}$$

$$\sigma_{t_{r=10}} = \frac{p2b^2}{c^2-b^2} = \frac{1846 \times 2 \times 8^2}{10^2-8^2} = 6{,}560 \text{ psi}$$

These initial stresses produced by the shrink-fit are shown in Fig. 2.23*a*. The stresses caused by the internal pressure are the same as for a solid-wall cylinder of thickness 10 in. − 6 in. = 4 in. and are shown in Fig. 2.23*b* as determined by Eq. 2.8.19.

$$\sigma_{t_{r=6}} = \frac{p_i(a^2+c^2)}{c^2-a^2} = \frac{20{,}000(6^2+10^2)}{10^2-6^2} = 42{,}500 \text{ psi}$$

$$\sigma_{t_{r=8}} = \frac{p_i a^2}{c^2-a^2}\left(1+\frac{c^2}{b^2}\right) = \frac{20{,}000 \times 6^2}{10^2-6^2}\left(1+\frac{10^2}{8^2}\right) = 28{,}830 \text{ psi}$$

$$\sigma_{t_{r=10}} = \frac{p_i 2a^2}{c^2-a^2} = \frac{20{,}000 \times 2 \times 6^2}{10^2-6^2} = 22{,}500 \text{ psi}$$

Superposition of these two stress patterns gives the final total stress distribution, Fig. 2.23*c*, from which it is seen that the initial assembly stresses reduce the maximum stress in the cylinder when it is subjected to internal pressure from 42,500 psi to 37,240 psi.

Shrink-fit stresses do not influence the bursting pressure of such builtup cylinders which is equal to that of a simple solid wall cylinder of the same overall diameter ratio, Par. 2.10.3.

2.10 Autofrettage of Thick Cylinders

1. *Theory*

It is also possible to obtain a favorable initial stress pattern, analogous to that mentioned in paragraph 2.9 for builtup shrink-fit cylinders, by applying a sufficiently high internal pressure to produce plastic flow in the inner part of the cylinder. After removing this internal pressure, residual stress persists due to the plastic flow or deformation that has taken place, with the inner part in compression and the outer part in tension. This is called "autofrettage."

Yielding of inner surface due to internal pressure will occur when the maximum shearing stress, Eq. 2.8.21, at this point becomes equal to the yield point stress in shear of the material $\tau_{Y.P.}$*. Substituting this value in Eq. 2.8.21 gives the pressure at which yielding begins as:

$$p_{Y.P.} = \tau_{Y.P.}\frac{b^2 - a^2}{b^2} \qquad (2.10.1)$$

As the pressure is further increased, the plastic deformation penetrates farther into the vessel wall until it reaches the outside surface at a pressure p_u when the entire wall of the cylinder has yielded. The stress distribution under these conditions can be determined by assuming the material to be perfectly plastic and to yield under the action of constant shearing stress $\tau_{Y.P.}$. Then for every point in the plastic region

$$\frac{\sigma_t - \sigma_r}{2} = \tau_{Y.P.} \qquad (2.10.2)$$

A second equation involving the principal stresses σ_t and σ_r can be obtained from the equilibrium of an element of the wall as shown in Fig. 2.16, which gave the equation (Eq. 2.8.3).

$$\sigma_t - \sigma_r - r\frac{d\sigma_r}{dr} = 0 \qquad (2.10.3)$$

and substituting the value of $\sigma_t - \sigma_r$ from Eq. 2.10.2 into Eq. 2.10.3 gives

$$\frac{d\sigma_r}{dr} = \frac{2\tau_{Y.P.}}{r} \qquad (2.10.4)$$

*See Chapter 5 for a discussion on the yielding and failure of materials under stress.

Upon integration this gives

$$\sigma_r = 2\tau_{\text{Y.P.}} \log_e r + C \tag{2.10.5}$$

and the constant of integration C can be determined from the condition that at the outer surface of the cylinder, $r = b$, the radial stress becomes zero, $\sigma_r = 0$, as

$$C = -2\tau_{\text{Y.P.}} \log_e b \tag{2.10.6}$$

Placing this constant of integration in Eq. 2.10.5 gives

$$\sigma_r = 2\tau_{\text{Y.P.}} \log_e \frac{r}{b} \tag{2.10.7}$$

or at the inner surface,

$$\sigma_{r_a} = 2\tau_{\text{Y.P.}} \log_e \frac{a}{b} \tag{2.10.8}$$

The pressure that is required to bring the entire wall of the cylinder into a state of plastic flow is then

$$p_u = -\sigma_{r_a} = -2\tau_{\text{Y.P.}} \log_e \frac{a}{b} \tag{2.10.9}$$

The tangential stress can be found by substituting the value of σ_r from Eq. 2.10.7 into Eq. 2.10.2, which gives

$$\sigma_t = 2\tau_{\text{Y.P.}} \left(1 + \log_e \frac{r}{b}\right) \tag{2.10.10}$$

If the internal pressure is removed after the cylinder material has been brought to a plastic condition, a residual stress will remain in the wall. This can be calculated by assuming that during unloading the material follows Hooke's law, and the stresses which are to be subtracted while unloading are those given by Eqs. 2.8.18 and 2.8.19 when $-p_u$ is substituted for p_i in these expressions. This is an effect superposing a radial tension, or negative pressure, on the inside to cancel that pressure causing the initial plastic flow condition. This is best illustrated by a particular case, say, $b = 2.2a$, for which the stresses are shown in Fig. 2.24. These were determined as follows:

(1) The distribution of radial stress σ_r through the wall is shown in Fig. 2.24a as calculated from Eq. 2.10.7

$$\sigma_{r_a} = 2\tau_{\text{Y.P.}} \log_e \left(\frac{1}{2.2.}\right) = -0.79(2\tau_{\text{Y.P.}}) \tag{a}$$

$$\sigma_{r_b} = 2\tau_{\text{Y.P.}} \log_e 1 = 0 \tag{b}$$

and likewise the tangential stress σ_t distribution calculated by Eq. 2.10.10:

$$\sigma_{t_a} = 2\tau_{\text{Y.P.}}\left[1+\log_e\left(\frac{1}{2.2.}\right)\right] = 0.21(2\tau_{\text{Y.P.}}) \qquad (c)$$

$$\sigma_{t_b} = 2\tau_{\text{Y.P.}}\,(1+\log_e 1) = 2\tau_{\text{Y.P.}} \qquad (d)$$

(2) The distribution of radial stress σ_r resulting from release of the internal pressure is shown in Fig. 2.24*b* as calculated from Eqs. 2.10.9 and 2.8.18,

$$p_u = -\,2\tau_{\text{Y.P.}}\log_e\left(\frac{1}{2.2.}\right) = 0.79(2\tau_{\text{Y.P.}}) \qquad (a)$$

$$\sigma_{r_a} = -\,\frac{a^2 p_u}{b^2-a^2}\left(1-\frac{b^2}{a^2}\right) = 0.79(2\tau_{\text{Y.P.}}) \qquad (b)$$

$$\sigma_{r_b} = 0 \qquad (c)$$

and the tangential stress σ_t distribution by Eq. 2.8.19:

$$\sigma_{t_a} = -\,\frac{a^2 p_u}{b^2-a^2}\left(1+\frac{b^2}{a^2}\right) = -\,1.52p_u = -\,1.52\times 0.79(2\tau_{\text{Y.P.}})$$
$$= -\,1.20(2\tau_{\text{Y.P.}}) \qquad (d)$$

$$\sigma_{t_b} = -\,\frac{2a^2 p_u}{b^2-a^2} = -\,0.52p_u = -\,0.52\times 0.79(2\tau_{\text{Y.P.}}) = -\,0.41(2\tau_{\text{Y.P.}}) \qquad (e)$$

(3) The final residual stress distribution is obtained by superposition to give that shown in Fig. 2.24*c*. It is very favorable to the re-application of internal pressure since the accompanying tangential tensile stress must now first overcome the residual compressive stress. For instance, if an internal pressure equal to p_u is again applied, the tangential stresses produced by this pressure must be superposed on the residual stresses, solid curve of Fig. 2.24*c*, with a resultant maximum stress of 2 $\tau_{\text{Y.P.}}$ now occurring at the outside wall of the vessel; no yielding will occur during the second application of this internal pressure. Comparing this to the pressure at which yielding will begin at the inside wall of a stress-free cylinder which is given by Eq. 2.10.1 for $b = 2.2a$ as $p_{\text{Y.P.}} = 0.79\ \tau_{\text{Y.P.}}$, it is seen that the pressure can be increased two and a half times while maintaining elastic behavior of the cylinder.

In this manner it is possible to increase considerably the pressure which a cylinder can contain elastically, and the method is frequently used in the design of accumulators, hydraulic ram cylinders, gun

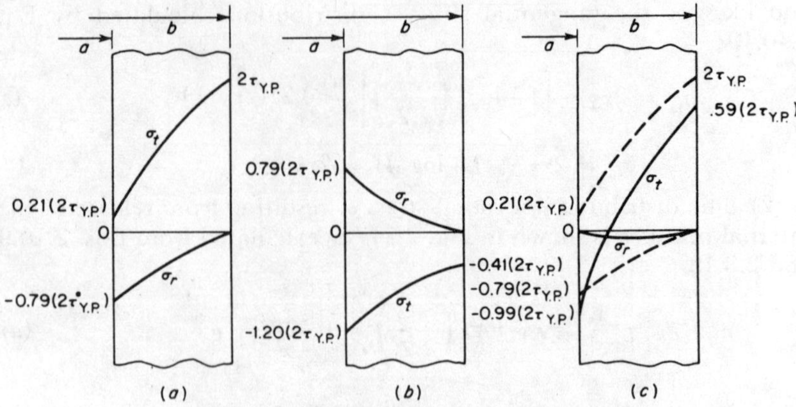

Fig. 2.24. Residual Stresses in an Autofrettaged Thick Cylinder

barrels, etc. Not only does it increase the elastic static strength, but it also has a favorable effect in increasing fatigue life, paragraph 5.18.4. This process is applicable only to thick-walled cylinders where a variation in pressure stress exists across the wall thickness. It is not applicable to those with thin walls since it is impossible to build up a favorable residual stress pattern.

The foregoing analysis considered plastic flow of the entire cylinder wall, but a similar analysis can be made when only a portion of the wall is made to yield while the rest remains elastic.[20] The residual stresses discussed above were induced by pressure, or so-called mechanical means. They can also be created by temperature changes,[21] such as the upsetting accompanying quenching or the volume change occurring during metallurgical phase changes during cooling. These stresses can become particularly important in large heavy forgings or castings where their values, not being selectively controlled as in the case above, can become large and influence the life of the part, paragraph 5.17.3.

2. *Potential of The Autofrettaged Pressure Vessel*

Equation 2.8.21 gives the maximum shearing stress occurring at the inside wall of a cylindrical vessel. This equation may be written in the form

$$\frac{p_{\text{elastic}}}{\tau} = \frac{\left(\frac{b}{a}\right)^2 - 1}{\left(\frac{b}{a}\right)^2} = \frac{K^2 - 1}{K^2} \qquad (2.10.11)$$

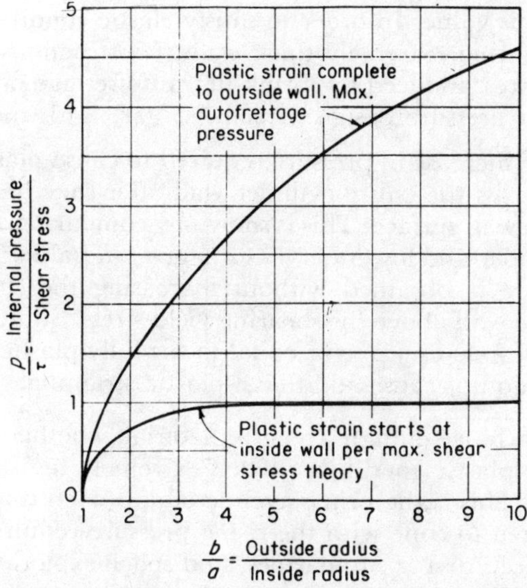

Fig. 2.25. Influence of Autofrettage on a Cylinder

and is plotted as the lower curve in Fig. 2.25. If τ is the shearing yield strength of the material, the ordinates of this lower curve represent values of the pressure, according to the maximum shear stress theory, which are the maximum that can be sustained elastically. Beyond this pressure the inside surface of the wall is strained plastically. The effect of further straining to the point where the outside surface just becomes plastic is given by Eq. 2.10.9. This equation can be written in the form

$$\frac{p_{\text{plastic}}}{\tau} = 2 \log_e \left(\frac{b}{a}\right) = 2 \log_e K \tag{2.10.12}$$

and is plotted as the upper curve in Fig. 2.25. These equations apply to either open-end or closed-end cylinders. This is due to the fact that they are based on the maximum shearing stress which is one-half the largest algebraic difference between any two of the three principal stresses or $1/2(\sigma_t - \sigma_r)$, and are therefore not influenced by the longitudinal stress which is a value intermediate between σ_t and σ_r. A comparison of these two curves indicates:

1. The limited merit of using extremely thick wall cylinders of $K = b/a$ ratios in excess of about 3.0 when an elastic stress condition must be maintained throughout the cylinder wall. This is shown by the flatness of the lower curve, Fig. 2.25, with K ratios

beyond this value. In order to satisfy elastic conditions for pressures requiring extremely thick walls, constructions embodying residual stress induced by shrink-fits, autofrettage or other novel ultra-high pressure design principles, Par. 2.15, must be used.

2. The large increase in pressure required to cause plastic strains to spread across the entire cylinder wall after they have started at the inside wall surface. This is shown by comparing the ordinates for the upper and lower curves for a given K value. This increase in pressure is obtained without increasing the shearing stress within the wall above the shearing yield stress. Subsequent applications of pressure up to or equal to the fully plastic autofrettage pressure do not cause additional plastic straining.
3. The importance of high strength material, whether designing for elastic or plastic conditions, when extremely high pressures are involved. This is the prime avenue of approach that is presently being taken to cope with the rising pressure requirements of industry, deep diving submarines, and space exploration vehicles.

3. *The Bursting Strength of Thick-Wall Cylindrical Vessels*

Predicting the maximum or bursting pressure that a thick-wall cylindrical vessel can withstand is an important consideration in its design. There have been a multitude of formulas used or proposed for establishing bursting strength.[22] These have ranged from entirely emperical ones to completely theoretical ones based on theories of plasticity and the true strain behavior of the material.

A simple formula for determining bursting pressure can be derived from Eq. 2.10.9. This equation gives the pressure in the cylinder when the shearing stress across the entire thickness has reached the shearing yield stress. If the assumption is made that upon reaching the bursting pressure the shearing stresses are uniform over the entire thickness and equal to the ultimate shearing strength of the material, the bursting pressure is given by

$$p_{\text{burst}} = 2\tau_{\text{ult.}} \log_e K \tag{2.10.13}$$

If the value $2\tau_{\text{ult.}}$ is assumed equal to and replaced by the ultimate tensile strength of the material, $\sigma_{\text{ult.}}$, since the former value is difficult to ascertain, this becomes

$$p_{\text{burst}} = \sigma_{\text{ult.}} \log_e K \tag{2.10.14}$$

A more accurate value of the bursting pressure has been developed by Svensson[23,50] as

$$p_{\text{burst}} = \left[\left(\frac{0.25}{n + 0.227}\right)\left(\frac{e}{n}\right)^n\right]\sigma_{\text{ult.}} \log_e K \qquad (2.10.15)$$

This formula embodies the strain hardening exponent, n, of the material and gives predictions in excellent agreement with experimental burst tests.

The term in brackets on the right hand side of Eq. 2.10.15 is a modifier of the basic Eq. 2.10.14 to account for this specific material property. Table 2.2 shows how this modifier varies with n. This shows that for a given vessel made from materials having the same ultimate tensile strength, the bursting pressure decreases as the material strain hardening exponent n increases. This decrease in bursting pressure is due to the greater reduction in the vessel wall thickness and accompanying increase in vessel diameter over which the pressure acts before the bursting pressure is reached. An indication of the variation of n as a function of the ratio of yield strength to ultimate tensile strength for low and intermediate tensile strength carbon and low alloy steels is shown in Fig. 2:26.

Equation 2.10.16 by Faupel,[55] which directly incorporates the yield and ultimate tensile strength, also gives good agreement with experiment.

$$p_{\text{burst}} = \frac{\sigma_{\text{Y.P.}}}{\sqrt{3}} \log_e K \left(2 - \frac{\sigma_{\text{Y.P.}}}{\sigma_{\text{ult.}}}\right) \qquad (2.10.16)$$

Equations 2.10.14, 2.10.15, and 2.10.16 give a basis for establishing the bursting strength of cylindrical vessels using the material specification properties, or those readily obtainable from simple tension tests. Material imperfections such as flaws, weld defects, etc., can alter bursting pressures. Accordingly, it is customary to subject vessels for the nuclear and other high pressure critical services to a complete nondestructive examination to assure soundness of material and weldments. Vessel designs based on burst pressure should use a minimum factor of safety of 1.5–2.0 because of the difficulty of accurately predicting burst pressures. Just as shrink-fit stresses do not influence

TABLE 2.2. EFFECT OF STRAIN HARDENING EXPONENT ON BURSTING STRENGTH OF CYLINDRICAL VESSELS[23]

n	0	0.10	0.20	0.30	0.40	0.50
Modifier (Bracketed Term, Eq. (2.10.15)	1.10	1.06	0.99	0.92	0.86	0.80

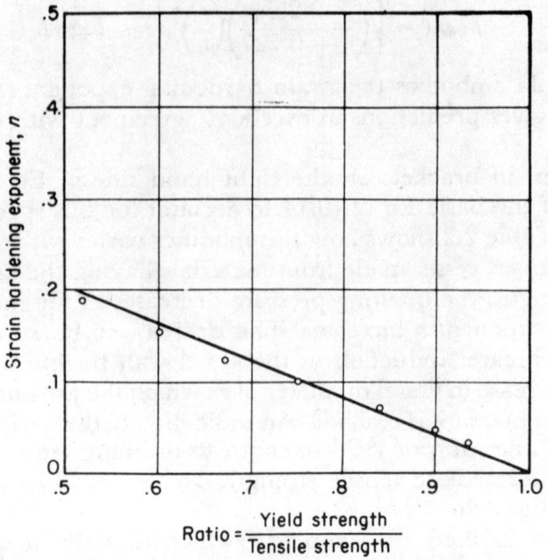

Fig. 2.26. Variation of Strain Hardening Exponent for Low and Intermediate Tensile Strength Carbon and Low Alloy Stress[23]

the bursting pressure of builtup cylinders, Par. 2.9, neither does the prestress autofrettage residual stress.[57] The bursting pressure remains equal to that of a solid wall cylinder of the same overall diameter ratio.

2.11 Thermal Stresses and Their Significance

Stresses which result from restricting the natural growth or contraction of a material due to a temperature change are called thermal stresses. If the bar in Fig. 2.27*a* is uniformly heated from an initial temperature T_1 to a new temperature T_2, the unit change in dimension is $\alpha(T_2 - T_1)$, where α is the coefficient of thermal expansion, and no thermal stresses are produced since it is free to expand. If, however, the bar is restricted from expanding in the y direction, Fig. 2.27*b*, but free to expand laterally due to the Poisson effect, the resulting uniaxial thermal stress becomes:

$$\frac{\sigma}{E} = -\alpha(T_2 - T_1) \tag{2.11.1}$$

$$\sigma = -E\alpha(T_2 - T_1) \tag{2.11.2}$$

If the bar is further restricted in the x direction also, Fig. 2.27*c*, the principal strains are equal, $e_1 = e_2 = \alpha(T_2 - T_1) = \alpha\Delta T$, and the

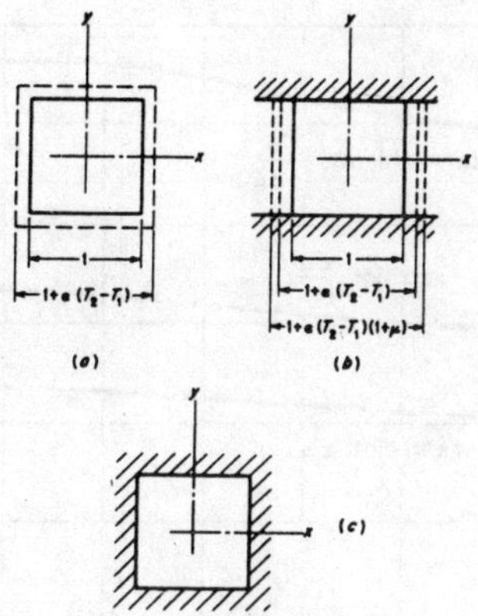

Fig. 2.27. Thermal Strains

principal stresses from Eqs. 2.3.4 and 2.3.5 are:

$$\sigma_1 = \sigma_2 = -\frac{\alpha E(T_2 - T_1)}{1 - \mu} = -\frac{\alpha E \Delta T}{1 - \mu} \qquad (2.11.3)$$

When a third restraint is imposed, perpendicular to the x–y plane of Fig. 2.27*c*, the stress becomes:

$$\sigma = -\frac{\alpha E(T_2 - T_1)}{1 - 2\mu} = -\frac{\alpha E \Delta T}{1 - 2\mu} \qquad (2.11.4)$$

These values of thermal stress are for full restraint and hence are the maximum that can be created. Fig. 2.28 gives the value of these thermal stresses for carbon steel per degree Fahrenheit temperature change, $\Delta T = 1°F$, at elevated temperatures for one, two, and three degrees of restraint. Most pressure vessel conditions involve two-dimensional restraint, Eq. 2.11.3 and it is noted that these are higher than those for simple uniaxial restraint in the ratio of $1/(1 - \mu)$, or 43 per cent for steel with $\mu = 0.3$.

The minus sign in the above equations indicates the bar is in compression since its expansion has been restricted. If the bar is prevented from contracting, a tensile stress is produced. Figure 2.27 has depicted the situation of uniform change in temperature with constraint from without so that the sign and magnitude of the stress remains constant throughout. Thermal stresses can also be induced by a temperature variation within the member creating a differential expansion such

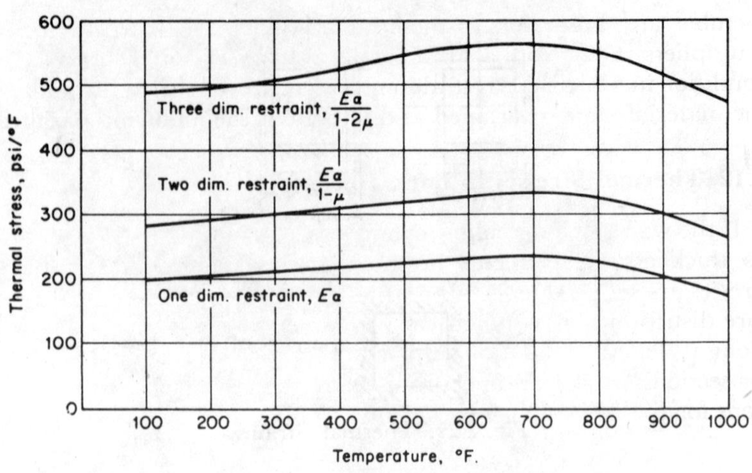

Fig. 2.28. Thermal Stress Values, Carbon Steel

that the natural growth of one fiber is influenced by the different growth requirements of adjacent fibers. The result is that fibers at high temperatures are compressed and those at lesser temperatures stretched, and herein lies the fundamental difference between thermal and mechanical stresses. The differential expansion requires only that a prescribed strain pattern be satisfied, and the accompanying stress pattern need only satisfy the requirements for equilibrium of the internal forces; hence, yielding produces relaxation of the thermal stress.[24] On the other hand, if the member is loaded mechanically, such as by pressure, to a stress beyond its yield strength, yielding will continue until it breaks, unless the deflection is limited by strain hardening or stress redistribution, Chapter 5. The internal stress must be in equilibrium with the external load which remains constant, hence the internal stresses cannot relax. Accordingly, thermal stresses are secondary stresses because they are "self-limiting." They will not cause failure by rupture in ductile materials upon their first application irrespective of their magnitude, but they will cause failure by repeated cycling, i.e., fatigue, paragraph 5.19. They may also produce essential failure by rendering the structure inoperative due to large deflections or distortions as would occur in the case of rotating devices such as turbines. It is most important to recognize this distinction in appraising the significance of thermal stresses.

The above equations are basic ones for determining the maximum thermal stress, and all others for the thermal stress in various shape members and with various shape thermal gradients throughout the member are multipliers of these basic equations. The value of these multipliers range from 1 for a totally restrained condition, to $\frac{1}{2}$ for a condition in which the restraining material is of the same rigidity as the material being restrained as discussed in the following paragraph.

2.12 Thermal Stresses in Long Hollow Cylinders

If the wall of a long hollow cylinder is heated nonuniformly through its thickness, its elements do not expand uniformly and thermal stresses are set up due to this mutual interference. When the temperature distribution is symmetrical with respect to the axis and constant along its length, a solution similar to that developed in paragraph 2.8 for sections a short distance from the ends* gives the following equations for the principal stresses.[25,26,27]

$$\sigma_r = \frac{\alpha E}{(1-\mu)r^2}\left[\frac{r^2-a^2}{b^2-a^2}\int_a^b Trdr - \int_a^r Trdr\right] \tag{2.12.1}$$

$$\sigma_t = \frac{\alpha E}{(1-\mu)r^2}\left[\frac{r^2+a^2}{b^2-a^2}\int_a^b Trdr + \int_a^r Trdr - Tr^2\right] \tag{2.12.2}$$

$$\sigma_z = \frac{\alpha E}{(1-\mu)}\left[\frac{2}{b^2-a^2}\int_a^b Trdr - T\right] \tag{2.12.3}$$

If the thermal gradient over the thickness of the wall is known, the integrals in these equations can be evaluated and the thermal stresses for the particular case determined.

1. *Steady-State Thermal Stresses, Logarithmic Thermal Gradient*

One of the most prevalent cases of thermal stress occurs in the cylindrical vessel when heat is flowing through the sides in a steady state, causing the equilibrium temperature difference between the inner and outer surfaces to remain constant. Under these conditions the flow is radial through a flow cross section proportional to the radius which gives rise to a logarithmic temperature distribution throughout the wall thickness,[28] paragraph 7.3.2. The temperature at any point is then given as a function of the temperature of the inner wall T_a by

$$T = T_a \frac{\log_e\left(\frac{b}{r}\right)}{\log_e\left(\frac{b}{a}\right)} \tag{2.12.4}$$

*See footnote 25 for a discussion of the stresses at the ends of a cylinder.

The temperature distribution is dependent only upon the ratio of the outer and inner radii, and, although it is independent of the thickness, it must be mentioned that normally greater total temperature differences are associated with increased thickness. Hence, thick wall vessels are more susceptible to failure due to thermal stresses than are thin ones. It is the temperature drop through the vessel thickness that gives rise to the thermal stress, and it is convenient to call the temperature at the outer surface zero, realizing that any other surface temperature conditions may be obtained by superposing on this condition a uniform heating or cooling which produces no stresses, Fig. 2.29. Substituting Eq. 2.12.4 into Eqs. 2.12.1, 2.12.2 and 2.12.3 gives

$$\sigma_r = \frac{\alpha E T_a}{2(1-\mu)\log_e\left(\frac{b}{a}\right)}\left[-\log_e\left(\frac{b}{r}\right) - \frac{a^2}{b^2-a^2}\left(1-\frac{b^2}{r^2}\right)\log_e\left(\frac{b}{a}\right)\right] \tag{2.12.5}$$

$$\sigma_t = \frac{\alpha E T_a}{2(1-\mu)\log_e\left(\frac{b}{a}\right)}\left[1-\log_e\left(\frac{b}{r}\right) - \frac{a^2}{b^2-a^2}\left(1+\frac{b^2}{r^2}\right)\log_e\left(\frac{b}{a}\right)\right] \tag{2.12.6}$$

$$\sigma_z = \frac{\alpha E T_a}{2(1-\mu)\log_e\left(\frac{b}{a}\right)}\left[1-2\log_e\left(\frac{b}{r}\right) - \frac{2a^2}{b^2-a^2}\log_e\left(\frac{b}{a}\right)\right] \tag{2.12.7}$$

When T_a is positive, the radial stress σ_r is compressive throughout the thickness and becomes zero at the inner and outer surfaces. The tangential stress σ_t and longitudinal stress σ_z have their largest numerical values at the inner and outer surfaces of the cylinder which can be found by substituting $r = a$ and $r = b$ into Eqs. 2.12.6 and 2.12.7 to obtain

$$\sigma_{t_a} = \sigma_{z_a} = \frac{\alpha E T_a}{2(1-\mu)\log_e\left(\frac{b}{a}\right)}\left[1-\frac{2b^2}{b^2-a^2}\log_e\left(\frac{b}{a}\right)\right] \tag{2.12.8}$$

$$\sigma_{t_b} = \sigma_{z_b} = \frac{\alpha E T_a}{2(1-\mu)\log_e\left(\frac{b}{a}\right)}\left[1-\frac{2a^2}{b^2-a^2}\log_e\left(\frac{b}{a}\right)\right] \tag{2.12.9}$$

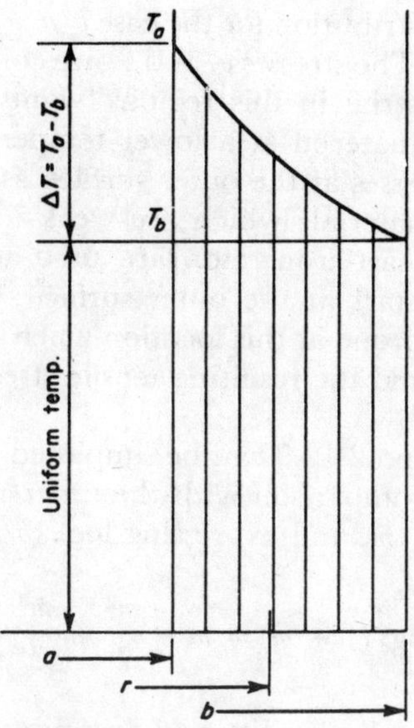

Fig. 2.29. Thermal Gradient in Hollow Cylinder

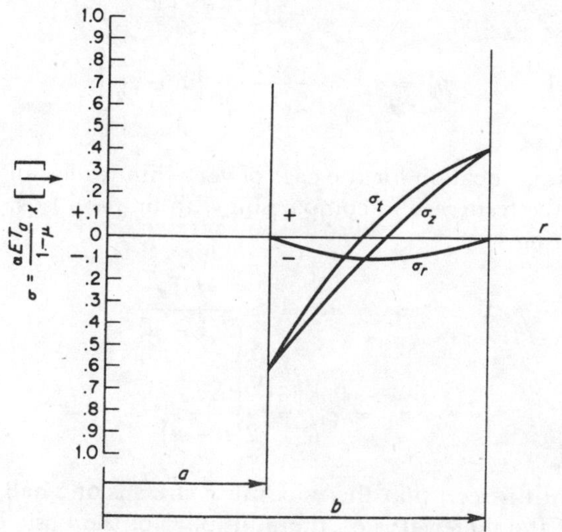

Fig. 2.30. Steady-State Thermal Stress in a Cylinder of $b/a = 2.0$ for a Logarithmic Thermal Gradient

The thermal stress distribution for the case $b/a = 2.0$ and T_a is positive is shown in Fig. 2.30. The stresses σ_t and σ_z are compressive at the inner surface; i.e., the material in this region "wants to grow" but is restricted by adjacent material at a lower temperature, and gradually changes to tensile stresses at the outer surface as the reverse situation takes place. When materials which are weak in tension, such as refractories, concrete, cast iron, etc., are used under this condition, cracks are likely to start at the outer surface; likewise failure from fatigue will be more prone at this location when the vessel is subjected to internal pressure and the resulting tensile stresses become additive, paragraph 5.19.

Equations 2.12.8 and 2.12.9 can be simplified when the thickness of the wall is small in comparison with the inner radius of the cylinder by writing $b/a = 1 + m$, and expressing $\log_e (b/a)$ as the series

$$\log_e \left(\frac{b}{a}\right) = \log_e (1 + m) = m - \frac{m^2}{2} + \frac{m^3}{3} - \frac{m^4}{4} + \ldots \quad (2.12.10)$$

Considering m as a small quantity and dropping terms of higher order gives

$$\sigma_{t_a} = \sigma_{z_a} = -\frac{\alpha E T_a}{2(1-\mu)}\left[1 + \frac{m}{3}\right] \quad (2.12.11)$$

$$\sigma_{t_b} = \sigma_{z_b} = \frac{\alpha E T_a}{2(1-\mu)}\left[1 - \frac{m}{3}\right] \quad (2.12.12)$$

A further simplification for the case of very thin walls can be made by neglecting the term $m/3$ in comparison with unity in Eqs. 2.12.11 and 2.12.12, giving

$$\sigma_{t_a} = \sigma_{z_a} = -\frac{\alpha E T_a}{2(1-\mu)} \quad (2.12.13)$$

$$\sigma_{t_b} = \sigma_{z_b} = \frac{\alpha E T_a}{2(1-\mu)} \quad (2.12.14)$$

from which it is seen that the maximum stress is one half that for full restraint of the material; i.e., the multiplier of the basic full restraint thermal stress equation, Eq. 2.11.3 is $\frac{1}{2}$ for this case.

2. *Steady-State Thermal Stresses, Linear Thermal Gradient*

If the thickness of the vessel wall is small in comparison with the outside radius, the logarithmic temperature gradient, Eq. 2.12.4, can be replaced by a linear one,

$$T = T_a \frac{b - r}{b - a} \tag{2.12.15}$$

and substituting the value of T from Eq. 2.12.15 into Eqs. 2.12.1, 2.12.2, and 2.12.3 gives

$$\sigma_r = \frac{\alpha E T_a}{(1 - \mu) r^2} \left[\frac{(r^2 - a^2)(b + 2a)}{6(b + a)} + \frac{2(r^3 - a^3) - 3b(r^2 - a^2)}{6(b - a)} \right] \tag{2.12.16}$$

$$\sigma_t = \frac{\alpha E T_a}{(1 - \mu) r^2} \left[\frac{(r^2 + a^2)(b + 2a)}{6(b + a)} - \frac{2(r^3 - a^3) - 3b(r^2 - a^2)}{6(b - a)} - \frac{(b - r) r^2}{(b - a)} \right] \tag{2.12.17}$$

$$\sigma_z = \frac{\alpha E T_a}{(1 - \mu)} \left[\frac{b + 2a}{3(b + a)} - \frac{b - r}{b - a} \right] \tag{2.12.18}$$

The tangential stress σ_t and longitudinal stress σ_z again, as for a logarithmic thermal gradient, have their greatest numerical value at the inner and outer surfaces of the cylinder which can be found by substituting $r = a$ and $r = b$ into Eqs. 2.12.17 and 2.12.18 to give

$$\sigma_{t_a} = \sigma_{z_a} = -\frac{\alpha E T_a}{1 - \mu} \left[\frac{2b + a}{3(b + a)} \right] \tag{2.12.19}$$

$$\sigma_{t_b} = \sigma_{z_b} = \frac{\alpha E T_a}{1 - \mu} \left[\frac{b + 2a}{3(b + a)} \right] \tag{2.12.20}$$

For a thin wall vessel, $a \approx b$ and the equations can be further simplified to

$$\sigma_{t_a} = \sigma_{z_a} = -\frac{\alpha E T_a}{2(1 - \mu)} \tag{2.12.21}$$

$$\sigma_{t_b} = \sigma_{z_b} = \frac{\alpha E T_a}{2(1 - \mu)} \tag{2.12.22}$$

and the thermal stress is the same as for a logarithmic thermal gradient, Eqs. 2.12.13 and 2.12.14. The thermal stress distribution over the thickness of the wall is also the same as that of a flat plate with clamped edges, so that bending of this plate is prevented when it is subjected to a linear thermal gradient through its thickness.

When the vessel wall is not thin, the exact shape of the thermal gradient becomes more influencing. Figure 2.31 is a plot of the logarithmic, Eq. 2.12.4, and linear, Eq. 2.12.15, temperature gradients which shows that, for a value of $b/a = 1.2$, the linear assumption is within 7 percent of the logarithmic gradient; whereas for an appreciably thicker wall vessel of $b/a = 2.0$, the difference is as high as 23 percent. The shape of the temperature gradient is correspondingly reflected in the magnitude of the thermal stresses it causes, although to a lesser extent, with the steeper gradient giving rise to the higher stresses. This can be verified by comparing the values of a maximum stress obtained from Eq. 2.12.8 for a logarithmic gradient to that obtained from Eq. 2.12.19 for a linear one. For a value of $b/a = 2.0$, the steeper logarithmic gradient gives a 10 percent higher stress than the linear variation with its constant minimum slope throughout the entire wall; i.e., the mutual interference of one fiber on the natural expansion of an adjacent one has been minimized by distributing the total temperature drop ΔT equally over the entire wall thickness. This is a major pressure vessel design consideration emphasizing the importance of avoiding

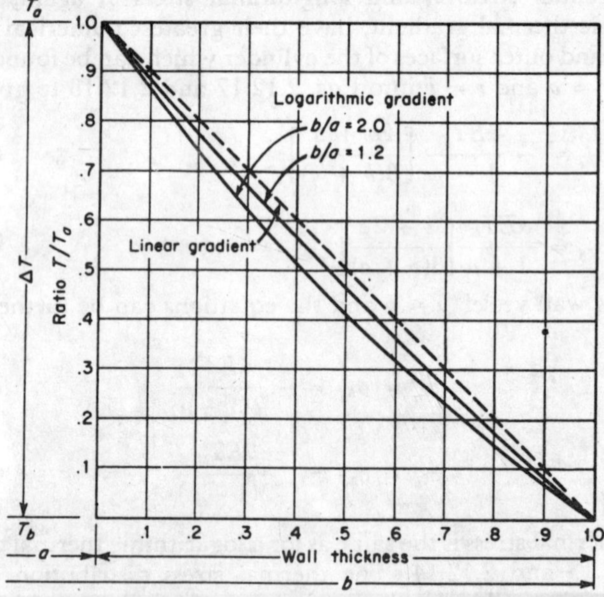

Fig. 2.31. Steady-State Thermal Gradient in Cylindrical Vessel Wall

construction shapes or contours which introduce steep thermal gradients,[56] Chapter 6.

2.13 Graphic Determination of Thermal Stress in a Cylindrical Vessel for Any Thermal Gradient

Frequently the thermal gradient throughout the cylindrical vessel wall cannot be simply expressed, and an analytical solution of Eqs. 2.12.1, 2.12.2, and 2.12.3 is not readily obtainable. Under these circumstances the integration can be done graphically by observing, for instance, that the tangential stress σ_t given by Eq. 2.12.2 may be written:

$$\sigma_t = \frac{\alpha E}{(1-\mu)}\left[\frac{r^2+a^2}{r^2(b^2-a^2)}\int_a^b Trdr + \frac{1}{r^2}\int_a^r Trdr - T\right] \quad (2.13.1)$$

The first integral can be expressed as

$$\frac{1+\dfrac{a^2}{r^2}}{b^2-a^2}\int_a^b Trdr = \frac{\displaystyle\int_a^b 2\pi Trdr}{\pi(b^2-a^2)} \quad (2.13.2)$$

since $1 + a^2/r^2 \approx 2$ (it is noted that $a \leqslant r \leqslant b$ and $b - a$ is small in comparison to r) and is the mean value of the temperature throughout the entire wall thickness. The second integral can be written:

$$\frac{1}{r^2}\int_a^r Trdr = \frac{\displaystyle\int_0^r 2\pi Trdr}{2\pi r^2} \quad (\text{if } T = 0 \text{ for } 0 \leqslant r \leqslant a) \quad (2.13.3)$$

and is one half of the mean value of the temperature distribution within the cylinder of radius r, where the integral is zero over most of the range of integration. The tangential stress at any point can then be written:

$$\sigma_t = \frac{\alpha E}{1-\mu}\left[\begin{pmatrix}\text{Mean temperature}\\ \text{of the entire}\\ \text{cylindrical wall}\\ \text{thickness}\end{pmatrix} + \begin{pmatrix}\tfrac{1}{2}\text{ the mean tem-}\\ \text{perature within}\\ \text{the cylinder of}\\ \text{radius } r\end{pmatrix} - \begin{pmatrix}\text{Temperature}\\ \text{of desired}\\ \text{stress location}\end{pmatrix}\right] \quad (2.13.4)$$

This may be further simplified by noting that, when the wall thickness is small compared with the inside radius of the cylinder, the mean value integral of Eq. 2.13.3 is considerably smaller than that of Eq. 2.13.2 and for an approximate solution may be dropped giving the approximation:

$$\sigma_t = \frac{\alpha E}{1-\mu}\left[\begin{pmatrix}\text{Mean temperature}\\ \text{of the entire}\\ \text{cylindrical wall}\\ \text{thickness}\end{pmatrix} - \begin{pmatrix}\text{Temperature of}\\ \text{desired stress}\\ \text{location}\end{pmatrix}\right] \tag{2.13.5}$$

The significance of Eq. 2.13.5 is that it is an algebraic expression showing that the thermal stress σ_t at any radius r is the value of this stress for full restraint, $\alpha E/(1 - \mu)$, multiplied by a factor proportional to the difference between the mean temperature of the whole cylinder and that of location r. This factor ranges from $\frac{1}{2}$ to 1 which is in agreement with the discussion of paragraphs 2.11 and 2.12. For instance, when the temperature distribution is linear, Eq. 2.13.5 gives a value of $\frac{1}{2}$ for this factor, whereas if the temperature distribution is a sharp one affecting the mean temperature of the entire cylindrical wall very little, a value of 1.0 for this factor is given by Eq. 2.13.5. Expressed in practical working terms, Eq. 2.13.5 states that, in order to find the thermal stress σ_t in a cylinder for a given thermal gradient, Fig. 2.32, the following should be done:

A. Plot the temperature distribution as a function of the square of the radius and determine the mean value of the plot.
B. Translate the abscissa axis to the mean temperature value determined in *A*.
C. Invert the ordinate scale and multiply the ordinate scale by $\alpha E/(1 - \mu)$. Inverting the scale gives the correct sign to the induced stress. That temperature below the minimum temperature occurring throughout the wall produces no stress since it is uniform throughout the entire wall thickness.

This curve is the approximate stress distribution in the new coordinate system. The cooler material will be in tension, and at least one surface will be in tension. A similar procedure can be followed to determine graphically the radial and longitudinal stresses.

Nuclear reactor vessels are subject to internal heat generation within their walls caused by the absorption of gamma radiation from the nuclear core. The conversion of such energy to heat behaves as a decaying exponential function through the wall of the cylindrical vessel.[29,30]

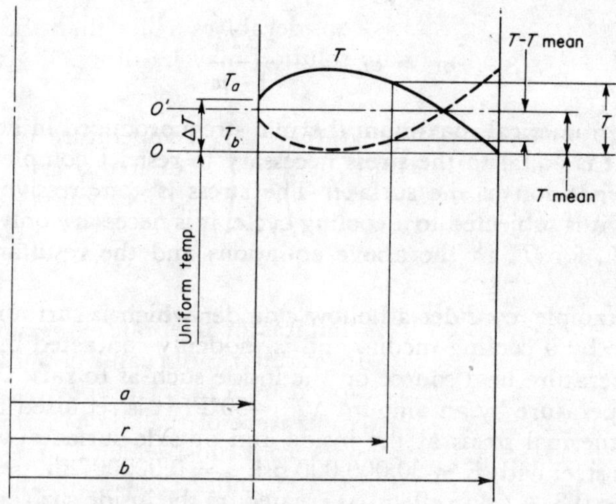

Fig. 2.32. Nature of the Thermal Gradient in a Cylindrical Vessel Due to Internal Heat Generation, and the Graphical Stress Evaluation Procedure

When the outside of the cylindrical reactor vessel is insulated, the temperature gradient is much like that shown in Fig. 2.32 which adapts itself well to the graphic integration solution above. A tabulation of many thermal gradients and resulting stress distribution equations for various heat generating geometries is given by Hankel.[31]

2.14 Thermal Stresses Due to Thermal Transients

In the preceding cases the thermal gradient was that which existed under steady-state conditions; i.e., it was independent of time. Of course, in order to reach this equilibrium thermal condition from an initial uniform temperature, a transient thermal gradient, or one changing with time, first occurs.[32] For instance, if the cylinder had an initial uniform temperature of zero, and beginning with time $t = 0$ the inside surface is maintained at a temperature T_a, the transient thermal gradients throughout the wall after various time intervals t_n are represented by the dotted curves of Fig. 2.33 as they approach the steady-state condition. From such curves the mean temperature of the whole cylindrical wall and also that of an inner portion of radius r can be determined. Then from Eq. 2.13.5, having these temperatures, the thermal stresses can be found for any time interval. For a very small time interval $t \approx 0$, the mean temperature approaches zero, and at the surface

$$\sigma_{t_a} = \sigma_{t_z} = -\frac{\alpha E T_a}{1-\mu} \tag{2.14.1}$$

This is the numerical maximum thermal stress produced in heating a cylinder. It is equal to the stress necessary to restrict completely the thermal expansion of the surface. The stress is compressive during heating. If it is subjected to a cooling cycle, it is necessary only to substitute $-T_a$ for T_a in the above equations and the resultant stress is tensile.

As an example, consider a hollow cylinder which is surrounded on the outside by a cooling media, and is suddenly contacted by a constant temperature heat source on the inside such as to raise its inside metal temperature by an amount $\Delta T = 50°F$. It is required to determine the thermal stress at the inside and outside surfaces when the material is steel with $E = 30{,}000{,}000$ psi, $\alpha = 0.000007$ in. per in. per °F, and $\mu = 0.3$. A "skin effect" is created on the inside surface at time zero when this transient is just started, because there has been insufficient time for heat conduction to take place, and a simple essentially vertical thermal gradient is assumed, Fig. 2.33 ($t \approx 0$). Using Eq.

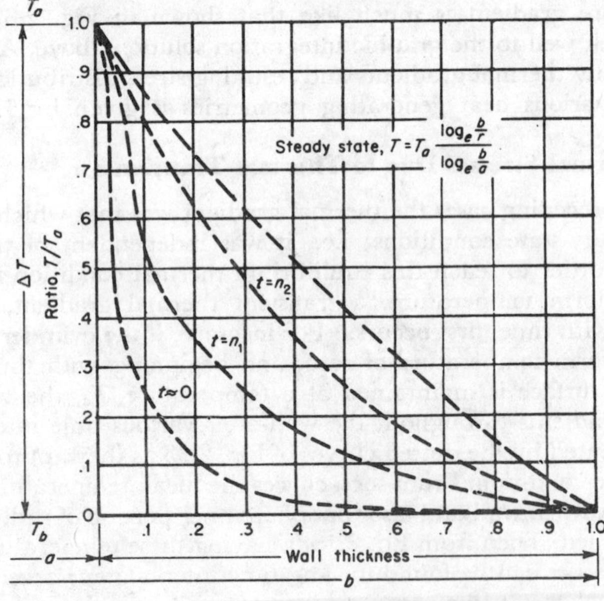

Fig. 2.33. Transient Thermal Gradients in Cylindrical Vessel Wall

2.13.5, and noting that the mean temperature of the entire cylindrical wall thickness is substantially unchanged, the thermal stress at the inside wall surface is

$$\sigma_{t_a} = \frac{\alpha E}{1-\mu}[0 - T_a] = -\frac{\alpha E T_a}{1-\mu} \qquad (2.14.2)$$

and at the outside surface is

$$\sigma_{t_b} = \frac{\alpha E}{1-\mu}[0 - 0] = 0 \qquad (2.14.3)$$

substituting the values of α, μ, E and T_a gives

$$\sigma_{t_a} = -\frac{.000007 \times 30{,}000{,}000 \times 50}{1 - .3} = -15{,}000 \text{ psi}$$

$$\sigma_{t_b} = 0$$

The shape of the thermal gradient progresses from one having a very steep slope to a logarithmic one after sufficient time elapses for a steady state condition of heat flow to prevail, Fig. 2.33. Approximating this logarithmic gradient by a linear one, the thermal stress at the inside wall surface is again found from Eq. 2.13.5, noting that the mean temperature of the entire cylindrical wall thickness is $T_a/2$

$$\sigma_{t_a} = \frac{\alpha E}{1-\mu}\left[\frac{T_a}{2} - T_a\right] = -\frac{\alpha E T_a}{2(1-\mu)} \qquad (2.14.4)$$

and at the outside surface is

$$\sigma_{t_b} = \frac{\alpha E}{1-\mu}\left[\frac{T_a}{2} - 0\right] = \frac{\alpha E T_a}{2(1-\mu)} \qquad (2.14.5)$$

Substituting the values of α, μ, E and T_a gives

$$\sigma_{t_a} = -\frac{.000007 \times 30{,}000{,}000 \times 50}{2(1 - .3)} = -7{,}500 \text{ psi}$$

$$\sigma_{t_b} = \frac{.000007 \times 30{,}000{,}000 \times 50}{2(1 - .3)} = 7{,}500 \text{ psi}$$

Maximum thermal stresses are associated with the maximum slope of the thermal gradient; hence, a linear thermal gradient will give minimum thermal stresses throughout since it has the minimum possible thermal gradient. When the transient thermal gradient can be expressed by simple analytical equations, the transient thermal stresses can be readily evaluated from Eqs. 2.12.1, 2.12.2, and 2.12.3; however, when they cannot be so expressed it is more convenient to solve for them by Eq. 2.13.4 and Eq. 2.13.5.

In order to reduce these maximum stresses in boiler drums, turbine rotors, process equipment, and nuclear vessels, it is customary practice to heat or cool them gradually to reduce the thermal gradient by beginning with a temperature much lower than the final temperature, and very slowly increasing the temperature when starting up, and reversing the procedure when shutting down.

2.15 Ultra-High Pressure Vessel Design Principles

Hydraulic and extrusion presses utilize very high fluid pressures to produce large forces, which in turn require extremely thick-walled cylinders. In such cylinders the hoop stress at the outside of the wall thickness is appreciably less than that at the inside surface; hence, the wall material is not used uniformly to its fullest stress and economic potential, Fig. 2.22 and Eq. 2.8.19. Several design principles that have been successfully used to overcome this situation follow.

1. *Wedge Principle*

Cylinders to withstand in the order of 200,000 psi are required in the synthetic gem and powder metallurgy industries.[33] One method that is used in the construction of these cylinders employs the wedge principle. This is shown in Fig. 2.34 and consists of placing a multitude of radial wedges inside a thick cylinder. The wedge surfaces are ground or fitted with membrane gaskets so as to preclude leakage along their mating surfaces; hence, the contained media is in contact with only the inside surface. Since the number of wedges is large the tangential strain in each one can be neglected and a contact pressure p^1 is created between each wedge and the unsplit integral cylinder equal to

$$p^1 = p\frac{c}{a} \tag{2.15.1}$$

This principle is based on these observations. In a cylinder under internal pressure both the tangential and radial stresses are a maximum at the inside surface, and fall off rapidly with distance into the wall of the cylinder, Eqs. 2.8.18 and 2.8.19. Since the shearing stress, to

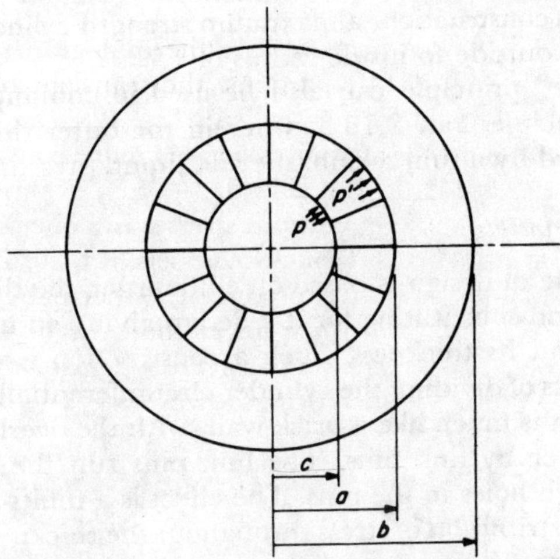

Fig. 2.34. Cross-section of a Thick-Walled Cylinder Showing Method of Sub-Dividing Wall for Wedge Construction Principle

which the failure of ductile material subscribes, Par. 5.15, is proportional to the difference of these two stresses, it can be reduced by moving the point where the highest tangential stress occurs away from that where the highest radial stress occurs. Hence, in this construction the maximum tangential stress occurs at radius "a" whereas the maximum radial stress occurs at radius "c".

The maximum shearing stress occurs at the inside surface, radius "a", of the integral cylinder and is found by substituting the value of the contact pressure p^1 from Eq. 2.15.1 in Eq. 2.8.21 as

$$\tau = p^1 \frac{b^2}{(b^2 - a^2)} = p\frac{c}{a}\frac{b^2}{(b^2 - a^2)} \tag{2.15.2}$$

The optimum dimensions of the cylinder of constant outside radius "b" to obtain minimum shear stress can be found by setting the derivative of this stress relative to the inside radius "a" of the cylinder equal to zero

$$\frac{d\tau}{da} = -pcb^2 \frac{(b^2 - 3a^2)}{a^2(b^2 - a^2)^2} = 0 \tag{2.15.3}$$

$$\frac{b}{a} = \sqrt{3} = 1.7 \tag{2.15.4}$$

Thus, with this construction, a maximum strength cylinder is obtained with a ratio of outside to inside radius of 1.7.

The "wedge" principle can also be used in conjunction with the "cascade" principle, Par. 2.15.3, wherein the outer thick-walled cylinder is replaced by a thin membrane and liquid pressure.

2. *Segment Principle*

This principle of design is based on eliminating the thick-wall hoop stress, per se, and substituting for it a design giving an uniform tensile stress throughout its thickness. Such a construction is shown in Fig. 2.35 and consists of dividing the cylinder circumferentially into a series of short links. It is much like a brick wall, with the overlapping bricks fastened together by link pins. The link pins run the length of the cylinder through holes in the pins. The effect is a multi-sided polygon with an even distribution of stress throughout the thickness of the polygon link members. A thin inner liner or seal membrane Fig. 2.35*a* can be employed to prevent fluid leakage. The width of the segments and diameter of hinge pins are based on the shearing, tensile, and bearing strength properties of the materials used just as with the design of bolted or riveted joints. When closure heads are required the hinge pins are used to fasten the head to the cylinder under this condition they also take longitudinal stress.

Neglecting the effect of the seal membrane liner, and equating the applied load on the elemental area *abcd* due to the internal pressure p to the resisting force in a segment uniformly spaced on a longitudinal pitch of twice the segment width, Fig. 2.35*b*, gives

$$2wp \int_0^{\theta/2} 2r \cos \phi \, d\phi = 2F_1 = 2F \tan\left(\frac{\theta}{2}\right) \qquad (2.15.5)$$

$$4wpr \sin\left(\frac{\theta}{2}\right) = 2F \tan\left(\frac{\theta}{2}\right) \qquad (2.15.6)$$

but since for small angles the sine, tangent and angle in radians are approximately equal, Eq. 2.15.6 gives

$$F = 2wpr \qquad (2.15.7)$$

Dividing this force, Eq. 2.15.7, by the cross-sectional area of the segment, Sect. A-A. Fig. 2.35*c*, gives the segment tensile stress

$$\sigma_2 = \frac{2wrp}{wh} = \frac{2rp}{h} \qquad (2.15.8)$$

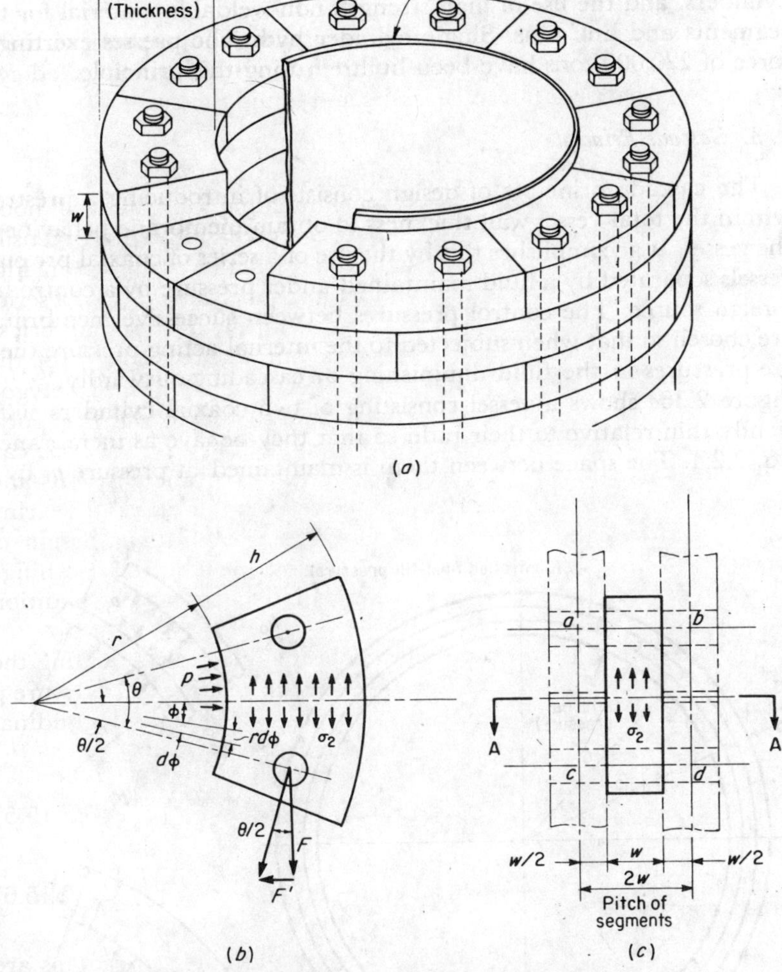

Fig. 2.35. Segmental Cylinder Construction

This is twice the hoop membrane stress in a cylinder of the same dimensions, Eq. 2.2.4, but since the stress is uniform across the segment, large thicknesses of material can be used to their maximum potential.

The segment principle permits the construction of single large diameter pressure cylinders to accomplish the task usually done by a multitude of small diameter cylinders acting in parallel. This type of construction also permits the field erection of extremely large heavy cylinders, and the use of high strength non-weldable material for the segments and link pins. Single cylinder hydraulic presses exerting a force of 220,000 tons have been built[34,35] using this principle.

3. *Cascade Principle*

The cascade principle of design consists of introducing a prestress within the total vessel wall thickness to obtain membrane behavior in the vessel. It accomplishes this by the use of a series of coaxial pressure vessels separated by a fluid maintained under pressure by a controlled foreign source. The control pressures between successive membranes are chosen so that when subjected to the internal acting pressure there are pressures in the fluid diminishing or cascading outwardly.[36,37,51] Figure 2.36*a* shows a vessel consisting of two coaxial cylinders sufficiently thin relative to their radii so that they behave as membranes, Eq. 2.2.4. The space between them is maintained at pressure p_1 by a

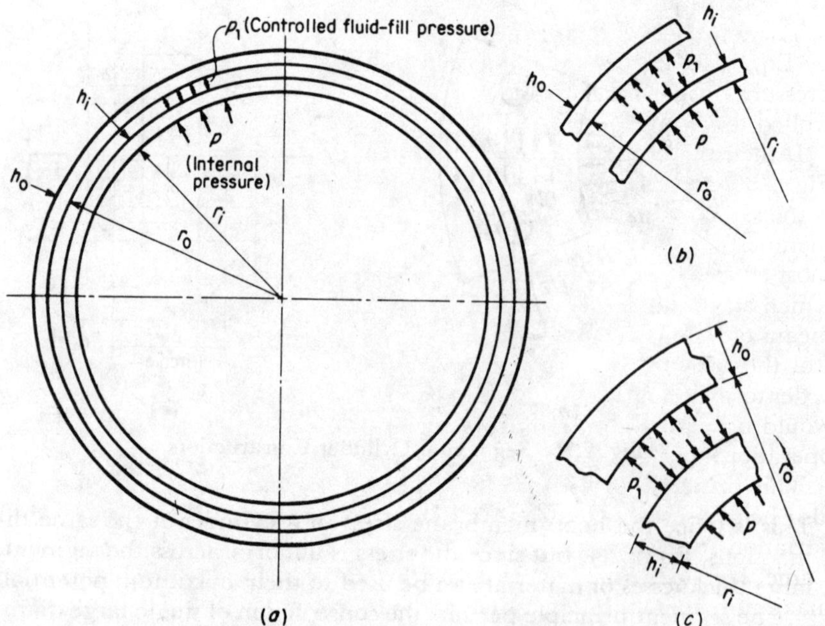

Fig. 2.36. Coaxial Cylindrical Vessel with Cascade Controlled Fluid-Fill Pressure

separate source so that the inner cylinder is placed in compression with a hoop stress of

$$\sigma_{2(i)} = \frac{-p_1 r_i}{h_i} \tag{2.15.9}$$

and the outer cylinder is placed under a hoop tension stress of

$$\sigma_{2(o)} = \frac{p_1 r_o}{h_o} \tag{2.15.10}$$

Application of the acting internal pressure only on the inner cylinder produces a hoop tension stress of

$$\sigma_{2(i)} = \frac{p r_i}{h_i} \tag{2.15.11}$$

The total hoop stress in the inner cylinder is obtained by adding Eq. 2.15.9 and Eq. 2.15.11 to give

$$\sigma_{2(i)} = \frac{-p_1 r_i}{h_i} + \frac{p r_i}{h_i} = \frac{(p - p_1) r_i}{h_i} \tag{2.15.12}$$

The total hoop stress in the outer cylinder remains the same as given by Eq. 2.15.10. It is not influenced by the application of the internal pressure on the inner cylinder since the pressure acting on it is controlled to a value p_1.

If the radii of the inner and outer cylinders are substantially equal, Fig. 2.36*b*, it is seen that the sum of the thickness of the two cylinders is the same as that for a single thickness cylinder assumed to act as a membrane with uniform stress throughout the entire wall thickness. The most efficient vessel in terms of economic material utilization is one in which all the material acts as a membrane and resists the pressure by means of a uniform stress through its thickness. It is important to control the pressure relationship p/p_1 with constructions of this type in order to avoid subjecting the inner vessel to a collapse condition as would occur if the internal pressure p was reduced below p_1 during operation.

This principle of a controlled fluid-fill pressure between vessels is also helpful even when the wall thickness is large, Fig. 2.36*c*, and a variation in stress occurs across the thickness in accordance with the Lame' formula, Par. 2.8. It helps through reducing the variation of maximum to minimum radial and tangential stress throughout the thickness, Equations 2.8.15 and 2.8.16, and is particularly advantageous in lowering the shear stress, Eq. 2.8.17, by reducing the value of the radial compressive stress occurring at the point of maximum tangential stress.

4. *Yoke Principle*

A version of the cylindrical monobloc, shrink-fit, or segmented vessel is one in which the axial force is not transmitted to the cylindrical vessel but instead is imposed on an external yoke via a floating cylindrical closure head. Tightness is accomplished by a deflection compensating seal, Fig. 6.50*b*, in which the internal pressure forces the seal carrier to dilate and follow the ID of the vessel. Hence, the O-ring gaskets do not see a clearance and cannot extrude. Therefore, they remain leakproof under pressure. The separate external structural yoke may consist of:

1. A series of needle beams supported by pinned ties, Fig. 2.37*a*, or
2. A series of continuous wire or strap wrappings, Fig. 2.37*b*.

The advantages claimed for this type of construction are its use of inexpensive structural material for the yoke, and a minimum amount of machining. After depressurizing, quick opening of the vessel is facilitated by sliding it out of the yoke for access and removal of end closures. In the case of beam tie-pins, removal of one set permits hinging the support beams. Alternatively, the yoke may be slid away from the vessel.[61]

This principle has been used for the construction of high pressure vessels in metallurgical sintering and bonding processes. It is well adapted to high pressures and large diameters, and there are no size limitations on the use of this type of end closure.

The strap wrappings, Fig. 2.37*b*, are in membrane tension and must sustain the total end load, so the force in the wrapping on one side of the cylinder is

$$F = \frac{\pi r^2 p_i}{2} \qquad (2.15.13)$$

and the average membrane stress in the wrapping is

$$\sigma_{\text{avg.}} = \frac{F}{A} = \frac{\pi r^2 p_i}{2A} \qquad (2.15.14)$$

where A is the cross-sectional area of the wrapping on one side of the cylinder (Fig. 2.37*b*). The stress in the wrapping will vary dependent

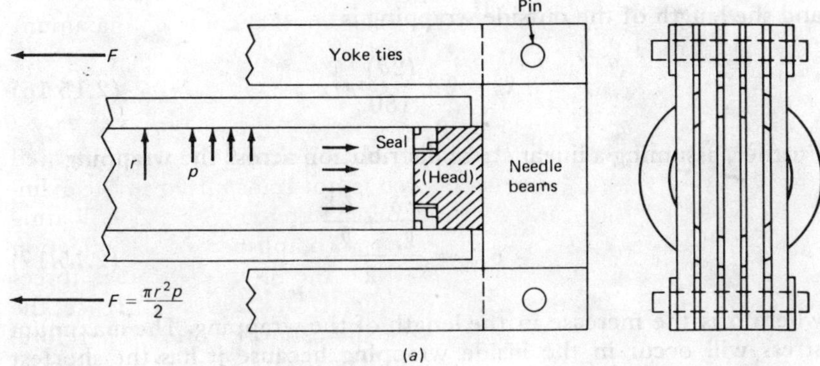

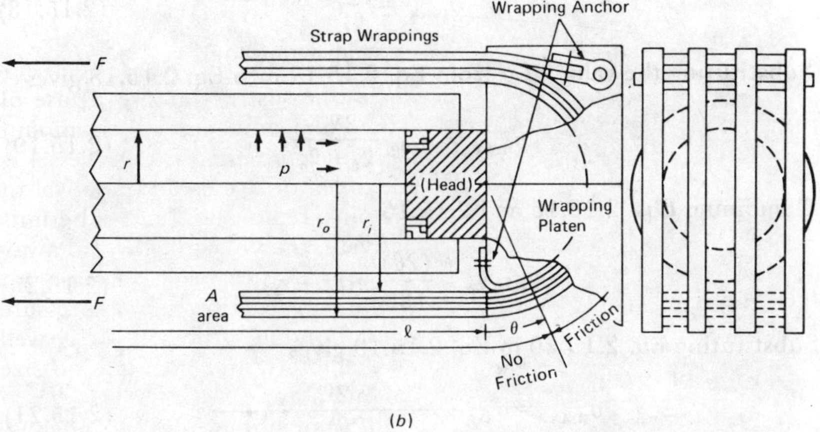

Fig. 2.37. Pressure Vessel with External Yoke. (*a*) Beam and Ties (*b*) Continuous Strap Wrapping.

upon the assumption made concerning friction between the individual wrappings. If no friction is present for $\theta < \theta_f$ and the wrappings are free to move relative to each other, the length of the inside wrapping is

$$\ell_i = \ell + \frac{(2\theta)}{180} \pi r_i \tag{2.15.15}$$

and the length of the outside wrapping is

$$\ell_o = \ell + \frac{(2\theta)}{180} \pi r_o \tag{2.15.16}$$

Further, assuming a linear stress distribution across the wrapping

$$\sigma_{avg.} = \frac{\dfrac{E\delta}{\ell_i} + \dfrac{E\delta}{\ell_o}}{2} \tag{2.15.17}$$

where δ is the increase in the length of the wrapping. The maximum stress will occur in the inside wrapping because it has the shortest length.

$$\sigma_{max.} = \frac{E\delta}{\ell_i} \tag{2.15.18}$$

Substituting the value of δ from Eq. 2.15.17 into Eq. 2.15.18 gives

$$\sigma_{max.} = \sigma_{avg.} \frac{2\ell_o}{\ell_i + \ell_o} \tag{2.15.19}$$

Combining Eqs. 2.15.15 and 2.15.16

$$\ell_i = \ell_o - \frac{(2\theta)}{180} \pi (r_o - r_i) \tag{2.15.20}$$

Substituting Eq. 2.15.20 in Eq. 2.15.19 gives

$$\sigma_{max.} = \sigma_{avg.} \frac{2\ell_o}{2\ell_o - \dfrac{(2\theta)}{180} \pi (r_o - r_i)} \tag{2.15.21}$$

The maximum stress condition occurs on the inside wrapping for $\theta = 90°$; that is, no friction exists between the wrappings, and from Eq. 2.15.21

$$\sigma_{max.} = \sigma_{avg.} \frac{2\ell_o}{2\ell_o - \pi(r_o - r_i)} \tag{2.15.22}$$

The maximum stress possible on the outer wrapping occurs when friction prevents sliding throughout the circular portion, $\theta = 0$; and from Eq. 2.15.21 is equal to the average stress.

PROBLEMS

1. What is the required thickness of a 6 ft. inside diameter cylinder, considering it as a thin wall vessel, to withstand an internal pressure of 1000 psi if the allowable tangential stress is 20,000 psi?

Ans.: 1.8 in.

2. What is the thickness of a spherical head for the cylinder of Problem 1 if the dilation of the head and cylinder is to be equal?

Ans.: $1.8\ (1 - \mu)/(2 - \mu)$

3. What are the numerically maximum and minimum thermal stresses occurring in a carbon steel cylinder of $b = 3a$ for steady-state heat flow condition in which the temperature of the inside wall is 10°F higher than the outside wall ($\alpha E/(1 - \mu) =$ 300 psi per °F) (hint, use Eqs. 2.12.8 and 2.12.9)?

Ans.: Max. $\sigma_{t_a} = \sigma_{z_a} = -\ 2{,}010$ psi
Min. $\sigma_{t_b} = \sigma_{z_b} = +\ 990$ psi

4. A point in a body is acted upon by three normal tensile stresses, σ_1, σ_2 and σ_3, on mutually perpendicular planes so as to form a triaxial stress condition. What are the strains in the direction of these stresses?

Ans.:
$$e_1 = 1/E(\sigma_1 - \mu\sigma_2 - \mu\sigma_3)$$
$$e_2 = 1/E(\sigma_2 - \mu\sigma_3 - \mu\sigma_1)$$
$$e_3 = 1/E(\sigma_3 - \mu\sigma_1 - \mu\sigma_2)$$

5. A cylinder with a 48 in. inside diameter, and a 60 in. outside diameter is subjected to an internal pressure of 5,000 psi. Determine the value and place of occurrence of: (*a*) the maximum tangential stress, (*b*) the maximum radial stress; and (*c*) the maximum shear stress.

Ans.: (*a*) 22,778 psi, tangent to inside surface.
(*b*) −5,000 psi, normal to inside surface.
(*c*) 13,889 psi, 45° plane at inside surface.

6. In the cylinder of Problem 5, (*a*) what is the average tangential stress; and (*b*) what per cent is this of the maximum tangential stress?

Ans.: (*a*) 20,000 psi
(*b*) 87.8 per cent

7. A 100 psi air accumulator consists of two 60 in. inside diameter spheres constructed to intersect at an angle of 30° with their common axis. (*a*) How thick must their wall be for an allowable design stress of 20,000 psi? (*b*) If a ring is used to reinforce this intersection, what is its required cross-sectional area to maintain the natural radial growth of these spheres at their intersection?

Ans.: (*a*) 0.075 in.
(*b*) 2.798 sq. in.

8. An 8 in. inside diameter ½ in. thick tube is bent into a 10 ft. diameter torus which is to be subjected to an internal pressure of 2,000 psi. What is the tangential stress at: (*a*) the centerline of the torus bend, (*b*) the outside of the bend cross section, (*c*) the crotch or inside of the bend cross section; and (*d*) by what per cent does the tangential crotch stress exceed the average tangential stress in a straight cylinder of the same cross-sectional dimension and subjected to the same internal pressure?

Ans.: (*a*) 16,000 psi
(*b*) 15,500 psi
(*c*) 16,571 psi
(*d*) 3.6 per cent

9. If a long sleeve of inside radius b and outside radius c is shrunk on a solid shaft, the external radius, of which in the unstressed condition is larger than the internal radius of the sleeve by an amount δ, establish an expression for the uniform contact pressure p.

Ans.: $p = \delta E(c^2 - b^2)/2bc^2$

10. In the sleeve given in Problem 9, (*a*) establish an expression for the maximum shearing stress in sleeve; and (*b*) if the shrink-fit is obtained by heating the sleeve to a uniform temperature T above the solid shaft, how hot must the sleeve be heated to produce a maximum shear stress in the sleeve of 20,000 psi if the diameter of the shaft is 6 in. (assume the materials to be steel with $E = 30{,}000{,}000$ psi, and $\alpha = 0.000007$ in. per in. per °F)?

Ans.: (*a*) $\tau_{max.} = E\delta/2b$
(*b*) 190°F

11. A thick cylinder is subjected to only an external pressure directed inward. Derive an expression and reduce it to its simplest form for the hoop stress, radial stress, and shear stress at the inside and outside surface.

Ans.:

$$\sigma_{t_a} = \frac{-2p_o b^2}{b^2 - a^2}, \qquad \sigma_{t_b} = \frac{-p_o(b^2 + a^2)}{b^2 - a^2}$$

$$\sigma_{r_a} = 0 \qquad \sigma_{r_b} = -p_o$$

$$\tau_a = \frac{-p_o b^2}{b^2 - a^2} \qquad \tau_b = \frac{-p_o a^2}{b^2 - a^2}$$

12. The ASME Nuclear Power Code uses the maximum shear stress theory of failure and requires the minimum thickness, h, of a thin cylindrical vessel of inside radius, r, which is closed on each end and subject to an internal pressure, p, to be:

$$h = \frac{p_r}{2\tau - 0.5p}$$

Develop this formula. (τ is the allowable maximum shear stress.)
[Hint: See Par. 5.15.1 (b).]

13. In a thin vessel of revolution of thickness "h," subject to an internal pressure "p," the two principal stresses are equal ($\sigma_1 = \sigma_2$). Prove that this vessel is a sphere. (Hint: start with Eq. 2.6.5.)

14. A banded cylindrical vessel, 30″ inside radius, has hemispherical heads and is subjected to an internal pressure of 1,000 psi, Fig. 2.38. Assume there is no friction between the band and the inner shell and also no prestress exists. (*a*) What is the required thickness of the hemispherical heads for a maximum membrane stress of 20,000 psi? (*b*) If the inner shell of the cylindrical portion is made the same thickness as the hemispherical head, what is the required thickness of the wrapper (or band) to maintain a maximum membrane hoop stress of 20,000

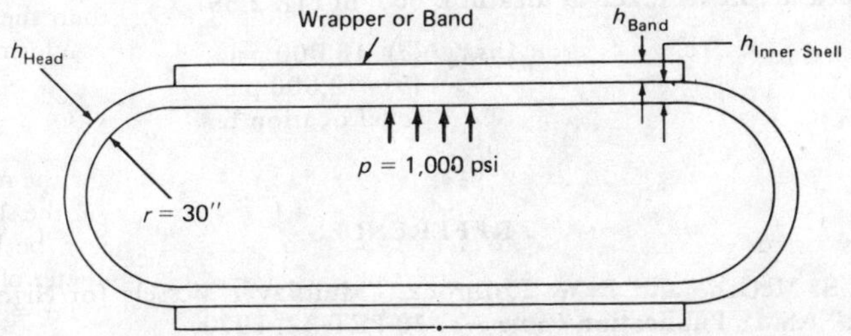

Fig. 2.38. Cross Section through the Banded Vessel of Problem 14.

psi in the inner shell? (assume $\mu = 0.3$) (Hint: The inner shell is subjected to both a hoop and longitudinal stress, whereas the outer band is subject to only a hoop stress. See Par. 7.5.4).

Ans.: (*a*) .75 in.
(*b*) 1.07 in.

15. The cylindrical portion of the vessel of Problem 14 is constructed of a single solid wall. (*a*) What is the required thickness to maintain a hoop stress of 20,000 psi? (*b*) How does this compare to the combined thickness of the cylindrical inner shell and band of Problem 14?

Ans.: (*a*) 1.5 in.
(*b*) 17% thinner

16. In practice it is frequently difficult to form perfectly circular vessels, but due to the fabricating process they may have a very small noncircularity as shown in Fig. 2.39. Under the action of internal pressure it tends to deform to a truly circular shape, thereby introducing a small local bending stress in the wall of the vessel.

Stress measurements on a thin wall cylinder showed a tangential stress of 20,000 psi on the inside surface and a tangential stress of 16,000 psi on the outside surface adjacent to the inside stress measurement. What is (*a*) the value

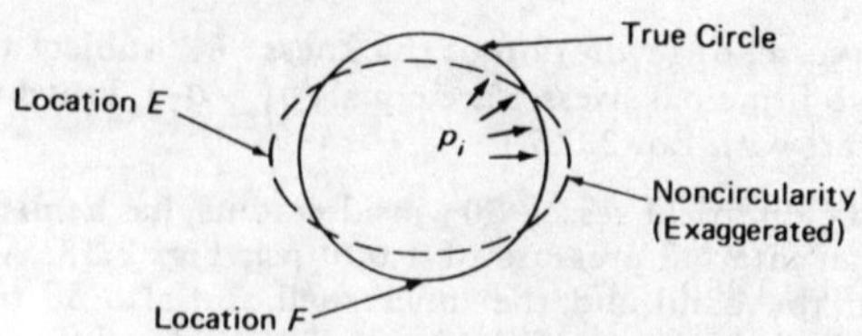

Fig. 2.39. Noncircular Vessel

of the direct or hoop stress, (*b*) the value of the bending stress and (*c*) were these stress measurements taken at locatin E or F in Fig. 2.39?

Ans.: (*a*) 18,000 psi
(*b*) ±2,000 psi
(*c*) Location E

REFERENCES

1. J. S. McCabe and E. W. Rothrock, "Multilayer Vessels for High Pressure Service," ASME Publication *Paper No.* 70-PET-32, 1970.

2. R. L. Davis and H. D. Keith "Fatigue Analysis of a Cladded Multilayer Pressure Vessel," ASME Publication *Paper No.* 71/WA/HT-21, 1971.

3. E. G. Menkes, G. N. Sandor and L. R. Wang, "Optimum Design of Composite Multilayer Shell Structures," ASME Publication *Paper No.* 71-WA/DE-12, 1971.

4. A. K. Shoemaker, T. Melville and J. E. Steiner, "Fracture Resistance of Wire-Wrapped Cylinders," ASME Publication *Paper No.* 72-PET-9, 1972.

5. C. Garland, "Design and Fabrication of Deep-Diving Submersible Pressure Hulls," *Transactions, The Society of Naval Architects and Marine Engineers,* Vol. 76, 1968.

6. H. Krause, "Elastic Stresses in Pressure Vessel Heads," Welding Research Council Bulletin 129, April, 1968.

7. R. L. Cloud, "Interpretive Report on Pressure Vessel Heads," Welding Research Council Bulletin 119, January, 1967.

8. F. A. Simonen and D. T. Hunter, "Elastic-Plastic Deformations in Pressure Vessel Heads," Welding Research Council Bulletin 163, July, 1971.

9. Z. Zudans and F. H. Gregory, "Analysis of Shells of Revolution Formed of Closed Box Section," ASME Publication Paper No. 70-PVP-12, 1970.

10. H. Kraus, G. G. Bilodeau, and B. F. Langer, "Stresses in Thin-Walled Pressure Vessels with Ellipsoidal Heads," *Transactions ASME* 83, 1961.

11. E. P. Esztergar and H. Kraus, "Analysis and Design of Ellipsoidal Pressure Vessels Heads," ASME Publication Paper No. 70-PVP-26, 1970.

12. E. O. Jones, "Thin-Shell Pressure Vessel Toroidal Region and Ellipsoidal Head Stresses," ASME Publication Paper No. 67-PET-19, 1967.

13. P. Rotondo and H. Kraus, "Buckling of an Ellipsoid Due to Internal Pressure," ASME Publication Paper No. 68-WA/PVP-12, 1968.

14. J. C. Gerdeen and D. N. Hutula, "Plastic Collapse of ASME Ellipsoidal Head Pressure Vessels," ASME Publication Paper No. 70-PVP-13, 1970.

15. J. Adachi and M. Benicek, "Buckling of Torispherical Shells Under Internal Pressure," *Journal of the Society for Experimental Stress Analysis*, August, 1964.

16. G. J. Schoessow and E. A. Brooks, "Analysis of Experimental Data Regarding Certain Design Features of Pressure Vessels," *Transactions of ASME*, Vol. 72, No. 5, July, 1950.

17. H. Fessler and P. Stanley, "Stresses in Torispherical Drumheads: A Critical Review," Welding Research Abroad, Welding Research Council, Vol. XII, No. 4, April, 1966.

18. G. D. Gallety, "Torospherical Shells–A Caution to Designers," ASME Publication Paper No. 58-PET-3, 1958.

19. R. J. Crisp, "A Computer Survey of the Behavior of Torispherical Drum Heads Under Internal Pressure Loading," *Nuclear Engineering and Design*, Vol. 11, 1969, No. 3, April 1970, pp. 457–495, North-Holland Publishing Co., Amsterdam, Netherlands.

20. S. Timoshenko, *Strength of Materials*, Part II, D. Van Nostrand Co. Inc., New York, 1956.

21. I. Berman and D. H. Pai, "Elevated Temperature Autofrettage," ASME Publication Paper No. 66-WA/PVP-3, 1966.

22. J. Marin and Tu-Lung Weng, "Strength of Thick-Walled Cylindrical Pressure Vessels," ASME Publication Paper No. 62-WA-227, 1962.

23. N. L. Svensson, "The Bursting Pressure of Cylindrical and Spherical Vessels," ASME Publication Paper No. 57-A-15, 1957.

24. B. F. Langer, "Design Values for Thermal Stress in Ductile Materials," *The Welding Journal*, Research Supplement, September, 1958.

25. S. Timoshenko and J. N. Goodier, *Theory of Elasticity*, McGraw-Hill Book Co. Inc., New York, 1951.

26. J. C. Heap, "Thermal Stresses in Concentrically Heated Hollow Cylinders," ASME Publication Paper No. 62-WA-228, 1962.

27. D. Burgreen, *Elements of Thermal Stress Analysis*, C. P. Press, Jamaica, New York, 1971.

28. W H. McAdams, *Heat Transmission*, McGraw-Hill Book Co., Inc., New York, 1942.

29. H. J. Honohan, "Steady State Stress Distribution in Long Hollow Cylinders Experiencing a Radial Exponential Internal Heat Generation," Master of Science Thesis, University of Pittsburgh, Pa., 1956

30. H. Krause and G. Sonneman, "Stresses in Hollow Cylinders Due to Asymmetrical Heat Generation," *ASME Transactions*, October, 1959.

31. R. Hankel, "Stress and Temperature Distribution," *Nucleonics*, Vol. 18, No. 11, November, 1960.

32. Z. Zudans, T. C. Yen and W. H. Steigelmann, *Thermal Stress Techniques*, American Elsevier Publishing Co., New York, 1965.

33. R. H. Wentorf, *Modern Very High Pressure Techniques*, Butterworths, Washington, D. C., 1962.

34. G. Birman, "Large Steel Vessels for Deep-Submergence Simulation," ASME Publication Paper No. 66-WA/UNT-13, 1966.

35. A. Zeitlin, "High-Pressure Technology," *Scientific American*, Vol. 212, No. 5, May, 1965.

36. I. Berman, "Design and Analysis of Commercial Pressure Vessels to 500,000 psi," ASME Publication Paper No. 65-WA/PT-1, 1965.

37. L. W. Hu and J. C. Schutzer, "Cascade Arrangement in Spherical Vessel

Design for Nuclear Power Reactors," *Nuclear Engineering and Design*, 1966, pp. 412–420, North-Holland Publishing Co., Amsterdam.

38. D. Bushnell and G. D. Galletly, "Stress and Buckling of Internally Pressurized, Elastic-Plastic Torispherical Vessel Heads—Comparison of Test and Theory," ASME Publication Paper No. 74-PVP-23, 1976.

39. G. D. Galletly, "Elastic and Elastic-Plastic Buckling of Internally-Pressurized 2:1 Ellipsoidal Shells," ASME Publication Paper No. 78-PVP-47, 1978.

40. I. Berman, R. Henschell and J. M. Horowitz, "Elastic-Plastic Behavior to Burst of Commerically Fabricated Pressure Vessel Heads with Attached Cylinders," Third International Conference on Pressure Vessel Technology, Tokyo, Japan, 1977.

41. E. P. Esztergar, "Development of Design Rules for Dished Pressure Vessel Heads," *Welding Research Bulletin* 215, May, 1976.

42. J. C. Gerdeen, "The Effect of Geometrical Variations on the Limit Pressures for 2:1 Ellipsoidal Head Vessels Under Internal Pressure," *Welding Research Bulletin* 215, May, 1976.

43. C. E. Washington, R. J. Clifton and B. W. Costerus, "Tests of Torispherical Pressure Vessel Heads Convex to Pressure," *Welding Research Bulletin* 227, June, 1977.

44. A. Biron and J. Veillon, "Influence of Head Thickness on Yield Pressure for Cylindrical Pressure Vessels," ASME Publication Paper No. 74-PVP-4.

45. W. L. Greenstreet, "Experimental Study of Plastic Responses of Pipe Elbows," Oak Ridge National Laboratory Report ORNL/NUREG-24, February, 1978.

46. J. Middleton and D. R. J. Owen, "Automated Design Optimization to Minimize Shearing Stress in Axisymmetric Pressure Vessels," *Nuclear Engineering and Design,* Vol. 44, No. 3, December 1977, North-Holland Publishing Co.

47. C. P. Wright, "Design and Analysis of Dished Covers Under External Pressure," ASME Publication Paper No. 73-WA/Oct-16, 1973.

48. S. S. Gill, *The Stress Analysis of Pressure Vessels and Pressure Vessel Components,* Pergamon Press, London, 1970.

49. J. P. Carter, "The Expansion of a Cylinder Under Conditions of Finite Plane Strain," *Nuclear Engineering and Design,* North-Holland Publishing Co., May, 1978.

50. C. P. Boyer and S. T. Rolfe, "Effect of Strain-Hardening Exponent and Strain Concentrations on the Bursting Behavior of Pressure Vessels," ASME Publication Paper No. 74-Mat-1, 1974.

51. A. I. Soler, "On Seal Forces in Removable Closures in Very High Pressure Test Chambers," ASME Publication Paper No. 74-WA/PVP-2, 1974.

52. J. T. S. Wang, "Cylindrical Panels of Various Shapes for Pressure Vessels," ASME Publication Paper No. 79-PVP-110, 1979.

53. G. D. Galletly and S. K. Radhamohan, "Elastic-Plastic Buckling of Internally-Pressurized Thin Torispherical Shells," ASME Publication Paper No. 79-PVP-52, 1979.

54. D. Bushnell, "Elastic-Plastic Buckling of Internally Pressurized Ellipsoidal Pressure Vessel Heads," Welding Research Council Bulletin 267, May 1981.

55. J. H. Faupel and F. E. Fisher, *Engineering Design*, McGraw-Hill Book Co., New York, 1982.

56. M. S. M. Rao, T.V. Narayanan, and G. P. Gupta, "Inelastic Analysis of Nonaxisymmetrically Heated Thick Cylindrical Shells,"ASME Publication Paper No. 79-PVP-8, 1979.

57. R. Labbens and R. Bovagne, "Mechanical Background of Pressure Vessel Calculation with Special Application to Layered Vessels," ASME Publication 83-PVP-44, 1983.

58. E. L. Thomas, "Design of Rectangular Pressure Vessels with Chamfered Corners," ASME Publication Paper 83-PVP-9, 1983.

59. R. L. Roche and B. Autrusson, "Experimental Tests on Buckling of Torispherical Heads—Comparison with Plastic Bifurcation Analysis," ASME Publication Volume PVP-Vol. 89, 1984.

60. W. Szyszkowski and P. Glockner, "Design of Buckle-Free Shapes in Pressure Vessels," ASME Publication Paper 84-PVP-105, 1984.

61. G. Härkegård, J. Liljeblad, and L. Ohlsson, "A Comparative Study of Various Design Concepts for High Pressure Vessels Subject to Cyclic Operation," ASME Publication Paper 84-PVP-116, 1984.

3

Stresses in Flat Plates

3.1 Behavior of Flat Plates

Flat plates may be thought of as two-dimensional beams. When a plate, such as a cylinder head or manway cover, bends under loads normal to its surface, the plate bends in two perpendicular planes rather than in only one plane as does a beam.

The behavior and failure of a plate may be depicted by Fig. 3.1. From 0 to *A* the deflection is proportional to the load and the deflection is due to bending only. This is the region that will be discussed in the succeeding paragraphs. In the region *A* to *B*, yielding has occurred over the entire plate thickness and direct tension carries a major part of the load as in a thin wall or membrane vessel, Chapter 2. The purely elastic strength of a plate is small compared to its total, and when its flatness must be maintained, such as for steam generator tube-sheets, the thickness must be adequate for the loading or supplemental stay-bolts, ties, or strong-backs employed.

Plates may be arbitrarily classified into three groups: (1) thick plates in which the shearing stress is important, much as with a short

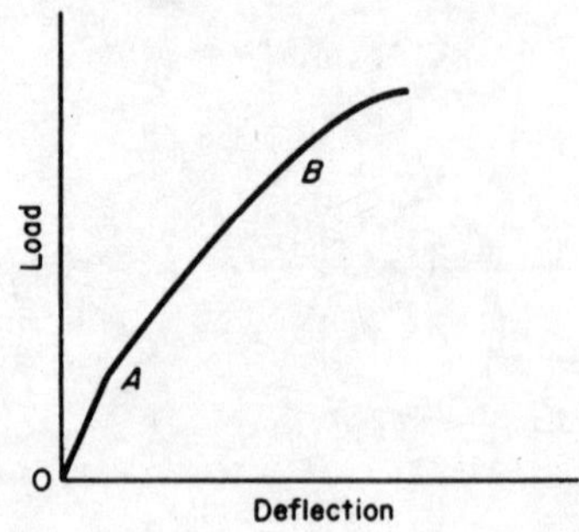

Fig. 3.1. General Behavior of a Flat Plate

deep beam; (2) medium thickness plates in which bending stresses are the most important and on which the major strength of the plate depends, corresponding to the usual beam, and about which this chapter is concerned; and (3) thin plates whose strength depends mainly on the direct tension accompanying stretching of its middle plane.

Circular flat plates are widely used in all pressure vessel design and construction, such as at the ends of cylinders or hemispheres, and for access and maintenance closures, as well as for the structural support of internal loads from catalysis beds in process vessels or nuclear cores in nuclear vessels.

3.2 Bending of a Plate in One Direction

When a simple beam is subjected to bending, not only are the fibers strained in the longitudinal direction, Fig. 3.2, equal to

$$e_x = \frac{s's_1}{nn'} = \frac{y}{r} \tag{3.2.1}$$

but they are also accompanied by lateral contraction on the convex (tensile) side, and lateral expansion on the concave (compressive) side due to the Poisson effect of

$$e_z = -\mu e_x = -\mu\frac{y}{r} \tag{3.2.2}$$

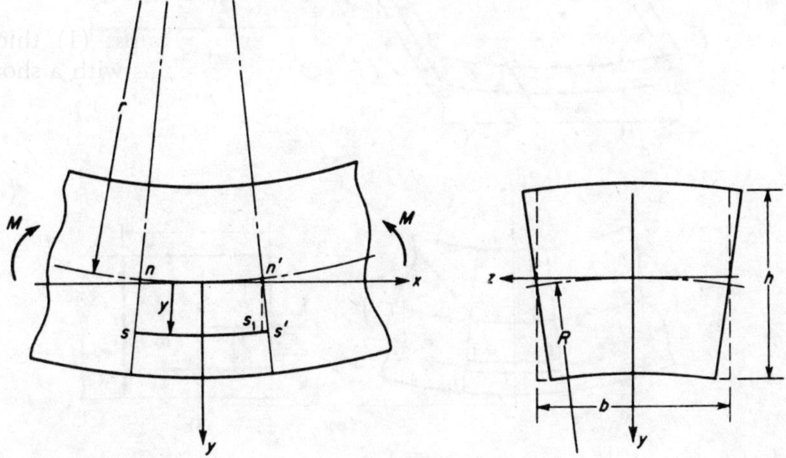

Fig. 3.2. Bending of a Beam

Due to this distortion all straight lines that are parallel to the z direction in a cross section curve so as to remain normal to the sides of the section, Fig. 3.2. Their radius of curvature R will be larger than r in the same proportion as e_x is larger than e_z, and equal to

$$R = -\frac{r}{\mu} \tag{3.2.3}$$

If a long rectangular plate of uniform thickness, h, is bent to a cylindrical surface by moments along its long sides or loads normal to its surface, Fig. 3.3a, it is sufficient to consider only a strip of unit width as a rectangular cross-section beam of length a. Since it can be concluded from the condition of continuity that there is no distortion of the strip cross section during bending, as shown in Fig. 3.3b, a longitudinal fiber in the strip ss is subjected to both a longitudinal tensile stress σ_x and a tensile stress σ_z in the lateral direction sufficient to prevent contraction of the fiber. Assuming that cross sections of the strip remain

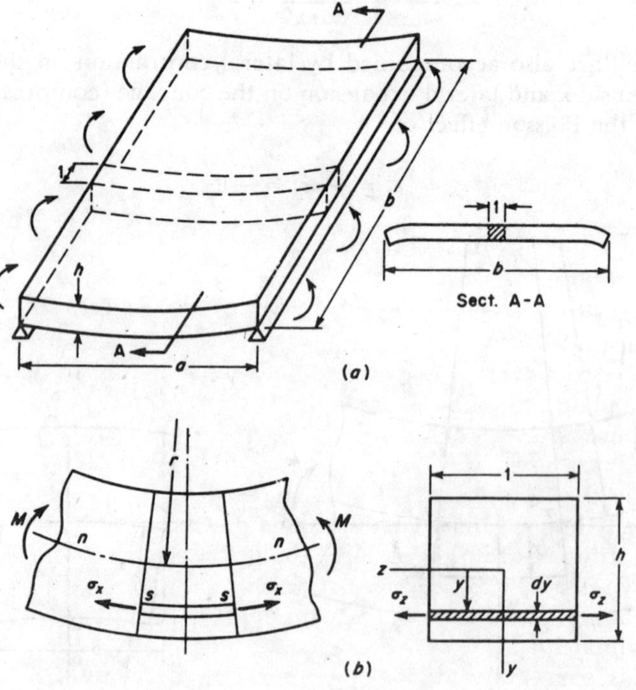

Fig. 3.3. Bending of Plate in One Plane

plane during bending, the unit elongation in the x and z directions, Fig. 3.3b, are:

$$e_x = \frac{y}{r} \quad \text{and} \quad e_z = 0 \tag{3.2.4}$$

The corresponding stresses are obtained from Eqs. 2.3.4 and 2.3.5 for the case of tension in two perpendicular directions as:

$$\sigma_x = \frac{e_x E}{1 - \mu^2} = \frac{Ey}{(1 - \mu^2)r} \tag{3.2.5}$$

$$\sigma_z = \frac{\mu e_x E}{1 - \mu^2} = \frac{\mu E y}{(1 - \mu^2)r} \tag{3.2.6}$$

The bending moment at any cross section of the strip is then

$$M = \int_{-h/2}^{+h/2} \sigma_x y\, dy = \frac{E}{(1 - \mu^2)r} \int_{-h/2}^{+h/2} y^2\, dy = \frac{Eh^3}{12(1 - \mu^2)r} \tag{3.2.7}$$

from which

$$\frac{1}{r} = \frac{M}{D} \tag{3.2.8}$$

where

$$D = \frac{Eh^3}{12(1 - \mu^2)} = \frac{EI}{1 - \mu^2} \tag{3.2.9}$$

The quantity D is called the "flexural rigidity" of the plate and takes the place of EI in the conventional beam formulas. Thus due to the lateral restriction and correspondingly induced lateral bending moment, a plate bent like a beam in one plane only is stiffer than it would be by pure beam action only in the ratio of $1/(1 - \mu^2)$, or about 10 per cent.

Along the unsupported edges, length a, of the plate there are no externally applied moments and none are induced by restricting the lateral strain at these edges. Here the plate edge also curls downward as shown in Fig. 3.3a Section A-A, in addition to deflecting in the normal manner The radius of curvature of this unsupported edge curl is approximately that given by Eq. 3.2.3. The remainder of the plate, i.e., that part removed from these unsupported edges, is bent into a cylindrical shape and Eq. 3.2.8 can be used for calculating deflections.

As with a beam, for small deflections the curvature $1/r$ can be replaced by d^2y/dx^2 and the differential equation for the deflection curve is

$$D\frac{d^2y}{dx^2} = -M \tag{3.2.10}$$

3.3 Bending of a Plate in Two Perpendicular Directions

When the rectangular plate of Fig. 3.4*a* is bent by uniform moments M_1 per unit of length along the edges parallel to the y axis and M_2 per unit of length along the edges parallel to the x axis, the middle plane does not undergo a deformation when the plate is slightly curved to give a small deflection w, and this surface is called the neutral surface. Since w is a function of both x and y, its derivatives which give the slopes when proceeding in the x and y directions are written $\partial w/\partial x$ and $\partial w/\partial y$, respectively. These correspond to the single plane slope dw/dx of a beam. Likewise, $\partial^2 w/\partial x^2$ and $\partial^2 w/\partial y^2$ are the corresponding curvatures, with their reciprocals giving the approximate radii of curvature:

$$\frac{1}{r_1} = -\frac{\partial^2 w}{\partial x^2} \quad \text{and} \quad \frac{1}{r_2} = -\frac{\partial^2 w}{\partial y^2} \tag{3.3.1}$$

r_1 is the radius of curvature of the neutral surface in sections parallel to the xz plane, and r_2 that in sections parallel to the yz plane.

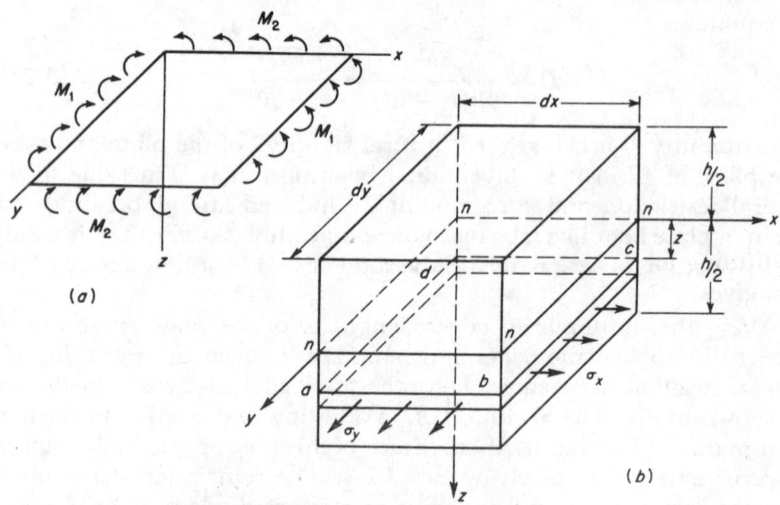

Fig. 3.4. Bending of Plate in Two Perpendicular Planes

The direction of the moments in Fig. 3.4*a* are considered positive and the middle plane of the plate is taken as the *xy* plane with the z axis positive downward. An element is cut out of the plate by two pairs of planes parallel to the *xz* and *yz* planes, and it is assumed that during bending its lateral sides remain plane and rotate about the neutral axes *n-n*, Fig. 3.4*b*. The applied moments shown in Fig. 3.4*a* put the upper part under compression and the lower part in tension. The unit elongations in the x and y directions of an elemental sheet *abcd*, at a distance z from the neutral surface are

$$e_x = \frac{z}{r_1} \qquad \text{and} \qquad e_y = \frac{z}{r_2} \tag{3.3.2}$$

From Eqs. 2.3.4 and 2.3.5 the stresses are

$$\sigma_x = \frac{zE}{1-\mu^2}\left(\frac{1}{r_1} + \mu\frac{1}{r_2}\right) \tag{3.3.3}$$

$$\sigma_y = \frac{zE}{1-\mu^2}\left(\frac{1}{r_2} + \mu\frac{1}{r_1}\right) \tag{3.3.4}$$

Equating the moments of the internal forces acting on the sides of the element, noting that the stresses are proportional to the distance from the neutral surface, to the external applied moments, gives the following equations:

$$\int_{-h/2}^{+h/2} \sigma_x z \, dy \, dz = M_1 \, dy \tag{3.3.5}$$

$$\int_{-h/2}^{+h/2} \sigma_y z \, dx \, dz = M_2 \, dx \tag{3.3.6}$$

substituting for σ_x and σ_y from Eqs. 3.3.3 and 3.3.4 into these equations gives

$$D\left(\frac{1}{r_1} + \mu\frac{1}{r_2}\right) = M_1 \tag{3.3.7}$$

$$D\left(\frac{1}{r_2} + \mu\frac{1}{r_1}\right) = M_2 \tag{3.3.8}$$

Equations 3.3.7 and 3.3.8 may be written in terms of the deflection w by substituting the values of r_1 and r_2 from Eq. 3.3.1, to become:

$$-D\left(\frac{\partial^2 w}{\partial x^2} + \mu\frac{\partial^2 w}{\partial y^2}\right) = M_1 \tag{3.3.9}$$

$$-D\left(\frac{\partial^2 w}{\partial y^2} + \mu\frac{\partial^2 w}{\partial x^2}\right) = M_2 \tag{3.3.10}$$

These equations correspond to the deflection curve for a straight beam.

In the special case of $M_1 = M_2 = M$, the curvatures of the surface in two perpendicular directions are equal and the deflection surface is spherical with the curvature given by Eq. 3.3.7 as

$$\frac{1}{r} = \frac{M}{D(1 + \mu)} \tag{3.3.11}$$

As long as the bending moment M is uniformly distributed along the plate edges, a spherical surface is obtained for any shape plate; i.e., it is independent of whether the plate is square, rectangular, or round.

3.4 Thermal Stresses in Plates

If a simply supported plate is heated uniformly throughout its entire thickness, no stresses are set up since its thermal expansion is not restricted externally or internally. If the plate is heated nonuniformly so that a linear thermal gradient exists through the plate thickness for a temperature difference between plate surfaces of ΔT, the thermal expansion corresponds to the moment elongations above, and, since the edges are free, the deflection surface produced by these expansions will be spherical. The difference between the maximum or minimum expansion, and the expansion at the middle surface is $\alpha\Delta T/2$, where α is the linear coefficient of thermal expansion, and the curvature resulting from this thermal gradient is:

$$\frac{\alpha\Delta T}{2} = \frac{h}{2r} \tag{3.4.1}$$

$$\frac{1}{r} = \frac{\alpha\Delta T}{h} \tag{3.4.2}$$

The curving of the plate creates no stresses because the edges are free to rotate and the deflection is small compared to the thickness.

However, if the edges of the plate are clamped so that they cannot rotate freely, bending moments will be induced along the edges of a magnitude sufficient to eliminate the curvature produced by the linear thermal gradient as given by Eq. 3.4.2; and accordingly, satisfy the condition of clamped edges. Substituting the value of the radius of curvature from Eq. 3.4.2 in Eq. 3.3.11 gives the bending moment per unit length of the clamped edge:

$$M = \frac{\alpha \Delta T(1 + \mu)D}{h} \tag{3.4.3}$$

Since M acts on a rectangular area of unit width and depth h, the maximum bending stress occurring on the outside surface fibers is

$$\sigma_{\text{max.}} = \frac{6M}{h^2} = \frac{6\alpha \Delta T(1 + \mu)D}{h^3} = \frac{\alpha E \Delta T}{2(1 - \mu)} \tag{3.4.4}$$

This is the same as that for the thermal stress in a cylinder, Eq. 2.12.22; and Eq. 3.4.4 developed for flat plates can also be used with sufficient accuracy for cylindrical and spherical vessels. As with a cylinder, the stress is proportional to the coefficient of thermal expansion α, the thermal drop ΔT across the plate thickness, and the modulus of elasticity. Also, although Eq. 3.4.4 shows that this thermal stress is independent of the plate thickness, in practice the total thermal drop ΔT is likely to be higher for thick plates than for thin ones.

3.5 Bending of Circular Plates of Constant Thickness

The deflection surface of a circular-symmetrically loaded circular plate is symmetrical about its central axis perpendicular to the plate, and hence depends on only one variable x. Figure 3.5 represents a diametrical section with the axis of symmetry $0z$, and w the deflection of any point A at a distance x from the axis. The slope at this point for small values of w is $\phi = -dw/dx$, and the curvature of the plate in the diametrical section xz is:

$$\frac{1}{r_1} = -\frac{d^2w}{dx^2} = \frac{d\phi}{dx} \tag{3.5.1}$$

The radius of curvature r_2 in a direction perpendicular to the xz plane can be found by noting that the original straight line mn remains a straight line after the plate is bent but is inclined to the central axis $0z$ at an angle ϕ; i.e., a cylindrical surface (mn vertical) in the unstressed plate having the line $0z$ for its geometric axis becomes a conical surface with its apex at point B. Then AB represents the radius r_2, and from Fig. 3.5 is

$$\frac{1}{r_2} = \frac{\phi}{x} \tag{3.5.2}$$

Neglecting the effect of shear on bending, and substituting the values of $1/r_1$ and $1/r_2$ from Eqs. 3.5.1 and 3.5.2 into Eqs. 3.3.7 and 3.3.8 give:

$$M_1 = D\left(\frac{d\phi}{dx} + \mu\frac{\phi}{x}\right) \tag{3.5.3}$$

$$M_2 = D\left(\frac{\phi}{x} + \mu\frac{d\phi}{dx}\right) \tag{3.5.4}$$

M_1 and M_2 are the bending moments per unit of length, with M_1 acting along cylindrical sections such as *mn*, Fig. 3.5, and M_2 acting along diametrical sections *xz*, Fig. 3.6.

Equations 3.5.3 and 3.5.4 contain only one variable, ϕ, which can be determined from the equilibrium equation for the forces on an element cut from the circular plate by two cylindrical sections, *mmnn* and $m_1m_1n_1n_1$, and two diametrical sections mnm_1n_1 and mnm_1n_1, Fig. 3.6. For the direction of forces shown, the upper portion of the element is in compression both in the radial and circumferential direction, and the lower portion in tension in each of these directions. The middle plane is the neutral plane; i.e., it is in an unstrained state.[1,2] The total moment acting on the side *mmnn* is

$$M_1 x\, d\theta \tag{3.5.5}$$

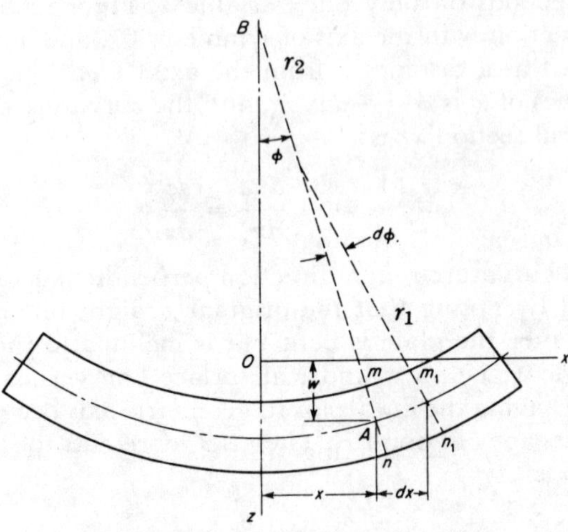

Fig. 3.5. Symmetrical Bending of Circular Plate

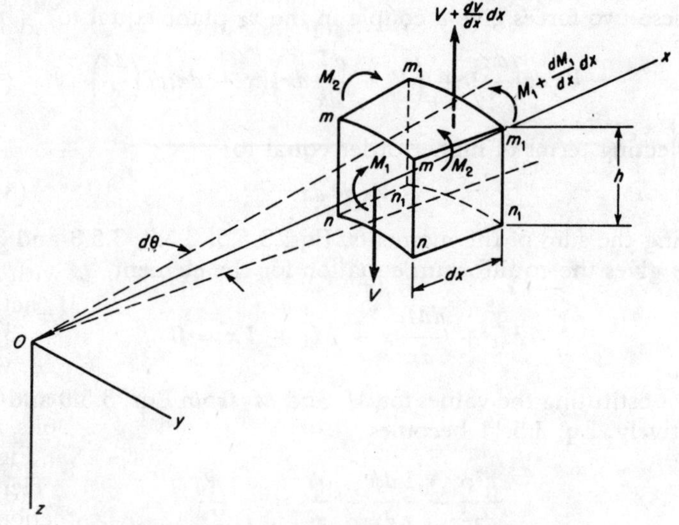

Fig. 3.6. Element of Circular Plate in Bending

and the total moment acting on the side $m_1m_1n_1n_1$ is

$$\left(M_1 + \frac{dM_1}{dx}dx\right)(x + dx)\,d\theta \tag{3.5.6}$$

or neglecting terms of higher order,

$$M_1x\,d\theta + \frac{dM_1}{dx}x\,dx\,d\theta + M_1\,dx\,d\theta \tag{3.5.7}$$

The total moments on the sides mnm_1n_1 are each equal to M_2dx, and these have a resultant in the xz plane, noting that for small angle the sine may be taken as the angle in radians, of

$$2M_2\,dx\,\sin\left(\frac{d\theta}{2}\right) = M_2\,dx\,d\theta \tag{3.5.8}$$

Due to symmetry there are no shearing forces on the sides mm_1nn_1; but if V is the shearing force per unit of length acting on the side $mmnn$, the total shearing force acting on this side of the element is $Vxd\theta$. The shearing force acting on the side $m_1m_1n_1n_1$ is

$$\left(V + \frac{dV}{dx}dx\right)(x + dx)\,d\theta$$

and these two forces give a couple in the xz plane equal to:

$$V x \, d\theta\left(\frac{dx}{2}\right) + \left(V + \frac{dV}{dx}dx\right)(x + dx)\,d\theta\left(\frac{dx}{2}\right) \tag{3.5.9}$$

or neglecting terms of higher order equal to:

$$V x \, d\theta \, dx \tag{3.5.10}$$

Equating the sum of the moments, Eqs. 3.5.5, 3.5.7, 3.5.8 and 3.5.10 to zero gives the equilibrium equation for the element.

$$M_1 + \frac{dM_1}{dx}x - M_2 + Vx = 0 \tag{3.5.11}$$

Upon substituting the values for M_1 and M_2 from Eqs. 3.5.3 and 3.5.4, respectively, Eq. 3.5.11 becomes

$$\frac{d^2\phi}{dx^2} + \frac{1}{x}\frac{d\phi}{dx} - \frac{\phi}{x^2} = -\frac{V}{D} \tag{3.5.12}$$

A second equilibrium equation stating that the algebraic sum of all the forces in a given direction is equal to zero can be written. From this V can be found and Eq. 3.5.12 used to determine the slope ϕ and deflection w of the plate. For instance, if a circular plate is subjected to a uniform load of intensity q and a centrally applied concentrated load P, the shear per unit of circumferential length on a circular section of radius x must be equal to the load within this radius divided by its circumference or

$$V = \frac{\pi x^2 q + P}{2\pi x} = \frac{qx}{2} + \frac{P}{2\pi x} \tag{3.5.13}$$

Placing the value of V in Eq. 3.5.12 gives

$$\frac{d^2\phi}{dx^2} + \frac{1}{x}\frac{d\phi}{dx} - \frac{\phi}{x^2} = -\frac{1}{D}\left(\frac{qx}{2} + \frac{P}{2\pi x}\right) \tag{3.5.14}$$

or

$$\frac{d}{dx}\left[\frac{1}{x}\frac{d}{dx}(x\phi)\right] = -\frac{1}{D}\left(\frac{qx}{2} + \frac{P}{2\pi x}\right) \tag{3.5.15}$$

The first integration of Eq. 3.5.15 gives

$$\frac{1}{x}\frac{d}{dx}(x\phi) = -\frac{1}{D}\left(\frac{qx^2}{4} + \frac{P}{2\pi}\log_e x\right) + C_1 \tag{3.5.16}$$

where C_1 is a constant of integration, and a second integration gives

$$x\phi = -\frac{qx^4}{16D} - \frac{P}{2\pi D}\left(\frac{x^2 \log_e x}{2} - \frac{x^2}{4}\right) + C_1\frac{x^2}{2} + C_2 \quad (3.5.17)$$

or

$$\phi = -\frac{qx^3}{16D} - \frac{Px}{8\pi D}(2\log_e x - 1) + \frac{C_1 x}{2} + \frac{C_2}{x} \quad (3.5.18)$$

where C_2 is the second constant of integration. For small deflections $\phi = -dw/dx$ and the equation becomes

$$\frac{dw}{dx} = \frac{qx^3}{16D} + \frac{Px}{8\pi D}(2\log_e x - 1) - \frac{C_1 x}{2} - \frac{C_2}{x} \quad (3.5.19)$$

and by integrating again gives

$$w = \frac{qx^4}{64D} + \frac{Px^2}{8\pi D}(\log_e x - 1) - \frac{C_1 x^2}{4} - C_2 \log_e x + C_3 \quad (3.5.20)$$

This is the general deflection equation for a symmetrically loaded flat circular plate. The constants of integration C_1, C_2, and C_3 are determined in each particular case of loading by the edge conditions of the plate.

3.6 Bending of Uniformly Loaded Plates of Constant Thickness

1. *Clamped Edges*

When the edges of a plate are prevented from rotating but are not otherwise restrained, i.e., there is no strain in the neutral plane of the plate, the edge condition is called a "clamped" one. Equation 3.5.18 gives the slope and Eq. 3.5.20 gives the deflection for this case when $P = 0$ is put into these equations. The value of the constants can be found by introducing the physical conditions that satisfy the plate. If a is the edge radius of the plate, then $\phi = 0$ at $x = 0$ and $x = a$, and from these two conditions Eq. 3.5.18 gives the following equations for determining the constants C_1 and C_2:

$$\left(\frac{qx^3}{16D} - \frac{C_1 x}{2} - \frac{C_2}{x}\right)_{x=a} = 0 \quad (3.6.1)$$

$$\left(\frac{qx^3}{16D} - \frac{C_1 x}{2} - \frac{C_2}{x}\right)_{x=0} = 0 \quad (3.6.2)$$

from which

$$C_1 = \frac{qa^2}{8D} \quad \text{and} \quad C_2 = 0 \tag{3.6.3}$$

Substituting these values into Eq. 3.5.18 gives

$$\phi = \frac{qx}{16D}(a^2 - x^2) \tag{3.6.4}$$

The deflection is established from Eq. 3.5.20 by setting $P = 0$ and introducing the values of C_1 and C_2 from Eq. 3.6.3 obtaining

$$w = \frac{qx^4}{64D} - \frac{qa^2x^2}{32D} + C_3 \tag{3.6.5}$$

The constant C_3 can be found from the condition that at the edge of the plate the deflection is zero, so

$$0 = \frac{qa^4}{64D} - \frac{qa^4}{32D} + C_3 \tag{3.6.6}$$

from which

$$C_3 = \frac{qa^4}{64D} \tag{3.6.7}$$

Substituting the value of C_3 in Eq. 3.6.5 gives the deflection

$$w = \frac{q}{64D}(a^2 - x^2)^2 \tag{3.6.8}$$

The maximum deflection is at the center of the plate equal to

$$\delta = \frac{qa^4}{64D} \tag{3.6.9}$$

The bending moments can be found from Eqs. 3.5.3 and 3.5.4 by substituting the expression for the slope ϕ from Eq. 3.6.4, thereby giving

$$M_1 = \frac{q}{16}[a^2(1 + \mu) - x^2(3 + \mu)] \tag{3.6.10}$$

$$M_2 = \frac{q}{16}[a^2(1 + \mu) - x^2(1 + 3\mu)] \tag{3.6.11}$$

The moments at the edge of the plate, $x = a$, are

$$M_1 = -\frac{qa^2}{8} \tag{3.6.12}$$

$$M_2 = -\frac{\mu qa^2}{8} \tag{3.6.13}$$

and at the center, $x = 0$, these moments are

$$M_1 = M_2 = \frac{1+\mu}{16}qa^2 \tag{3.6.14}$$

The maximum stress is at the edge of the plate and equal to

$$\sigma_{x_{\max.}} = \frac{6}{h^2}\frac{qa^2}{8} = \frac{3}{4}\frac{qa^2}{h^2} \tag{3.6.15}$$

This stress is 3/8 that of the bending stress of a like thickness beam clamped at the ends and length equal to the diameter of the plate.

2. *Simply Supported Edges*

In the case of a clamped-edge plate there are negative edge bending moments, Fig. 3.7*a*, of magnitude $M_1 = -qa^2/8$ from Eq. 3.6.12. Using the method of superposition, this can be combined with the case of pure bending, Fig. 3.7*b*, thereby eliminating the edge bending mo-

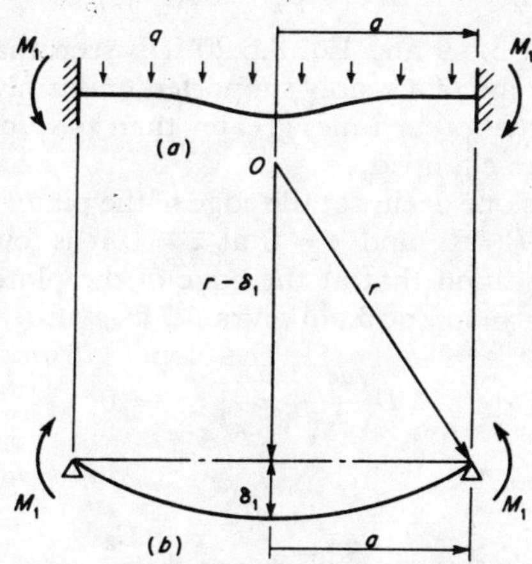

Fig. 3.7. Circular Plate (*a*) Clamped Edge, (*b*) Simply Supported Edge

ment and obtaining the bending of a plate simply supported at the edge. The deflection due to pure bending can be found from Eq. 3.3.11 by substituting in this equation $M = qa^2/8$ to give

$$\frac{1}{r} = \frac{qa^2}{8D(1 + \mu)} \tag{3.6.16}$$

The corresponding deflection at the center for a spherical surface can be found by noting in Fig. 3.7*b* that r is the hypotenuse of right triangle and a is one leg, so

$$a^2 = r^2 - (r - \delta_1)^2 = 2r\delta_1 - \delta_1^2 \tag{3.6.17}$$

and since δ_1 is small compared to r, the quantity δ_1^2 can be neglected to give

$$\delta_1 = \frac{a^2}{2r} \tag{3.6.18}$$

or

$$\delta_1 = \frac{qa^4}{16D(1 + \mu)} \tag{3.6.19}$$

This deflection is added to that for a clamped edge plate as given by Eq. 3.6.9 to obtain the deflection of a plate simply supported at the edge and gives

$$\delta = \frac{qa^4}{64D} + \frac{qa^4}{16D(1 + \mu)} = \frac{5 + \mu}{64(1 + \mu)D}qa^4 \tag{3.6.20}$$

By comparing Eq. 3.6.9 and Eq. 3.6.20 it is seen that, for steel with $\mu = 0.3$, the deflection of a simply supported uniformly loaded circular plate is approximately four times greater than that for the same plate when the edges are clamped.

The maximum slope occurs at the edge of the plate. From Eq. 3.5.18 for the condition $P = 0$, and $\phi = 0$ at $x = 0$ it is found that $C_2 = 0$. Also from the condition that at the edge of the plate the radial moment $M_1 = 0$ at $x = a$, Eq. 3.5.3 gives

$$D\left(\frac{d\phi}{dx} + \mu\frac{\phi}{x}\right)_{x=a} = 0 \tag{3.6.21}$$

From Eq. 3.5.18

$$\phi = -\frac{qx^3}{16D} + \frac{C_1 x}{2} + \frac{C_2}{x} \tag{3.6.22}$$

and

$$\frac{d\phi}{dx} = -\frac{3qx^2}{16D} + \frac{C_1}{2} - \frac{C_2}{x^2} \tag{3.6.23}$$

Substituting Eqs. 3.6.22 and 3.6.23 into Eq. 3.6.21, recalling $C_2 = 0$, gives

$$C_1 = \frac{qa^2(3+\mu)}{8D(1+\mu)} \tag{3.6.24}$$

from which

$$\phi_{x=a} = \frac{qa^3}{16D}\left[\frac{2}{1+\mu}\right] = \frac{3qa^3(1-\mu)}{2Eh^3} \tag{3.6.25}$$

The bending moments can likewise be found by superposing on the moments from Eqs. 3.6.10 and 3.6.11 for the case of a clamped edge, the constant bending moment $qa^2/8$ (Eq. 3.6.12) to satisfy the edge condition that the radial moment be zero, $M_{1_{x=a}} = 0$; hence,

$$M_1 = \frac{q}{16}(3+\mu)(a^2 - x^2) \tag{3.6.26}$$

and

$$M_2 = \frac{q}{16}[a^2(3+\mu) - x^2(1+3\mu)] \tag{3.6.27}$$

The maximum bending moment occurs at the center where

$$M_1 = M_2 = \frac{3+\mu}{16}qa^2 \tag{3.6.28}$$

and corresponding maximum stress is

$$\sigma_{x_{\max.}} = \sigma_{y_{\max.}} = \frac{6}{h^2}\frac{3+\mu}{16}qa^2 = \frac{3(3+\mu)}{8}\frac{qa^2}{h^2} \tag{3.6.29}$$

Figure 3.8 gives a graphic comparison of the bending stresses σ_x and σ_y at the lower surface of the plate for a condition of clamped edges, and simply supported edges. It is seen that the stress varies parabolically with radial distance from the center of the plate and the maximum stress occurring in the clamped edge plate is much lower than that occurring in a simply supported plate by approximately 40 per cent. It is also noted that while at the edge of a simply supported plate

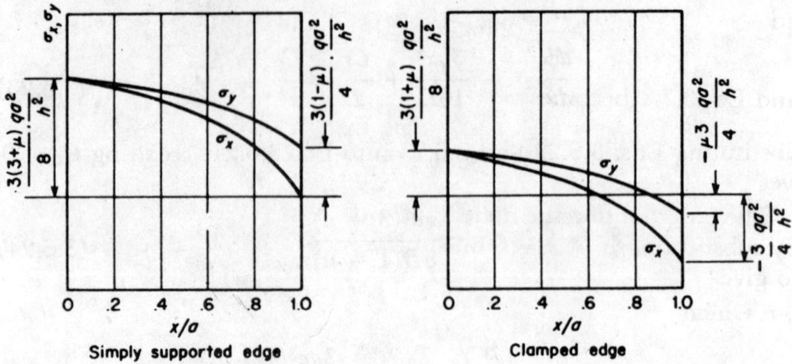

Fig. 3.8. Comparison of Bending Stresses in Simply Supported and Clamped Edge Uniformly Loaded Circular Plates

the radial stress σ_x is zero, the circumferential stress σ_y at this location has a value of $3(1 - \mu)qa^2/4h^2$, Eq. 3.6.27.

The effect of shearing strain on the deflection is usually negligible in the size and thicknesses generally encountered in pressure vessel construction and has been neglected in the above discussion. When the thickness of the plate is not small in comparison with its radius, the additional deflection due to shear may be found by the same method as used for beams.[1,2]

3.7 Bending of Centrally Loaded Circular Plate of Constant Thickness

1. *Clamped Edges*

In this case of a load placed at the center of the plate, its slope is given by Eq. 3.5.18 when substituting $q = 0$ as

$$\phi = -\frac{Px}{8\pi D}(2\log_e x - 1) + \frac{C_1 x}{2} + \frac{C_2}{x} \tag{3.7.1}$$

The constants of integration C_1 and C_2 can be found from the physical conditions that $\phi = 0$ at $x = 0$ and $x = a$; i.e.,

$$\left[-\frac{Px}{8\pi D}(2\log_e x - 1) + \frac{C_1 x}{2} + \frac{C_2}{x}\right]_{x=0} = 0 \tag{3.7.2}$$

$$\left[-\frac{Px}{8\pi D}(2\log_e x - 1) + \frac{C_1 x}{2} + \frac{C_2}{x}\right]_{x=a} = 0 \tag{3.7.3}$$

From these expressions it is found that

$$C_1 = \frac{P}{4\pi D}(2 \log_e a - 1) \qquad \text{and} \qquad C_2 = 0 \tag{3.7.4}$$

and Eq. 3.7.1 becomes

$$\phi = \frac{Px}{4\pi D} \log_e \frac{a}{x} \tag{3.7.5}$$

The equation for the deflection surface is obtained from Eq. 3.5.20 by substituting in it $q = 0$ and the values of C_1 and C_2 from Eq. 3.7.4 to give

$$w = \frac{Px^2}{8\pi D}\left(\log_e \frac{x}{a} - \frac{1}{2}\right) + C_3 \tag{3.7.6}$$

The constant C_3 is determined from the condition that at the clamped edge the deflection is zero, thereby giving $C_3 = Pa^2/16\pi D$. Substituting this into Eq. 3.7.6 gives

$$w = \frac{Px^2}{8\pi D} \log_e \frac{x}{a} + \frac{P}{16\pi D}(a^2 - x^2) \tag{3.7.7}$$

The maximum deflection occurs at the center of the plate and is

$$\delta = \frac{Pa^2}{16\pi D} \tag{3.7.8}$$

Comparing this equation with Eq. 3.6.9 shows that the deflection produced by a concentrated central load is four times that of a uniformly distributed load of the same magnitude. The bending moments can be determined from Eqs. 3.5.3 and 3.5.4 through the use of Eq. 3.7.5 which gives

$$M_1 = \frac{P}{4\pi}\left[(1 + \mu) \log_e \frac{a}{x} - 1\right] \tag{3.7.9}$$

$$M_2 = \frac{P}{4\pi}\left[(1 + \mu) \log_e \frac{a}{x} - \mu\right] \tag{3.7.10}$$

The moments at the edge, $x = a$, become

$$M_1 = -\frac{P}{4\pi} \tag{3.7.11}$$

$$M_2 = -\mu\frac{P}{4\pi} \tag{3.7.12}$$

and give the corresponding maximum bending stresses

$$\sigma_{x_{max.}} = \frac{6M_1}{h^2} = \frac{6}{h^2}\frac{P}{4\pi} = \frac{3}{2}\frac{P}{\pi h^2} \tag{3.7.13}$$

$$\sigma_{y_{max.}} = \frac{6M_2}{h^2} = \frac{6}{h^2}\frac{\mu P}{4\pi} = \frac{3\mu}{2}\frac{P}{\pi h^2} \tag{3.7.14}$$

It is seen that these stresses for a concentrated central load on a clamped-edge circular plate are twice as great as the stresses produced in the same plate by the same total load uniformly distributed over the entire plate.

Equations 3.7.9 and 3.7.10 give infinitely large bending moments and stresses at the center of the plate, but this is a result of the assumption that the load is concentrated at a point.[1] If the load is distributed over a small area, the stresses become finite as discussed in paragraph 3.8. In practice, infinite stresses are not produced for two reasons: first, loads are physically applied by lugs, brackets, skirts, piping, etc., over a relatively large area; and second, load or stress redistribution via local plastic flow of the material takes place to alleviate this condition, Chapter 5. These facts point out, however, the importance of mitigating the effects of concentrated loads by spreading or applying the load over the largest possible area.

2. *Simply Supported Edge*

The deflection of a circular plate simply supported at the edge can be obtained by the method of superposition. Thus, if to the deflection for a clamped-edge plate, Eq. 3.7.7, is superposed on that produced by the uniformly distributed radial edge moments $M_1 = P/4\pi$, the case of a simply supported plate is obtained. The curvature produced by these edge moments is found from Eq. 3.3.11 to be

$$\frac{1}{r} = \frac{P}{4\pi(1+\mu)D} \tag{3.7.15}$$

and the corresponding deflection at the middle is (Eq. 3.6.18)

$$\delta_1 = \frac{a^2}{2r} = \frac{Pa^2}{8\pi(1+\mu)D} \tag{3.7.16}$$

Adding this to the deflection of Eq. 3.7.8 gives the deflection at the middle of a simply supported circular plate with a central concentrated load as

$$\delta = \frac{Pa^2}{16\pi D} + \frac{Pa^2}{8\pi(1+\mu)D} = \frac{3+\mu}{1+\mu}\frac{Pa^2}{16\pi D} \tag{3.7.17}$$

This deflection is approximately 2.5 times as great as that for the case of a clamped-edge plate.

The bending moments for this case of a simply supported edge plate are found by adding the moment $P/4\pi$ to the moments of Eqs. 3.7.9 and 3.7.10 obtained for the case of a clamped edge. Likewise, the maximum stress is obtained by adding $(6/h^2)(P/4\pi)$ to the stress obtained for a clamped-edge plate.

3.8 Bending of a Circular Plate Concentrically Loaded

1. *Clamped Edge*

In this case in which the load is distributed along a concentric circle of radius b, Fig. 3.9, the portion of the plate inside this circle b is considered separately from that portion outside. Using the general deflection equation Eq. 3.5.20, letting $q = 0$ for both portions, $P = 0$ for the inner portion, and P be the total load for the outer portion, the six arbitrary constants can be found from the following edge conditions, and continuity at the circle $x = b$,

(*a*) $\left[\dfrac{dw}{dx}\right]_{\substack{\text{inner}\\ x=0}} = 0$

(*b*) $\left[\dfrac{dw}{dx}\right]_{\substack{\text{outer}\\ x=a}} = 0$

(*c*) $\left[\dfrac{dw}{dx}\right]_{\substack{\text{inner}\\ x=b}} = \left[\dfrac{dw}{dx}\right]_{\substack{\text{outer}\\ x=b}}$

(*d*) $\left[\dfrac{d^2w}{dx^2}\right]_{\substack{\text{inner}\\ x=b}} = \left[\dfrac{d^2w}{dx^2}\right]_{\substack{\text{outer}\\ x=b}}$

(*e*) $[w]_{\substack{\text{inner}\\ x=b}} = [w]_{\substack{\text{outer}\\ x=b}}$

(*f*) $[w]_{\substack{\text{outer}\\ x=a}} = 0$

and the deflection for the inner portion ($x < b$) has been found[1] to be

$$w = \frac{P}{8\pi D}\left[-(x^2 + b^2)\log_e\frac{a}{b} + (x^2 - b^2) + \frac{1}{2}\left(1 + \frac{b^2}{a^2}\right)(a^2 - x^2)\right] \tag{3.8.1}$$

and for the outer portion ($x > b$),

$$w = \frac{P}{8\pi D}\left[-(x^2 + b^2)\log_e\frac{a}{x} + \frac{1}{2}\left(1 + \frac{b^2}{a^2}\right)(a^2 - x^2)\right] \tag{3.8.2}$$

2. *Simply Supported Edge*

In a like manner the deflection for a simply supported edge condition is found for the inner portion ($x < b$) to be

$$w = \frac{P}{8\pi D}\left[-(x^2+b^2)\log_e\frac{a}{b} + (x^2-b^2) + \frac{(3+\mu)a^2-(1-\mu)b^2}{2(1+\mu)a^2}(a^2-x^2)\right] \quad (3.8.3)$$

and for the outer portion ($x > b$)

$$w = \frac{P}{8\pi D}\left[-(x^2+b^2)\log_e\frac{a}{x} + \frac{(3+\mu)a^2-(1-\mu)b^2}{2(1+\mu)a^2}(a^2-x^2)\right] \quad (3.8.4)$$

3. *Clamped Edge Concentrically Loaded*

These equations are useful in solving other cases of bending of symmetrically loaded circular plates. For instance, one occurring frequently in practice is that in which the load is uniformly distributed over the inner part of the plate bounded by a circle of radius c, Fig. 3.10. Using Eq. 3.8.1 and letting $P = 2\pi bqdb$, the center deflection produced by this elemental ring load, db, is

$$dw = \frac{q}{4D}\left[-b^2\log_e\frac{a}{b} - b^2 + \frac{1}{2}(a^2+b^2)\right]bdb \quad (3.8.5)$$

The center deflection of the plate produced by the entire load is

$$\delta = \int_0^c dw = \frac{q}{4D}\int_0^c\left[-b^2\log_e\frac{a}{b} - b^2 + \frac{1}{2}(a^2+b^2)\right]bdb$$

$$= \frac{q}{4D}\left[-\frac{c^4}{4}\log_e\frac{a}{c} - \frac{3c^4}{16} + \frac{a^2c^2}{4}\right] \quad (3.8.6)$$

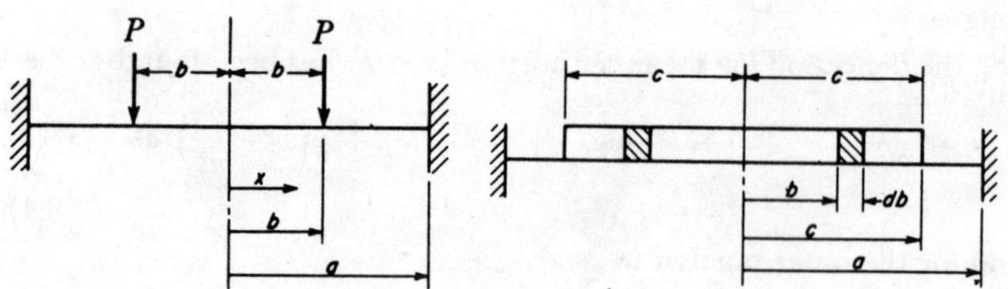

Fig. 3.9. Concentrically Loaded Circular Plate

Fig. 3.10. Concentrically Loaded Circular Plate

When $c = a$, this equation coincides with Eq. 3.6.9 for a uniformly loaded plate. Also, when substituting $c = 0$ and $\pi c^2 q = P$ into Eq. 3.8.6, the deflection for a concentrated central load is obtained which coincides with Eq. 3.7.8.

The bending moment at the center of the plate can be found by first determining the curvature at this location. This can be calculated from Eq. 3.8.1 by taking the second derivative with respect to x, and setting $x = 0$ and $P = 2\pi bq db$ into this derivative, which gives

$$\frac{d^2w}{dx^2}_{x=0} = \frac{q}{4D}\int_0^c\left(-2\log_e\frac{a}{b} + 1 - \frac{b^2}{a^2}\right)b\,db$$

$$= -\frac{qc^2}{4D}\left(\log_e\frac{a}{c} + \frac{c^2}{4a^2}\right) \tag{3.8.7}$$

The corresponding bending moment at the center is found from Eqs. 3.3.9 and 3.3.10 to be

$$M_1 = M_2 = -D(1+\mu)\frac{d^2w}{dx^2} = \frac{1+\mu}{4}qc^2\left(\log_e\frac{a}{c} + \frac{c^2}{4a^2}\right) \tag{3.8.8}$$

The maximum bending stresses at the center are

$$\sigma_{x_{\text{max.}}} = \sigma_{y_{\text{max.}}} = \frac{3}{2}(1+\mu)\frac{qc^2}{h^2}\left(\log_e\frac{a}{c} + \frac{c^2}{4a^2}\right) \tag{3.8.9}$$

This equation can also be written as follows, letting P equal the entire load $\pi c^2 q$,

$$\sigma_{x_{\text{max.}}} = \sigma_{y_{\text{max.}}} = \frac{3}{2}(1+\mu)\frac{P}{\pi h^2}\left(\log_e\frac{a}{c} + \frac{c^2}{4a^2}\right) \tag{3.8.10}$$

From this it is seen that as the radius c of the circle over which the load is distributed diminishes, the condition of a single concentrated load is approached, and the stresses increase as c diminishes but remain finite as long as c is finite. Investigations by Timoshenko[1,2] for the case of a single central concentrated load indicate that the proper formula for calculating the tensile stress is

$$\sigma_{x_{\text{max.}}} = \sigma_{y_{\text{max.}}} = \frac{P}{h^2}(1+\mu)\left(0.485\log_e\frac{a}{h} + 0.52\right) \tag{3.8.11}$$

The compressive stress at the top of the plate may be several times the value of the tensile stress at the bottom of the plate but does not necessarily represent a direct failure potential because of its highly localized nature, paragraph 3.7.1.

3.9 Deflection of a Symmetrically Loaded Circular Plate of Uniform Thickness with a Circular Central Hole

Circular holes are frequently used in pressure vessel construction, such as for an access opening; consequently, their effect is important in all pressure vessel shapes. It is particularly so when these openings are placed in flat plates which derive their major strength from bending action, and hence cannot readily be reinforced in the same manner as openings in curved pressure vessel shapes depending on membrane tensile action for their strength, Chapter 6.

1*a*. *Bending by Couples, Edge*

In Fig. 3.11*a*, if M_{1a} and M_{1b} represent the bending moments per unit length on the outer and inner edges, respectively, Eqs. 3.5.18 and 3.5.20 give for the case $P = q = 0$,

$$\phi = \frac{C_1 x}{2} + \frac{C_2}{x} \tag{3.9.1}$$

$$w = -\frac{C_1 x^2}{4} - C_2 \log_e x + C_3 \tag{3.9.2}$$

The constants C_1, C_2, and C_3 can be determined from the physical conditions at the edge of the plate. Substituting from Eq. 3.9.1 into Eq. 3.5.3 gives

$$M_1 = D\left[\frac{C_1}{2} - \frac{C_2}{x^2} + \mu\left(\frac{C_1}{2} + \frac{C_2}{x^2}\right)\right] \tag{3.9.3}$$

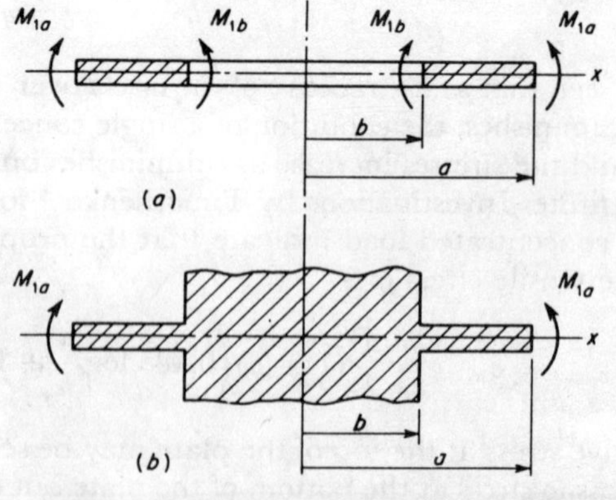

Fig. 3.11. Circular Plate with a Central Circular Hole

and for the conditions $x = a$ and $x = b$, Eq. 3.9.3 presents the following expressions for finding the constants C_1 and C_2

$$M_{1a} = D\left[\frac{C_1}{2}(1 + \mu) - \frac{C_2}{a^2}(1 - \mu)\right] \tag{3.9.4}$$

$$M_{1b} = D\left[\frac{C_1}{2}(1 + \mu) - \frac{C_2}{b^2}(1 - \mu)\right] \tag{3.9.5}$$

from which

$$C_1 = \frac{2(a^2M_{1a} - b^2M_{1b})}{(1 + \mu)D(a^2 - b^2)} \quad \text{and} \quad C_2 = \frac{a^2b^2(M_{1a} - M_{1b})}{(1 - \mu)D(a^2 - b^2)} \tag{3.9.6}$$

The constant C_3 is determined by considering the deflection of the plate which is zero for a condition of a simply supported outer edge $x = a$, so from Eq. 3.9.2,

$$-\frac{C_1a^2}{4} - C_2 \log_e a + C_3 = 0 \tag{3.9.7}$$

or

$$C_3 = \frac{a^2}{2}\frac{(a^2M_{1a} - b^2M_{1b})}{(1 + \mu)D(a^2 - b^2)} + \frac{a^2b^2(M_{1a} - M_{1b})}{(1 - \mu)D(a^2 - b^2)} \log_e a \tag{3.9.8}$$

The radial bending moment is then found by placing into Eq. 3.9.3 the value of the constants of integration C_1 and C_2 from Eq. 3.9.6, to be

$$M_1 = \frac{1}{a^2 - b^2}\left[a^2M_{1a} - b^2M_{1b} - \frac{a^2b^2(M_{1a} - M_{1b})}{x^2}\right] \tag{3.9.9}$$

In like manner the slope of the plate can be determined by substituting the values of C_1 and C_2 into Eq. 3.9.1, and the deflection surface of the plate determined by substituting the value of C_1, C_2, and C_3 into Eq. 3.9.2.

1b. Bending by Couples, Inner Edge Restrained From Rotation

Figure 3.11*b*, illustrates the case of bending by a couple M_{1a} along the outer edge when the inner edge is restrained from rotating. In this case the constants C_1 and C_2 in Eq. 3.9.1 are determined from the edge conditions $\phi = 0$ at $x = b$, and $M_1 = M_{1a}$ at $x = a$, which from Eqs. 3.9.1 and 3.9.3 gives

$$0 = \frac{C_1 b}{2} + \frac{C_2}{b} \tag{3.9.10}$$

and

$$M_{1a} = D\left[\frac{C_1(1+\mu)}{2} - \frac{C_2(1-\mu)}{a^2}\right] \tag{3.9.11}$$

Therefore,

$$C_1 = \frac{2a^2 M_{1a}}{D[a^2(1+\mu) + b^2(1-\mu)]} \tag{3.9.12}$$

$$C_2 = -\frac{a^2 b^2 M_{1a}}{D[a^2(1+\mu) + b^2(1-\mu)]} \tag{3.9.13}$$

Substituting these values of C_1 and C_2 into Eq. 3.9.3 gives

$$M_1 = \frac{a^2 M_{1a}}{a^2(1+\mu) + b^2(1-\mu)}\left[1 + \mu + (1-\mu)\frac{b^2}{x^2}\right] \tag{3.9.14}$$

Again in a similar manner, the slope of the plate can be determined by substituting the values of C_1 and C_2 into Eq. 3.9.1, and the deflection surface determined by substituting into Eq. 3.9.2 the value of C_1 and C_2, and C_3 found from the condition that $w = 0$ at $x = b$, thus yielding

$$w = \frac{-M_{1a}a^2}{2D[a^2(1+\mu) + b^2(1-\mu)]}\left[x^2 - b^2 + 2b^2 \log_e\frac{b}{x}\right] \tag{3.9.15}$$

2. *Bending by a Load Uniformly Distributed Along the Inner and Outer Edges*

When bending is produced by a load uniformly distributed along the inner and outer edges as shown in Fig. 3.12*a*, the value of $q = 0$ and P equals the total load in Eqs. 3.5.18 and 3.5.20. This gives the slope from Eq. 3.5.18 as

$$\phi = -\frac{Px}{8\pi D}(2\log_e x - 1) + \frac{C_1 x}{2} + \frac{C_2}{x} \tag{3.9.16}$$

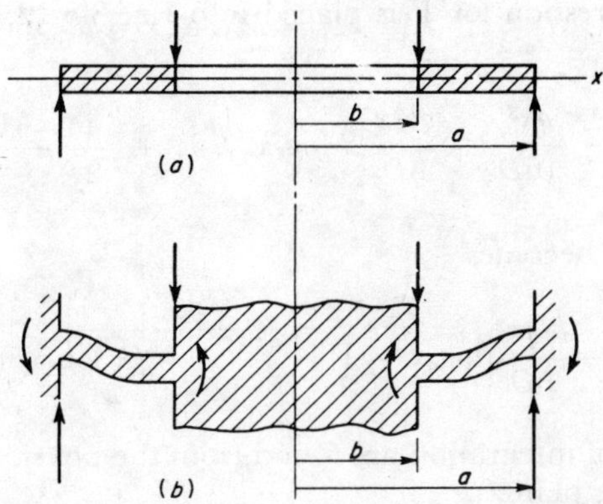

Fig. 3.12. Circular Plate with Uniform Edge Loading

In the case of a simply supported edge, Fig. 3.11a, the constants of integration C_1 and C_2 are obtained from Eq. 3.5.3 for the conditions that $M_1 = 0$ at $x = a$ and $x = b$; thus,

$$D\left(\frac{d\phi}{dx} + \mu\frac{\phi}{x}\right)_{x=a} = 0 \tag{3.9.17}$$

$$D\left(\frac{d\phi}{dx} + \mu\frac{\phi}{x}\right)_{x=b} = 0 \tag{3.9.18}$$

The slope ϕ is determined by placing the constants so found in Eq. 3.9.16, and the bending moments from Eqs. 3.5.3 and 3.5.4.

If the plate is clamped at both inner and outer edges, Fig. 3.12b, the constants of integration are found from the condition that $\phi = 0$ at $x = a$ and $x = b$; thus

$$0 = -\frac{Pa}{8\pi D}(2 \log_e a - 1) + \frac{C_1 a}{2} + \frac{C_2}{a} \tag{3.9.19}$$

$$0 = -\frac{Pb}{8\pi D}(2 \log_e b - 1) + \frac{C_1 b}{2} + \frac{C_2}{b} \tag{3.9.20}$$

and the slope and moments determined as described above.

When the load consists of an uniform pressure over the plate, Fig. 3.13, instead of a load distribution over the edges of the plate, the shearing force V at a point x distant from the center is

$$V = \frac{1}{2\pi x}\pi q(x^2 - b^2) = \frac{qx}{2} - \frac{qb^2}{2x} \tag{3.9.21}$$

When this expression for V is placed into Eq. 3.5.12, Eq. 3.5.18 becomes

$$\phi = -\frac{qx^3}{16D} + \frac{qb^2x}{8D}(2\log_e x - 1) + \frac{C_1x}{2} + \frac{C_2}{x} \tag{3.9.22}$$

and Eq. 3.5.20 becomes

$$w = \frac{qx^4}{64D} - \frac{b^2qx^2}{8D}(\log_e x - 1) - \frac{C_1x^2}{4} - C_2\log_e x + C_3 \tag{3.9.23}$$

The constants of integration are found from the physical conditions at the edges of the plate.

Constructions involving the solution of such problems are frequently encountered in pressure vessels. For instance, the flat circular head with central access opening on the end of a cylinder, Fig. 3.14, combines the cases of edge loading and uniform pressure loading. Several cases[4,5,7,17,18] of practical importance are shown in Fig. 3.15, in which the maximum stress can be represented by an expression of the type

$$\sigma_{\text{max.}} = k\frac{qa^2}{h^2} \quad \text{or} \quad \sigma_{\text{max.}} = k\frac{P}{h^2} \tag{3.9.24}$$

depending on whether the load is uniformly applied over the surface or concentrated along the edges. The numerical values of the factor k, calculated for several values of the ratio a/b and Poisson's ratio $\mu = 0.3$, are given in Table 3.1. The maximum deflection for the same cases are given by expressions of the type

$$w_{\text{max.}} = k_1\frac{qa^4}{Eh^3} \quad \text{and} \quad w_{\text{max.}} = k_1\frac{Pa^2}{Eh^3} \tag{3.9.25}$$

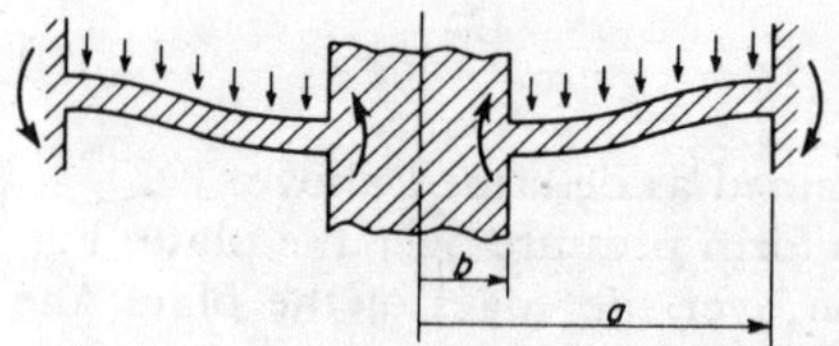

Fig. 3.13. Circular Plate with Uniformly Distributed Load

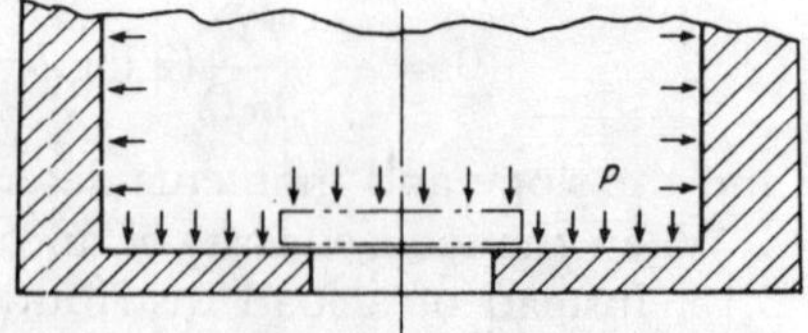

Fig. 3.14. Flat Head with Central Access Opening that Closes the End of a Cylindrical Pressure Vessel

TABLE 3.1. COEFFICIENTS k AND k_1 FOR USE IN EQS. 3.9.24 AND 3.9.25 FOR THE BENDING CASES SHOWN IN FIG. 3.15

a/b =	1.25		1.5		2		3		4		5	
Case	k	k_1	k	k_1	k	k_1	k	k_1	k	k_1	k	k_1
1	1.10	0.341	1.26	0.519	1.48	0.672	1.88	0.734	2.17	0.724	2.34	0.704
2	0.66	0.202	1.19	0.491	2.04	0.902	3.34	1.220	4.30	1.300	5.10	1.310
3	0.135	0.00231	0.410	0.0183	1.04	0.0938	2.15	0.293	2.99	0.448	3.69	0.564
4	0.122	0.00343	0.336	0.0313	0.74	0.1250	1.21	0.291	1.45	0.417	1.59	0.492
5	0.090	0.00077	0.273	0.0062	0.71	0.0329	1.54	0.110	2.23	0.179	2.80	0.234
6	0.115	0.00129	0.220	0.0064	0.405	0.0237	0.703	0.062	0.933	0.092	1.13	0.114
7	0.592	0.184	0.976	0.414	1.440	0.664	1.880	0.824	2.08	0.830	2.19	0.813
8	0.227	0.00510	0.428	0.0249	0.753	0.0877	1.205	0.209	1.514	0.293	1.745	0.350

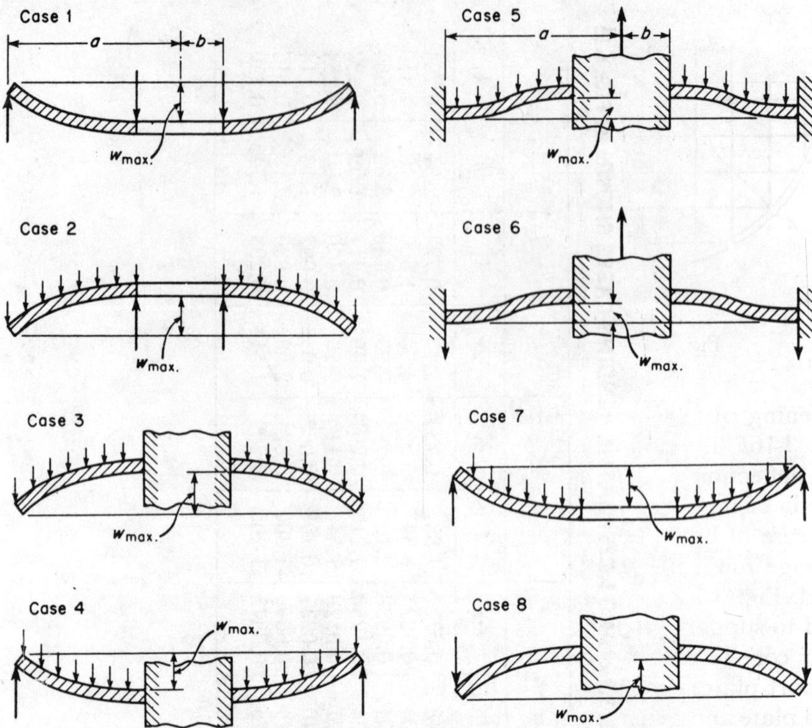

Fig. 3.15. Cases of Bending of Circular Plates with Central Holes

3.10 Reinforced Circular Plates

1. *Orthogonal Grillage Reinforcement*

Plates may be reinforced by an equidistant orthogonal system of ribs to form a grillage type of reinforcement, Fig. 3.16. If the reinforcing ribs are symmetrical about the middle plane of the plate, Fig. 3.16*a*, the composite structure may be treated as a circular plate with the beam stiffening effect of the ribs averaged over the entire plate and added to the plate rigidity to give[2,7,8]

$$D = \frac{Eh^3}{12(1 - \mu^2)} + \frac{EI}{d} \tag{3.10.1}$$

where d is the spacing between ribs and I is the moment of inertia of a rib with respect to the middle axis of the plate cross section. If the

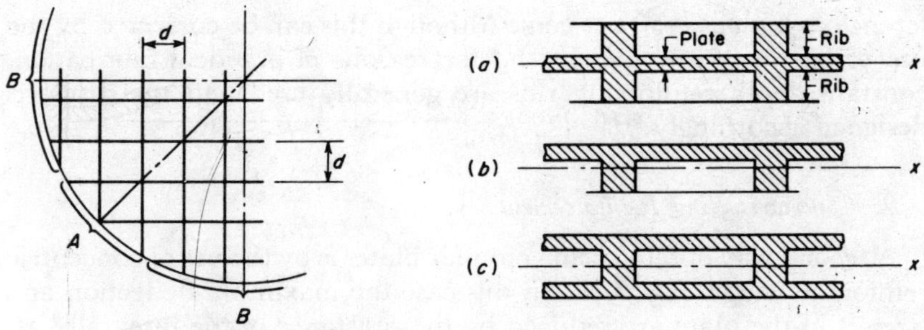

Fig. 3.16. Orthogonally Reinforced Circular Plates

stiffening ribs are not symmetrical with the middle plane of the plate, Fig. 3.16*b*, the circular flat plate formulas give an approximation of the deflection and stresses if D is taken as the average moment of inertia of the Tee-section of width d about its centroid, $D = EI/d$. In the case of both a top and bottom cover plate construction, I is the average moment of inertia of the composite structure about its centroid, Fig. 3.16*c*. The orthogonal rib reinforcing system is extensively used to support internal loads within pressure vessels, such as the nuclear core in reactor vessels, Fig. 3.17. The loading and size of these support plates are frequently such that they cannot be procured as a solid plate and hence must be built up by welding together a web-plate system with a top and/or bottom cover plate. It has been found in such grillage plates that there is a tendency for the vertical boundary edge reaction to concentrate in the region A and to be less in region B, Fig. 3.16, with the variation in average load per unit of boundary circum-

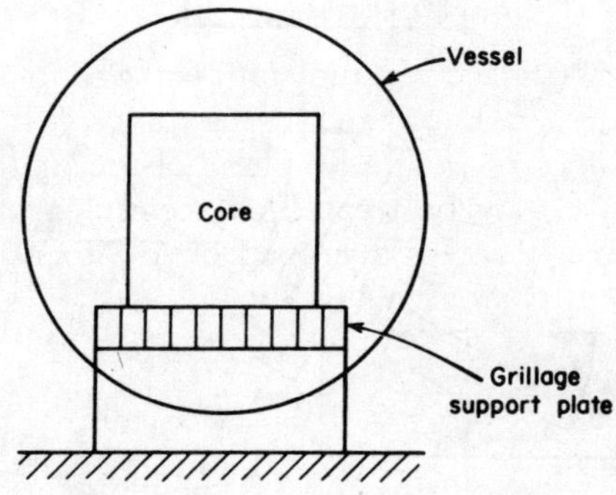

Fig. 3.17. Nuclear Core Support

ference as high as ±50 per cent. Although this can be corrected by the use of beams of varying depth,[6] for reasons of practical fabrication, constant depth reinforcing ribs are generally used and the supports designed accordingly.[9,22]

2. *Concentric Ring Reinforcement*

A second way of reinforcing circular plates is by the use of concentric reinforcing rings, Fig. 3.18. In this case the maximum deflection and stresses in the plate are reduced by the resistance of the integrally attached ring to "turning inside out." In effect such a reinforced plate consists of an inner plate, a ring, and an outer plate which can be solved by equating the slope of the short reinforcing ring, paragraph 4.9, to that of a flat plate, paragraph 3.8, at the juncture, Fig. 3.18*c*.

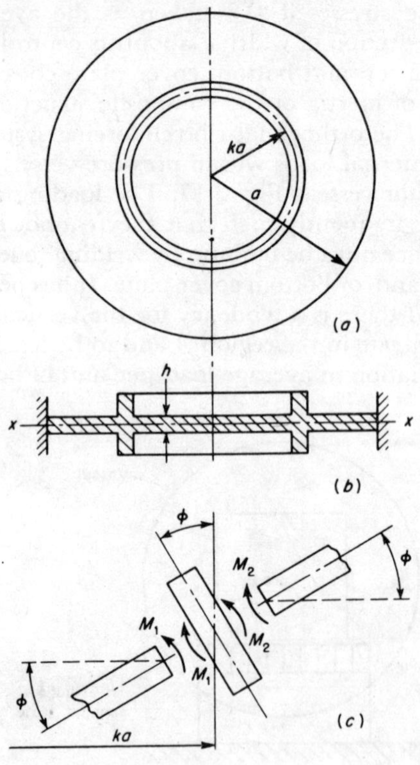

Fig. 3.18. Circular Plate Reinforced by Concentric Ring

The effect of such a reinforcing ring for various locations and reinforcing amounts[3] on the maximum deflection, and the maximum radial stress for a circular clamped-edge plate subject to uniform normal load q are shown in Figs. 3.19 and 3.20, respectively. These show that reinforcing rings located in the region of greatest slope of the plate are the most advantageous in reducing both deflection and stress.

The exact location for minimizing the deflection can be found from the curves of Fig. 3.19, in which the vertical deflection of any point, ka, is expressed as the product of the maximum (central) deflection of an unreinforced clamped-edge plate of the same size, and a coefficient which is the function of the position of the ring and a dimensionless ratio EI/aD, where I is the moment of inertia of the ring cross section about the middle axis of the plate. As in the unreinforced plate, the maximum vertical deflection of any reinforced plate is at the center, $k = 0$. When the location of the reinforcing ring approaches the center or edge of the plate, its restoring moment vanishes and Eq. 3.6.9 for a circular clamped-edge plate holds. Also the effectiveness of the ring reinforcing is not directly proportional to the size of the ring but has a diminishing effect as the amount of reinforcing is increased. For instance, from Fig. 3.19 it is seen that a "moderately stiff" reinforcing ring of $EI/aD = 10$ will reduce the maximum deflection to about 38 per cent of its central value in an unreinforced plate when the ring is most effectively placed. It is only reduced further to 30 per cent when the rigidity of the ring is increased two and a half times to $EI/aD = 25$.

The radial stress at any point in the plate can also be expressed as a product of the maximum value of that stress in the same size unreinforced clamped-edge plate and a coefficient which is a function of the

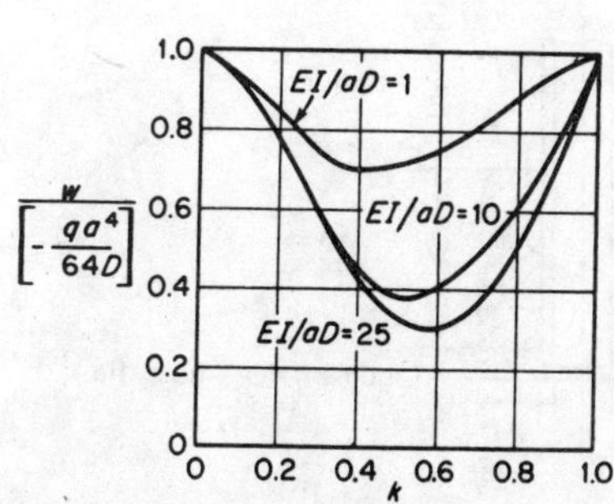

Fig. 3.19. Effect of Reinforcing Ring on Maximum Deflection of Reinforced Plate[3]

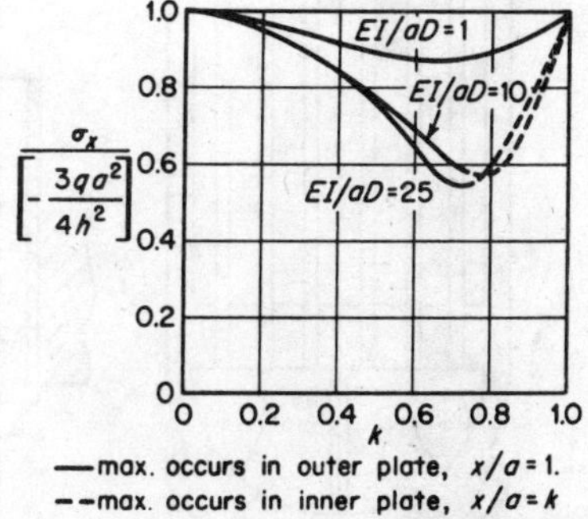

Fig. 3.20. Effect of a Reinforcing Ring on Maximum Radial Stress in Reinforced Plate[3]

location of the ring, and a dimensionless ratio EI/aD, Fig. 3.20. From these figures the relative position a ring of given size should occupy to be most effective can be determined. This position is not the same for the radial stress as it is for the deflection; hence, must be selected on the basis of primary concern; however, a general compromise value of $k = 0.6$ may be taken with satisfactory results. This method of reinforcement lends itself well to flat plates which form an integral part of a pressure vessel, such as the end closure of a cylinder, since its edges are readily welded to the remainder of the vessel.

3. *Flat Perforated Plates*

Tube-sheets, or tube-plates, are flat plates which have been drilled with a multitude of holes into which tubes are "rolled-in," and frequently welded on one side to ensure tightness. Such perforated tube-plates are used in the construction of boiler feed-water heaters, heat exchangers, and in nuclear steam generators,[23] Fig. 3.21. The applied pressure load is uniform and is resisted mainly by the flexural strength of a circular plate, weakened from its original condition by the holes required to receive the tubes. These tubes are in the form of "hairpins"

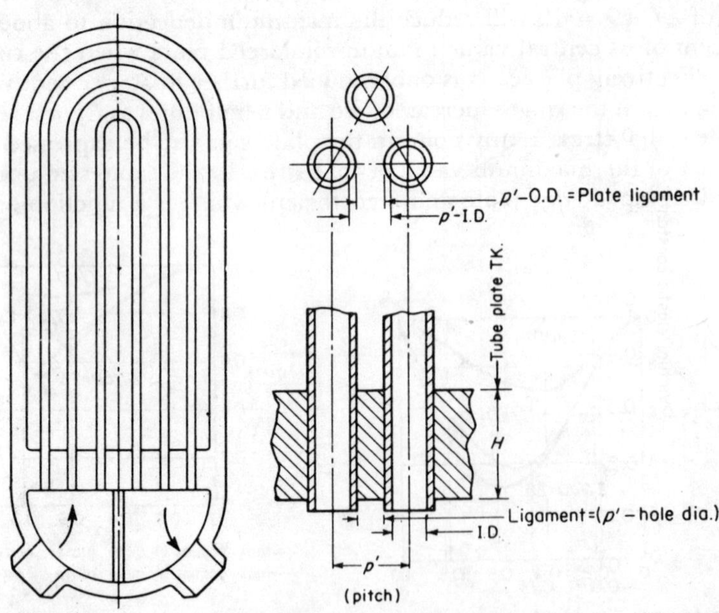

Fig. 3.21. Arrangement of a Typical Nuclear Steam Generator with a Perforated Tube Plate

and so offer no direct ties between tube-plates. The formulas for like size unperforated flat circular plates can be satisfactorily used if:

A. A virtual modulus of elasticity E^1 is used in place of E, and also a virtual Poisson's ratio μ^1 is used in place of μ to compensate for the loss of rigidity and bi-axial effect due to the perforations.

B. The effect of the rolled-in tubes on the deflection and ligament stress of the tube-sheet is considered.

Figure 3.22 shows the manner in which E^1 and μ^1 vary with the tube hole ligament efficiency for an equilateral triangular pattern of tube holes.[10] Similar values for square and rectangular tube hole patterns have also been established. The stiffening and strengthening effect of rolled-in-tubes on a tube-sheet is appreciable, especially when the tubes are very closely spaced. For instance, the results of tests on the tube-sheets of nuclear steam generators have shown that when this effect of rolled-in-tubes is neglected, the tube-plate calculated deflection and strains were 75 percent higher than those measured. However, when the entire thickness of the rolled-in-tubes was assumed

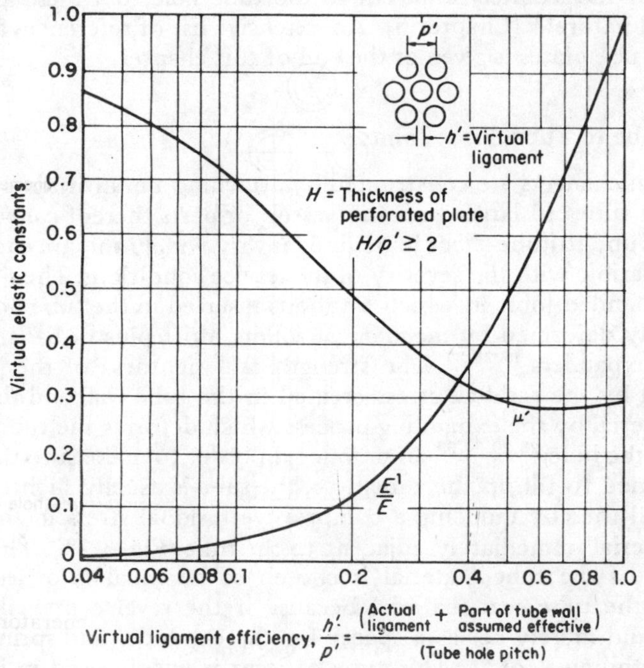

Fig. 3.22. Virtual Elastic Constants for Perforated Plates

effective in stiffening the tube-plate and reducing ligament strains, agreement within 5 percent of the actual was obtained.[25] Similar tests on thinner heat exchanger tubesheets also confirm this stiffening effect.[10] This stiffening effect may be accounted for in the calculation by including part or all of the tube wall and establishing a virtual ligament efficiency, (actual ligament plus portion of tube wall assumed effective)/(tube hole pitch), used to obtain the virtual elastic constants of the plate. The amount of tube wall thickness that may be credited toward the ligament efficiency is:

A. A minimum of 50 percent to a maximum of 95 percent for tube joints expanded the full depth of the tubesheet.
B. No credit for joints attached by welding only, Fig. 3.27*a, b,* and *c,* with no tube expanding.

Ligament stress is higher than that calculated for an unperforated plate inversely proportional to the actual ligament efficiency, (p' − dia. hole)/(p'), so that $\sigma_{\text{avg.}} = \sigma/$(actual ligament efficiency). This is the average stress across the ligament. Peak stresses higher than the average occur immediately adjacent to the tube hole, but these are of a localized nature[10] (Chapter 6). An extensive list of references for perforated tube plates is given at the end of this chapter.

3.11 Tube to Tube-Sheet Joints

Heat exchangers are constructed by attaching a mulititude of small diameter tubes to larger pressure vessels or parts thereof called tubesheets. Tube to tube-sheet joint designs vary widely and are chosen to be compatible with the severity of the service conditions. The simplest is an expanded joint in which a tube is inserted in the tube hole and plastically deformed by mechanical rollers, drift pins or balls, or hydraulic expanders.[19,25,27] The strength and tightness of the joint is obtained by the residual stress created in the tube wall and the tube seat material by the expanding process which deforms the tube to fill up the tube hole.[28,29,30,33] In practice, in order to make sure the tube is deformed to fill up the tube hole, the tube is usually slightly overexpanded thereby inducing a compressive residual stress in the tube seat material immediately adjacent to the tube, Fig. 3.23*a*. The yield strength of the tube material is chosen to be equal to or less than that of the tube seat material because if the reverse prevailed the tube would merely act as a spacer for the expander and spring back upon withdrawal of it. This type of joint is widely used in boilers, condensers, heaters, etc., when it is not subject to thermal transients

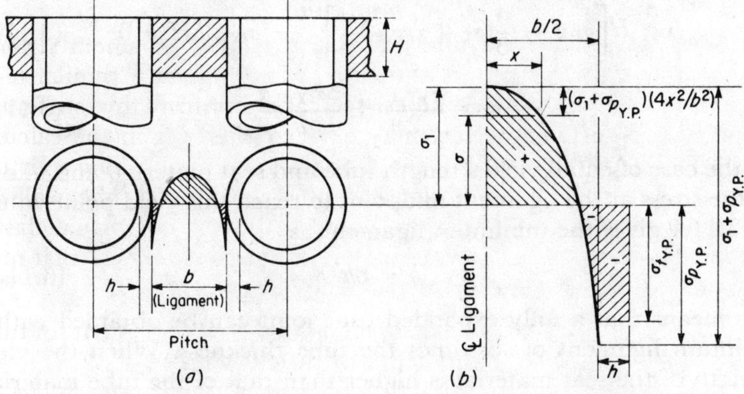

Fig. 3.23. Expanded Tube to Tube-sheet Joint

which would tend to shrink the tube away from the seat, or otherwise grossly alter the residual stress pattern upon which it depends for strength and tightness. The minimum ligament for an expanded tube joint can be established by the following analysis. The direct residual stress in the ligament varies from yield point compression at the edge of the hole to tension at the center of the ligament, Fig. 3.23*b*. The summation of the residual forces must equal zero; which means that the area under the tension curve must equal the area under the compression curve plus that of the tube wall.

Representing the stress curve by a parabola and letting—

σ = stress at any point
σ_1 = stress at ligament mid point
$\sigma_{tY.P.}$ = yield strength of tube material
$\sigma_{pY.P.}$ = yield strength of tube seat material
h = thickness of tube
b = width of tube seat
H = thickness of tube seat
F = total force

$$\sigma = \sigma_1 - (\sigma_1 + \sigma_{pY.P.})\frac{4x^2}{b^2} \tag{3.11.1}$$

$$\Sigma F_y = H\int_0^{b/2} \sigma dx - \sigma_{tY.P.}hH = 0 \tag{3.11.2}$$

Substituting the value of σ from Eq. 3.11.1 in Eq. 3.11.2 and integrating gives

$$H\left[\sigma_1 x - \frac{4}{3}(\sigma_1 + \sigma_{pY.P.})\frac{x^3}{b^2}\right]_0^{b/2} - \sigma_{tY.P.}hH = 0 \qquad (3.11.3)$$

$$\sigma_1 = \frac{\sigma_{pY.P.}}{2} + \frac{3\sigma_{tY.P.}h}{b} \qquad (3.11.4)$$

For the case of equal yield strength tube and seat material, and allowing the stress at the ligament midpoint to reach this yield point value, Eq. 3.11.4 gives the minimum ligament as

$$b = 6h \qquad (3.11.5)$$

This means that a fully expanded tube joint can be obtained with a minimum ligament of six times the tube thickness. When the yield strength of the seat material is higher than that of the tube material, the minimum ligament b is less than that given by Eq. 3.11.5 and can be found by substituting the respective material yield strength values in Eq. 3.11.4.

Expanded tube joints are made with a plain interface (seat) or are circumferentially grooved to increase their holding ability, Fig. 3.24. The total design axial allowable mechanical-frictional holding force, P, of an expanded tube joint is established by experimental testing and applying a suitable factor of safety, Fig. 3.25. Bending and torque may also be applied to these tube joints, in which case it is necessary to compute equivalent loads to analyze the joint. Referring to Fig. 3.26 and using the following nomenclature:

F = mechanical-frictional force
M_R = resisting moment of expanded tube joint
P = total axial allowable tube joint holding force
$P^1 = P/2\pi r$ = axial allowable tube joint holding force per unit of circumference
T = resisting torque of expanded tube joint

$$M_R = \int_0^{2\pi} Fy = 4\int_0^{\pi/2} P^1 r^2 \sin^2\phi \, d\phi \qquad (3.11.6)$$

Substituting the value of P^1 gives

$$M_R = \frac{2Pr}{\pi}\int_0^{\pi/2} \sin^2\phi \, d\phi \qquad (3.11.7)$$

$$M_R = \frac{Pr}{2} \qquad (3.11.8)$$

(*a*) (*b*)

Fig. 3.24. (*a*) Photograph of Expanded Tube Joint With Two Circumferential Grooves. The Tube End has also Been Flared to Further Increase the Pull-Out Resistance. (*b*) Tube Removed From Tube-Seat to Show Grooving and Flaring

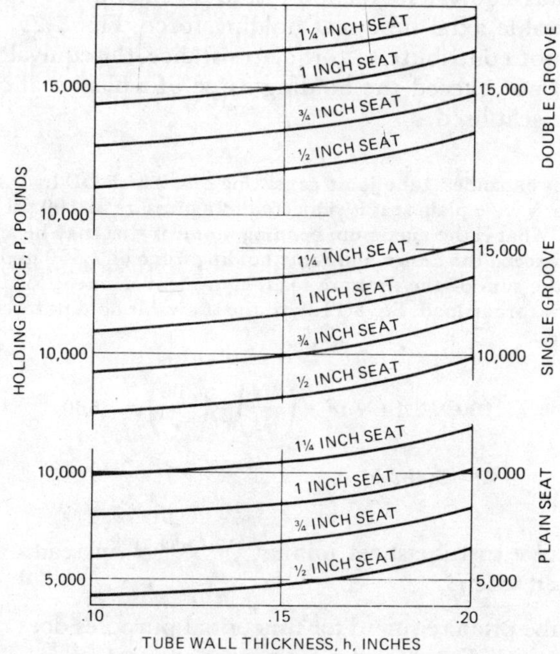

Fig. 3.25 Typical Design Allowable Holding Force of a 2 Inch Outside Diameter Expanded Tube Joint

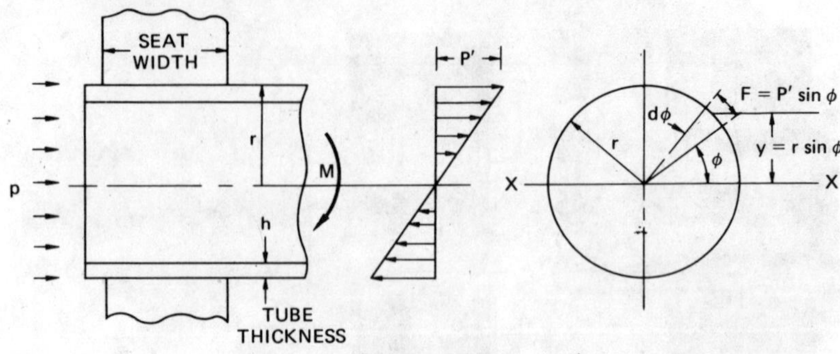

Fig. 3.26. Expanded Tube Joint

Similarly,

$$T = P^1 (2\pi r) r = Pr \tag{3.11.9}$$

The safety of the joint requires that the total equivalent axial force, including the pressure load $p\pi r^2$, plus equivalent load for bending moment, plus equivalent load for torque, shall not exceed the safe design allowable axial tube seat holding force, Fig. 3.25. Also, since grooves do not contribute to torque resistance, the equivalent load for torque shall not exceed the holding force of a plain seat of the same width as the seat used.

Example: An expanded tube joint consisting of a 2 inch OD by 0.10 inch thick tube in a 1 inch wide plain seat is subjected to a pressure of 100 psi and a torque of 200 in. lb. What is the maximum bending moment that may be applied to this seat and not exceed the design allowable holding force of 7,500 lb per Fig. 3.25?

Equating the sum of the pressure load, equivalent moment load, Eq. 3.11.8, and equivalent torque load, Eq. 3.11.9; to the allowable holding force of 7,500 lb, Fig. 3.25, gives:

$$100 \times 3.14 \times 1^2 + \left(\frac{2M}{r}\right) + \left(\frac{200}{r}\right) = 7{,}500$$

$$M = 3{,}493 \text{ in. lb}$$

Welded tube to tube-sheet joints,[11,12,13,20,21] instead of expanded ones, are used when:

1. The tube pitch required for functional purposes does not provide an adequate ligament for an expanded joint.
2. The joint is subject to thermal transients that will loosen it.

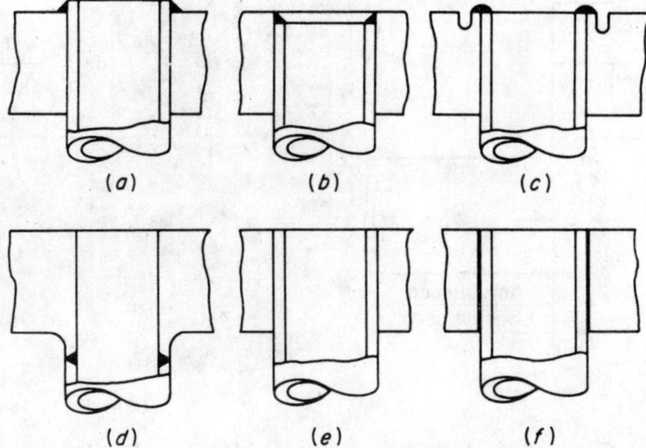

Fig. 3.27. Typical Welded Tube to Tube-Sheet Joints. (*a*) Projected Tube Fillet Weld, (*b*) Recessed Tube Fillet Weld, (*c*) Trepanned Seat Weld, (*d*) Butt Weld, (*e*) Explosive or High Energy Weld, (*f*) Brazed or Adhesive Joint

3. The potential crevice between a portion of the tube and tube sheet can become a place of hideout and concentration of injurious chemicals from the contacting media.
4. Maintenance accessibility is limited or not available. This occurs with some nuclear and chemical vessels.

Welded joints vary widely in construction details, Fig. 3.27. They may consist of simple fillet welds, *a* and *b*, or trepanned junctures, *c*, to those butt welded, *d*, explosively welded, *e*, brazed or adhesively bonded, *f*, etc. The choice is based upon the service and environment.

3.12 Local Flexibility at the Supports of Clamped-edge Beams and Plates

Beams and plates are frequently considered as being rigidly built-in and hence undergo no edge rotation, Fig. 3.28*a*. However, because of the elasticity in the support, a local distortion occurs which allows them to rotate at their built-in ends even if there is no deflection in the support, Fig. 3.28*b*. This rotation produces a deflection in addition to that caused by bending and shear stresses in the beam or plate itself, and in the case of statically indeterminate ones the bending stress is also affected. This additional deflection can be accounted for by using an effective length equal to the actual span length plus a distance L from the face of the support in the applicable beam and plate equations,[14,15,16] Fig. 3.29. This local flexibility effect is nil for long thin

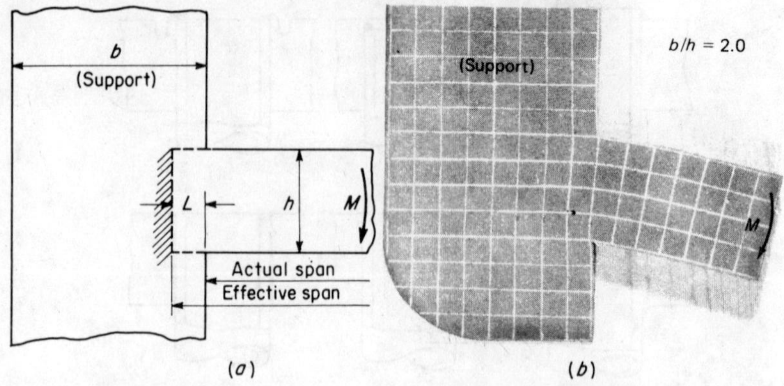

Fig. 3.28. (*a*) Geometry of Built-in Support. (*b*) Local Distortion of Built-in Edge Supports Depicted by the Strain Distribution in a Flat Rubber Model Showing the Slope at the Elastic Deflection Curve is not Zero at the Face of the Support but Extends Slightly Within the Support. (See Chapter 7 for a Discussion of this Method of Experimental Stress Analysis).

members of span/thickness ratios over 10; that is, it is only a major contributor to deflection for short thick beams or plates. Fillets and haunches are effective means of reducing the local flexibility of supports.

In the construction of many heat exchangers, including nuclear steam generators, extremely thick tubesheets are welded at their periphery to comparatively thin cylindrical shells, Fig. 3.30. In this case the thin cylindrical shell has little influence on the thick tubesheet and

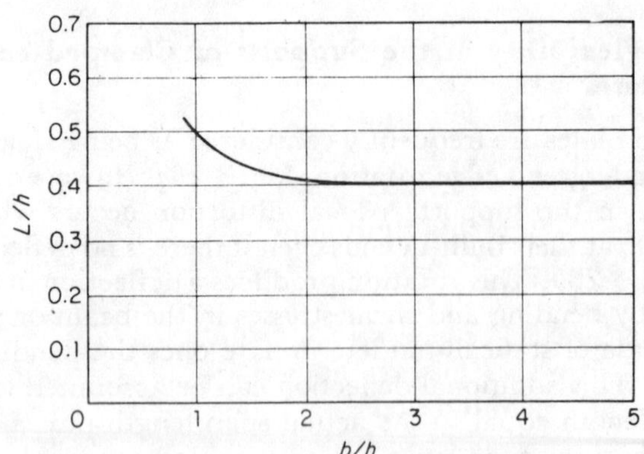

Fig. 3.29. Location of Effective Built-in Support

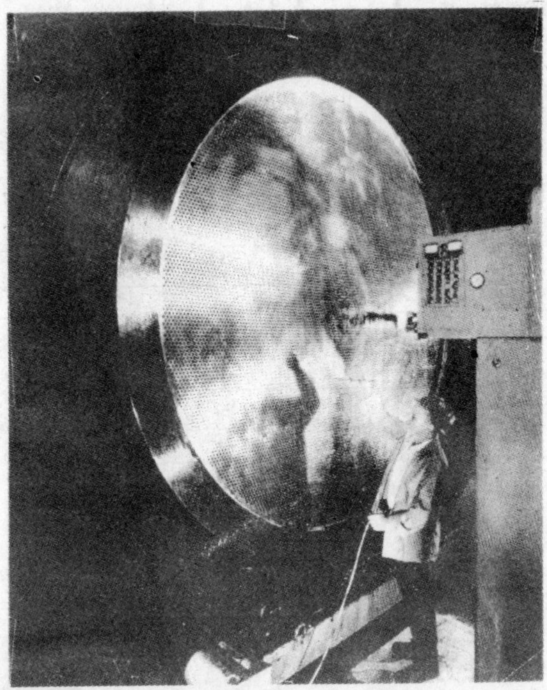

Fig. 3.30. Deep-drilling the Perforations in the Extremely Thick Tube-sheet of a Nuclear Steam Generator. The Small Diameter Closely Spaced Holes Will Be Fitted with Heat Exchanger Tubes as Shown in Fig. 3.27 (*Courtesy The Babcock & Wilcox Company*)

it approaches free-edge behavior. However, thick tubesheets are often costly and difficult material procurement items in which case optimizing their thickness by introducing adjacent short thick cylinders or heads whose interactions[24,31,32] can result in reduced thickness requirements, Chapter 4.

3.13 Stacked Plates and Built-up Plates

1. *Stacked Plates*

Flat plates stacked one on top of the other are frequently used to support loads. This may be done because the construction arrangement does not permit a single plate of equivalent thickness to be used, or because a single plate of sufficient thickness is not obtainable. Fig. 3.31*a* shows three stacked circular plates of different thickness simply

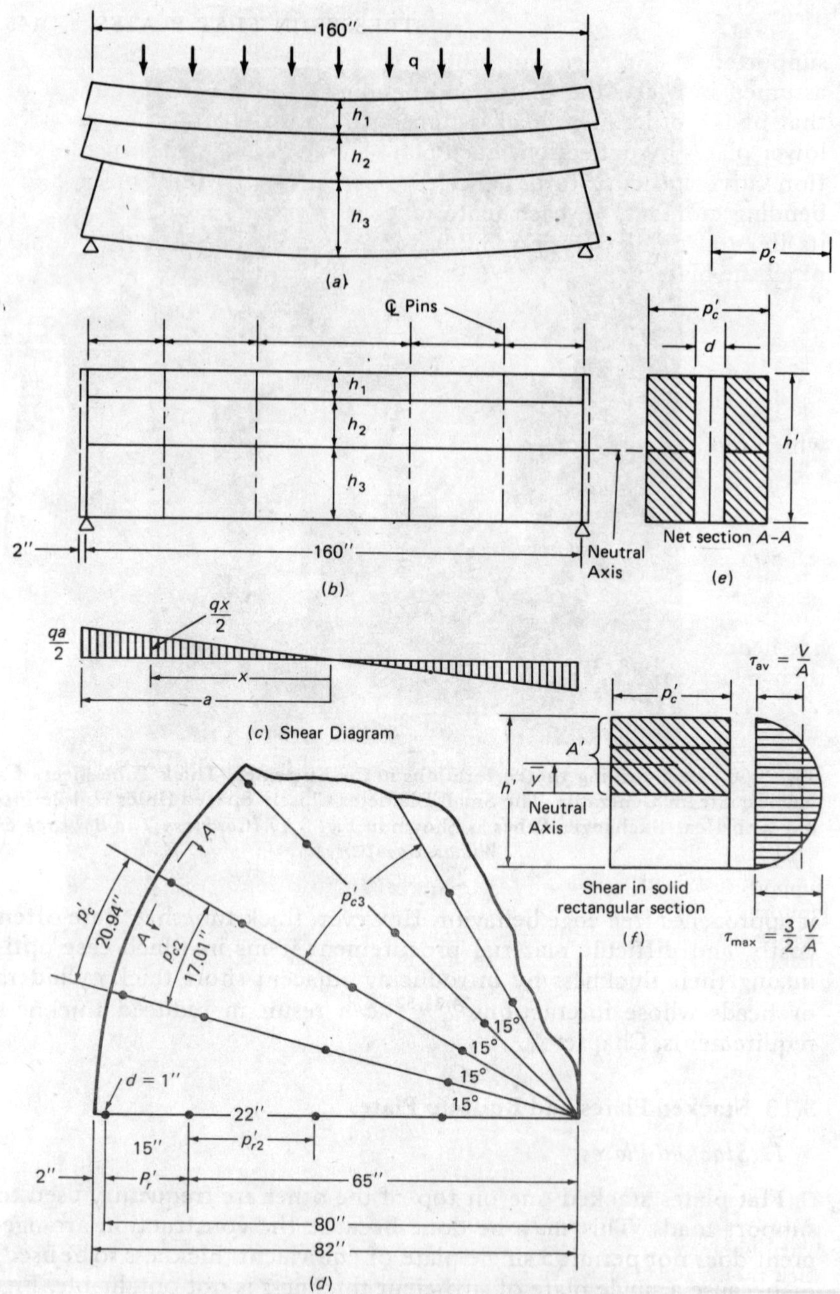

Fig. 3.31. Stacked Plates and Built-Up Plates

supported at the edge and subject to a uniform load. No friction is assumed between the plates and bending of each is independent of that of the other. The load is transmitted from the top to successive lower plates by deflection. Each plate has essentially the same deflection curve and curvature; hence, the percentage of the total applied bending moment, M, each plate takes is proportional to the ratio of its flexural rigidity D_n, Eq. 3.2.9, to the sum of the flexural rigidities of all the plates:

$$M_n = \frac{M\,D_n}{\sum_1^n D_n} \tag{3.13.1}$$

which reduces to

$$M_n = \frac{M\,h_n^3}{\sum_1^n h_n^3} \tag{3.13.2}$$

The section modulus of each plate varies as the square of its thickness, h_n^2; hence, the maximum stresses in each plate are in direct ratio to their relative thicknesses:

$$\sigma_n = \frac{6M}{\sum_1^n h_n^3}\,\frac{h_n^3}{h_n^2} = \frac{6M\,h_n}{\sum_1^n h_n^3} \tag{3.13.3}$$

Example 1. Three stacked 160-in.-diameter flat circular plates are simply supported around the edge of the lower plate, and a total load of 40,000 lb is uniformly applied to the top plate. The thickness of the top plate is 0.5 in., the middle plate is 1.0 in., and the bottom plate is 1.5 in. Assuming no sliding friction between plates, what is the maximum bending stress in each of the plates?

From Eq. 3.6.28

$$M = \left(\frac{3+\mu}{16}\right) qa^2, \qquad P = \pi a^2 q, \qquad a^2 q = \frac{40{,}000}{\pi}$$

$$M = \left(\frac{3+.3}{16}\right)\frac{40{,}000}{3.1416} = 2626 \text{ in. lb. (Total)}$$

This total moment applied to the three stacked plates is proportional to each plate per Eq. 3.13.2, and the stress in each plate determined as shown in Table 3.2. When one plate is extremely thin relative to the others, it is customary to sandwich the thin plate between the thicker ones in order to forestall buckling of the thin plate between attachment pins, Chapter 8.

TABLE 3.2. STRESS IN STACKED PLATES OF EXAMPLE 1

Thickness, h, in.	h^2	h^3	*% Mom. to Each Plate*	*Mom. to each Plate, in. lb.*	*Stress σ, psi*
0.5	.25	.125	.125/4.5 = .028	.028 × 2626 = 74	74 × 6/.25 = 1,776
1.0	1.00	1.00	1.00 /4.5 = .222	.222 × 2626 = 583	583 × 6/1 = 3,500
1.5	2.25	3.375	3.375/4.5 = .750	.750 × 2626 = 1,969	1969 × 6/2.25 = 5,250
Σ	—	4.5	1.000	2,626	

2. *Built-Up Plates*

If the differential sliding of one stacked plate on another is prevented through the use of pins, rivets, bolts, or keys to take inplane shear at the plate interfaces; the result is a built-up plate which is stiffer and stronger than the same stacked plates, Fig. 3.31*b*. The stresses in a built-up plate are calculated on the assumption that its individual plates are rigidly connected; that is, (1) it is a solid plate, and (2) the size, shape, and spacing of the uniting elements (pins, rivets, bolts, and keys) fulfill this requirement.

A. Bending Stresses. The formulas for the bending stresses in solid plates are applicable but allowance must be made for the pin holes, etc. by the use of reduced sections, Fig. 3.31. When the reduced sections occur at regular intervals, the resulting efficiency, similar to ligament efficiency, Par. 1.15, is employed; i.e., the value of I and Z must be multiplied by the circumferential plate efficiency $(p_c' - d)/p_c'$ so

$$\sigma = \frac{\sigma_{\text{solid}}}{\text{Efficiency}} = \sigma_{\text{solid}} \frac{p_c'}{p_c' - d} \qquad (3.13.3)$$

B. Shearing Stresses. The shearing stresses in the shear restraints, here assumed to be pins, can be approximated by comparing them to those at the same location in a solid rectangular beam whose height is the sum of the individual plate thicknesses, h', and width is the circumferential pitch, p_c', of the pins, Fig. 3.31*b* and *d*. Here the shear stress at any location in a cross section is perpendicular to the plane of the plate and numerically equal to the inplane shearing stress at the same location, and is given by*

$$\tau = \frac{V}{p_c'}\frac{A'\bar{y}}{I} = \frac{V}{p_c'}\frac{Q}{I} \qquad (3.13.4)$$

where V is the shearing force acting on the cross section $h'p_c'$, Q is the statical moment about the composite neurtal axis of that portion of the whole area, A', between the horizontal plane at which the shearing stress is desired and the outside surface, and I is the moment of inertia of the cross section about its neutral axis, Fig. 3.31*f*. The shearing stresses are not uniformly distributed from top to bottom. From Eq. 3.13.4 for a rectangular cross section, they vary parabolically from zero at the top and bottom surfaces to a maximum at the

*See any text on strength of materials.

neutral axis which is 50 percent larger than the average:

$$\tau_{\max} = 1.5 \frac{V}{A} = 1.5 \frac{V}{h'p'_c} \tag{3.13.5}$$

where A is the area of the cross section, $h'p'_c$, Fig. 3.31*f*.

The weakening effect of the pin holes can be accounted for in the same manner as that used for calculating the bending stress, namely, through the use of the plate efficiency of these holes. The cross-sectional area is reduced by the pin holes in the ratio of $(p'_c - d)/p'_c$, which is the plate efficiency, Fig. 3.31*e*. Hence, the shearing stress is increased by the presence of these pin holes and its value is found by dividing the right side of Eqs. 3.13.4 and 3.13.5 by the efficiency. This gives

$$\tau = \frac{V}{p'_c}\frac{Q}{I}\frac{p'_c}{p'_c - d} = \frac{V}{p'_c - d}\frac{Q}{I} \tag{3.13.4a}$$

and

$$\tau_{\max.} = \frac{1.5V}{h'p'_c}\frac{p'_c}{p'_c - d} = \frac{1.5V}{h'(p'_c - d)} \tag{3.13.5a}$$

The shearing force, F_s, acting on each pin is

$$F_s = \tau(p'_c - d)p'_r \tag{3.13.6}$$

and substituting Eq. 3.13.4*a* in Eq. 3.13.6 gives

$$F_s = \frac{VQ}{I(p'_c - d)}(p'_c - d)p'_r = \frac{VQ}{I}p'_r \tag{3.13.7}$$

If a plate interface coincides with the neutral axis, Fig. 3.31*b*, the shear force on the pins is a maximum at this interface, and when substituting Eq. 3.13.5*a* in Eq. 3.13.6, becomes

$$F_s = 1.5 \frac{V}{h'}p'_r \tag{3.13.8}$$

If the neutral axis of the built-up plate does not coincide with a plate interface, the shearing force at the interface will be less than that when it does, and its value is found from Eq. 3.13.7.

Equations 3.13.7 and 3.13.8 for the shearing force, F_s, acting on each pin can also be developed by assuming it to be equal to the shearing force distributed in a solid plate over an area $p'_c p'_r$, where p'_c is the circumferential pitch and p'_r is the radial pitch of the pins. Then by using Eq. 3.13.4 and 3.13.5 for the average shearing stress over the same area ($d = 0$) Eqs. 3.13.7 and 3.13.8 are obtained. This assumes the shearing force on a pin acts over a rectangular area $p_c'p_r'$, whereas the area is actually a trapezoidal one. The difference in area is nil and is usually not considered in practice.

The dimensions of the shear pins and their pitch or spacing must be chosen so as to insure strength against shearing through the pin, and against crushing of the material on the lateral sides of the pin. Here the assumption is made that the shearing stresses are uniformly distributed over the cross sectional area of the pin of diameter d:

$$\tau_{\text{pin}} = \frac{F_s}{\pi d^2/4} \tag{3.13.9}$$

The further simplifying assumption is made that the compressive bearing area of the pin on the plate is $h'd$, so the bearing stress on the plate is

$$\sigma = \frac{F_s}{h''d} \tag{3.13.10}$$

where h'' is the thinner of the two interfacing plates.

The shear pins, rivets, bolts or keys used in built-up plates should have a tight fit in order to prevent high local deformation and initial deflection.

Example 2. The stacked plates of the example given in Par. 3.13.1 and shown in Fig. 3.31*a* are fastened together with 1-in.-diameter pins. For calculation purposes, the pin diameter and hole diameter are assumed to be the same d. These are located on radii spaced at 15 degree intervals and the circumferential pitch of the 24 pins in the outer 160-in.-diameter pitch circle is $p_c' = 20.94$ inches. (Note: For small angles, the arc may be taken as equal to the chord.) What is the required radial pitch of the pins if the allowable shearing stress on the pins is 16,000 psi, and the allowable bearing stress on the plate is 36,000 psi? (The maximum shear occurs just prior to the support edge at an assumed diameter of 160 inches.)

A. Allowable shear and bearing forces on pins. The allowable shear force per pin from Eq. 3.13.9 is

$$F_{\text{S.al.}} = 16{,}000\ \pi \frac{d^2}{4} = 12{,}600 \text{ lb}$$

and the allowable bearing force on the thinner of the two adjacent plates (1 inch) from Eq. 3.13.10 is

$$F_{\text{br.al.}} = 36{,}000 \times 1 \times 1 = 36{,}000 \text{ lb}$$

so the shearing of the pin governs in establishing the radial pitch or spacing.

B. Shear acting on pins. The maximum shear stress occurs at the neutral axis, Eq. 3.13.5*a* and 3.13.8; hence, the pin size and spacing that provides for the shearing force at this location will also be more than adequate to take the lesser shear at the interface between the middle 1-in.-thick plate and the top $\frac{1}{2}$-in.-thick plate.

The shear diagram for a uniformly loaded simply edge-supported circular plate is given by Eq. 3.5.13 and shown in Fig. 3.31*c*. The shear force per circumferential pin spacing p_c' is

$$V = \frac{qx}{2} p_c'$$

The maximum shear force occurs in the edge row of pins and for

$$q = \frac{40,000}{3.1416 \times 80^2} = 2 \text{ psi}$$

is equal to

$$V = \frac{2 \times 80}{2} p_c' = 80\, p_c'$$

The shearing force on one pin per Eq. 3.13.8 is

$$F_s = \frac{1.5 \times 80}{3} p_c' p_r' = 40\, p_c' p_r'$$

C. Required radial pitch of pins. Equating this to the allowable shearing force of 12,600 lb per pin, and for $p_c' = 20.98$ inches the maximum radial pitch is

$$p_r' = \frac{12,600}{40 \times 20.98} = 15 \text{ inches}$$

Since the shear force is decreasing to a value of zero at the center of the plates, the radial pitch of the pins can be increased and/or the circumferential pitch increased. If the radial pin pitch is continued along the same radii, the pitch may be increased. The new pitch can be found by taking the total shear force at the last circumferential pitch circle, $X_2 = 80 - 15 = 65$ inches, and from Eq. 3.13.8 for

$$p_{c2}' = p_c' \times \frac{65}{80} = 20.94 \times \frac{65}{80} = 17.01 \text{ inches}$$

$$V = \frac{2 \times 65}{2} p_{c2}' = 65\, p_{c2}' \text{ lb}$$

$$F_s = \frac{1.5 \times 65}{3} p_{c2}' p_{r2}' = 32.5\, p_{c2}' p_{r2}'$$

$$p_{r2}' = \frac{12,600}{32.5 \times 17.01} = 22.8 \text{ inches (say 22 inches)}$$

In this manner, the pin spacing throughout the built-up plate is established. As the center is approached, both the circumferential and radial pitches are increased, Fig. 3.31. Though not required by calculations, it is good engineering practice to add pins in the central region to position the plates with respect to each other and the center of the stack. Likewise, in practice the outer row of pins is usually placed beyond the edge and over the support. The example was chosen to simplify the calculations and procedure.

REFERENCES

1. S. Timoshenko, *Strength of Materials*, Part II, D. Van Nostrand Co., Inc., Princeton, N.J., 1956.

2. S. Timoshenko and S. Woinowsky-Krieger, *Theory of Plates and Shells*, McGraw-Hill Book Co., Inc., New York, 1959.

3. W. A. Nash, "Effect of a Concentric Reinforcing Ring on Stiffness and Strength of a Circular Plate," *ASME Journal of Applied Mechanics*, Paper No. 47-A-15, 1947.

4. A. M. Wahl and G. Lobo, Jr., "Stresses and Deflections in Flat Circular Plates with Central Holes," *ASME Transactions*, Paper APM-52-3, 1929.

5. W. E. Trumpler, Jr., "Design Data for Flat Circular Plates with Central Holes," *ASME Journal of Applied Mechanics*, Vol. 10, p. 173, 1943.

6. H. D. Conway, "The Bending Symmetrically Loaded Circular Plates of Variable Thickness," *ASME Journal of Applied Mechanics*, Paper 47-A35, 1957.

7. N. J. Huffington, Jr., "Theoretical Determination of Rigidity Properties of Orthogonally Stiffened Plates," *ASME Journal of Applied Mechanics*, March, 1956.

8. W. H. Hoppmann, N. J. Huffington, Jr., and L. S. Magness, "A Study of Orthogonally Stiffened Plates," *ASME Journal of Applied Mechanics*, September, 1956.

9. R. K. Livesley, "Structural Problems in Reactor Engineering," *Chemical and Process Engineering*, August, 1960.

10. See list of references at end of Chapter.

11. K. S. Brundige, H. Phillips, E. J. Galvanek and E. J. Lachner, "Innovations in Carbon-Steel Tubed Heater Design, Fabrication, and Operation," ASME Publication Paper No. 66-WA/PWR-6, 1966.

12. E. B. Norris, P. D. Watson and R. D. Wylie, "Consideration in Design of Tube-to-Tubesheet Joints in High-Temperature Heat-Exchange Equipment," ASME Publication Paper No. 68-PVP-11, 1968.

13. A. M. Impagliazzo, "Tube-to-Tubesheet Attachment Welds," ASME Publication Paper 68-PVP-16, 1968.

14. W. J. O'Donnell, "Stresses and Deflections in Built-in Beams," ASME Publication Paper No. 62-WA-16, 1962.

15. R. D. Cook, "Deflections of a Series of Cantilevers Due to Elasticity of Supports," Transactions of the ASME *Journal of Applied Mechanics*. September, 1967.

16. D. P. Jones and W. J. O'Donnell, "Local Flexibility Coefficients for Asisymmetric Junctures," ASME Publication Paper No. 70-PVP-16, 1970.

17. B. Alami and D. G. Williams, *Thin Plate Design for Transverse Loading*, John Wiley & Sons, New York, 1975.

18. C. C. Chiang, "Closed Form Design Solutions for Box Type Heat Exchangers," ASME Publication Paper 75-WA/DE-15, 1975.

19. H. Krips and M. Podhorsky, "Hydraulic Expansion—A New Method for

the Anchoring of Tubes," 4th International Conference on Structural Mechanics in Reactor Technology, San Francisco, August 15-19, 1977.

20. L. J. Wolf and R. M. Mains, "Heat Exchanger Expansion Joints—Failure Modes, Analysis and Design Criteria," ASME Publication Paper No. 74-PVP-7, 1974.

21. W. R. Apblett, E. D. Montrone and W. Wolowodiuk, "Thermal Shock Testing of Fillet-Type Tube-to-Tubesheet Welds," ASME Publication Paper No. 76-PVP-35, 1976.

22. L. K. Chang and A. H. Marchertas, "Stress Analysis of a Nuclear Reactor Grid Plenum Assembly and Its Application to the EBR-II," *Nuclear Engineering and Design*, May, 1977.

23. J. P. Callahan and J. W. Bryson, "Stress Analysis of Perforated Flat Plates Under In-Plane Loadings," Oak Ridge National Laboratory Report ORNL/NUREG-2, August, 1976.

24. D. D. Dueck, "The Effect of Primary Head Shape on Tubesheet Thickness," ASME Publication Paper No. 78-PVP-30, 1978.

25. T. Terakawa, A. Imai, K. Yagi, Y. Fukada, and K. Okada, "Stiffening Effects of Tubes in Heat Exchanger Tube Sheet," ASME Publication Paper 83-PVP-3, 1983.

26. D. A. Scott, G. A. Wolgemuth, and J. A. Aiken, "Hydraulically Expanded Tube to Tube Sheet Joints," ASME Publication Paper No. 81-PVP-35, 1981.

27. S. Yokell, "Hydroexpanding: The Current State of the Art,"ASME Publication Paper No. 82-JPGC-Pwr-1, 1982.

28. K. P. Singh and A. I. Soler, "A Design Concept for Minimizing Tubesheet Stresses and Tubejoint Load in Fixed Tubesheet Heat Exchangers," ASME Publication Paper 82-PVP-34, 1982.

29. K. Uragami, M. Sugino, S. Urushibata, T. Kodama, and Y. Fujiwara, "Experimental Residual Stress Analysis of Tube to Tube Sheet Joints During Expansion," ASME Publication Paper 82-PVP-61, 1982.

30. B. Kasraie, J. S. Porowski, W. J. O'Donnell, and A. Selz, "Elastic-Plastic Analysis of Tube Expansion in Tubesheets," ASME Publication Paper 83-PVP-71, 1983.

31. C. A. Laberge and R. Baldur, "The Effect of Corner Radius on Plate-Cylinder Intersections," ASME Publication Paper No. 79-PVP-15, 1979.

32. D. P. Jones and J. L. Gordon, "Local Flexibility Coefficients for Built-in Ends of Beams and Plates Including the Effects of Foundation Fillet Radii," ASME Publication Paper No. 79-PVP-4, 1979.

33. A. I. Soler and Xu Hong, "Analysis of Tube-Tubesheet Joint Loading Including Thermal Loading," ASME Publication Paper 84-APM-15, 1984.

PROBLEMS

1. In a clamped-edge circular plate that is uniformly loaded, determine the location along a radius at which (*a*) the radial bending stress $\sigma_x = 0$, and (*b*) the circumferential stress $\sigma_y = 0$. (Take $\mu = 0.3$)

Ans.: (*a*) $= 0.628a$
Ans.: (*b*) $= 0.827a$

2. In the plate of Fig. 3.11*a*, establish the equation for (1) the radial moment, and (2) the circumferential moment when $M_{1b} = 0$. (Hint: Follow the procedure of para-

graph 3.9.1*a*, and utilize Eq. 3.5.4 for the circumferential stress.)

$$\textit{Ans.:}\quad (1)\ M_1 = \frac{a^2 M_{1a}}{a^2 - b^2}\left(1 - \frac{b^2}{x^2}\right)$$

$$(2)\ M_2 = \frac{a^2 M_{1a}}{a^2 - b^2}\left(1 + \frac{b^2}{x^2}\right)$$

3. A flat circular steel plate 160 in. in diameter is located inside a reactor to support a nuclear core. It is simply supported around its edge and is loaded uniformly with a total core load of 40,000 lb. (*a*) How thick must the plate be for a design stress limit of 8,000 psi; and (*b*) what is the maximum deflection of this plate if $E = 26{,}000{,}000$ psi and $\mu = 0.3$?

Ans.: (*a*) 1.4 in.
(*b*) 0.795 in.

4. If it becomes necessary to pierce the core support plate in Problem 3 with a 40 in. diameter central hole, (*a*) to what value does the maximum stress increase; and (*b*) to what value does the corresponding maximum deflection increase? (Hint: Use Eqs. 3.9.24 and 3.9.25.) (Assume unit uniform load is the same as in Problem 3.)

Ans.: (*a*) 13,512 psi
(*b*) 0.948 in.

5. A 20 ft. diameter nuclear core support plate for a nuclear reactor is composed of a solid plate 2 in. thick to which an orthogonal grillage consisting of 10 in. deep by 2 in. thick webs on 10 in. centers is integrally welded to one side of the flat cover plate to form a monolithic structure as shown in Fig. 3.32. It supports uniformly spaced nuclear fuel elements of 320,000 lb total weight. Assuming the structure to act as a simply supported plate with uniform edge reactions, determine: (*a*) the flexural rigidity *D* of the structure, (*b*) the bending stress at the top of the flat cover plate, and (c) the bending stress at the bottom of the grillage web at the center of the support plate.

Ans.: (*a*) 53.3 *E*
(*b*) − 1,576 psi
(*c*) 3,153 psi

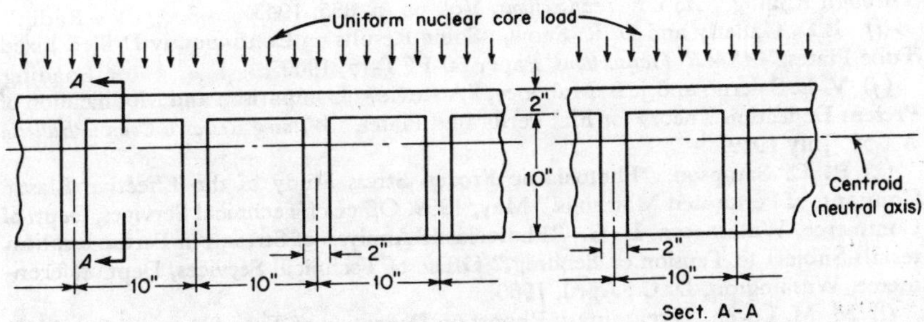

Fig. 3.32. Cross Section through the Nuclear Core Support Grillage of Problem 5

6. Relatively low values of bending stress, as found in Problem 5, are frequently associated with nuclear core support plates which are often designed to minimize

deflections in order to obviate misalignment, possible binding of nuclear fuel elements, control rods, or other moving parts within the reactor. What is the maximum deflection of the core support in Problem 5? (Assume $E = 28{,}000{,}000$ and $\mu = 0.3$.)

Ans.: 0.0626 in.

7. The single flat circular plate in Problem 3 is replaced by two equal thickness plates of the same diameter placed one on top of the other and supported in the same manner. What is the required thickness of these plates to maintain the same stress limit?

Ans.: 0.99 in.

REFERENCES FOR PERFORATED TUBE PLATES

(*a*) K. A. Gardner, "Heat-Exchanger Tube-Sheet Design," *ASME Journal of Applied Mechanics,* Vol. 15, p. 377, 1948; "Heat-Exchanger Tube-Sheet-2, Fixed Tube-Sheets," *ibid.,* Vol. 19, p. 573, 1952.

(*b*) G. Horvay, "Bending of Honeycombs and of Perforated Plates," *ASME Journal of Applied Mechanics,* Vol. 19, p. 122, 1952; "The Plane-Stress Problem of Perforated Plates," *ibid.,* Vol. 19, p. 355, 1952.

(*c*) I. Malkin, "Notes on a Theoretical Basis for Design of Tube-Sheets of Triangular Layout," *ASME Transactions,* Vol. 74, p. 387, 1952.

(*d*) K. A. G. Miller, "The Design of Tube-Plates in Heat-Exchangers," *Proc. of the Inst. of Mech. Engrs.,* Vol. 1B, p. 215, 1952.

(*e*) J. P. Duncan, "The Structural Efficiency of Tube-Plates of Heat Exchangers," *Proc. of the Inst. of Mech Engrs.,* Vol. 169, No. 39, p. 789, 1955; "Heat Exchanger Tube Sheet Design-3, U-Tube and Bayonet Tube Sheets," *ASME Journal of Applied Mechanics,* Vol. 27, p. 25, 1960.

(*f*) J. N. Goodier, and G. J. Schoessow, "The Holding Power and Hydraulic Tightness of Expanded Tube Joints," *ASME Transactions,* Vol. 65, p. 489, 1943.

(*g*) G. Sachs, "Notes on the Tightness of Expanded Tube Joints," *ASME Journal of Applied Mechanics,* Vol. 69, p. A285, 1947.

(*h*) A. Nadai, "Theory of the Expanding of Boiler and Condenser Tube Joints Through Rolling," *ASME Transactions,* Vol. 65, p. 865, 1943.

(*i*) G. D. Galletly and D. R. Snow, "Some Results on Continuously Drilled Fixed Tube Plates," *ASME Transactions,* Paper 60-PET-16, 1960.

(*j*) V. L. Salerno and J. B. Mahoney, "A Review, Comparison and Modification of Present Deflection Theory for Flat Perforated Plates," *Welding Research Council Bulletin No.* 52, July 1959.

(*k*) R. C. Sampson, "Photoelastic Frozen Stress Study of the Effective Elastic Constants of Perforated Materials," May, 1959, Office of Technical Services, Dept. of Commerce, Washington, D. C.; "Photoelastic Analysis of Stresses in Perforated Materials Subject to Tension or Bending," Office of Technical Services, Dept. of Commerce, Washington, D. C., April, 1960.

(*l*) M. M. Leven, "Preliminary Report on Deflection of Tube Sheets," May, 1959, and "Photoelastic Determination of stresses in Tube Sheets and Comparison with Calculated Values," Office of Technical Services, Dept. of Commerce, Washington, D. C., April, 1960.

(*m*) W. J. O'Donnell, "The Effect of the Tube on Stresses and Deflections in U-Tube Steam Generator Tube Sheets," Office of Technical Services, Dept. of Commerce, Washington, D. C., November, 1960.

(*n*) O. Tamate, "Transverse Flexure of a Thin Plate Containing Two Circular Holes," ASME Publication Paper No. 58-A-35, 1958.

(*o*) W. A. Bassali and M. Nassif, "Stresses and Deflections in an Elastically Restrained Circular Plate Under Uniform Normal Loading over a Segment," ASME Publication Paper No. 58-A-27, 1958.

(*p*) W. J. O'Donnell and B. F. Langer, "Design of Perforated Plates," ASME Publication Paper No. 61-WA-115, 1961.

(*q*) H. Kraus, "Flexure of a Circular Plate with a Ring of Holes," ASME Publication Paper No. 62-WA-9, 1962.

(*r*) J. B. Mahoney and V. L. Selerno, "Analysis of a Circular Plate Containing a Rectangular Array of Holes," Welding Research Council Bulletin No. 106, July 1965.

(*s*) H. M. S. Abdul-Wahab and J. Harrup, "The Rigidity of Perforated Plates with Reinforced Holes," *Nuclear Engineering and Design*, Vol. 5, No. 2, 1967, North-Holland Publishing Co., Amsterdam.

(*t*) W. J. O'Donnell, "Effective Elastic Constants for the Bending of Thin Perforated Plates with Triangular and Square Penetration Patterns," ASME Publication Paper No. 72-PVP-9, 1972.

(*u*) W. J. O'Donnell, "A Study of Perforated Plates with Square Penetration Patterns," Welding Research Council Bulletin No. 124, 1967.

(*v*) T. Slot, "Stress Analysis of Thick Perforated Plates," 1972, Technomic Publishing Co., Westport, Conn.

(*w*) M. D. Bernstein and A. I. Soler, "The Tubesheet Analysis Method in the New HEI Condenser Standards," ASME Publication Paper No. 77-JPGC-NE-18, 1977.

(*x*) K. P. Singh, "Analysis of Vertically Mounted Through-Tube Heat Exchangers," ASME Publication Paper No. 77-JPGC-NE-19, 1977.

(*y*) K. J. Tong, "Inelastic Stress Distribution in Tubesheet Liagments Under Thermal and Mechanical Loading," ASME Publication Paper No. 77-JPGC-NE-22, 1977.

(*z*) J. S. Porowski and W. J. O'Donnell, "Elastic Design Methods for Perforated Plates," ASME Publication Paper No. 77-JPGC-NE-20, 1977.

(*aa*) A. I. Solar, "Analysis of Closely Spaced Double Tubesheets Under Mechanical and Thermal Loading," ASME Publication Paper No. 77-HPGC-NE-21, 1977.

(*bb*) L. E. Hulbert and F. A. Simonen, "Analysis of Stresses in Shallow Spherical Shells with Periodically Spaced Holes," ASME Publication Paper No. 70-PVP-11, 1970.

(*cc*) J. Porowski and W. J. O'Donnell, "Effective Plastic Constants for Perforated Materials," Report to the Pressure Vessel Research Committee, April 1971, New York.

(*dd*) W. J. O'Donnell, J. S. Porowski, and R. D. Kichko, "Plastic Design of Ligaments," ASME Publication Paper No. 79-PVP-37, 1979.

(*ee*) D. P. Jones, "Axisymmetric Finite Element Analysis of Plates Containing Penetrations Arranged in a Square Pattern with Experimental Qualification," ASME Publication Paper No. 79-PVP-79, 1979.

(*ff*) T Terakawa, A. Imai, K. Yagi, Y. Fukada, and K. Okada, "Stiffening Effects of Tubes in Heat Exchanger Tubesheet," ASME Publication Paper 83-PVP-3, 1983.

(*gg*) B. Kasraie, J. S. Porowski, W. J. O'Donnell, and A. Selz, "Elastic-Plastic Analysis of Tube Expansion in Tubesheets," ASME Publication Paper 83-PVP-71, 1983.

4

Discontinuity Stresses in Pressure Vessels

4.1 Introduction

In Chapter 2 it was shown that the principal membrane stresses in a vessel subjected to internal or external pressure are produced by this pressure and remain as long as it is applied. Likewise, in Chapter 3 the bending stresses in plates subjected to pressure or structural loads are produced by these loads and remain as long as they are applied. These are called primary stresses. Primary stresses may be defined as those stresses developed by the imposed loading which are necessary to satisfy the laws of equilibrium of external and internal forces and moments. The basic characteristic of primary stresses is that they are not self-limiting; hence, when they exceed the yield point of the material they can result in failure or gross distortion. Another example of primary stress is that produced by wind, snow, or other specified live loads. These may produce either tension, compression, or bending, and must be combined with those produced by pressure in determining the total primary stress.

Secondary stresses, on the other hand, are those stresses developed by the constraint of adjacent parts or by self-constraint of a structure. The basic characteristic of secondary stresses is that they are self-limiting. Local yielding or minor distortion can satisfy the conditions causing the stress to occur and failure is not expected in one application. For example, all thermal stresses produced by thermal gradients within the structure are secondary. Another source of secondary stresses occurring in vessels is that occurring at the juncture of a cylindrical vessel and its closure head resulting from the differential growth or dilation of these parts under pressure. This effect is not uniform over the entire vessel, nor is it unrelenting in the sense that the primary membrane stresses are, since these remain as long as the pressure is applied. In fact, these stresses are relatively local in extent and self-limiting in magnitude since once the differential deflection is satisfied by plastic flow of the material a more favorable stress distribution results. This behavior is much like that of a beam on an elastic foundation, and this concept is used to evaluate the extent and magnitude of such disturbances occurring at shell-to-head junctures, support skirt, etc. Although these stresses are secondary ones and do not effect the static or bursting strength of the vessel, they are none the less important when: (1) clearances for moving parts, such as control rod drives for the cores of nuclear reactors, will not permit any gross local deflection, and (2) the vessel is subject to repetitive loading such that high local stresses can seriously limit its fatigue life, Chapters 5 and 6.

4.2 Beam on an Elastic Foundation

When a straight prismatic beam rests on a continuous supporting elastic foundation, Fig. 4.1, and is subjected to a concentrated load P in the principal plane of the symmetrical cross section, the beam will deflect producing a continuous distributed reaction force q in the foundation proportional to the deflection y of the beam at that point; i.e., $q = ky$ per unit length. This force opposes the deflection of the beam; hence, when the deflection is downward (positive) the foundation is in compression, whereas the reverse is true when the deflection is negative and the foundation is placed in tension. In some problems the supporting foundation is not actually continuous but consists of a series of closely spaced individual supports such as a railroad rail resting on closely spaced cross-ties. This may be considered continuous for practical purposes; in fact, the development and application of this theory was first concerned with rails,[1] and it is strongly recommended that in order to "get a feel" of the elastic foundation behavior, the nature of the deflection of a rail as a railroad car passes over it be observed. Here the deflection is relatively large and can be seen with the naked eye, whereas this analogous action in pressure vessels is not so visible.

The force with which the foundation resists the deflection of the beam is proportional to its deflection, and its "spring constant" or foundation modulus, as it is called, is equal to the force required per unit area to cause unit deflection, k pounds per square inch; hence

$$q = ky \tag{4.2.1}$$

Cutting a small element of length dx from the beam, Fig. 4.2, and applying the equilibrium equation that the summation of the vertical forces equal zero ($\Sigma F_v = 0$) to this element gives

$$V - (V + dV) + kydx = 0 \tag{4.2.2}$$

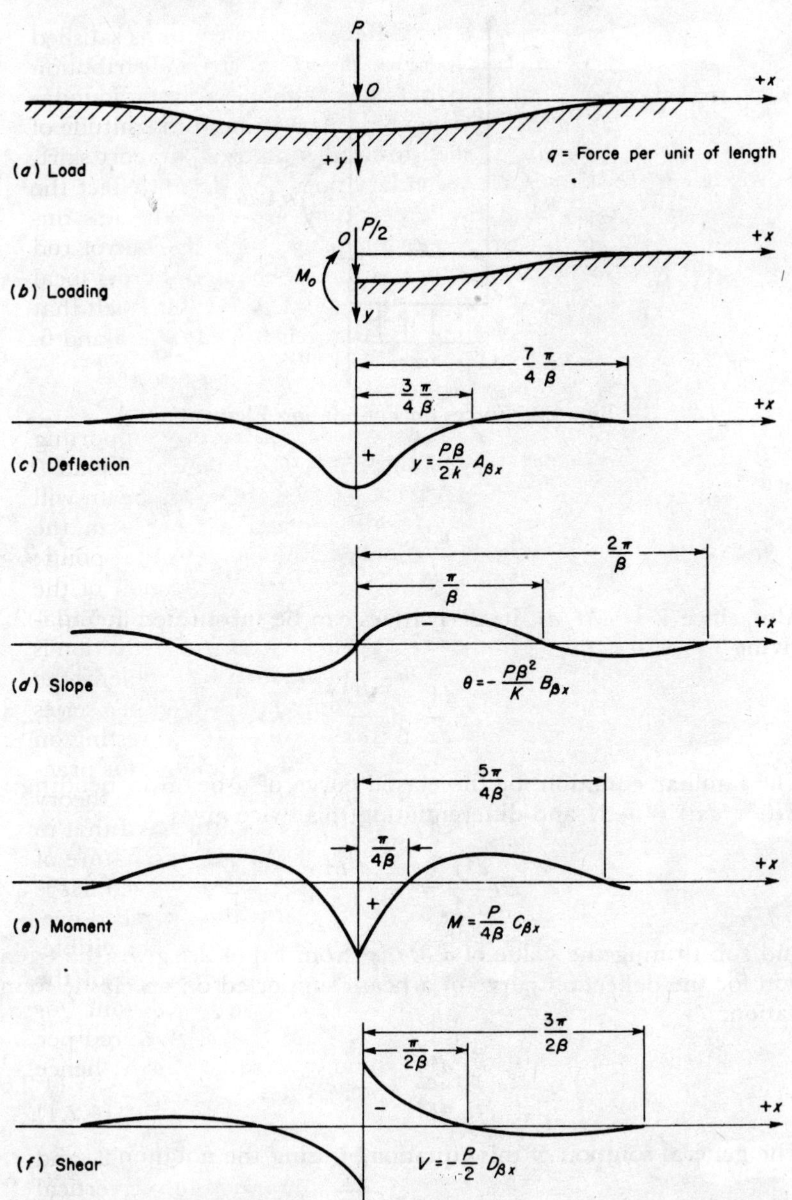

Fig. 4.1. Loading, Deflection, Slope, Moment, and Shear in a Beam on an Elastic Foundation

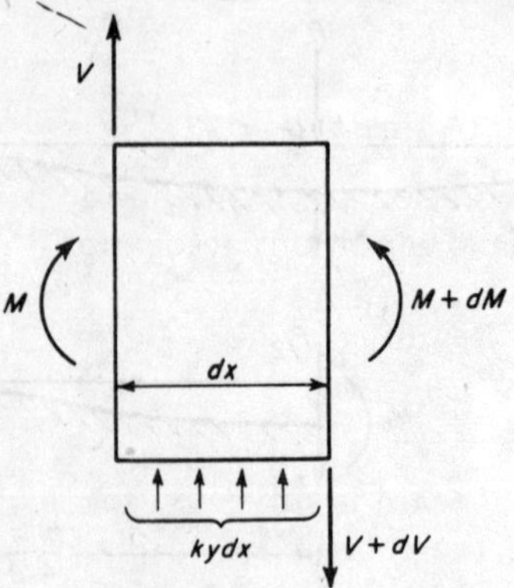

Fig. 4.2. Forces on Foundation Element

or

$$\frac{dV}{dx} = ky \tag{4.2.3}$$

Also, since $V = dM/dx$, its derivatives can be substituted in Eq. 4.2.3 giving

$$\frac{dV}{dx} = \frac{d^2M}{dx^2} = ky \tag{4.2.4}$$

The familiar equation for the elastic curve of a beam in bending is $EI(d^2y/dx^2) = -M$ and differentiating this twice gives

$$EI\frac{d^4y}{dx^4} = -\frac{d^2M}{dx^2} \tag{4.2.5}$$

and substituting the value of d^2M/dx^2 from Eq. 4.2.4 gives the equation for the deflection curve of a beam supported on an elastic foundation,

$$EI\frac{d^4y}{dx^4} = -ky \tag{4.2.6}$$

The general solution of this equation,[2,3] using the notation

$$\beta = \sqrt[4]{\frac{k}{4EI}} \tag{4.2.7}$$

is

$$y = e^{\beta x}(C_1 \cos \beta x + C_2 \sin \beta x) + e^{-\beta x}(C_3 \cos \beta x + C_4 \sin \beta x) \quad (4.2.8)$$

The constants of integration C_1, C_2, C_3, and C_4 must be determined from the known physical conditions at locations throughout the beam.

4.3 Infinitely Long Beam

1. *Single Concentrated Load*

If an infinitely long beam is subjected to a concentrated load P at point 0, the origin of the coordinate system, Fig. 4.1*a*, the deflection curve is symmetrical about the origin and only that part of the curve to the right need be considered, Fig. 4.1*b*. It is reasonable to assume that at a point infinite distance from the load the deflection is zero; hence, in Eq. 4.2.8 those terms involving $e^{\beta x}$ must vanish, which requires that $C_1 = C_2 = 0$. Therefore, the deflection curve for the right portion of the beam becomes

$$y = e^{-\beta x}(C_3 \cos \beta x + C_4 \sin \beta x) \quad (4.3.1)$$

Another physical condition of the beam is that at the origin, $x = 0$, its slope is zero, so

$$\frac{dy}{dx}_{x=0} = -\beta(C_3 - C_4) = 0 \quad (4.3.2)$$

or

$$C_3 = C_4 = C \quad (4.3.3)$$

and Eq. 4.3.1 becomes

$$y = Ce^{-\beta x}(\cos \beta x + \sin \beta x) \quad (4.3.4)$$

The value of C may be determined by noting that the summation of the foundation reaction forces must equal the applied force ($\Sigma F_v = 0$), or

$$2\int_0^\infty q\,dx = 2\int_0^\infty ky\,dx = P \quad (4.3.5)$$

When the value of y from Eq. 4.3.4 is substituted into Eq. 4.3.5 and the integration performed, C is found to be

$$C = \frac{P\beta}{2k} \quad (4.3.6)$$

Placing this value of C into Eq. 4.3.4, the deflection curve for the right portion of the beam becomes, Fig. 4.1*c*,

$$y = \frac{P\beta}{2k} e^{-\beta x}(\cos \beta x + \sin \beta x) \tag{4.3.7}$$

Expressions for the slope, bending moment, and shear at any positive value of x (on the right portion of the beam) can be found by taking successive derivatives of Eq. 4.3.7, giving

$$\frac{dy}{dx} = \theta = -\frac{P\beta^2}{k} e^{-\beta x} \sin \beta x \tag{4.3.8}$$

$$-EI\frac{d^2y}{dx^2} = M = \frac{P}{4\beta} e^{-\beta x}(\cos \beta x - \sin \beta x) \tag{4.3.9}$$

$$-EI\frac{d^3y}{dx^3} = V = -\frac{P}{2} e^{-\beta x} \cos \beta x \tag{4.3.10}$$

These equations may be written

Deflection $$y = \frac{P\beta}{2k} A_{\beta x} \tag{4.3.7a}$$

Slope $$\theta = -\frac{P\beta^2}{k} B_{\beta x} \tag{4.3.8a}$$

Moment $$M = \frac{P}{4\beta} C_{\beta x} \tag{4.3.9a}$$

Shear $$V = -\frac{P}{2} D_{\beta x} \tag{4.3.10a}$$

where the following notation has been used, and their value as a function of βx tabulated in Table 4.1:

$$A_{\beta x} = e^{-\beta x}(\cos \beta x + \sin \beta x)$$

$$B_{\beta x} = e^{-\beta x} \sin \beta x$$

$$C_{\beta x} = e^{-\beta x}(\cos \beta x - \sin \beta x)$$

$$D_{\beta x} = e^{-\beta x} \cos \beta x$$

The nature of the variation with distance from the point of applied load of the deflection, slope, moment, and shear are shown in Figs.

TABLE 4.1. FUNCTIONS $A_{\beta x}$, $B_{\beta x}$, $C_{\beta x}$, AND $D_{\beta x}$

βx	$A_{\beta x}$	$B_{\beta x}$	$C_{\beta x}$	$D_{\beta x}$	βx	$A_{\beta x}$	$B_{\beta x}$	$C_{\beta x}$	$D_{\beta x}$
0	1.0000	0	1.0000	1.0000	3.6	−0.0366	−0.0121	−0.0124	−0.0245
0.1	0.9907	0.0903	0.8100	0.9003	3.7	−0.0341	−0.0131	−0.0079	−0.0210
0.2	0.9651	0.1627	0.6398	0.8024	3.8	−0.0314	−0.0137	−0.0040	−0.0177
0.3	0.9267	0.2189	0.4888	0.7077	3.9	−0.0286	−0.0140	−0.0008	−0.0147
0.4	0.8784	0.2610	0.3564	0.6174	$5\pi/4$	−0.0278	−0.0140	0	−0.0139
0.5	0.8231	0.2908	0.2415	0.5323	4.0	−0.0258	−0.0139	0.0019	−0.0120
0.6	0.7628	0.3099	0.1431	0.4530	4.1	−0.0231	−0.0136	0.0040	−0.0095
0.7	0.6997	0.3199	0.0599	0.3798	4.2	−0.0204	−0.0131	0.0057	−0.0074
$\pi/4$	0.6448	0.3224	0	0.3224	4.3	−0.0179	−0.0125	0.0070	−0.0054
0.8	0.6354	0.3223	−0.0093	0.3131	4.4	−0.0155	−0.0117	0.0079	−0.0038
0.9	0.5712	0.3185	−0.0657	0.2527	4.5	−0.0132	−0.0108	0.0085	−0.0023
1.0	0.5083	0.3096	−0.1108	0.1988	4.6	−0.0111	−0.0100	0.0089	−0.0011
1.1	0.4476	0.2967	−0.1457	0.1510	4.7	−0.0092	−0.0091	0.0090	0.0001
1.2	0.3899	0.2807	−0.1716	0.1091	$3\pi/2$	−0.0090	−0.0090	0.0090	0
1.3	0.3355	0.2626	−0.1897	0.0729	4.8	−0.0075	−0.0082	0.0089	0.0007
1.4	0.2849	0.2430	−0.2011	0.0419	4.9	−0.0059	−0.0073	0.0087	0.0014
1.5	0.2384	0.2226	−0.2068	0.0158	5.0	−0.0046	−0.0065	0.0084	0.0019
$\pi/2$	0.2079	0.2079	−0.2079	0	5.1	−0.0033	−0.0057	0.0080	0.0023
1.6	0.1959	0.2018	−0.2077	−0.0059	5.2	−0.0023	−0.0049	0.0075	0.0026
1.7	0.1576	0.1812	−0.2047	−0.0235	5.3	−0.0014	−0.0042	0.0069	0.0028
1.8	0.1234	0.1610	−0.1985	−0.0376	5.4	−0.0006	−0.0035	0.0064	0.0029
1.9	0.0932	0.1415	−0.1899	−0.0484	$7\pi/4$	0	−0.0029	0.0058	0.0029
2.0	0.0667	0.1230	−0.1794	−0.0563	5.5	0.0000	−0.0029	0.0058	0.0029
2.1	0.0439	0.1057	−0.1675	−0.0618	5.6	0.0005	−0.0023	0.0052	0.0029
2.2	0.0244	0.0895	−0.1548	−0.0652	5.7	0.0010	−0.0018	0.0046	0.0028
2.3	0.0080	0.0748	−0.1416	−0.0668	5.8	0.0013	−0.0014	0.0041	0.0027
$3\pi/4$	0	0.0671	−0.1342	−0.0671	5.9	0.0015	−0.0010	0.0036	0.0026
2.4	−0.0056	0.0613	−0.1282	−0.0669	6.0	0.0017	−0.0007	0.0031	0.0024
2.5	−0.0166	0.0492	−0.1149	−0.0658	6.1	0.0018	−0.0004	0.0026	0.0022
2.6	−0.0254	0.0383	−0.1019	−0.0636	6.2	0.0019	−0.0002	0.0022	0.0020
2.7	−0.0320	0.0287	−0.0895	−0.0608	2π	0.0019	0	0.0019	0.0019
2.8	−0.0369	0.0204	−0.0777	−0.0573	6.3	0.0019	+0.0001	0.0018	0.0018
2.9	−0.0403	0.0132	−0.0666	−0.0534	6.4	0.0018	0.0003	0.0015	0.0017
3.0	−0.0423	0.0070	−0.0563	−0.0493	6.5	0.0018	0.0004	0.0012	0.0015
3.1	−0.0431	0.0019	−0.0469	−0.0450	6.6	0.0017	0.0005	0.0009	0.0013
π	−0.0432	0	−0.0432	−0.0432	6.7	0.0016	0.0006	0.0006	0.0011
3.2	−0.0431	−0.0024	−0.0383	−0.0407	6.8	0.0015	0.0006	0.0004	0.0010
3.3	−0.0422	−0.0058	−0.0306	−0.0364	6.9	0.0014	0.0006	0.0002	0.0008
3.4	−0.0408	−0.0085	−0.0237	−0.0323	7.0	0.0013	0.0006	0.0001	0.0007
3.5	−0.0389	−0.0106	−0.0177	−0.0283	$9\pi/4$	0.0012	0.0006	0	0.0006

4.1*c*, *d*, *e*, and *f*, respectively. The maximum values of the deflection, moment, and shear occur at the point of application of the load, $x = 0$, and are found from Eqs. 4.3.7*a*, 4.3.9*a*, and 4.3.10*a* to be

$$y_{max.} = \frac{P\beta}{2k} \tag{4.3.7b}$$

$$M_{max.} = M_0 = \frac{P}{4\beta} \tag{4.3.9b}$$

$$V_{max.} = -\frac{P}{2} \tag{4.3.10b}$$

The effect of multiple concentrated loads may be handled by the principle of superposition and the theorem of reciprocity. The principle of superposition merely means that the effect of each load may be determined independent of all others, and the overall result arrived at by adding together the individual effects. The theorem of reciprocity states that a reflex (such as deflection) at point 1 due to an action (such as load) at point 2, is the same as a reflex at point 2 produced by the action at point 1. As an example, if two equal loads, P_1 and P_2, spaced 60 in. apart rest on a beam on an elastic foundation of $\beta = 1/40$, and the origin of coordinates is taken at the first load, Table 4.2 gives the value of the functions $A_{\beta x}$, and $C_{\beta x}$, as taken from Table 4.1.

The total deflection under the load P_1 is from Eq. 4.3.7a,

$$y = \frac{P\beta}{2k}(1 + 0.2384) = 1.2384\frac{P\beta}{2k}$$

TABLE 4.2

Function	Load	
	P_1	P_2
βx	0	1.5
$A_{\beta x}$	1	0.2384
$C_{\beta x}$	1	−0.2068

or 24 percent more than that produced by a single load. The total moment under the load P_1 is from Eq. 4.3.9a

$$M = \frac{P}{4\beta}(1 - 0.2068) = 0.7932\frac{P}{4\beta}$$

or 20 percent less than that produced by a single load P.

2. *Uniform Load*

The principal of superposition can also be used to solve the problem of an uniformly distributed load over a portion of the beam length, Fig. 4.3, by considering that the distributed load is equivalent to a series of closely spaced concentrated loads qdx. The deflection at point O produced by such an element is obtained by substituting qdx for P in Eq. 4.3.7 which gives

$$\delta_y = \frac{qdx\beta}{2k}e^{-\beta x}(\cos\beta x + \sin\beta x) \tag{4.3.11}$$

where x is the distance from the element qdx to the point of origin O. The total deflection can be found by integrating Eq. 4.3.11 between the limits $O - a$ on the left side of the origin and $O - b$ on the right side of the origin giving

$$y = \int_0^a \frac{qdx\beta}{2k}e^{-\beta x}(\cos\beta x + \sin\beta x) + \int_0^b \frac{qdx\beta}{2k}e^{-\beta x}(\cos\beta x + \sin\beta x) \tag{4.3.12}$$

The value of a in Eq. 4.3.12 is negative, but since the basic equation has been set up to give the deflection for positive values of x only, it is used with a positive sign in Eq. 4.3.12. This is in order because the deflection of a beam under a single concentrated load has the same value

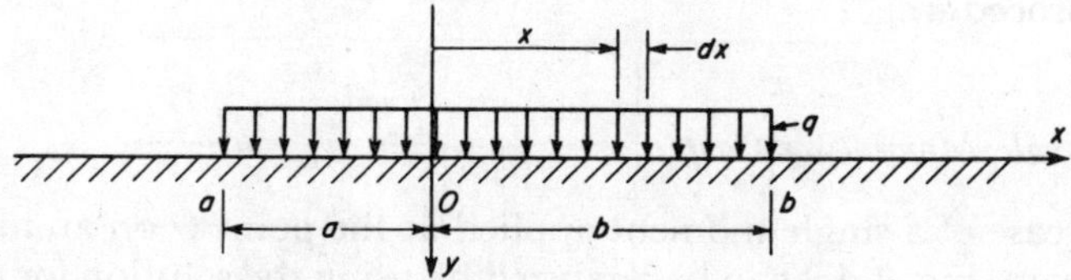

Fig. 4.3. Uniformly Distributed Load Over a Portion of a Beam on an Elastic Foundation

at equal distances in the positive and negative directions due to symmetry. This integration gives

$$y = \frac{q}{2k}[(1 - e^{-\beta a} \cos \beta a) + (1 - e^{-\beta b} \cos \beta b)] \tag{4.3.13}$$

or

$$y_0 = \frac{q}{2k}(2 - D_{\beta a} - D_{\beta b}) \tag{4.3.14}$$

The maximum deflection occurs at the mid-point of the loaded portion of the beam. If $O - a$ and $O - b$ are large, the values $e^{-\beta a}$ and $e^{-\beta b}$ will be small and the deflection from Eq. 4.3.14 becomes q/k; i.e., the bending of the bar can be neglected and it can be assumed that the uniform load q is transmitted directly to the elastic foundation. In a pressure vessel in which the uniformly distributed load is the pressure, $q = p$; then the deflection is proportional to the membrane stress which is uniform throughout the entire vessel.

In a similar manner, by substituting qdx for P in Eqs. 4.3.8*a*, 4.3.9*a* and 4.3.10*a* and integrating between the assigned limits, the values of the slope moment and shear are found to be, respectively,

$$\theta = \frac{q\beta}{2k}(A_{\beta a} - A_{\beta b}) \tag{4.3.15}$$

$$M = \frac{q}{4\beta^2}(B_{\beta a} + B_{\beta b}) \tag{4.3.16}$$

$$V = \frac{q}{4\beta}(C_{\beta a} - C_{\beta b}) \tag{4.3.17}$$

When the point of origin occurs at one end of the distributed load, or beyond the distributed load, these expressions can be found by the same procedure.

3. *Single Moment, or Couple*

The case of a single moment applied at the point O on an infinitely long beam, Fig. 4.4*a*, can be analyzed by using the solution for a single load, Eq. 4.3.7. The single moment can be considered as equivalent to two forces P, Fig. 4.4*b*, a distance a apart if it is assumed that Pa

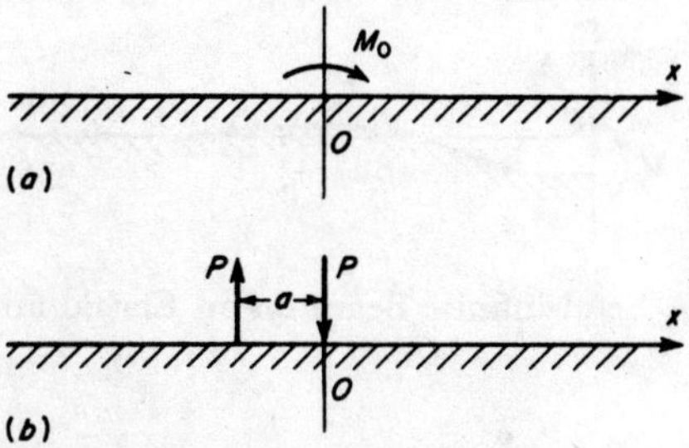

Fig. 4.4. Single Moment Acting on a Beam on an Elastic Foundation

approaches M_0 as a approaches zero. The deflection at a distance x from the origin is from Eq. 4.3.7a,

$$y = \frac{P\beta}{2k}(A_{\beta x} - A_{\beta(x+a)}) \tag{4.3.18}$$

$$= \frac{M_0\beta\,(A_{\beta x} - A_{\beta(x+a)})}{2k \qquad a} \tag{4.3.19}$$

but from Eq. 4.3.8,

$$-\left[\frac{A_{\beta(x+a)} - A_{\beta x}}{a}\right]_{a\to 0} = -\frac{d}{dx}A_{\beta x} = 2\beta B_{\beta x} \tag{4.3.20}$$

hence the deflection curve produced by the moment M_0 is

$$y = \frac{M_0\beta^2}{k}B_{\beta x} \tag{4.3.21}$$

The successive derivatives of this equation give

$$\frac{dy}{dx} = \theta = \frac{M_0\beta^3}{k}C_{\beta x} \tag{4.3.22}$$

$$-EI\frac{d^2y}{dx^2} = M = \frac{M_0}{2}D_{\beta x} \tag{4.3.23}$$

$$-EI\frac{d^3y}{dx^3} = V = -\frac{M_0\beta}{2}A_{\beta x} \tag{4.3.24}$$

The deflection, slope, bending moment and shear can be found from these equations, with the aid of Table 4.1 for any location on the beam.

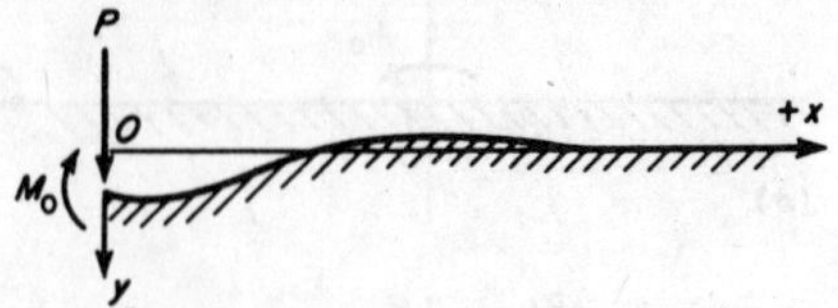

Fig. 4.5. Semi-infinite Beam on an Elastic Foundation

4.4 Semi-infinite Beam

A semi-infinite beam refers to one that has unlimited extension in one direction, but also has a finite end at point O. Figure 4.5 shows such a beam bent by a force P and moment M_0 applied at the finite end. The general solution, Eq. 4.2.8, for the case of an infinite beam can also be used for that of the semi-infinite one. Since the deflection and bending moment approach zero as distance x from the loaded end increases, the constants C_1 and C_2 must equal zero giving the same deflection curve as that for an infinite beam; namely,

$$y = e^{-\beta x}(C_3 \cos\beta x + C_4 \sin\beta x) \tag{4.4.1}$$

The constants of integration C_3 and C_4 are found from the conditions at the origin, taken at the point of application of the load and moment,

$$EI\left(\frac{d^2y}{dx^2}\right)_{x=0} = -M_0 \tag{4.4.2}$$

$$EI\left(\frac{d^3y}{dx^3}\right)_{x=0} = -V = P \tag{4.4.3}$$

Substituting from Eq. 4.4.1 into Eqs. 4.4.2 and 4.4.3 and performing the differentiation permits determining the constants as

$$C_3 = \frac{1}{2\beta^3 EI}(P - \beta M_0) \qquad \text{and} \qquad C_4 = \frac{M_0}{2\beta^2 EI} \tag{4.4.4}$$

Introducing these constants into Eq. 4.4.1 gives the deflection curve as

$$y = \frac{e^{-\beta x}}{2\beta^3 EI}[P\cos\beta x - \beta M_0(\cos\beta x - \sin\beta x)] \tag{4.4.5}$$

or, using the foregoing notation,

$$y = \frac{2P\beta}{k}D_{\beta x} - \frac{2M_0\beta^2}{k}C_{\beta x} \tag{4.4.6}$$

Likewise, taking the successive derivatives of Eq. 4.4.5 gives the slope, moment, and shear as

$$\theta = -\frac{2P\beta^2}{k}A_{\beta x} + \frac{4M_0\beta^3}{k}D_{\beta x} \quad (4.4.7)$$

$$M = -\frac{P}{\beta}B_{\beta x} + M_0 A_{\beta x} \quad (4.4.8)$$

$$V = -PC_{\beta x} - 2M_0\beta B_{\beta x} \quad (4.4.9)$$

The deflection and slope are a maximum at the end, $x = 0$, where

$$y_{\text{max.}} = \frac{2P\beta}{k} - \frac{2M_0\beta^2}{k} \quad (4.4.10)$$

$$\theta_{\text{max.}} = -\frac{2P\beta^2}{k} + \frac{4M_0\beta^3}{k} \quad (4.4.11)$$

These equations, in conjunction with the principle of superposition, can be used in the solution of problems involving discontinuity stresses at the juncture of heads and shells in vessels, etc., paragraph 4.7.

4.5 Cylindrical Vessel Under Axially Symmetrical Loading

One of the most important applications of the theory of beams on elastic foundations is to thin-walled pressure vessels. Considering the cylinder of Fig. 4.6*a* which is subject to rotationally symmetrical loading, but variable along its length, sections through the cylinder

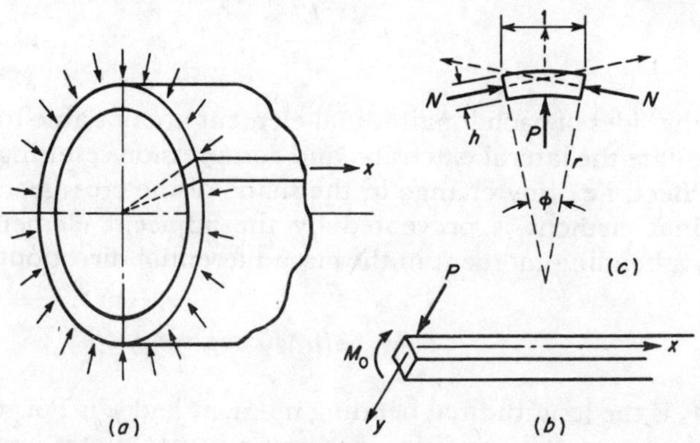

Fig. 4.6. Cylindrical Vessel Under Axially Symmetrical Loading

normal to the axis will remain circular because of symmetry, and the radius will undergo a displacement $\Delta r = y$ which will be different for each cross section, Fig. 4.6*b*. This radial displacement can be considered as deflection of a longitudinal element of unit width of the cylinder setting up bending so that the element acts as a "beam" on an "elastic foundation" created by the remainder of the supporting cylinder. The radial displacement y at any cross section of the longitudinal element results in a corresponding shortening of the radius of the cylinder at this location giving rise to a compressive strain in the hoop direction of y/r (Eq. 2.4.1). The accompanying hoop stress is Ey/r, or a hoop force per unit length of the longitudinal element of (Fig. 4.6*c*)

$$N = \frac{Eh}{r} y \tag{4.5.1}$$

The angle ϕ subtended by this unit width longitudinal element is $1/r$, so the radial resultant of these forces is

$$P = N\frac{1}{r} = \frac{Eh}{r^2} y \tag{4.5.2}$$

This reactive force P opposes the deflection and is distributed along the length of the longitudinal element in proportion to the deflection, where Eh/r^2 is the proportionality factor. Hence, a longitudinal element of a cylindrical vessel loaded symmetrically with respect to the axis behaves as a beam on an elastic foundation, the modulus of which is

$$k = \frac{Eh}{r^2} \tag{4.5.3}$$

Since the sides of each longitudinal element are not able to rotate to accommodate the lateral extension and compression resulting from the Poisson effect, i.e., any change in the shape of the cross section of the longitudinal element is prevented by the adjacent elements of the cylinder, a bending moment in the circumferential direction is created equal to

$$M_c = \mu M_x \tag{4.5.4}$$

where M_x is the longitudinal bending moment and μ is Poisson's ratio. This is similar to the condition occurring in flat plates and is taken into account by using $D = EI = Eh^3/12(1 - \mu^2)$ for its flexural rigidity, paragraph 3.2. Introducing this value for EI, and that for k from Eq. 4.5.3 in Eq. 4.2.7 gives the factor β as

$$\beta = \sqrt[4]{\frac{k}{4EI}} = \frac{\sqrt[4]{3(1 - \mu^2)}}{\sqrt{rh}} \tag{4.5.5}$$

For steel with $\mu = 0.3$ this becomes

$$\beta = \frac{1.285}{\sqrt{rh}} \tag{4.5.6}$$

4.6 Extent and Significance of Load Deformations on Pressure Vessels

1. *Attenuation Factors*

It is seen from Fig. 4.1 that the value of the deflection, slope, bending moment, and shear all have the characteristic damped wave form of rapidly diminishing amplitude. The length of this wave is given by the period of the functions cos βx and sin βx which is equal to

$$a = \frac{2\pi}{\beta} = 2\pi\sqrt[4]{\frac{4EI}{k}} \tag{4.6.1}$$

and the factor β is called the "damping factor." It is noted that these values are all very small at about a distance $x = \pi/\beta$ on either side of the load. This means that a beam of length $2\pi/\beta$ (π/β on each side of the point of loading) will have essentially the same deflection curve as an infinitely long beam, and Eqs. 4.3.7a, 4.3.8a, 4.3.9a, and 4.3.10a may be used without appreciable error. A cylindrical steel vessel then, of length greater than $2\pi\sqrt{rh}/1.285 = 4.9\sqrt{rh}$ acts as if it were infinitely long. This is particularly helpful since the solution of finite beams becomes more complicated and time consuming because the constants of integration in Eq. 4.2.8 are not so readily determined.[3]

2. *Equivalent Elastic Foundation—Cantilever Beam Length*

Another observation that can be made from the nature of the curves of Fig. 4.1 is an appraisal of the extent to which the influence of the elastic foundation characteristics may be considered of primary concern; as for example, the distance beyond the application of the load at which structural reinforcing of a pressure vessel may be assumed to

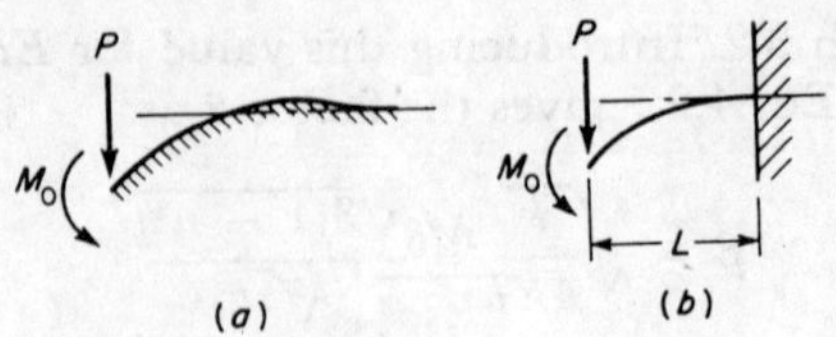

Fig. 4.7. Elastic Foundation—Cantilever Beam Comparison

have no significant effect. Obviously, it is not at a distance $x = \pi/\beta$ where all effects are nil as discussed above, but is somewhat closer to the point of load application. One simple approach is to approximate the elastic foundation deflection and slope characteristics with those of a similarly loaded equivalent length cantilever beam, which has similar deflection and slope curves, to give the same maximum values of these characteristics. The equivalent length cantilever beam is taken as the farthest distance from the point of application of the load at which a significant effect is registered. It is found, referring to Fig. 4.7, by equating the end deflections and slopes of a beam on an elastic foundation loaded by a force P and moment M_0 with those for a similarly loaded cantilever; hence,

A. Deflection Consideration:

(*a*) Load P only

Elastic beam from Eqs. 4.4.6 and 4.2.7

$$y = \frac{P}{2EI\beta^3} D_{\beta x} \tag{4.6.2}$$

Cantilever beam

$$y = \frac{PL^3}{3EI} \tag{4.6.3}$$

and substituting the value of y at $x = 0$ from Eq. 4.6.2 into Eq. 4.6.3 gives

$$L = \frac{1.11}{\beta} \tag{4.6.4}$$

when it is noted that the I of the elastic beam is equal to the I of the cantilever beam divided by $1 - \mu^2$, paragraph 3.2, and a value $\mu = 0.3$ for steel is used.

(*b*) Moment M_0 only
Elastic beam from Eqs. 4.4.6 and 4.2.7

$$y = \frac{M_0}{2EI\beta^2} C_{\beta x} \tag{4.6.5}$$

Cantilever beam

$$y = \frac{M_0 L^2}{2EI} \tag{4.6.6}$$

and substituting the value of y at $x = 0$ from Eq. 4.6.5 into Eq. 4.6.6 gives

$$L = \frac{0.95}{\beta} \tag{4.6.7}$$

B. Slope Consideration:

(*a*) Load P only
Elastic beam from Eqs. 4.4.7 and 4.2.7

$$\theta = \frac{P}{2EI\beta^2} A_{\beta x} \tag{4.6.8}$$

Cantilever beam

$$\theta = \frac{PL^2}{2EI} \tag{4.6.9}$$

and substituting the value of θ at $x = 0$ from Eq. 4.6.8 into Eq. 4.6.9 gives

$$L = \frac{0.95}{\beta} \tag{4.6.10}$$

(*b*) Moment M_0 only
Elastic beam from Eqs. 4.4.7 and 4.2.7

$$\theta = \frac{M_0}{EI\beta} D_{\beta x} \tag{4.6.11}$$

Cantilever beam

$$\theta = \frac{M_0 L}{EI} \tag{4.6.12}$$

and substituting the value of θ at $x = 0$ from Eq. 4.6.11 into Eq. 4.6.12 gives

$$L = \frac{0.91}{\beta} \tag{4.6.13}$$

It is seen from Eqs. 4.6.4, 4.6.7, 4.6.10, and 4.6.13 that the average value is approximately $L = 1/\beta$. This factor is useful in two ways:

1. It gives a good "feel" of the elastic foundation beam behavior via a comparison with familiar simple cantilever beam action.
2. It can be used conveniently for determining the end deflection and slope of elastic foundation beams when the equivalent cantilever length is taken as $1/\beta$, and it establishes a practical limit for its primary effect, such as the reinforcing limits around vessel openings, paragraph 6.6.

4.7 Discontinuity Stresses in Vessels

In Chapter 2, when a vessel was subjected to internal pressure only the direct tensile stresses, called membrane stresses, occurring over the entire wall thickness were considered. Differential displacements due to membrane stresses of varying magnitudes throughout the vessel can also occur causing bending of the wall, and, even though these bending stresses are local in extent, they may become very high in magnitude. One such location would be at the juncture of the cylindrical shell with its closure head, Fig. 4.8 and Fig. 4.10 (given subsequently) where the radial growth of the cylindrical portion of the vessel is not the same as that of the head when the vessel is pressurized; hence, at the juncture of these parts local bending takes place to preserve the continuity of the vessel wall. The additional stresses set up at these locations are called "discontinuity stresses."

In problems of this type the deformation and stress in the longitudinal or meridian elements can be determined from the elastic

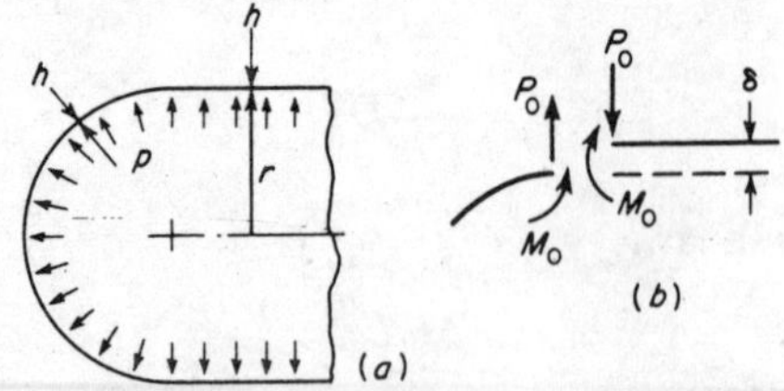

Fig. 4.8. Discontinuity at Hemispherical Head and Cylindrical Shell Juncture

foundation beam formulas, and then adding the longitudinal pressure stress σ_1. The stress in the circumferential direction is obtained by adding to the normal hoop pressure stress σ_2, that due to direct compression (shortening) or tension (extension) of the radius σ_c, and that caused by circumferential bending, σ_b. Thus the total circumferential stress in a cylindrical vessel is

$$\sigma = \sigma_2(\pm)\sigma_c \pm \sigma_b = \frac{pr}{h}(\pm)\frac{E}{r}y_x \pm \frac{6}{h^2}\mu M_x \tag{4.7.1}$$

When h is small in comparison with r, as it is in thin-walled vessels the deflection and bending becomes very local in extent and affects the stresses only in the immediate vicinity of the juncture. This narrow zone at the edge of the head can be considered as nearly cylindrical in shape; hence, the equations developed for the cylindrical portion of the vessel can be used for approximate calculations of the deflections in spherical, elliptical, or conical shape heads[4,5,6,9,10] at the juncture. In fact, these local effects can be determined for any shell of revolution by approximating the actual shell with an "equivalent cylinder" that has a radius equal to the radius of curvature r_2 in the hoop direction of the actual shell.

1. *Cylindrical Vessel with Hemispherical Heads*

This method of evaluating local bending stresses can be used for the case of a cylindrical vessel with hemispherical heads subjected to internal pressure, Fig. 4.8*a*. The hoop and longitudinal stresses in the cylindrical portion are from Eqs. 2.2.4 and 2.2.6, respectively.

Hoop

$$\sigma_2 = \frac{pr}{h} \tag{4.7.2}$$

Longitudinal

$$\sigma_1 = \frac{pr}{2h} \tag{4.7.3}$$

where r is the radius and h the thickness of the wall. These stresses in the spherical portion are from Eq. 2.2.7

Hoop

$$\sigma_2 = \frac{pr}{2h} \tag{4.7.4}$$

Longitudinal

$$\sigma_1 = \frac{pr}{2h} \tag{4.7.5}$$

The radial growth under internal pressure of the cylindrical portion from Eq. 2.4.3 is

$$\delta_c = \frac{pr^2}{2hE}(2 - \mu) \tag{4.7.6}$$

and for the spherical portion from Eq. 2.4.4 is

$$\delta_s = \frac{pr^2}{2hE}(1 - \mu) \tag{4.7.7}$$

If the cylindrical and spherical portions are disjointed, Fig. 4.8*b*, the difference in radial growth produced by the membrane stresses in the two portions would be

$$\delta = \delta_c - \delta_s = \frac{pr^2}{2hE} \tag{4.7.8}$$

However, in the actual vessel the head and cylinder are kept together at this juncture by shearing forces P_0 and bending moments M_0 per unit length of the circumference. These discontinuity forces produce local bending stresses in the adjacent parts of the vessel. One of the simplest cases which frequently occurs in practice is that in which the cylindrical wall and spherical head are of the same thickness. In this case the deflections and slopes induced at the edges of the cylindrical and spherical parts by the forces P_0 are equal; hence, the conditions of continuity at the juncture are satisfied if $M_0 = 0$ and P_0 is of a magnitude to create a deflection at the edge of the cylinder equal to $\delta/2$. Substituting $M_0 = 0$ at $x = 0$ in Eq. 4.4.6 gives an equation from which the value of P_0 can be found,

$$\frac{\delta}{2} = \frac{2P_0\beta}{k}D_{\beta x} \tag{4.7.9}$$

or using the value of k from 4.5.3, δ from Eq. 4.7.8 and noting $D_{\beta x} = 1$ at $x = 0$ gives

$$P_0 = \frac{p}{8\beta} \tag{4.7.10}$$

The deflection and bending moment at any distance from the point of juncture can be found from Eqs. 4.4.6 and 4.4.8, respectively, when P_0 is known; hence, the total longitudinal stress at any point x from the point of juncture of the cylinder and hemisphere is, in the cylinder,

$$\sigma = \frac{pr}{2h} \pm \frac{6}{h^2} \cdot \frac{p}{8\beta^2} B_{\beta x} \tag{4.7.11}$$

and the total hoop stress from Eq. 4.7.1 is, in the cylinder,

$$\sigma = \frac{pr}{h} - \frac{E}{r} \cdot \frac{p}{8\beta} \cdot \frac{2\beta}{k} D_{\beta x} \pm \frac{6}{h^2} \cdot \mu \cdot \frac{p}{8\beta^2} B_{\beta x} \tag{4.7.12}$$

substituting $k = Eh/r^2$ from Eq. 4.5.3, gives

$$\sigma = \frac{pr}{h} - \frac{pr}{4h} D_{\beta x} \pm \frac{3\mu p}{4h^2\beta^2} B_{\beta x} \tag{4.7.13}$$

As an example, the discontinuity stresses in the vessel of Fig. 4.8 for the conditions of $p = 300$ psi, $r = 50$ in., $h = 1$ in., and $\mu = 0.3$ can be found as follows:

A. From Eq. 4.5.6,

$$\beta = \frac{1.285}{\sqrt{rh}} = \frac{1.285}{\sqrt{50 \times 1}} = 0.182, \quad \beta^2 = 0.033$$

B. From Eq. 4.7.11 the longitudinal stress in the cylindrical portion is

$$\begin{aligned} \sigma &= \frac{pr}{2h} \pm \frac{6}{h^2} \cdot \frac{p}{8\beta^2} B_{\beta x} \\ &= \frac{300 \times 50}{2 \times 1} \pm \frac{6 \times 300}{1^2 \times 8 \times 0.033} B_{\beta x} \\ &= 7{,}500 \pm 6{,}820 B_{\beta x} \end{aligned}$$

The first quantity in this equation, membrane stress, remains constant along the length of the cylinder, while the second quantity in this equation, the bending stress, varies along the length of the vessel reaching a maximum numerical value at $\beta x = \pi/4$ as observed from an inspection of Table 4.1. The variation in stress is plotted in Fig. 4.9.

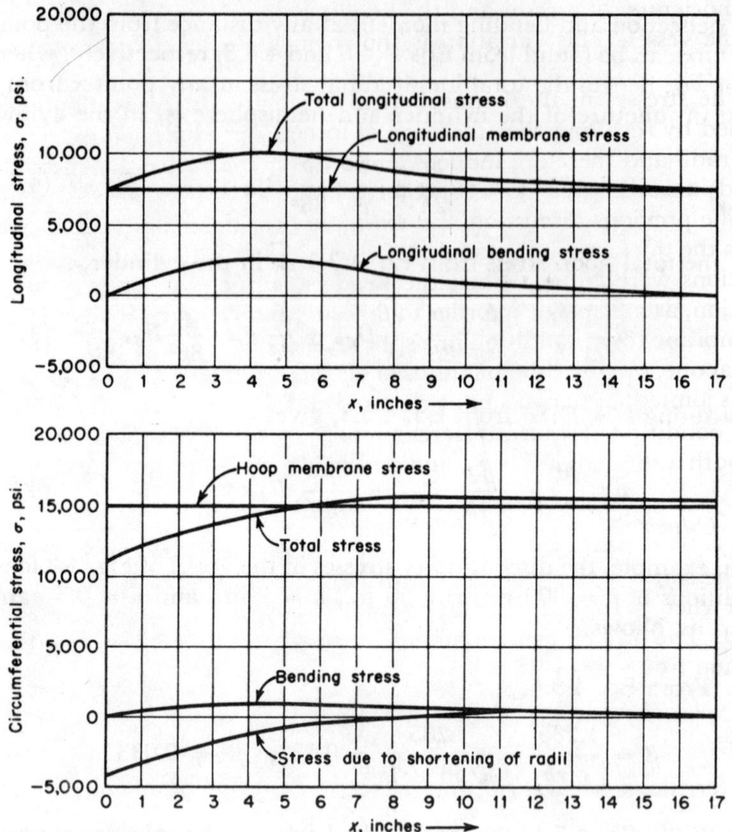

Fig. 4.9. Stress in Cylindrical Portion of Vessel of Fig. 4.8

C. From Eq. 4.7.13 the total hoop stress in the cylindrical portion is

$$\sigma = \frac{pr}{h} - \frac{pr}{4h}D_{\beta x} \pm \frac{3\mu p}{4h^2\beta^2}B_{\beta x}$$

$$= \frac{300 \times 50}{1} - \frac{300 \times 50}{4 \times 1}D_{\beta x} \pm \frac{3 \times 0.3 \times 300}{4 \times 1^2 \times 0.033}B_{\beta x}$$

$$= 15{,}000 - 3{,}750D_{\beta x} \pm 2{,}040B_{\beta x}$$

The first quantity in this equation, membrane stress, remains constant along the length of the cylinder, while the direct compression stress due to shortening of a radii and the bending stress varies along the length of the vessel. This variation is shown in the plot of these stresses in Fig. 4.9.

The stresses in the hemispherical head are correspondingly determined by recalling that the edge of force P_0 produces an extension of the radii and therefore induces tension over the entire thickness of the head; accordingly, the second term in Eq. 4.7.1 is taken as positive.

The previous discussion and example considered the case in which both the thickness and modulus of elasticity of the head and cylindrical portions were equal. When the head is thinner than the cylindrical portion, as may occur in order to take advantage of the more favorable membrane stress condition in a spherical shape as compared to a cylindrical one, or when the modulus of elasticity, E, is not the same for the parts jointed, as may occur at elevated temperatures when special design conditions require different materials for these parts, there will be both a shearing force P_0 and a moment M_0 at the juncture.[14, 15]

2. *Cylindrical Vessel with Ellipsoidal Heads*

The same method of calculating discontinuity stresses can also be used when the end closure to a cylindrical shell is an ellipsoidal shaped head. The radial growth under internal pressure of the cylindrical portion from Eq. 2.4.3 is

$$\delta_c = \frac{pr^2}{2hE}(2 - \mu) \tag{4.7.14}$$

and for the ellipsoidal portion from Eq. 2.4.7 is

$$\delta_e = \frac{pr^2}{hE}\left(1 - \frac{a^2}{2b^2} - \frac{\mu}{2}\right) \tag{4.7.15}$$

The difference in radial growth produced by the membrane stresses in the two portions of equal thickness is

$$\delta = \delta_c - \delta_e = \frac{pr^2}{2hE}\left(\frac{a^2}{b^2}\right) \tag{4.7.16}$$

Comparing this to Eq. 4.7.8 for a spherical closure head shows that this difference in dilation at the juncture, which is what causes the discontinuity stresses, is greater than that for a spherical head in the ratio of a^2/b^2. The shearing force P_0 and discontinuity stresses are also increased in the same proportion. The membrane stresses in the ellipsoidal head are obtained from Eqs. 2.6.25 and 2.6.26, and those in the cylindrical shell from Eqs. 2.2.4 and 2.2.6. As in the case of the spherical head the membrane and discontinuity stresses are added to give the total acting stress. These quantities can be computed from the continuity conditions that: (1) the sum of the edge deflections of head and cylindrical portions must equal δ, Fig. 4.8*b*, and; (2) the angle of rotation, or slope, of the two edges must be equal.

3. *Cylindrical Vessel with a Flat Head*

Flat plates are often used for the heads of cylindrical pressure vessels,[7] Fig. 4.10*a*. In this case the head may be considered as a flat circular plate uniformly loaded by the internal pressure, p, and hence bends to a spherical surface with a corresponding change in slope at its juncture with the cylinder. This change in slope sets up a bending moment M_0 which makes the cylindrical shell slope agree with the slope of the head. The other continuity condition requires that the radial growth of the cylindrical shell under pressure be restricted at the head juncture by a force P_0 and moment M_0. Referring to Fig. 4.10*b* and using the subscripts H for the head and C for the cylinder, the slope continuity equation at the juncture is

$$\theta_{H,p} - \theta_{H,M_0} = \theta_{C,M_0} - \theta_{C,P_0} \tag{4.7.17}$$

The head edge slope due to the uniform pressure p is from Eq. 3.6.25

$$\theta_{H,p} = \frac{3pa^3}{2Eh_H^3}(1 - \mu) \tag{4.7.18}$$

and that due to the edge moment M_0 is from Eqs. 3.3.11 and 3.5.1

$$\theta_{H,M_0} = \frac{rM_0}{D(1 + \mu)} \tag{4.7.19}$$

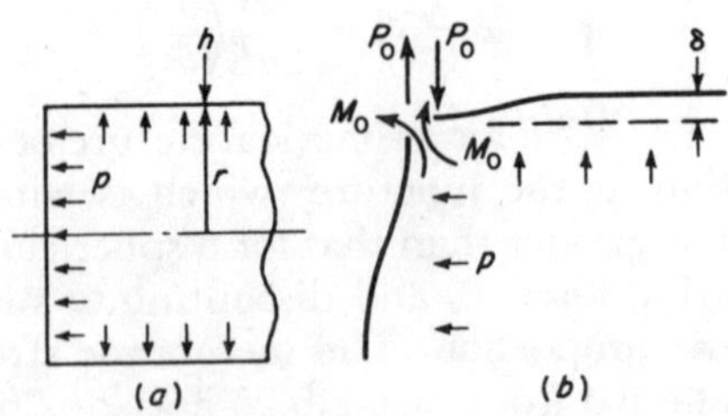

Fig. 4.10. Discontinuity at Flat Head and Cylindrical Shell Juncture

The slope of cylindrical portion at the juncture due to P_0 and M_0 is obtained from Eq. 4.4.7 giving for the right side of Eq. 4.7.14

$$\theta_{C,M_0} - \theta_{C,P_0} = \frac{4M_0\beta^3}{k}D_{\beta x} - \frac{2P_0\beta^2}{k}A_{\beta x} \tag{4.7.20}$$

The second continuity equation for the radial displacement can be written

$$\delta = \delta_{H,P_0} + \delta_{C,P_0M_0} \tag{4.7.21}$$

where δ is the unrestrained growth of the cylindrical portion due to internal pressure and is from Eq. 2.4.3

$$\delta = \frac{pr^2}{2hE}(2 - \mu) \tag{4.7.22}$$

The radial deflection of the flat head due to P_0 is

$$\delta_{H,P_0} = \frac{rP_0}{h_H E}(1 - \mu) \tag{4.7.23}$$

and the radial deflection of the cylindrical portion due to P_0 and M_0 is from Eq. 4.4.6

$$\delta_{C,P_0M_0} = \frac{2P_0\beta}{k}D_{\beta x} - \frac{2M_0\beta^2}{k}C_{\beta x} \tag{4.7.24}$$

By substituting these values of the individual terms into Eqs. 4.7.17 and 4.7.21 and solving simultaneously, P_0 and M_0 can be found. As an example, the magnitude of P_0 and M_0 for the vessel of Fig. 4.10 for the conditions of $r = 5$ in., $h_H = h_c = h = 3/8$ in., $p = 100$ psi, $\mu = 0.3$, $E = 25{,}000{,}000$ psi is found as follows:

A. From Eq. 4.7.18 noting $a = r$,

$$\theta_{H,p} = \frac{3pr^3(1 - \mu)}{2Eh^3} = \frac{3 \times 100 \times 5^3(0.7)}{2Eh^3} = \frac{13{,}125}{Eh^3}$$

B. From Eqs. 4.7.19 and 3.2.9,

$$\theta_{H,M_0} = \frac{12r(1 - \mu^2)M_0}{Eh^3(1 + \mu)} = \frac{12 \times 5 \times (0.7)M_0}{Eh^3} = \frac{42M_0}{Eh^3}$$

C. From Eqs. 4.7.20 and 4.5.3, at $x = 0$,

$$\theta_{C,M_0} - \theta_{C,P_0} = \frac{4r^2\beta^3M_0}{Eh} - \frac{2r^2\beta^2P_0}{Eh}$$

D. From Eq. 4.7.17,

$$\frac{13{,}125}{Eh^3} - \frac{42M_0}{Eh^3} = \frac{4r^2\beta^3 M_0}{Eh} - \frac{2r^2\beta^2 P_0}{Eh}$$

E. The growth of the cylindrical portion is from Eq. 4.7.22,

$$\delta = \frac{pr^2}{2hE}(2 - \mu) = \frac{100 \times 5^2 \times (1.7)}{2 \times hE} = \frac{2{,}125}{hE}$$

F. From Eq. 4.7.23,

$$\delta_{H,P_0} = \frac{rP_0}{hE}(1 - \mu) = \frac{3.5P_0}{hE}$$

G. From Eqs. 4.7.24 and 4.5.3,

$$\delta_{C,P_0M_0} = \frac{2r^2\beta P_0}{Eh} - \frac{2r^2\beta^2 M_0}{Eh}$$

H. From Eq. 4.7.21.

$$\frac{2{,}125}{hE} = \frac{3.5P_0}{hE} + \frac{2r^2\beta P_0}{Eh} - \frac{2r^2\beta^2 M_0}{Eh}$$

$$2{,}125 = (3.5 + 2r^2\beta)P_0 - 2r^2\beta^2 M_0$$

I. Solving simultaneously the equations from D and H after introducing the value $h^2 = 0.375^2 = 0.1406$, $\beta = 1.285/\sqrt{rh} = 1.285/\sqrt{0.375 \times 5} = 0.938$, $\beta^2 = 0.881$, $\beta^3 = 0.826$ gives

$P_0 = 284$ lb per in. of circumference

$M_0 = 277$ in. lb per in. of circumference

J. Total radial stress in the flat head at juncture due to P_0 and M_0 is:

1. Stress due to $P_0 = 284$ lb per inch circumference:

$$o = \frac{P_0}{h_H} = \frac{284}{.375} = 759 \text{ psi (tension)}$$

2. Stress due to $M_0 = 277$ in. lb per in circumference:

$$\sigma = \pm\frac{6}{h_H^2} M_0 = \frac{6 \times 277}{.375^2} = \pm 11{,}844 \text{ psi}$$

3. Total stress inside surface:

$$\sigma = 759 + 11{,}844 = 12{,}603 \text{ psi}$$

4. Total stress outside surface:

$$\sigma = 759 - 11{,}844 = -\ 11{,}085 \text{ psi}$$

K. Total radial stress at the center of the flat head:

1. Bending stress at center of free edge circular plate uniformly loaded by pressure p, Eq. 3.6.29:

$$\sigma = \pm \frac{3(3+\mu)}{8} p \frac{r^2}{h^2}$$

$$\sigma = \pm \frac{3(3+.3)}{8} 100 \frac{5^2}{.375^2} = \pm\ 22{,}000 \text{ psi}$$

Stress	*Inside Surface*	*Outside Surface*
Due to P_o	759	759 (same as J.1)
Due to M_o	11,844	-11,844 (same as J.2)
Due to p	-22,000	22,000
Total	-9,397	10,915

4.8 Stresses in a Bimetallic Joint

Special conditions often require that a pressure vessel be constructed of several materials of different metallurgical and physical properties whose incompatibility induces stresses when the vessel is subjected to its operating environment. This occurs in the piping of boilers and turbines where austenitic steels, such as 18 chrome–8 nickel, are required in the high-temperature zone; whereas in the cooler zones the more economical ferritic steels are used. It also occurs in nuclear reactor vessels where cleanliness requirements of the fluid necessitates the use of stainless steel clad vessels and stainless steel piping. In either case local discontinuity stresses are produced in the region where these dissimilar materials are welded together due to the fact that the coefficient of thermal expansion of the austenitic steel is about 50 per cent greater than that of the ferritic steel; hence, the free growth of each portion under a temperature change is restricted, Fig. 4.11*a*. These stresses can be evaluated in the same manner as those at the juncture of a cylindrical vessel with hemispherical heads, paragraph 4.7.1, but in this case the differential dilation of the two parts is due to their different thermal expansion and is

$$\delta = r\Delta T(\alpha_s - \alpha_f) \tag{4.8.1}$$

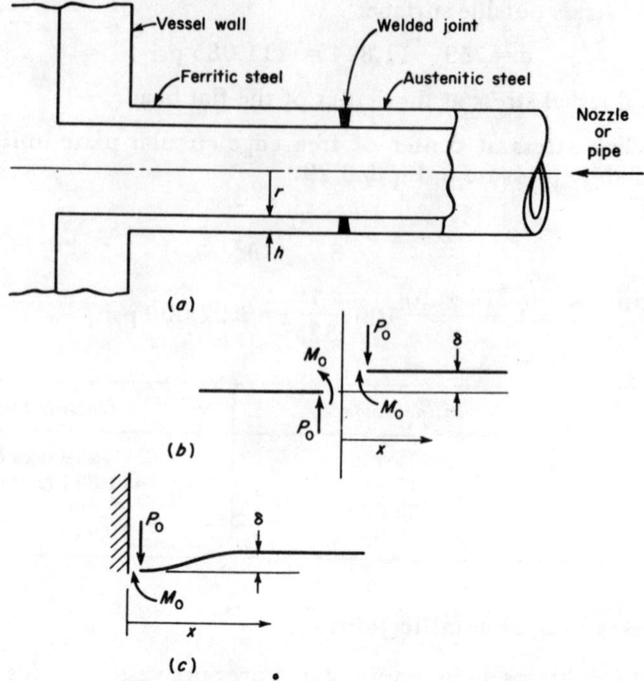

Fig. 4.11. Bimetallic Joints in Nozzles

where ΔT is the temperature change, r is the radius, and α_s and α_f are the linear coefficient of thermal expansion of the austenitic and ferritic steels, respectively. The stresses produced by a bimetallic joint can be minimized by the choice of its location in the vessel. Consider the attachment of a stainless steel nozzle to a heavy wall ferritic steel vessel by (1) locating the joint at relatively large distance from the vessel wall, and (2) locating the joint at the vessel wall.

1. *Location of a Bimetallic Welded Joint in a Nozzle of Uniform Thickness in a Region Away from the Vessel Wall*

In this case, Fig. 4.11*b*, the shearing force P_0 and bending M_0 per unit length of circumference at the joint necessary to preserve continuity of deflection and slope in the two portions is obtained by equating the sum of the deflection of the stainless steel portion and the ferritic steel portion resulting from P_0 and M_0, Eq. 4.4.6, to that created by the differential thermal expansion, Eq. 4.8.1,

$$r\Delta T(\alpha_s - \alpha_f) = \delta_{s,P_0} - \delta_{s,M_0} + \delta_{f,P_0} + \delta_{f,M_0} \tag{4.8.2}$$

$$= \frac{2P_0\beta_s}{k_s}D_{\beta x} - \frac{2M_0\beta_s^2}{k_s}C_{\beta x} + \frac{2P_0\beta_f}{k_f}D_{\beta x} + \frac{2M_0\beta_f^2}{k_f}C_{\beta x} \tag{4.8.3}$$

$$= 2P_0\left[\frac{\beta_s D_{\beta x}}{k_s} + \frac{\beta_f D_{\beta x}}{k_f}\right] + 2M_0\left[\frac{\beta_f^2 C_{\beta x}}{k_f} - \frac{\beta_s^2 C_{\beta x}}{k_s}\right] \tag{4.8.4}$$

and by equating the edge rotations for the two parts from Eq. 4.4.7,

$$\theta_s = \theta_f \tag{4.8.5}$$

$$-\frac{2P_0\beta_s^2}{k_s}A_{\beta x} + \frac{4M_0\beta_s^3}{k_s}D_{\beta x} = -\frac{2P_0\beta_f^2}{k_f}A_{\beta x} - \frac{4M_0\beta_f^3}{k_f}D_{\beta x} \tag{4.8.6}$$

$$-2P_0\left[\frac{\beta_s^2 A_{\beta x}}{k_s} - \frac{\beta_f^2 A_{\beta x}}{k_f}\right] = -4M_0\left[\frac{\beta_s^3 D_{\beta x}}{k_s} + \frac{\beta_f^3 D_{\beta x}}{k_f}\right] \tag{4.8.7}$$

When the temperature of the metal is sufficiently high so that their moduli of elasticity, E_s and E_f, are not the same, both Eqs. 4.8.4 and 4.8.7 are required to solve for P_0 and M_0. When these two values are equal, then $\beta_s = \beta_f$, $k_s = k_f$, and the left side of Eq. 4.8.7 vanishes, indicating that M_0 equals zero and the condition of slope continuity is provided by the action of the forces P_0 only. This is equivalent to saying that the slope and deflection induced at the edges of the two parts by the forces P_0 are equal; hence, the conditions of continuity are met by $M_0 = 0$, and Eq. 4.8.4 gives

$$\frac{r\Delta T(\alpha_s - \alpha_f)}{2} = \frac{2P_0\beta}{k}D_{\beta x} \tag{4.8.8}$$

This agrees with Eq. 4.7.9 when it is noted that the left side of this equation is $\delta/2$. The stresses can then be computed in the same manner as described in paragraph 4.7.1.

2. *Location of a Bimetallic Welded Joint in a Nozzle of Uniform Thickness at Its Juncture with the Vessel Wall*

In this case the wall of the ferritic steel vessel is assumed to be rigid and will not deflect or twist, so that the austenitic steel nozzle must absorb the complete deflection, δ, due to the difference in thermal expansion of the two parts, and it must be prevented from rotating at the juncture; i.e., $\theta_{x=0} = 0$, Fig. 4.11*c*. The deflection continuity condition then becomes from Eqs. 4.8.1 and 4.4.6

$$r\Delta T(\alpha_s - \alpha_f) = \frac{2P_0\beta}{k}D_{\beta x} - \frac{2M_0\beta^2}{k}C_{\beta x} \tag{4.8.9}$$

and equating the edge slope from Eq. 4.4.7 to zero gives

$$0 = -\frac{2P_0\beta^2}{k}A_{\beta x} + \frac{4M_0\beta^3}{k}D_{\beta x} \tag{4.8.10}$$

Substituting the values of $A_{\beta x}$, $C_{\beta x}$ and $D_{\beta x}$ from Table 4.1 at $x = 0$ into Eqs. 4.8.9 and 4.8.10 gives

$$r\Delta T(\alpha_s - \alpha_f) = \frac{2P_0\beta}{k} - \frac{2M_0\beta^2}{k} \tag{4.8.11}$$

and

$$0 = -\frac{2P_0\beta^2}{k} + \frac{4M_0\beta^3}{k} \tag{4.8.12}$$

from which the simultaneous solution gives

$$M_0 = \frac{k[r\Delta T(\alpha_s - \alpha_f)]}{2\beta^2} \tag{4.8.13}$$

and

$$P_0 = \frac{k[r\Delta T(\alpha_s - \alpha_f)]}{\beta} \tag{4.8.14}$$

With these values of M_0 and P_0, the bending moment at any distance x from the juncture can be found from Eq. 4.4.8, and the deflection from Eq. 4.4.6.

It can be deduced from a comparison of the forces P_0 and M_0 at the juncture for these two cases that the optimum location for a bimetallic welded joint is away from points of rigidity so that both portions can bend to absorb the differential thermal growth of the two joined parts. For instance, comparing Eq. 4.8.14 with Eq. 4.8.8 shows the shearing force P_0 at a juncture so located is 1/4 of that when the juncture is located at a point of fixity; and the accompanying juncture moment M_0 is zero as compared with that given by Eq. 4.8.13, Table 4.3. These are local discontinuity effects which attenuate rapidly, and a distance $x = \pi/\beta = 2.45\sqrt{rh}$ is sufficient to remove the joint from the effects of a point of rigidity or fixity. The stress variation in the region

TABLE 4.3

Location of Bimetallic Welded Joint	P_0	M_0
1. Away from Point of Fixity	$\dfrac{k[r\Delta T(\alpha_s - \alpha_f)]}{4\beta}$	0
2. At Point of Fixity	$\dfrac{k[r\Delta T(\alpha_s - \alpha_f)]}{\beta}$	$\dfrac{k[r\Delta T(\alpha_s - \alpha_f)]}{2\beta^2}$

of a bimetallic welded joint for the two locations discussed previously is shown in the following examples.

Example 1. Determine the longitudinal stress (pressure and bending) in the wall of the nozzle of Fig. 4.11*b* in the region of the weld joining an austenitic material (alloy steel 18 cr–8 ni) to a ferritic material (carbon steel) for the conditions of $r = 5$ in., $h = 3/8$ in., $p = 750$ psi, $\Delta T = 500°F$, E for both materials = 25,000,000 psi,

$$\alpha_s = 0.000010 \text{ in./in.-°F}, \quad \text{and} \quad \alpha_f = 0.000008 \text{ in./in.-°F}$$

A. Longitudinal pressure stress, Eq. 2.2.6,

$$\sigma_1 = \frac{pr}{2h} = \frac{750 \times 5}{2 \times 0.375} = 5{,}000 \text{ psi}$$

B. Differential radial growth, Eq. 4.8.1,

$$\delta = r\Delta T(\alpha_s - \alpha_f) = 5 \times 500(0.00001 - 0.000008) = 0.005 \text{ in.}$$

C. Damping factor, Eq. 4.5.6.

$$\beta = \frac{1.285}{\sqrt{rh}} = \frac{1.285}{\sqrt{5 \times 0.375}} = 0.938, \qquad \beta^2 = 0.881$$

D. Shearing force P_0, Eq. 4.8.8 and Eq. 4.2.7,

$$\frac{\delta}{2} = \frac{P_0}{2EI\beta^3}$$

$$P_0 = 0.005EI\beta^3$$

E. The longitudinal bending moment, Eq. 4.4.8,

$$M = -\frac{P_0}{\beta}B_{\beta x}$$

$$= -0.005EI\beta^2 B_{\beta x}$$

F. Longitudinal bending stress, noting $I = h^3/12\,(1 - \mu^2)$ per Eq. 3.2.9,

$$\sigma_b = \pm M/Z = \frac{6M}{h^2}$$

$$= \pm \frac{0.0025Eh\beta^2 B_{\beta x}}{1 - \mu^2}$$

G. The total longitudinal stress σ is the sum of the pressure stress, A, and the bending stress, F, and is plotted in Fig. 4.12.

Example 2. Determine the longitudinal stress (pressure and bending) in the wall of the nozzle of Fig. 4.11*c* if the dissimilar metal weld joint connects the austenitic steel nozzle directly to the ferritic steel vessel, all other conditions remaining the same as in Example 1.

A. Longitudinal pressure stress, Eq. 2.2.6,

$$\sigma_1 = \frac{pr}{2h} = \frac{750 \times 5}{2 \times 0.375} = 5{,}000 \text{ psi}$$

B. Differential radial growth, Eq. 4.8.1,

$$\delta = r\Delta T(\alpha_s - \alpha_f) = 5 \times 500(0.00001 - 0.000008) = 0.005 \text{ in.}$$

C. Damping factor, Eq. 4.5.6,

$$\beta = \frac{1.285}{\sqrt{rh}} = \frac{1.285}{\sqrt{5 \times 0.375}} = 0.938, \qquad \beta^2 = 0.881$$

D. Juncture bending moment M_0, Eq. 4.8.13 and Eq. 4.2.7,

$$M_0 = \frac{k[r\Delta T(\alpha_s - \alpha_f)]}{2\beta^2} = 0.01EI\beta^2$$

E. Juncture shearing force P_0, Eq. 4.8.14 and Eq. 4.2.7,

$$P_0 = \frac{k[r\Delta T(\alpha_s - \alpha_f)]}{\beta} = 0.02EI\beta^3$$

F. Longitudinal bending moment, Eq. 4.4.8,

$$M = -\frac{P_0}{\beta}B_{\beta x} + M_0 A_{\beta x}$$

$$= -0.02EI\beta^2 B_{\beta x} + 0.01EI\beta^2 A_{\beta x}$$

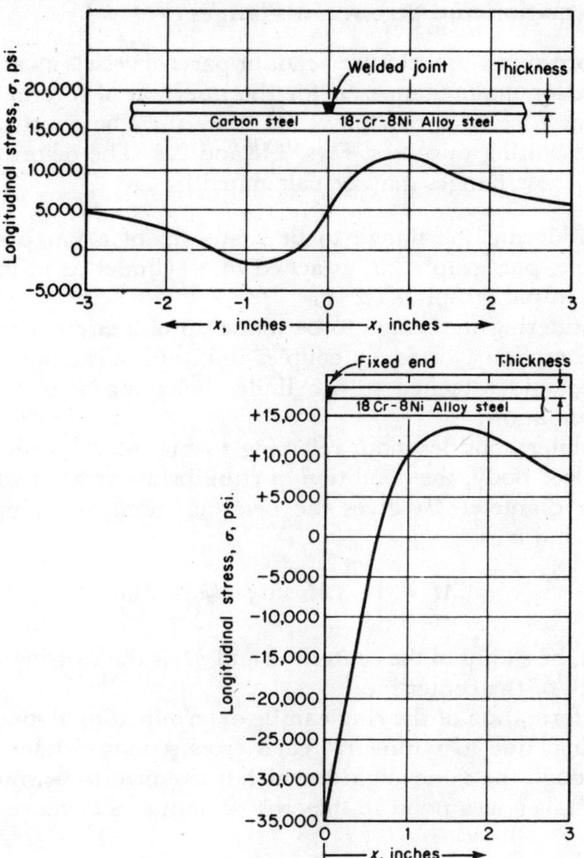

Fig. 4.12. Stress in Bimetallic Joint of Examples 1 and 2, Paragraph 4.8

G. Longitudinal bending stress, noting $I = h^3/12(1 - \mu^2)$ per Eq. 3.2.9,

$$\sigma_b = \pm \frac{6}{h^2}(-0.02EI\beta^2 B_{\beta x} + 0.01EI\beta^2 A_{\beta x})$$

$$= \pm \frac{hE\beta^2}{1 - \mu^2}(-0.01 B_{\beta x} + 0.005 A_{\beta x})$$

H. The total longitudinal stress σ is the sum of the pressure stress, A and the bending stress, G, and is plotted in Fig. 4.12. It is seen that the maximum longitudinal stress in Example 2 with the bimetallic weld located at a point of fixity is appreciably higher than when it is located away from such a point as in Example 1.

4.9 Deformation and Stresses in Flanges

When pressure vessel closure heads or parts of vessels must be readily removable for maintenance, or for the insertion of internals, such as the nuclear core of a reactor vessel, they may be constructed with flanges for bolting purposes, Figs. 1.18 and 2.2. The deformation and stresses in these flanges may be calculated by:

A. Considering the flange to be made up of a flat plate with a central hole, paragraph 3.9, attached to a cylinder as in the manner employed in paragraph 4.7.2; or

B. Considering the flange to be made up of a circular ring of uniform cross section twisted by couples uniformly distributed along its centerline,[1] and attached to a cylinder behaving as a beam on an elastic foundation.

In the later consideration, referring to Fig. 4.13 showing half the ring as a free body, the condition of equilibrium relative to moments about the diameter Ox gives the bending moment acting on each section m and n as

$$M = \int_0^{\pi/2} M_t \sin\phi a d\phi = M_t a \tag{4.9.1}$$

where a is the radius of the centerline and M_t is the twisting couple per unit length of the centerline.

The deformation of the ring can be determined by noting that during twisting, due to symmetry, each cross section rotates in its own plane through the same angle θ, which is assumed to be small. In Fig. 4.13, B is taken as a point in the cross section at a distance ρ from the

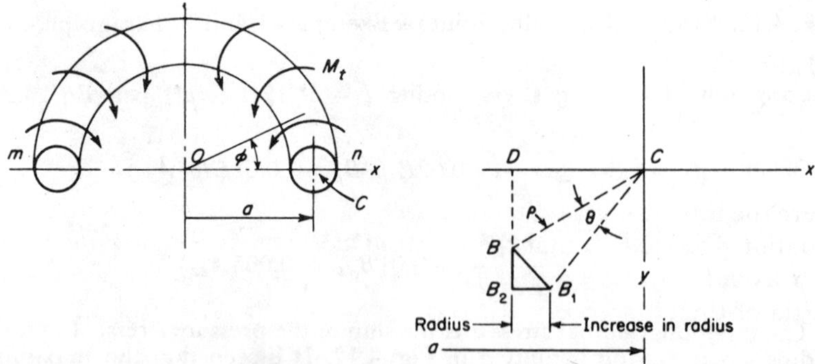

Fig. 4.13. Rotation of a Circular Ring by Uniformly Applied Couples

center of rotation C. During rotation of the cross section the point B describes an arc $\overline{BB_1} = \rho\theta$. Due to this deformation, the annular fiber of the ring (which is perpendicular to the section at point B) increases its radius by B_2B_1. Triangles BB_1B_2 and BDC are similar, so

$$\overline{B_1B_2} = \overline{BB_1}\left(\frac{\overline{DB}}{\overline{BC}}\right) = \rho\theta\frac{y}{\rho} = \theta y \tag{4.9.2}$$

1. *Case I, Ring Dimensions Small Compared with Centerline Radius*

When cross-sectional dimensions are small in comparison with the radius a of the centerline of the ring, the radius of any fiber may be taken equal to a without appreciable error, so that the unit elongation of the fiber B, due to the total displacement given in Eq. 4.9.2, is

$$e = \frac{\theta y}{a} \tag{4.9.3}$$

and the corresponding fiber stress is

$$\sigma = \frac{E\theta y}{a} \tag{4.9.4}$$

Just as in a simple beam, the sum of all the normal forces acting on the cross section of the ring must equal zero, and the moment of these forces about the x axis must equal the externally applied moment M, Eq. 4.9.1. If an elemental area of the cross section is denoted by dA, the first of these equilibrium equations becomes

$$\int_A \frac{E\theta y}{a} dA = 0 \tag{4.9.5}$$

and the second becomes

$$\int_A \frac{E\theta y^2}{a} dA = M \tag{4.9.6}$$

where the integration is extended over the entire cross-sectional area A. Equation 4.9.5 shows that the centroid of the cross section must be on the x axis. If it is noted that $\int y^2 dA$ about the x axis is the moment of inertia of the cross section I_x, Eq. 4.9.6 gives

$$\theta = \frac{Ma}{EI_x} = \frac{M_t a^2}{EI_x} \tag{4.9.7}$$

and substituting this value of θ in Eq. 4.9.4 gives the stress

$$\sigma = \frac{M_t a y}{I_x} \tag{4.9.8}$$

It is seen that the distribution of the normal stresses over the cross section of the ring is the same as in the bending of a straight bar; i.e., the stress is proportional to the distance from the neutral axis x, and the maximum stress occurs at the point farthest from this axis. It is well to emphasize that the twisting of a ring by uniform couples is not the same problem as the torsion of a straight bar. In this case there is no shearing stress on a diametrical section and the moment of inertia involved is the rectangular moment of inertia about the x axis and not the polar moment of inertia as used in problems of torsion.

2. *Case II, Ring Dimensions Not Small Compared to Centerline Radius*

When the cross-sectional dimensions of the ring are not small compared to the centerline radius, the simplifying assumptions of Case I cannot be made.[8,11,12,13] For instance, if we consider the rectangular cross-sectional ring of Fig. 4.14 whose width b is not small compared to the radius a of the centerline, and assume as before that the deformation of the ring consists of a rotation of its cross section through an angle θ, the elongation of a fiber at radius r is

$$e = \frac{\theta y}{r} \tag{4.9.9}$$

and the corresponding stress is

$$\sigma = \frac{E\theta y}{r} \tag{4.9.10}$$

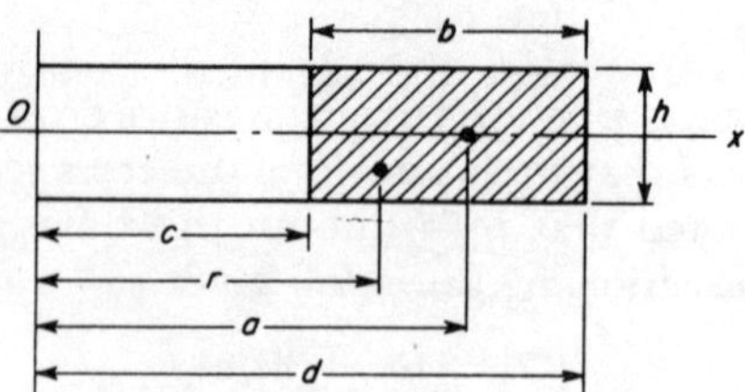

Fig. 4.14. Rectangular Flange

The moment equilibrium equation comparable to Eq. 4.9.6 becomes

$$\int_{-h/2}^{+h/2}\int_{c}^{d}\frac{E\theta y^2 dr\, dy}{r} = M \tag{4.9.11}$$

which upon integration yields

$$\frac{E\theta h^3}{12}\log_e\frac{d}{c} = M \tag{4.9.12}$$

Replacing M with its value from Eq. 4.9.1 gives the angle of rotation as

$$\theta = \frac{12M_t a}{Eh^3\log_e\dfrac{d}{c}} \tag{4.9.13}$$

Substituting the value of θ from Eq. 4.9.13 in Eq. 4.9.10 gives the bending stress

$$\sigma = \frac{12M_t ay}{h^3 r\log_e\dfrac{d}{c}} \tag{4.9.14}$$

The maximum stress occurs at the inner corners of the ring where $r = c$, and $y = h/2$,

$$\sigma_{\text{max.}} = \frac{6M_t a}{h^2 c\log_e\dfrac{d}{c}} \tag{4.9.15}$$

3. *Flange Stresses*

These equations are readily adaptable to calculating the stresses produced in a pipe flange or the closure flange of a vessel. Figure 4.15*a* shows a flange subjected to a force F per unit length of the inner circumference of the vessel. The force per unit length of the outer circumference is then $F(c/d)$. Under the action of these forces the flange rotates through an angle θ, and the wall of the vessel rotates a like amount at the juncture and behaves as a beam on an elastic foundation, Fig. 4.15*b*. Letting M_0 and P_0 be the bending moment and shearing force per unit length of the inner circumference of the flange respectively, their magnitude can be found by the conditions of continuity at the juncture of the flange and vessel. Since generally flanges are very rigid in a plane perpendicular to the axis of the vessel, the

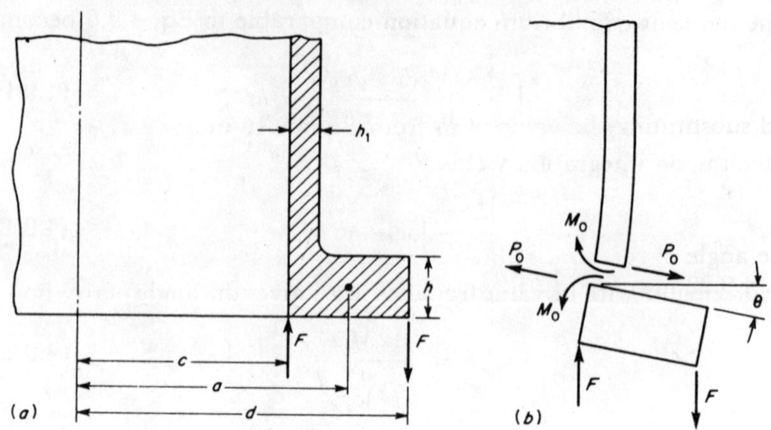

Fig. 4.15. Vessel Flange Rotation

radial displacement produced in the flange by P_0 is negligible and the radial deflection at the edge of the vessel can be considered zero. Then, from Eq. 4.4.6 and letting $D = EI = Eh^3/12(1 - \mu^2)$ from paragraph 4.5, equating the end deflection of the vessel to zero gives

$$\frac{1}{2\beta^3 D}(P_0 - \beta M_0) = 0 \tag{4.9.16}$$

and equating the angle of rotation of the edge of the vessel to the angle of rotation of the flange cross section, θ, gives

$$-\frac{1}{2\beta^2 D}(P_0 - 2\beta M_0) = \theta \tag{4.9.17}$$

From Eq. 4.9.16,

$$P_0 = \beta M_0 \tag{4.9.18}$$

and from Eq. 4.9.17,

$$M_0 = 2\beta D\theta \tag{4.9.19}$$

and

$$P_0 = 2\beta^2 D\theta \tag{4.9.20}$$

The value of β is obtained from Eq. 4.5.6, wherein c is the radius of the vessel and h_1 is its wall thickness. The twisting couple, M_t, per unit length of the centerline of the flange is

$$M_t = \frac{c}{a}\left[F(d - c) - M_0 - \frac{P_0 h}{2}\right] \tag{4.9.21}$$

and substituting the value of P_0 from Eq. 4.9.18 into 4.9.21 gives

$$M_t = \frac{c}{a}\left[F(d - c) - M_0 - M_0\frac{h\beta}{2}\right] \tag{4.9.22}$$

The angle of rotation θ is found by substituting the value of M_t from Eq. 4.9.22 into Eq. 4.9.13, and then from Eq. 4.9.19,

$$M_0 = 2\beta D \frac{12c}{Eh^3 \log_e \frac{d}{c}}\left[F(d - c) - M_0 - M_0\frac{h\beta}{2}\right] \tag{4.9.23}$$

Further, replacing D by its value $Eh_1^3/12(1 - \mu^2)$ gives

$$M_0 = F(d - c)\frac{1}{1 + \dfrac{\beta h}{2} + \dfrac{1 - \mu^2}{2c\beta}\left(\dfrac{h}{h_1}\right)^3 \log_e \dfrac{d}{c}} \tag{4.9.24}$$

When the dimensions of the flange and vessel, and the values of Poisson's ratio and the force F are given, the quantities P_0 and M_0 can be determined from Eqs. 4.9.18 and 4.9.24. The bending stresses in the vessel can be found as in paragraph 4.7, and in the flange from Eq. 4.9.14. The force F is established from the total pressure load over a cross section equal to the inside diameter of the sealing gasket plus the gasket precompression load. In practice this is obtained by measuring the stress in the bolts or studs and establishing an equivalent force per inch of circumference. As an example, the maximum bending stress in the vessel of Fig. 4.15 when $d = 6$ in., $c = 3$ in., $h = 1.25$ in., $h_1 = 0.75$ in., and $\mu = 0.3$ can be found as follows:

A. The damping factor β is from Eq. 4.5.6,

$$\beta = \frac{1.285}{\sqrt{ch_1}} = \frac{1.285}{\sqrt{3 \times 0.75}} = 0.86 \text{ in.}^{-1}$$

B. Substituting the above values in Eq. 4.9.24 gives

$$M_0 = F(6 - 3)\frac{1}{1 + \dfrac{0.86 \times 1.25}{2} + \dfrac{1 - 0.3^2}{2 \times 0.86 \times 3}\left(\dfrac{1.25}{0.75}\right)^3 \log_e \dfrac{6}{3}}$$

$$M_0 = 1.43F$$

C. From Eq. 4.9.18,

$$P_0 = \beta M_0 = 0.86 \times 1.43F = 1.23F$$

D. The maximum bending stress in the vessel is

$$\sigma = \pm \frac{6M_0}{h_1^2} = \frac{6 \times 1.43F}{0.75^2} = 15.25F$$

PROBLEMS

1. In the long thin tube of Fig. 4.16 subjected to a circumferential ring loading of P lb. per in. of circumference, determine the radial deflection directly under the load. Assume the tube is steel, $\mu = 0.3$.

$$Ans.\ \delta = 0.64\frac{P}{E}\left(\frac{r}{h}\right)^{3/2}$$

At what distance from the point of application of the load does the deflection first become zero?

$$Ans.\ 1.83\sqrt{rh}$$

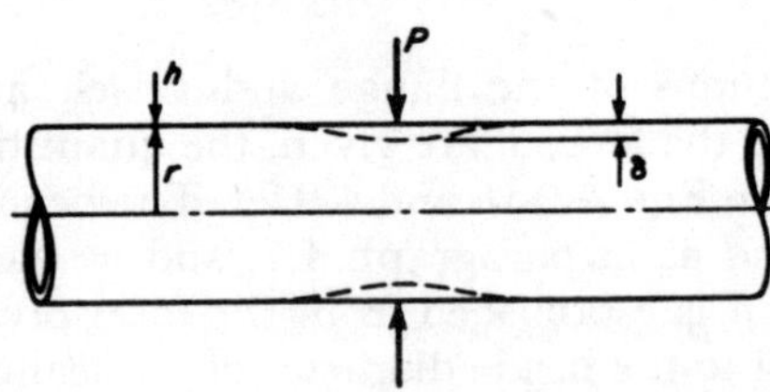

Fig. 4.16. Deflection of Thin Tube Under Circumferential Ring Loading

2. A flat circular head on a cylindrical pressure vessel, similar to Fig. 4.10, is subject to an internal $p = 100$ psi. A discontinuity analysis of the head-to-cylinder juncture gives a value $P_0 = 200$ lb and $M_0 = 100$ in. lb per inch of circumference. What is the total maximum stress occurring at the center of the flat head on the outside surface and inside surface of this head if it is $\frac{1}{2}$ in. thick and its radius is 6 inches?

Ans. +15,800 psi (outside at center)
- 15,000 psi (inside at center)

3. In the example given in paragraph 4.9.3, determine the maximum bending stress in this vessel wall if the flange is assembled with eight 3/4 in. diameter bolts stressed to 10,000 psi in the shank.

Ans. ± 28,600 psi

4. In order to prevent contamination of the coolant fluid by rust particles in nuclear reactor vessels fabricated of carbon steel, the inside of the vessel is frequently clad with

a very thin layer of stainless steel (which does not affect its strength or deflection), and solid wall stainless steel nozzles are welded into the vessel wall because of the difficulty or inability to clad relatively small-diameter nozzles internally. If a stainless steel nozzle (coefficient of thermal expansion $\alpha = 0.000009$) of $r = 20$ in., $h = 1/2$ in., $\mu = 0.3$ and $E = 26{,}000{,}000$ psi is welded into a thick carbon steel vessel (coefficient of thermal expansion $\alpha = 0.000007$) so as to simulate a built-in edge, what is: (*a*) the value of the nozzle characteristic β, and (b) the flexural rigidity D? What is: (*c*) the shearing force and bending moment per inch length of circumference at the built-in edge, and (*d*) what are the maximum thermal stresses at this location when the reactor vessel is operating to give a uniform increase in metal temperature of 300°F? The nozzle is free to expand in its axial direction.

Ans. (*a*) $\beta = 0.4063$ in.$^{-1}$
(*b*) $D = 297{,}600$ lb in.
(*c*) $P_0 = 960$ lb
$M_0 = 1{,}181$ in. lb
(*d*) $\sigma_1 = \pm 28{,}349$ psi (longitudinal)
$\sigma_2 = -24{,}105$ psi (circumferential)

5. Solve Problem 4 assuming that the edge at the juncture of the nozzle and reactor vessel is simply supported.

Ans. (*a*) $\beta = 0.4063$ in.$^{-1}$
(*b*) $D = 297{,}600$ lb in.
(*c*) $P_0 = 480$ lb
$M_0 = 0$ in. lb
(*d*) $\sigma_1 = 0$ psi (longitudinal)
$\sigma_2 = -15{,}600$ psi (circumferential)

6. As part of an external support system for a cylindrical nuclear reactor vessel, a narrow ring of cross-sectional area A is fastened snugly around the outside of the vessel at a distance well removed from the ends. Assuming zero clearance between the outside diameter of the vessel and the inside diameter of the ring, establish an expression for (*a*) the load P_0 per inch of circumference of the vessel, (*b*) the maximum bending moment M_0 in the vessel wall, and (*c*) the maximum bending stress produced in the vessel wall, due to the radial dilation of the vessel of radius r and thickness h resulting from an internal pressure p, all material being steel, and $\mu = 0.3$ [Hint: The total unrestrained outward dilation of the vessel due to internal pressure is given by Eq. 2.4.3 as $\delta = pr^2(2 - \mu)/2hE$. At the ring location this total amount must be absorbed by a local decrease in the radius of the vessel equal to $(P_0\beta/2k)$ (Eq. 4.3.7a and Eq. 4.5.3), and an increase in the radius of the ring equal to (P_0r^2/AE) (Eq. 2.2.3 and Eq. 2.4.1). The maximum bending moment can be found from (Eq. 4.3.9b)].

$$\textit{Ans.}\ (a)\ P_0 = \frac{1.7pA}{\beta A + 2h} = \frac{0.85Ap\sqrt{rh}}{0.6425A + \sqrt{rh^3}}$$

$$(b)\ M_{(\max.)} = 0.1945\sqrt{rh}\,P_0$$

$$= \frac{0.1653prhA}{0.6425A + \sqrt{rh^3}}$$

$$(c)\ \sigma b_{(\max.)} = \frac{1.167 P_0 \sqrt{r}}{\sqrt{h^3}}$$

$$= \frac{0.9919Apr}{0.6425Ah + h^2\sqrt{rh}}$$

REFERENCES

1. S. Timoshenko, *Strength of Materials*, Part II, D. Van Nostrand Co., Inc., New York, N.Y., 1956.

2. H. B. Phillips, *Differential Equations*, John Wiley & Sons, Inc., New York, 1934.

3. M. Hetényi, *Beams on Elastic Foundation*, The University of Michigan Press, Ann Arbor, 1958.

4. A. Pfluger, *Elementary Statics of Shells*, F. W. Dodge Corporation, New York, 1961.

5. G. D. Galletly, "Influence Coefficients and Pressure Vessel Analysis," *ASME Transactions, Journal of Engineering for Industry*, August, 1960.

6. G. W. Watts and H. A. Lang, "Stresses in a Pressure Vessel with a Conical Head," *ASME Transactions*, Paper 51-PET-8, April, 1952.

7. G. Watts and H. A. Lang, "The Stresses in a Pressure Vessel with a Flat Head Closure," *ASME Transactions*, Paper 51-A-146, August, 1952.

8. A. M. Wahl, "Stresses in Heavy Closely Coiled Helical Springs," *ASME Transactions*, Vol. 51, p. 185–200, 1929.

9. W. J. Graff, "Junction Stresses for a Conical Section Joining Cylinders of Different Diameter Subject to Internal Pressure," ASME Publication Paper No. 76-PET-31, 1976.

10. E. C. Rodabauch and S. E. Moore, "Stress Indices and Flexibility Factors for Concentric Reducers," Oak Ridge National Laboratory Report ORNL-TM-3795, February 1975.

11. H. H. Buchter, "Accurate Design Analysis of Flange Joints," ASME Publication Paper No. 76-PET-37, 1976.

12. H. Fessler and D. A. Perry, "Stresses in High-Pressure Taper-Hub Flanges with Recesses for Nuts," *Journal of Strain Analysis*, Vol. 10, No. 2, 1975.

13. H. Fessler and D. A. Perry, "Behavior of Brazed Pipe Flanges with Separate Clamping Rings," *Journal of Strain Analysis*, Vol. 10, No. 2, 1975.

14. D. Redekop and J. Schroeder, "Approximate Predictions of Hoop Stresses at Major Sections of Tee Intersections of Cylindrical Shells Subjected to Internal Pressure," ASME Publication Paper No. 79-PVP-1, 1979.

15. D. P. Updike and A. Kalnins, "Approximate Analysis of Intersecting Equal Diameter Cylindrical Shells Under Internal Pressure," ASME Publication Paper No. 79-PVP-2, 1979.

5

Fracture Control

5.1 Materials and Their Environment

The previous chapters dealt with the methods for analyzing the primary and secondary stresses in pressure vessels. These analyses usually assume that the material follows Hooke's law, so that it is enough to know the modulus of elasticity of the material for solving the problem based on elastic behavior. This is a good start for selecting the material and establishing the dimensions of the structure, but not enough to guarantee safety. The environmental limits under which the material remains elastic for various stress conditions is important. Equally important is the behavior of the material beyond these elastic limits, or the plastic range, which represents a stress range approximately one half the ultimate for annealed structural and pressure vessel steels. Engineering-wise, the structural merit of a material, especially when the member is a complex one and stress concentrations are present, is dependent not only upon its ultimate strength but also upon its plastic properties. It is the latter which permits local yielding in the presence of high peak stresses to give a more favorable stress state, thereby eliminating the danger of failure that would occur in more brittle materials which lack this property. Material specifications for pressure vessel materials recognize this factor and require minimum ductility, or plastic properties, as well as elastic and ultimate strength properties.

5.2 Ductile Material Tensile Tests

1. *Mechanical Properties*

Standard test procedures and size specimens are used to determine the physical properties of ductile materials, such as pressure vessel and structural steels. Of these, the simplest and most widely used is the tension test. This consists of pulling a $\frac{1}{2}$-in. diameter, 2-in. gage length coupon and noting proportional limit, yield point, ultimate strength, elongation in the 2-in. gage length, and reduction in cross-sectional area at failure. Figure 5.1 shows such a stress-strain curve for a mild carbon steel.

The *proportional limit* is the point at which the greatest stress value varies lineally with the corresponding strain. Sensitive extensometers are necessary to establish this point and accordingly this limit depends greatly on this sensitivity, Fig. 5.1. The *modulus of elasticity* is the slope of this curve in this region and, for instance, for this steel is $E = \sigma/e = 30{,}000/0.001 = 30{,}000{,}000$ psi.

The *yield point* is a most important characteristic since it represents the point at which elastic action ceases and plastic flow begins. In the usual vessel and structural steels this is frequently accompanied by an abrupt decrease in the stress, point *b*, Fig. 5.1, followed by considerable elongation, in the order of 2 per cent, with negligible increase in load,

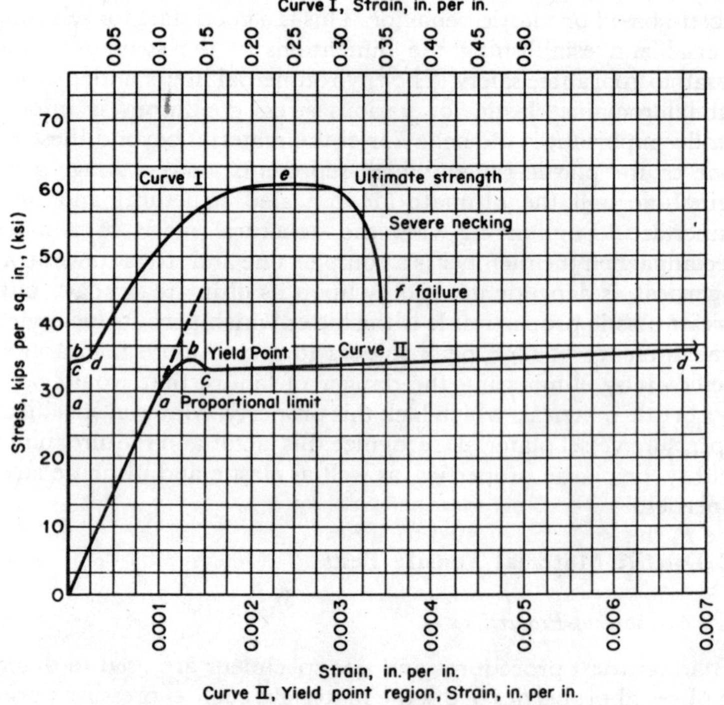

Fig. 5.1. Stress-Strain Curve for Mild Steel (0.25 carbon). Curve II is an enlargement of Curve I in Region of Yield Point

c to *d*. This justifies the simplified straight-line stress-strain diagram assumption of Fig. 1.5*c*. This point is readily recognized by a drop in beam or halt of the gage of the testing machine, or observing the gross stretching period. The upper stress limit in this region, point *b*, is called the *upper yield point* and the lower one, point *c*, the *lower yield point*. In the United States, reference to yield point implies the upper value.

The *ultimate strength* is the maximum stress obtained, point *e*, Fig. 5.1, computed on the basis of load and original cross-sectional area as is the premise for all standard physical test data evaluation. The entire area under the stress-strain curve *oabcdef* represents the amount of work required to produce failure—hence, is in great measure a characteristic of the material, since it depends both on its strength and ductility.

Ductility is measured by the *elongation* of the gage length, and *reduction of area* of the cross section at time of failure. Uniform elongation and reduction in area occur up to the ultimate strength, point *e*, Fig. 5.1, at which time necking begins and further elongation becomes localized (see Figs. 5.24 and 5.25 given subsequently). The increase in the gage length attributable to necking is appreciable, and since this is the same for all gage lengths, the percentage *elongation* (ratio of the total elongation of the gage length to its original length) will increase with decreasing gage lengths. The *reduction in area* is defined as a ratio of the cross-sectional area at time of failure, point *f*, Fig. 5.1, to the original cross-sectional area. Both of these properties obtained in this manner are dependent upon the proportions of the test coupon; hence for comparable results identical size coupons must be used. These properties are a measure of the ductility of the material, and in a sense represent an inherent material abuse safety factor. They are incorporated, along with yield point and ultimate strength, in all material specification requirements for vessel and structural steels.

2. *"True" and "Engineering" Values of Stress and Strain*

In a typical tension member as shown in Fig. 1.2, subjected to a force *P*, the force is assumed to act uniformly over the cross section so the stress is

$$\sigma^1 = \frac{P}{A} \qquad (5.2.1)$$

This is the "true" stress and is defined as the force divided by the instantaneous cross-sectional area *A*. When it is further assumed that the

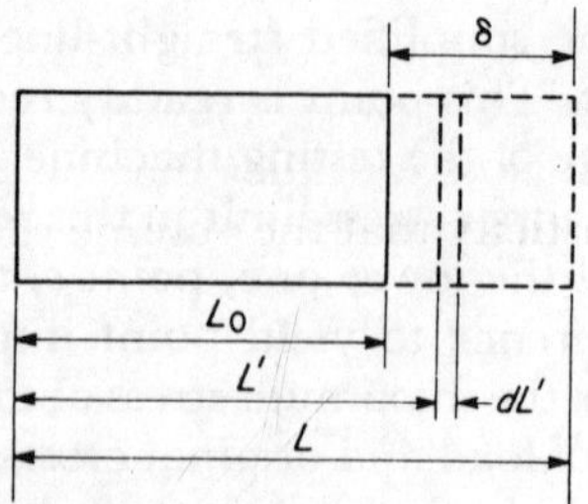

Fig. 5.2. True and Engineering Strain

cross-sectional area remains substantially constant and equal to the original area A_0 the stress becomes

$$\sigma = \frac{P}{A_0} \tag{5.2.2}$$

This is called "engineering" stress and is the force divided by the original cross-sectional area.

The stress causes an elongation δ; that is, an original length L_0 has grown to L as shown in Fig. 5.2, which results in a strain

$$e^1 = \int_{L_0}^{L} \frac{dL^1}{L^1} = \log_e \frac{L}{L_0} \tag{5.2.3}$$

This is called "true" strain and is defined as the integral of the ratio of the incremental change in length to the instantaneous length. Performing the integration over the limits of length from the original length L_0 to the final length L results in an alternate but equivalent definition of true strain as the natural logarithm of L/L_0. When deformations are large, such as occur in the plastic range, the quantity δ becomes large compared with L_0 and it is necessary to consider the change in length during stressing and base the strain on the actual length at any particular instant rather than the original length. In the elastic range the original length L_0 can be considered to remain constant regardless of the imposed stress; hence, Eq. 5.2.3 can be written

$$e = \int_{L_0}^{L} \frac{dL^1}{L_0} = \frac{L - L_0}{L_0} = \frac{\delta}{L_0} \tag{5.2.4}$$

This is called "engineering" strain. It is the ratio of the change in length to the original length.

Engineering stress, Eq. 5.2.2, and strain, Eq. 5.2.4, are used in design since most engineering structures are constructed to keep the applied stress within the elastic limit in which case the "true" and "engineering" values are very close.[1] However, in the plastic range the two values are no longer close, but they may be correlated by using the basic assumption of plasticity that the volume remains constant so that

$$A_0L_0 = AL \tag{5.2.5}$$

$$A = A_0\frac{L_0}{L} \tag{5.2.6}$$

But from Eq. 5.2.4

$$\frac{L}{L_0} = 1 + e \tag{5.2.7}$$

and substituting this value of L/L_0 in Eq. 5.2.6 gives

$$A = \frac{A_0}{1 + e} \tag{5.2.8}$$

Further placing the value of A from Eq. 5.2.8 in Eq. 5.2.1 gives the true stress as

$$\sigma^1 = \frac{P}{A_0}(1 + e) \tag{5.2.9}$$

and replacing the value of P/A_0 from Eq. 5.2.2

$$\sigma^1 = \sigma(1 + e) \tag{5.2.10}$$

which gives the "true" stress in terms of the "engineering" stress. Likewise, combining Eqs. 5.2.3 and 5.2.7 results in a correlation for "true" strain in terms of "engineering" strain.

$$e^1 = \log_e (1 + e) \tag{5.2.11}$$

Deformation may also be measured in terms of the unit reduction in area, a_e, instead of a change in length

$$a_e = -\int_{A_0}^{A} \frac{dA}{A} = \log_e \frac{A_0}{A} \tag{5.2.12}$$

but substituting the value of A from Eq. 5.2.8 in Eq. 5.2.12 gives

$$a_e = \log_e (1 + e) \tag{5.2.13}$$

and since the right side of Eq. 5.2.13 is equal to the true strain, Eq. 5.2.11 becomes

$$a_e = e^1 \tag{5.2.14}$$

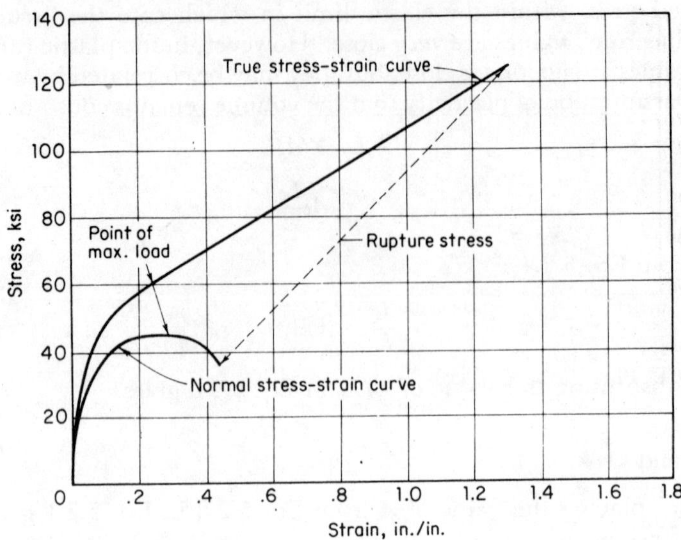

Fig. 5.3. True and Engineering Stress-Strain Curves for a 0.05% Carbon Steel[1]

Figure 5.3 is a comparison of the true and engineering stress-strain curve for a mild carbon steel. It shows an increasing discrepancy in the two values with extension into the plastic deformation process. For instance, the engineering stress-strain curve indicates that beyond the ultimate strength, the stress decreases with increase in strain. This is a false indication of material behavior since the true stress-strain curve shows that as straining progresses more stress develops.

3. *Shape of the Stress-Strain Curve*

The shape of the true stress-strain curve is of particular significance for large deformations encountered in metal forming, and also in appraising the behavior and properties of metals in the plastic range. It has been found experimentally that the true stress-strain curve in the plastic range for many metals plots as a straight line on log-log coordinates, Fig. 5.4, of which the equation is

$$\log \sigma' = \log C + n \log e' \tag{5.2.15}$$

In this equation σ' is the true stress in the plastic range starting at the yield point $\sigma_{Y.P.}$ and e' is the true strain. In Eq. 5.2.15, n is called the "strain hardening exponent" and is the slope of the plotted straight

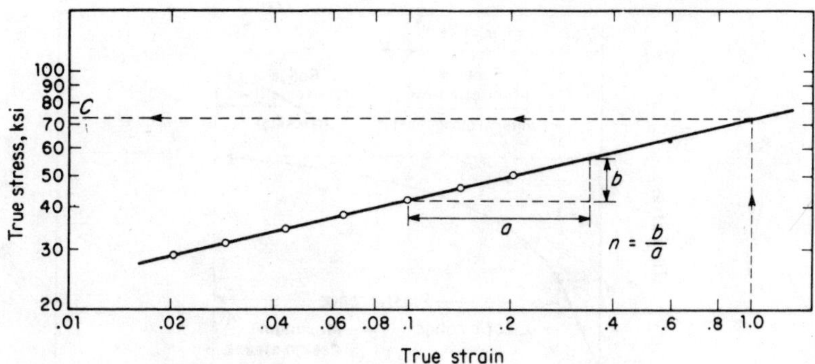

Fig. 5.4. Plot of True Stress-Strain Curve on Log-Log Coordinates for a Mild Steel (0.25% Carbon)

line, and C is the "strength coefficient" which is the true stress corresponding to true strain at 1.0 on this plot, Fig. 5.4. Equation 5.2.15 is the equation of a parabola which can also be written as

$$\sigma' = C(e')^n \tag{5.2.16}$$

A hypothetical stress-strain curve can be drawn as shown in Fig. 5.5. The elastic range is covered by Hooke's law, Eq. 1.1.3, and terminates at the yield point stress, $\sigma_{Y.P.} = \sigma'_{Y.P.}$. Beyond the yield point stress and up to the ultimate stress σ'_u is a range of "uniform elongation" in which the cross section of the member is under uniform elongation and is described by Eq. 5.2.16. Equation 5.2.16 is valid only in the region of uniform elongation. At the stress σ'_u the load-carrying capacity of the material reaches its ultimate; that is, the increase in strength due to strain-hardening just balances the decrease in strength due to the reduction in cross-sectional area. At this stress the material "necks" locally, Fig. 5.5, and becomes unstable with fracture occurring at the true stress σ'_f. A comparison of a true and an engineering stress-strain curve is shown in Fig. 5.3. In the region of instability the true stress continues to increase with strain but the rate of load-carrying capacity F, which is the product $\sigma' A$, decreases; accordingly, σ'_u is the maximum limit of design stress. The point at which this instability occurs is defined as

$$dF = 0 \tag{5.2.17}$$

where

$$F = \sigma' A \tag{5.2.18}$$

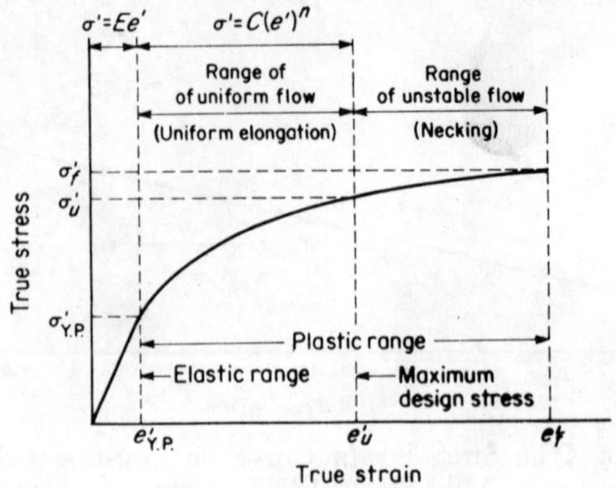

Fig. 5.5. Hypothetical True Stress-Strain Curve

From Eqs. 5.2.11, 5.2.12, 5.2.13 and 5.2.14

$$e' = \log_e \frac{A_0}{A} \tag{5.2.19}$$

and the instantaneous area A is then

$$A = A_0(e)^{-e'} \tag{5.2.20}$$

where (e) is the base of natural logarithms.
Substituting the value of A from Eq. 5.2.20 in Eq. 5.2.18 gives

$$F = \sigma' A_0(e)^{-e'} \tag{5.2.21}$$

Since the load F is a function of both the true stress and the true strain

$$dF = \frac{(\partial F)}{(\partial \sigma')} d\sigma' + \frac{(\partial F)}{(\partial e')} de' \tag{5.2.22}$$

From Eq. 5.2.21

$$\frac{\partial F}{\partial \sigma'} = A_0(e)^{-e'} \tag{5.2.23}$$

$$\frac{\partial F}{\partial e'} = -A_0 \sigma'(e)^{-e'} \tag{5.2.24}$$

Substituting the values given by Eqs. 5.2.23 and 5.2.24 in Eq. 5.2.22 and then equating Eq. 5.2.22 to Eq. 5.2.17

$$0 = A_0(e)^{-e'}(d\sigma' - \sigma' de') \tag{5.2.25}$$

In Eq. 5.2.25 the term $A_0(e)^{-e'}$ can not equal zero, therefore

$$d\sigma' - \sigma' de' = 0$$

or instability is defined by the relation

$$\sigma' = \frac{d\sigma'}{de'} \tag{5.2.26}$$

The value of the true stress is given by Eq. 5.2.16 as

$$\sigma' = C(e')^n \tag{5.2.27}$$

and the first derivative of this is

$$\frac{d\sigma'}{de'} = nC(e')^{n-1} \tag{5.2.28}$$

Substituting Eqs. 5.2.27 and 5.2.28 in Eq. 5.2.26 gives

$$C(e')^n = nC(e')^{n-1} \tag{5.2.29}$$

dividing both sides of Eq. 5.2.29 by $C(e')^{n-1}$

$$\frac{(e')^n}{(e')^{n-1}} = n \tag{5.2.30}$$

$$e' = n \tag{5.2.31}$$

Thus, from Eq. 5.2.31, the moment of instability of flow in uniaxial tension occurs when the true strain e' is numerically equal to the strain-hardening exponent. When the loading is the 2:1 biaxiality of the cylindrical vessel under internal pressure the instability occurs at a circumferential strain of $n/2$ or half of the strain to necking in the tensile test,[2,3] and for the 1:1 biaxiality of the sphere it occurs at a circumferential strain of $n/3$. Accordingly, in the instability (ductile failure) types of bursting the strain hardening exponent is a most significant property.

5.3 Structure and Strength of Steel

In approaching the study of the plastic flow of metals, it is well to consider first the structure and properties of a single crystal. Metals are composed of a random assembly of crystals or grains, which in turn are composed of atoms in a three-dimensional geometric lattice arrangement.[4,5] For instance, iron and steel consist of body-centered cubic cells, Fig. 5.6*a*, in an interlocking structure, Fig. 5.6*b*. The lattice dimensions, such as distance "*a*" in Fig. 5.6*a*, are constant for each material and are on the order of 2.5×10^{-8} cm. The atoms can

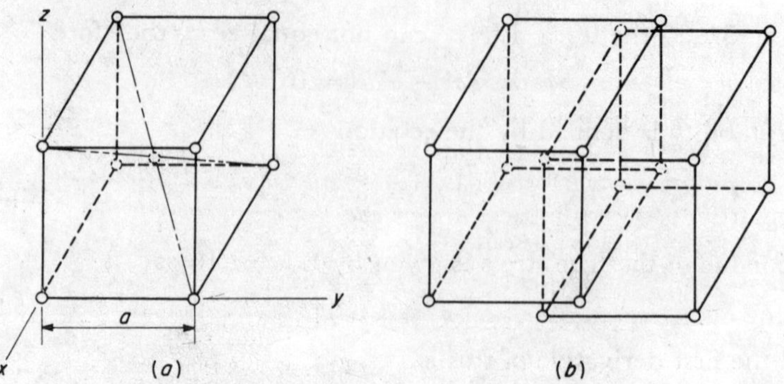

Fig. 5.6. Crystal of Iron Consisting of Unit Body-Centered Cubic Cells (*a*) in an Interlocking Body-Centered Structure (*b*)

be considered to lie in sheets or atomic planes, and external forces are resisted by internal stress components acting perpendicular to these planes (normal stress) and parallel to them (shear stress), Fig. 5.7*a*. Stress normal to the atomic plane tends to pull apart these planes and results in a cohesion or brittle type fracture, Fig. 5.7*b*. Shear stress tends to slide some of the planes relative to the others in the same direction, Fig. 5.7*c*. When a series of sliding planes occurs, Fig. 5.7*d*, a slip band

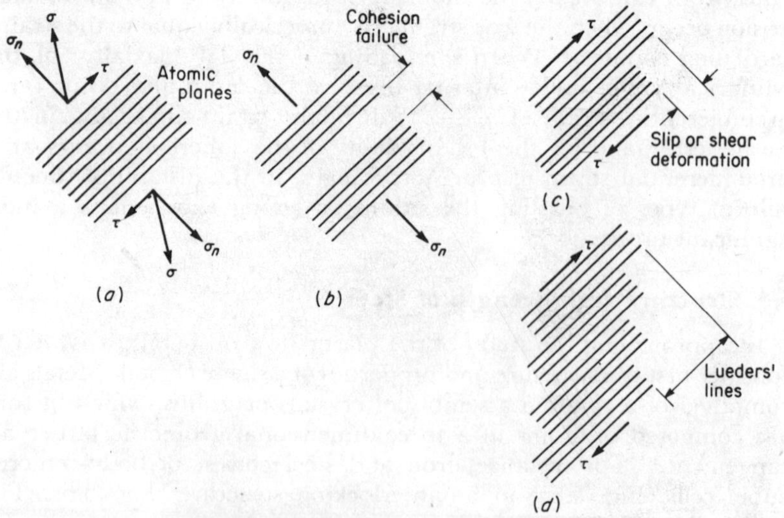

Fig. 5.7. Stresses Acting on a Crystal

is formed which is called a Lueders' Line. This may be visualized much as a stack of coins that is pushed sideways so as to expose part of the face of each coin. Metals contain dislocations or linear defects that move easily when small shear stresses are applied, and it is the movements of these dislocations that result in a shear deformation. The most common form of dislocation occurs when a plane of atoms is missing from a crystal lattice, and is represented by the black *T*-shape in Fig. 5.8*a*. Movement of dislocations accounts for the weakness of metals. When a small shearing stress is applied a simple slip in atomic bonding allows the dislocation to jump one cell to the right, Fig. 5.8*b*. Ultimately the dislocation reaches the edge of the crystal producing a unit slip, Fig. 5.8*c*. Figure 5.9 is a photograph of grain boundaries and dislocations. Many such slips will lead to a visible change in the

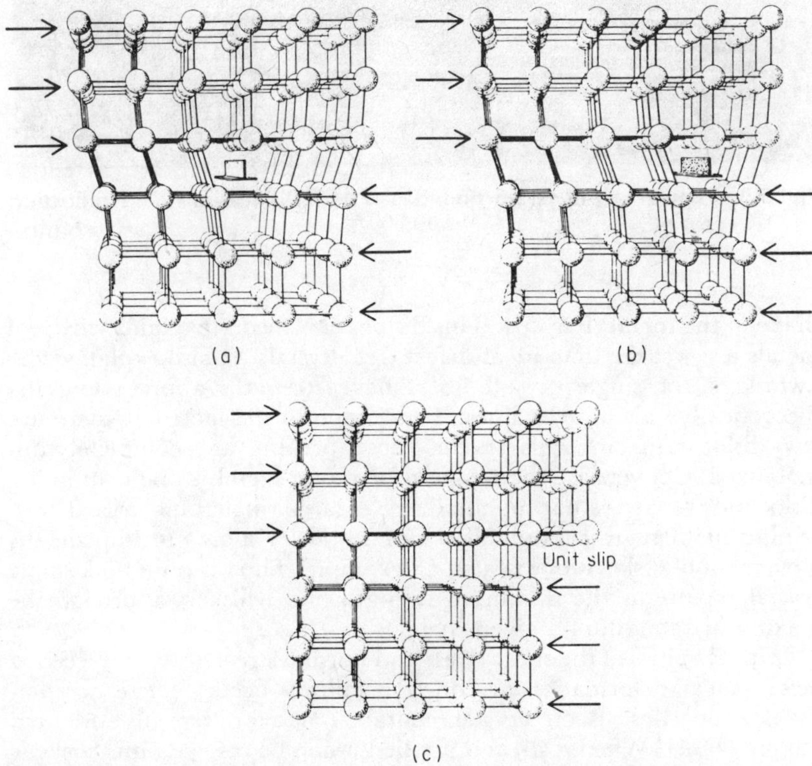

Fig. 5.8. Movement of Dislocation, Represented by the Black T-Shape, Under a Shear Stress to Produce Unit Slip [6]

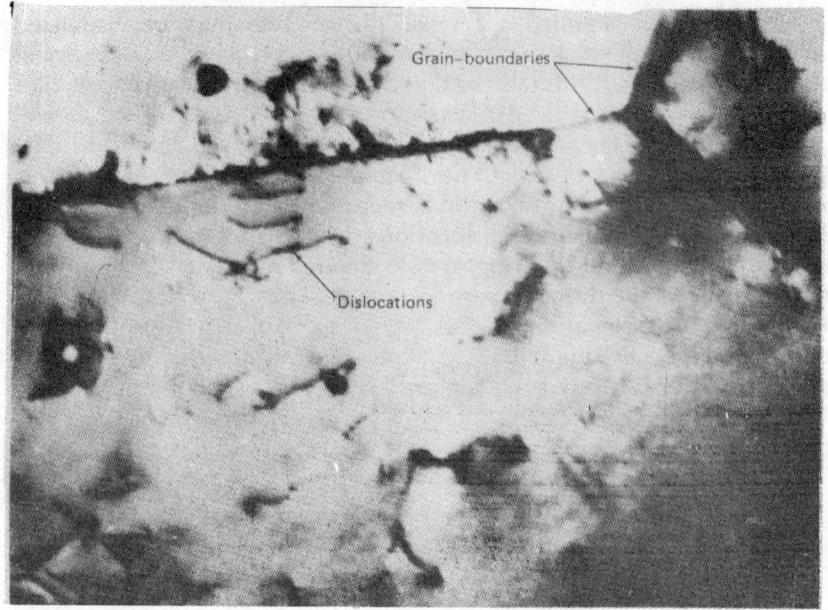

Fig. 5.9. Photograph of Grain Boundaries and Dislocations (Magnification = 40,000)

shape of the metal, Par. 5.4. This dislocation theory explains why real metals are weaker than ideal defect free crystals. It also explains why "whiskers" or single crystal fibers have strengths approaching the theoretically calculated values. This is due to the fact that there are few dislocations in whiskers or those present are completely immobilized. Conversely, the weakest form of a metal is one containing dislocations that are not immobilized. A large single bulk crystal containing neither grain boundaries nor impurity atoms to impede the movement of dislocations is such an example. Thus, a pure bulk single crystal represents the minimum strength and whiskers approach the maximum strength of a given metal.

Critical values of the shear stress and normal stress must be exceeded before shear deformation or cohesion failures occur. These are material properties. Each crystal contains many differently oriented atomic planes. When a stress is applied it can be resolved into a shear component and a normal component occurring on every plane.

Hence, shear deformation will occur when the acting shearing stress component exceeds the critical shear stress value. Likewise, cohesion failure will take place when the acting normal stress component exceeds the critical normal stress. The mode of failure is dependent upon which of these conditions is first reached. Failure seldom occurs without some shear deformation. As shear takes place on one plane it has the effect of increasing the stress on other planes. The same general considerations apply to a polycrystalline metal, with further modifications due to the many differently oriented grains. The relative magnitude of the applied shearing stress and normal stress is determined by the type and direction of the loading. Hence, the behavior of metals under load is governed by two conditions: (1) the material properties, and (2) the applied loading. High applied shear stresses and low critical shear material properties favor deformation and vice versa. Change in behavior from shear deformation to cohesion failure under a given loading condition results in a reduction in prefracture deformation and a decrease in the energy absorbed in the fracturing process. This behavior has been borne out by experiments with single large crystals which show these to respond elastically so long as they are loaded to produce a stress within the proportional limit; but, when this limit is exceeded, sliding occurs along crystallographic planes. Tests of single crystal specimens in the elastic range show considerable variation in the elastic properties, depending upon the orientation of

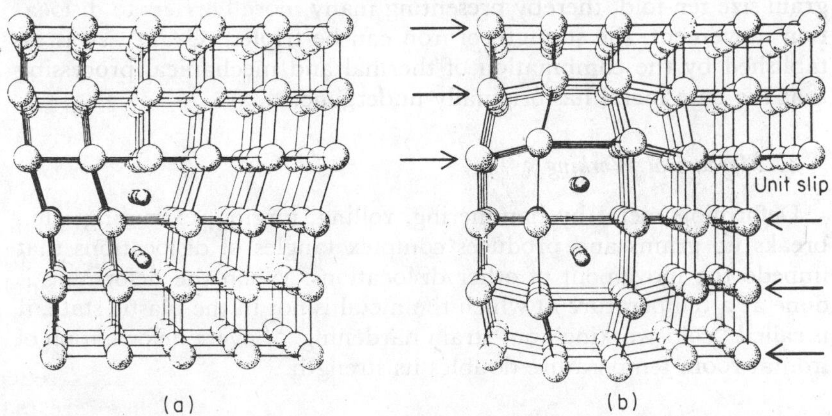

Fig. 5.10. Carbon Strengthening of Iron. (*a*) Carbon Atoms Dissolve at the Site of a Dislocation to Create, (*b*) Stressed Region which Impedes the Passage of Other Dislocations [6]

the crystal. Hence, the crystallographic plane on which the maximum shearing stress acts may not be a 45° plane with the applied load axis.

Metal properties depend basically upon their atomic structure, and the manner in which this structure is altered is the means by which their strengths are established.[6] For instance, steels can be strenghtened to the extent that the movement of atomic dislocations can be immobilized. Five effective methods of accomplishing this are:

1. *Addition of Carbon*

The carbon-strengthening of iron takes place in two steps. First, carbon atoms dissolve at the site of the dislocation, Fig. 5.10*a*. Second, after a unit slip occurs, the carbon atoms create a stressed region which interferes with the passage of other dislocations, Fig. 5.10*b*. The result is an increase in the applied stress required to move them. Increasing the amount of carbon dissolved in pure iron from .0001 to .005 per cent increases the strength of the metal four-fold.

2. *Reduction of Grain Size*

Decreasing the size of the crystal grains in the metal limits the movement of dislocations because, although a dislocation can move through a crystal in which it originates, it cannot easily jump across a grain boundary and propagate itself in adjacent crystals. By decreasing the grain size ten-fold, thereby presenting many more barriers to dislocation movement, the strength of iron can be tripled. Grain size is established by the combination of thermal and mechanical processing that the base metal has originally undergone.

3. *Mechanical Working*

Deforming metal by hammering, rolling, forging, extruding, etc., breaks up grains and produces complex tangles of dislocations that impede the movement of other dislocations. When the deforming is done at a temperature at which the metal is not in the plastic state, it is called "cold-working" or "strain hardening." Severe deformation of iron at room temperature doubles its strength.

4. *Inclusion of Hard Particles or Precipitates*

The dispersal of hard particles or precipitates in alloy steels blocks the movement of dislocations. Steel normally consists of hard iron carbides dispersed in a relatively soft matrix of pure iron ferrite. The closer the spacing of these carbides, the stronger the steel. These steels have useable tensile strengths of 30,000 to 150,000 psi.

5. *Quenching or Quick Cooling*

Strengths greater than those obtainable by Method 4 can be obtained by quenching from a high temperature. The rapid cooling prevents the formation of a carbide-ferrite microstructure of the type obtained in Method 4, and yields a metallurgical structure called martensite. This type of structure can contain many times more carbon in solution than ferrite. The strength of martensitic steels is directly proportional to the dissolved carbon. Their strength will be as high as 300,000 psi when the carbon is 0.4 to 0.6 percent. It is well to mention, however, that this strength is obtained at a sacrifice of ductility and efforts to rectify this loss by heat treatments and alloying are the source of many patented forming processes.

5.4 Lueders' Lines

Structural and pressure vessel steels are conglomerates of randomly located crystals such that the physical properties, as given by the ordinary tensile test specimen, represent the average of the physical properties in various directions of the crystals. Due to the small size and large number of crystals, these average values, provided the material has not undergone strain hardening, are independent of the direction in which the specimen is cut from the material; consequently, such a material may be considered isotropic in such analysis as it is involved. When the material has a pronounced yield point and the tensile stress in the specimen reaches this value, a large plastic flow takes place which is the sliding of a portion of the specimen along planes on which the shearing resistance has been overcome. In an individual crystal these slip bands are evidenced by the shadows of the ridges they form. If the surface has been given a mirror-like polish, these regions of sliding are visible to the eye as a dark band; or the wavy surface can be felt with the finger tips, since now the slip bands occur not only over a microscopic individual crystal but over the entire portion of the specimen subjected to this yield point stress.[7,8] These lines are known as Lueders' lines, Fig. 5.11.

Lueders' lines generally begin at points of stress concentration, since here the yield point stresses are first reached and their propagation direction is influenced by the direction and intensity of these localized high-stress zones. If that portion of a tensile test specimen unaffected by the location of fillets and changes of cross section were of perfect

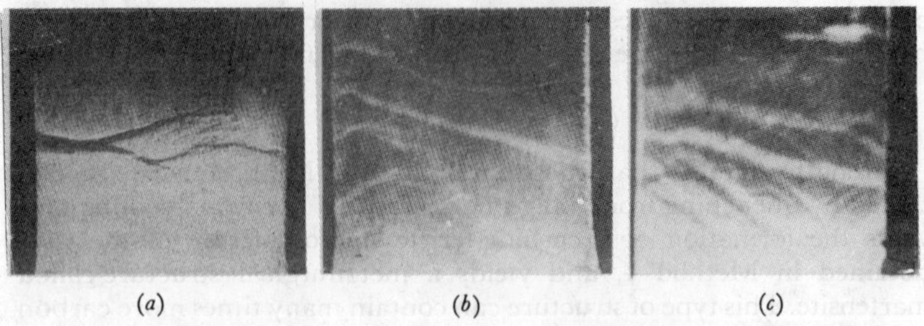

(a) (b) (c)

Fig. 5.11. Lueders' Lines on Polished Carbon Steel Vessel Plate in Tension. First Stage Shown in (a) and (b), and Second Stage in (c)

homogeneity, a yield point stress would be reached in all portions at the same time. Leuders' lines would appear over the entire portion in the same direction—namely, along the line of maximum shear which is 45° to the principal stress. This is not the case, however, but there is a tendency for them to start normal to the edge, Fig. 5.11*a* and *b*, due to the minute scratches and cracks left here by machining, which in turn give rise to high local stresses at these edges. The first lines occur at no regular interval, but once removed from the edge spread like the branches of a tree, first seeking those directions which offer the least resistance to the shearing component on that plane. This first irregular stage of propagation is due to the lack of material homogeneity, and it is this action that causes the first deviation from Hooke's law. On the stress-strain curve this is known as the proportional limit, point *a*, Fig. 5.1. As these planes of weakness are used up, the specimen becomes uniformly resistant and Lueders' lines, taking the direction of maximum shear stress, occur over the entire section, point *b*, Fig. 5.1. This point is known as the upper yield point on the stress-strain curve and is characterized by a drop in the applied load of the testing machine. The extension and widening of the flow lines in the first stage are checked by the strain hardening effect, whereas at the upper yield point or second stage, the extension is too rapid for the strain hardening effect, and one can easily watch the lines spread over the entire member as the stress drops from the upper yield point *b* to the lower yield point *c*, Fig. 5.1. During this stage neighboring portions seem to become unstable when deprived of support by the breakdown of the first, and an orthogonal system of lines which have the general direction of 45° with the load axis occurs, Fig. 5.11*c*. With further strain to

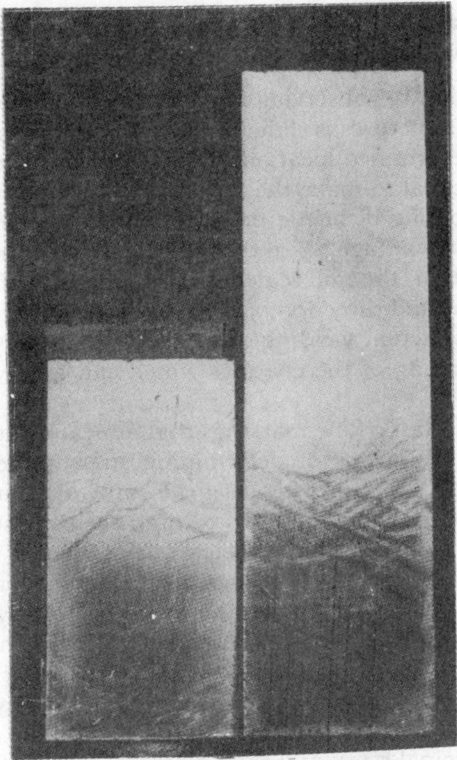

Fig. 5.12. Lueders' Lines on Polished Carbon Steel Plate in Compression

point *d* the line system becomes indistinguishable, indicating all crystals have undergone plastic deformation.

Lueders' lines also occur in the same manner and pattern under compression, as shown on the polished short mild steel column of Fig. 5.12. These surface observations are indicative only of the stress at the surface. If it can be concluded from the shape and loading of the member that the stress is not uniform throughout the cross section, such as occurs under bending or torsion, the internal flow lines may be made visible by cutting a section through the body and etching with Fry's reagent, a solution containing 110 cc of hydrochloric acid, 90 grams of cupric chloride, and 100 cc of water. The plastically deformed layers will etch darker and deeper than those elastically deformed.

5.5 Failure Analysis and Determination of Stress Patterns from Plastic Flow Observations.

The preparation for observing Lueders' lines on polished surfaces of large vessels or structures is difficult, and so such observations are restricted to predetermined locations of small extent. Consequently, it is often more practical to make the flow lines visible on rough and large surfaces by a coating of brittle paint; or the mill scale, if still intact, may be used to advantage.[9,10] In the region where the metal undergoes plastic deformation, the mill scale or paint will flake off, thereby indicating the extent and direction of the flow region. There will be a lag between the first actual yielding of the metal and the yielding as first evidenced by flaking of the coatings which can be evaluated by test coupons.

The nature of plastic flow markings that appear on structural members is a valuable means of determining stress patterns, and after failure offers a means of investigating the cause of failure. Since these flow lines follow the direction of maximum shearing stress, which bisects the angle between the principal stress planes, they form an orthogonal shear trajectory system oriented at 45° to each of the principal stresses. From such a shear stress trajectory system, the lines of principal stress can be drawn so that they intersect the shear lines at 45° and also form an orthogonal principal stress pattern. If the lines are equally spaced in regions of known stress uniformity, a qualitative picture is presented, with the orientation of the lines indicating the direction of principal stress and their closeness the relative stress mag-

Fig. 5.13. Flow Lines in Mill Scale in Carbon Steel Under Uniform Tension

nitude. The peculiar flow characteristics of typical structural details and loadings are helpful in interpretating these systems in composite members.[11]

1. *Plate Under Uniform Tension*

Figure 5.13 shows flow lines in mill scale in the vicinity of a butt weld in a uniformly thick tension member. They appear at 45° to the direction of loading and occur throughout the entire member. Figure 5.16*a*, given subsequently, shows the same oriented lines below the first row of rivets.

2. *Circular Hole in a Plate Under Uniform Tension*

Holes in members are frequent; in fact all riveted construction makes use of such means of fabrication, and all vessels must have openings. The stress distribution in the vicinity of a small circular hole of radius, a, in a plate stretched elastically by a uniform tensile stress, σ, in the direction of the polar axis $\theta = 0$, is given by the following stress components, Fig. 5.14.

$$\sigma_r = \frac{\sigma}{2}\left(1 - \frac{a^2}{r^2}\right) + \frac{\sigma}{2}\left(1 + \frac{3a^4}{r^4} - \frac{4a^2}{r^2}\right)\cos 2\theta \tag{5.5.1}$$

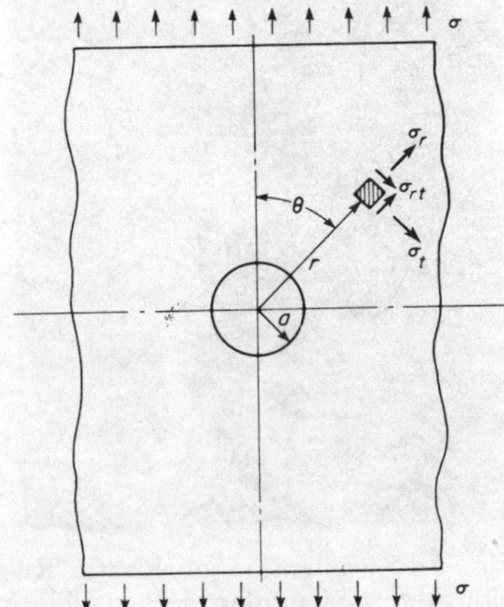

Fig. 5.14. Hole in a Plate Subjected to Tension

$$\sigma_t = \frac{\sigma}{2}\left(1 + \frac{a^2}{r^2}\right) - \frac{\sigma}{2}\left(1 + \frac{3a^4}{r^4}\right)\cos 2\theta \tag{5.5.2}$$

$$\sigma_{rt} = -\frac{\sigma}{2}\left(1 - \frac{3a^4}{r^4} + \frac{2a^2}{r^2}\right)\sin 2\theta \tag{5.5.3}$$

At the circumference of the hole, $r = a$ and $\sigma_r = 0$, $\sigma_t = \sigma(1 - 2\cos 2\theta)$, $\sigma_{rt} = 0$. The tangential stress is a maximum at the points $\theta = \pi/2$ and $3\pi/2$ located on the circumference of the hole, and on an axis perpendicular to the direction of the applied tension. At these points the stress $\sigma_t = 3\sigma$. For $r = a$, and $\theta = 0$ or $180°$, $\sigma_t = -\sigma$. Thus it is seen that a small hole in a plate subjected to tension in a given direction causes an increase in the stress in the vicinity of the hole to a maximum value of three times that in a normal undisturbed portion of the plate. As the plate is further stretched so as to cause yielding to spread, the flow area takes the form of two narrow strips symmetrically located at 45° with the tension axis, Fig. 5.15. This is the merger of two separate

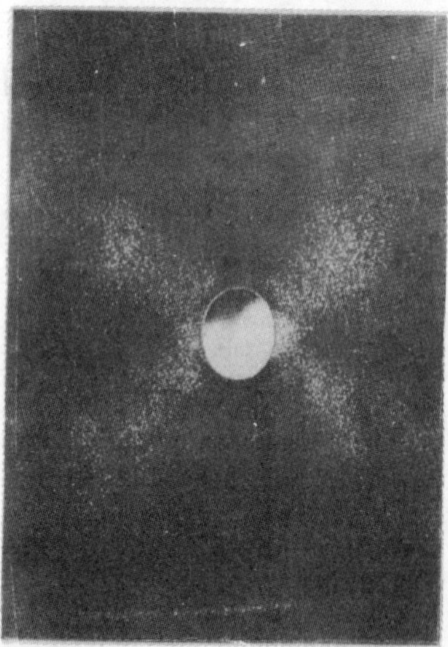

Fig. 5.15. Extension of Plastic Region About a Circular Hole in a Plate Subject to Uniform Tension in the Vertical Direction. Mill Scale on Mild Carbon Steel Vessel Material

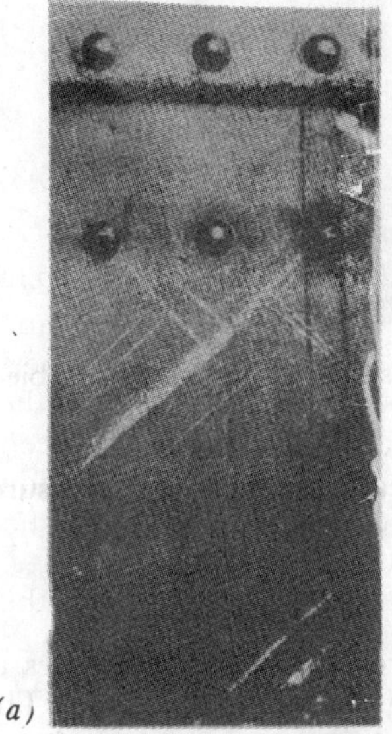
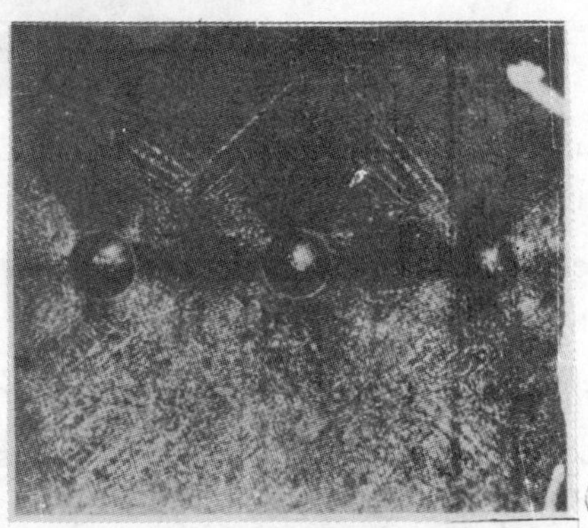

(a) (b)

Fig. 5.16. Flow Lines in a Large Riveted Joint. (*a*) First Appearance, and (*b*) Characteristic Spread Showing 45° Cross Pattern About Holes

flow regions. The first region is a small one at the edge of the hole and on an axis perpendicular to the direction of the load which is caused by the high tangential stress and is evidenced by the bright spots which extend only a short distance from the edge at these locations, Eq. 5.5.2. The second region is a result of the shearing stress and takes the form of a cross oriented at 45° with the direction of the principal stress, where it is a maximum, Eq. 5.5.3, and it extends much further into the surrounding plate. These patterns show up in riveted joints, Fig. 5.16, and around openings in pressure vessels that have been plastically deformed.

3. *Thick-walled Cylinder, or Plate with a Circular Hole, Subjected to Internal Pressure*

Figure 5.17 shows the flow lines in a thick-walled cylinder, or plate with a circular hole, that has been subjected to internal pressure sufficient to develop a plastic state. The principal stresses[7] for such a condition vary logarithmically with distance from the edge of the hole, and, since flow or slip lines make an angle of 45° to these, they too form an orthogonal system of logarithmic spirals. Similar markings may be

Figure 5.17. Logarithmic Spiral Lueders' Lines in a Thick Cylinder Subject to Internal Pressure

observed around rivets that have been headed under high pressure, Fig. 5.18, or in the region of parts that have a press or shrink fit.

4. *Uniform Pressure Along a Narrow Strip*

Loads are frequently applied to pressure vessels and structures in relatively local areas, such as by lifting lugs or support brackets. The principal stress trajectories[7] for such a condition are a system of confocal ellipses and hyperbolas. When the loading is near the free edge

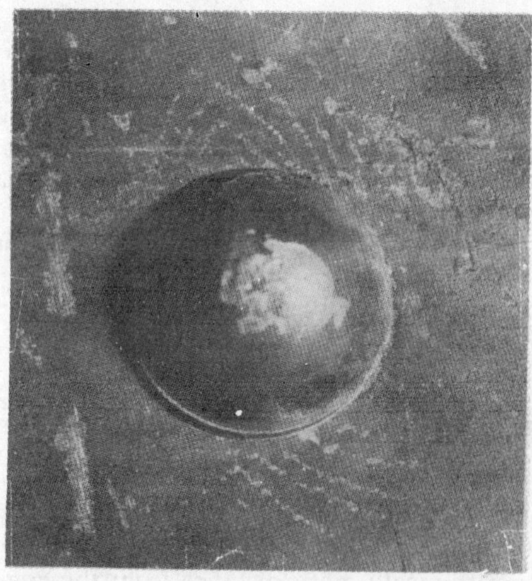

Fig. 5.18. Logarithmic Spiral Flow Lines About a Rivet

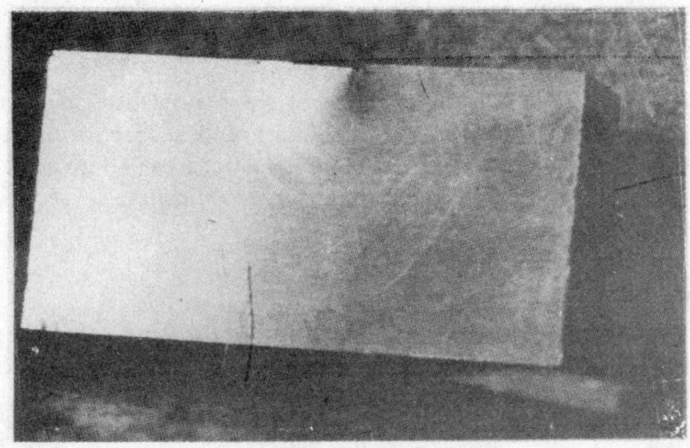

Fig. 5.19. Lueders' Lines Under the Pressure Zone of a Narrow Rigid Punch at Its Free Edge. Polished Mild Carbon Steel

Fig. 5.20. Flow Figures in a Riveted Joint Before Necking and Fracture

of the member, such as would give visible external flow markings, the material is free to flow in the unrestrained direction and a flow pattern similar to that shown in Fig. 5.19 occurs.

Figures 5.16 and 5.20 show some of these flow lines occurring in combination in a large riveted joint. From such observations the complete stress pattern in riveted and welded structures can be ascertained and used as a means of determining cause of failure, or securing design information.

5.6 Dynamic Loading

The shape of the stress-strain diagram at and beyond the yield point depends not only on the size of the specimen tested, paragraph 5.2, but also on the characteristics of the testing machine, of which the most important is the speed of testing.[12,13] Experiments show that not only is the yield point particularly affected, but also that the ultimate tensile strength and total elongation are greatly dependent upon the rate of strain. The curves of Fig. 5.21 show the stress-strain diagrams for mild steel for a wide range of strain rates ($u = de/dt = 9.5 \times 10^{-7}$ per sec to $u = 300$ per sec). In general these properties increase with the rate of strain; and accordingly, to ensure uniformity of material comparison

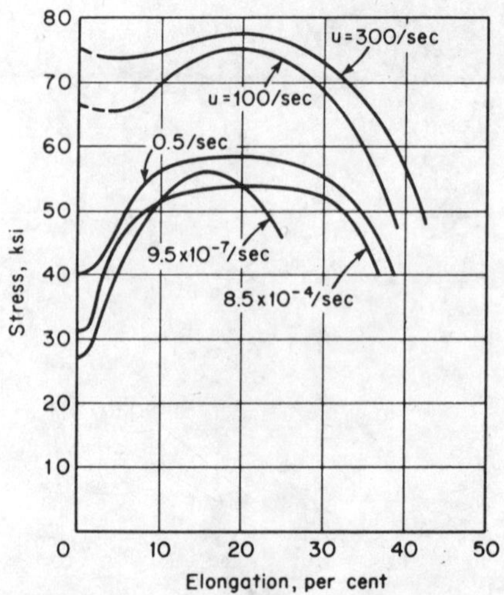

Fig. 5.21. Effect of Strain Rate on Mild Steel[12]

and acceptance tests, a standard testing strain rate is a part of all material specifications.

In the case of dynamic loading this is a very favorable attribute and explains the ability of materials to withstand impact or rapidly applied loading with less distress than would be anticipated from the magnitude of the resulting stresses. For instance, if a load which is just touching a structure is suddenly released, the stress and deflection are double those that would occur if the load were gradually applied. This may be illustrated by Fig. 5.22 in which OA is the tensile test diagram for the material. Then for any displacement OC, the area OAC under the diagram gives the corresponding internal strain energy. When an external load W is suddenly applied, its magnitude remains constant throughout the entire deformation and the work done by W is represented by the area of the rectangle $ODBC$ ($W\delta$); whereas the corresponding internal force increases from zero to a value such that its total energy represented by the triangle OAC ($P\delta/2$) is equal to the applied work. Figure 5.21 indicates that for very high rates of loading, $u = 300$/sec, the yield point more than doubles, which is helpful in preventing permanent structural distortion and malalignment. Most important, however, is the increase in ultimate strength and elongation which give maximum toughness or ability to absorb energy, since this ability is measured by the area under the stress-strain curve.

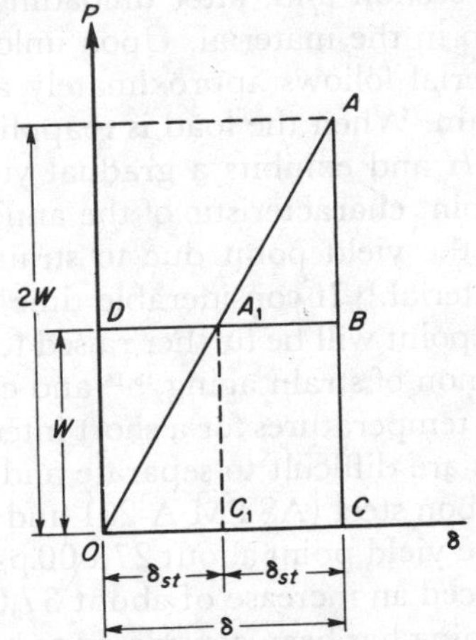

Fig. 5.22. Load-Deflection Diagram Under Impact

Hence, materials which show both an increase in strength and ductility are most useful for structures and vessels subject to dynamic loading.

5.7 Effect of Cold Work or Strain Hardening on the Physical Properties of Pressure Vessel Steels

In the manufacture of pressure vessels, initially flat plate material is frequently shaped into desired vessel contours by a cold-forming operation, and the resulting plastic deformation plays an important role in establishing the mechanical properties of the completed vessel. During this stretching of the material beyond its yield point it hardens and the stress required to continue the stretching increases as shown by the portion *de* of the stress-strain diagram, Fig. 5.1. This phenomena is called "strain or work hardening", and it is seen that the rate of strain hardening, as measured by the slope of the stress-strain curve, is high at low strains and decreases as the magnitude of strain increases. It is a result of redefinition of the crystalline structure proper to a stronger one, as mentioned in paragraph 5.3. During this elongation the work of stretching is not entirely transformed into heat, but part of it is retained in the form of strain energy.

1. *Effect on Yield Point and Ultimate Strength*

Due to the random orientation of the crystals, stresses are not uniform over the cross section and, after unloading, residual stress and strain energy remain in the material. Upon unloading from point *H*, Fig. 5.23*a*, the material follows approximately a straight line *HI*, as shown on the diagram. When the load is reapplied, the yield point is raised to the value *H* and exhibits a gradual yielding rather than a well-defined yield point characteristic of the annealed steel. This represents a raising of the yield point due to strain hardening by prestretching of the material.[14] If considerable time elapses after the first unloading, the yield point will be further raised to the point *H'*. This is due to the phenomenon of strain aging,[15,16] and can be accelerated by soaking at moderate temperatures for a short interval of time. In practice these two effects are difficult to separate and the combined result is used. On mild carbon steel (ASTM A-201 and 285) a prestrain of 5 percent increased the yield point about 27,000 psi, whereas a prestrain of 10 percent produced an increase of about 37,000 psi.[17]

So far the prestraining has been considered to have been in the same direction as the applied loading; if, however, the prestraining has been in the opposite or right-angle direction to the direction of the applied loading, it does not raise the property in this direction but tends to lower the yield point. This phenomenon, wherein a prior strain in one direction lowers the stress required to produce yielding under subsequent loading in the opposite direction, is known as the Bauschinger effect.[18] This is of special concern when the material is subject to reversal of stresses, paragraph 5.11. The ultimate tensile strength is affected by prior strain in the same manner as yield point.[19]

2. *Effect on Ductility*

Material that has undergone prestrain suffers a loss of ductility essentially equal to that consumed by the prestrain, Fig. 5.23. When the prestrain is done at a temperature higher than the final service

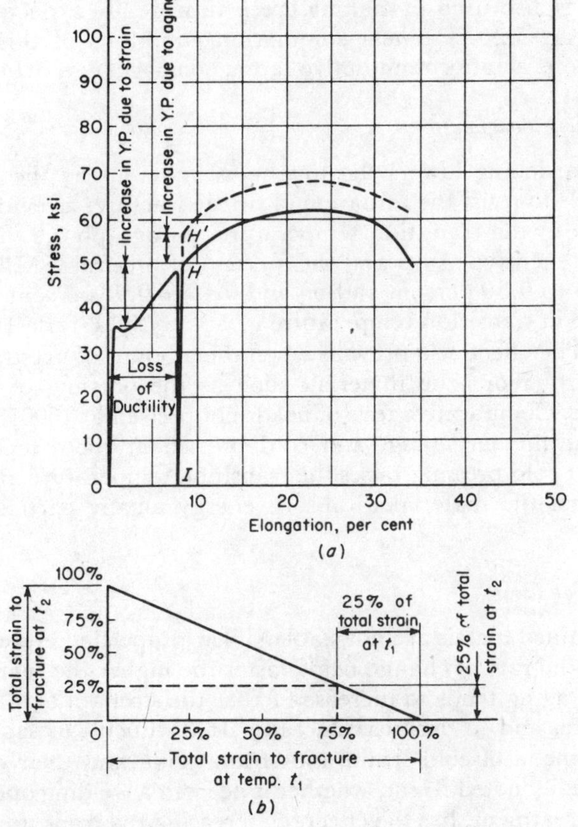

Fig. 5.23. Effect of Strain on Physical Properties

temperature, the percentage of remaining ductility is that which would have been observed if there had been no prior strain at the first temperature.[20] For example, if a strain is applied at temperature 1 corresponding to 75 percent of the fracture strain at that temperature then the additional strain at temperature 2 before fracture will be 25 per cent of the total ductility at that temperature without regard to the prior strain at temperature 1, Fig. 5.23*b*.

This ductility exhaustion can be expressed by

$$\sum \frac{e_i}{e_{Ii}} = 1 \tag{5.7.1}$$

where e_i is the strain for a given stress and temperature and e_{Ii} is the strain at rupture under the same stress and temperature. Equation 5.7.1 is also used in making creep rupture life expectancey predictions, Par. 5.20.2. Linear summation equations of this type are also used to evaluate cummulative fatigue damage, Par. 5.14.

3. *Effect on Toughness*

Strain hardening lowers the toughness and impairs the ability of ferritic steels to resist the initiation of brittle fracture, as evidenced by an increase in the transition temperature, paragraph 5.9. The effect varies with each type steel and the amount of prestrain. Mild carbon steels, 0.15 to 0.30 percent carbon and 0.40 to 0.90 percent Mn, show an increase in transition temperature of 50° to 70°F for a 10 percent prestrain. The effect also prevails when the amount of prestrain is not uniform throughout the material, such as the pressing of spherical vessel heads. Comparative tests of heads hot pressed at 1600°F, so as to eliminate strain hardening, and cold pressed at room temperature showed that cold pressing raises the transition temperature and lowers the ability of the material to absorb energy at any particular temperature.[21,22]

4. *Effect of Heat*

Cold-strained metals are not stable. Their properties tend to change with time, the rate of change being faster the higher the temperature. This strain aging tends to increase further the strength and transition temperature, and to decrease ductility. It is difficult to separate the two phenomena of cold strain and strain aging and their combined result is usually noted. Heat, whether it be from a welding operation or total heat treatment, has the effect of increasing the transition temperature with metal temperatures up to approximately 800°F. Beyond this temperature recovery begins to take place with relaxation of residual stresses, and acceleration of the recrystallization process of the metal.[20] Although complete removal of all strain hardening and strain aging effects can be guaranteed only at annealing or normalizing temperatures, many steels show a major recovery at stress relieving temperatures, 1050° to 1150°F.[23]

The properties of steel plate prior to fabrication into pressure vessels give no assurance of corresponding properties in the completed vessel because of the above effects, nor can the fabricated properties be relied upon, since these too are subject to modification by temperature whether it be from the welding process, thermal stress relieving, or service operating temperatures. Accordingly, it must be recognized that although in general strain hardening increases yield point, decreases ductility, and decreases toughness, they are unstable properties that depend upon the type of steel involved and its specific thermal history.

The use of strain hardened materials in vessels has both advantages and disadvantages, depending upon their application and service requirements. Probably the first criterion for determining the material suitability is the accuracy with which the amount and uniformity of the strain hardening is known. For instance, the cylinders of hydraulic presses are sometimes subjected to an initial internal pressure sufficient to produce permanent distortion in the walls, after they have first been annealed or heat treated to restore the desired uniform properties. The strain hardening and residual stresses that are produced help to prevent permanent set in service. Any pressure vessel subject to external shock loads or internal pressure surges must retain sufficient material ductility to absorb this energy, and have a material transition temperature below its service temperature to preclude brittle failure. It must also be mentioned that material which has been stretched beyond the yield point is more sensitive to corrosion in that region. This phenomenon is broadly termed "stress corrosion" and is of importance in boilers and other vessels subjected simultaneously to stress and chemical action. The preferential rusting readily observed adjacent to weld seams, punched holes, etc., illustrates this type of action.

5.8 Fracture Types in Tension

It has been mentioned in paragraph 5.3 that two kinds of fractures can be observed in a single crystal. A brittle fracture takes place without substantial plastic flow and occurs when the normal stress on one of the principal planes of the crystal reaches the critical value. This is called a "cohesive fracture"—rock salt at room temperature is an example. When large plastic deformation occurs, consisting of sliding along crystallographic planes, prior to fracture, the failure is called a "shear fracture"—copper, iron and most metals are examples.

The relationship of the resistance to these two types of failure does not remain constant for a given material, but is dependent upon the temperature and the speed at which the test is performed. There is evidence that sliding resistance increases with decrease in temperature, and increase in velocity of deformation; hence, the design of structures subject to low temperatures and impact loading is of particular concern, paragraphs 5.9 and 5.22. These two types of fracture are also characteristic of polycrystalline materials. The "brittle type" exhibits little deformation and has a flat fracture surface essentially perpendicular to the direction of maximum stress; whereas, the "shear type" exhibits gross deformation and the fracture surface has the familiar cup-and-cone or jagged pyramid form with shear-lips, Figs. 5.24 and 5.25.[284]

The ultimate strength of ductile metals is determined by a simple tension test. Here the specimen undergoes large plastic deformation and reduction in cross section during the necking stage just prior to rupture. Due to this necking effect, a three-dimensional stress condition exists[24] and the material near the center of the minimum cross section has its ductility so reduced that, during stretching, the crack

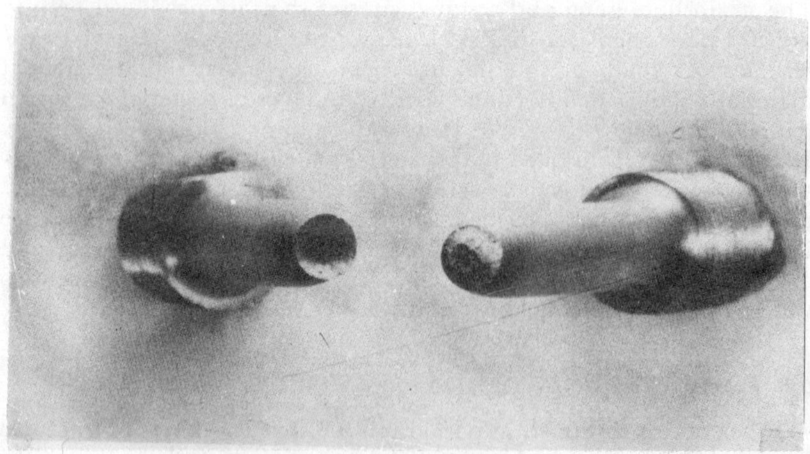

Fig. 5.24. Typical Cup-and-cone Fracture of a Half-Inch Round Test Specimen of Mild Carbon Steel

(a) (b)

Fig. 5.25. Fracture of Rectangular Specimen of Mild Carbon Steel.
(a) Crack Starting in Center of Broad Face
(b) Nature of Fracture Face Showing Curved Boundary

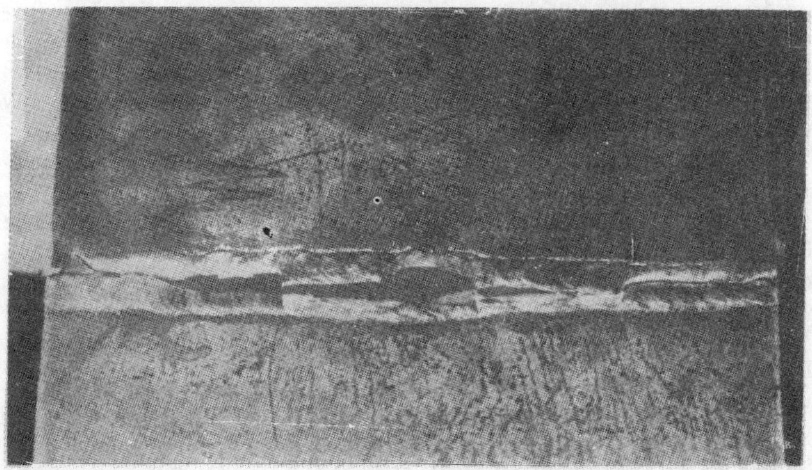

Fig. 5.26. Fracture of Welded Butt Joint in Large Mild Carbon Steel Plate. Fracture starts at Center and Works Outward, as Evidenced by the Larger Gap in the Central Region

starts in this region while the material near the surface continues to stretch plastically. Thus, the central portion of a cup-and-cone fracture exhibits a brittle likeness, while near the surface this is a shear type of failure. This same or like condition holds for the case of a wide thin plate wherein a two-dimensional stress condition is created in the center portion of the cross section and the material near the edges continues to stretch plastically. The curved boundary in the center part of the fractured rectangular bar of Fig. 5.25*b* is a result of this stress condition, as is the crack starting in the center of the broad face of this bar, Fig. 5.25*a*. The same condition prevails in large flat welded members; consequently, failure starts at the center of the member and works outward, Fig. 5.26.

5.9 Toughness of Materials

The "toughness" of a material is its ability to absorb energy during plastic deformation. In a static tensile test this is measured by the area under the tensile test diagram. Thus it is reasoned that in order for a material to have high toughness, it must possess both high tensile strength and large ductility. Impact tests are used to study and evaluate the toughness of materials. In practice, such tests consist of striking a standard notched specimen with a falling weight or pendulum and observing the energy absorbed in fracturing the specimen, or percentage of the fracture surface having a brittle appearance. Brittle materials have low toughness because they undergo negligible deformation before fracture. They are unsuitable for pressure vessel construction, since the material has no ability to redistribute high local stresses or distort under impact, and fracture may occur suddenly without noticeable deformation.

In discussing kinds of fracture in paragraphs 5.3 and 5.8, it was indicated that the same material may behave as a brittle or ductile material, depending upon the applied external stress conditions. The temperature of testing also has a similar effect. For instance, at room temperature a single crystal of salt will exhibit a brittle fracture, whereas the same crystal will deform plastically if tested in hot water. Pressure vessel and structural steels behave similarly; those which show a large plastic deformation in an ordinary tensile test may fracture in a brittle manner if tested at a lower temperature. The temperature at which the material changes from ductile to brittle type fracturing is called the "transition temperature." It can readily be determined from impact tests conducted over a wide range of temperatures, noting that since the amount of work required to produce failure in the case of a brittle fracture is considerably less than that for a shear fracture, a sharp change in the amount of energy absorbed at the transition temperature occurs, Fig. 5.27.[25] The transition temperature of pressure vessel and structural steels becomes less sharply defined as their ultimate tensile strengths increase, but it remains a good guide in appraising their use in fracture-safe design,[26] Par. 5.22.2. The use of brittle materials in structures is dangerous, as is also the use of ductile materials which are subjected to service temperatures below the transition temperature for the material. The numerous failures of welded World War II cargo ships,[27] and pressure vessels attest to this.[28,29]

It is important to determine the transition temperature of the material that is truly representative of that which is in the completed vessel or structure, and assure that it does not coincide with any portion of the service temperature cycle. In the case of a pressure vessel this means performing a series of impact tests on material that is representative of that in the completed vessel. This is done by using specimens cut from the actual material taken from the vessel via access opening cut-outs, etc., or from specimens that have undergone a simulated fabrication and heat treatment cycle duplicating that of the actual vessel. The result is then compared to the intended operating service temperature cycle, including not only the normal condition, but also that which the vessel will encounter from other sources, such as the hydrostatic test in cold weather. The latter may be overlooked in the case of vessels designed for operation at temperatures obviously above the transition temperature; however, they must undergo a hydrostatic or pneumatic acceptance proof pressure test prior to going

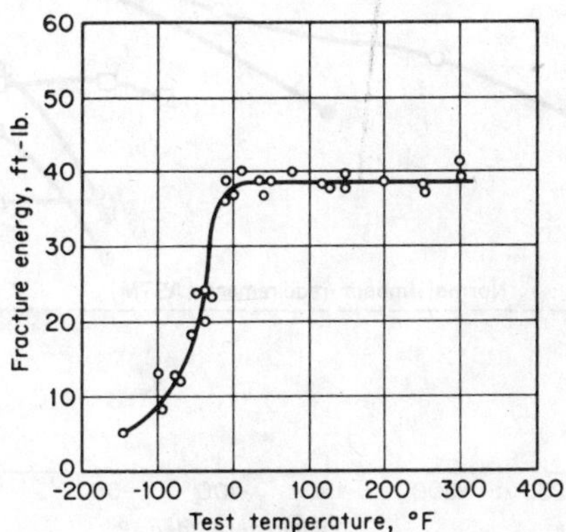

Fig. 5.27. Typical Impact Energy Transition Curve for Carbon Steel ASTM A-212B[25]

into service, and this condition has resulted in brittle failures.[30,31,32] Material toughness can be enhanced by basically two means; namely, changing the chemical composition of the material, and changing the grain structure (size, shape, orientation) by heat or other treatments. In steels, although alloying elements of silicon, copper, and manganese improve impact resistance,[33,34] nickel is the most satisfactory and the main alloying element for low-temperature service,[35] Fig. 5.28. Other materials, such as aluminum, are also used for extremely low temperatures. Fine grain steels have a lower transition temperature than do those of coarse grain;[36] hence, heat treatment to enhance grain refinement and toughness has been found advantageous.

It is current practice to specify minimum impact properties for pressure vessel materials; and as a further safety measure, to add an arbitrary value of 60° to 100° to the transition temperature as the minimum metal temperature for pressurizing (stressing) the vessel throughout its operating cycle and service life.[37] Cryogenic temperatures associated with the storage and transportation of liquified gases, and the neutron irradiation embrittlement,[38,39] par. 5.10, environment encountered in atomic power plants require the selection of material toughness for both initial use as well as degradation throughout its service life. This, together with the continuing use of higher pressures

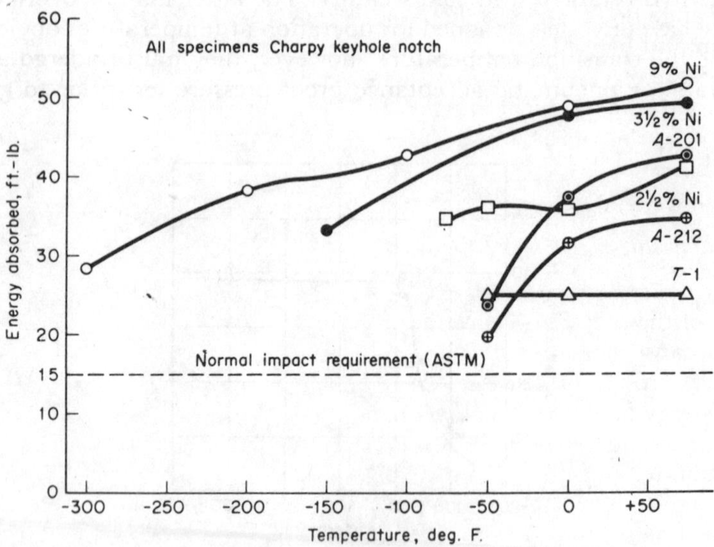

Fig. 5.28. Typical Impact Transition Curves for Low-Temperature Steels Showing the Effect of Nickel[35]

and stresses, has focused considerable attention on designing to prevent brittle fracture, par. 5.22.

5.10 Effect of Neutron Irradiation of Steel

Nuclear reactor vessels are subject to material irradiation induced damage by neutron bombardment from the core. Neutrons are classed as fast (energy greater than one million electron volts, mev) and thermal (energy less than one million electron volts, mev), and these produce two different effects. Fast neutrons cause damage by dislocation or displacement of the atomic structure of the metal, whereas the effect of thermal neutrons is one of transmutation of trace impurities that can materially change the properties of the reactor vessel structural material. The amount of neutron bombardment is measured by the integrated fast neutron flux, nvt (number of neutrons per cu cm × velocity in cm per sec × time in seconds), and is the accepted base parameter for correlating and measuring radiation damage effects.

The main concern with vessel material subjected to thermal neutron absorption is that of a gas-producing reaction that can lead to swelling or gross distortion in ductile materials, or to stress concentration inducing points that lead to embrittlement and fracture. Burst tests[40] of thin-walled inconel tubing have shown that the in-pile life was reduced by as much as a factor of two compared to unirradiated tubing, this change being attributable to transmutation of trace boron to form lithium and helium. In time, helium collects at the grain boundaries, the most likely vacancy source, to weaken the metal. This emphasizes the importance of specifying reactor vessel materials by strict chemical composition to eliminate impurities, and grain size to help reduce the available supply of vacancies, as well as the usual physical performance requirements.

1. *Embrittlement Damage*

Fast neutron irradiation causes damage through dislocations of the atoms of the metal and speeding the resulting vacancy diffusion reaction because it adds vacancies above the equilibrium number. The physical properties of vessel steels are altered by exposure to these high-energy neutrons. Typical changes are a marked increase in yield point, a smaller increase in tensile strength, increase in notch-impact transition temperature and decrease in ductility, and fracture energy. The magnitude of these changes is a function of the material, the total absorbed neutron irradiation, and the irradiation temperature. There is presently no suitable theory of brittle fracture in the presence of radiation, and reliance must be placed on empirical data for design.

TABLE 5.1. PHYSICAL PROPERTIES OF IRRADIATED CARBON STEEL PLATE AND WELD METAL[25]

Material	*Fast Neutron Absorption,* nvt	*Yield Strength, psi*	*Tensile Strength, psi*	*Total Elongation per cent*	*Uniform* Elongation per cent*
ASTM-A-212B (0.35 C. Steel)	0	41,300	75,400	36.0	27.0
	1.7×10^{19}	91,600	98,000	11.6	5.8
	1.0×10^{20}	108,500	115,800	7.0	4.0
E-7016 Weld Metal (C. Steel)	0	57,900	73,200	25.5	15.5
	5.0×10^{18}	69,300	77,500	18.5	10.5
	1.7×10^{19}	108,700	. . . †	8.0	—
	1.0×10^{20}	115,000	—	7.5	—

*Uniform elongation is that prior to onset of necking.
†Load decreased continuously after yielding.

The effect of irradiation on the usual physical properties of a commonly used carbon steel reactor vessel material, ASTM-A-212B and weld metal, is shown in Table 5.1.[25] The trend is typical of carbon and low alloy steels,[281,282] and in many respects resembles like properties created by mechanical strain hardening, whereby strength is obtained at a sacrifice of ductility and can be reversed in the same manner, namely, by annealing. Of particular note is the approach of the yield

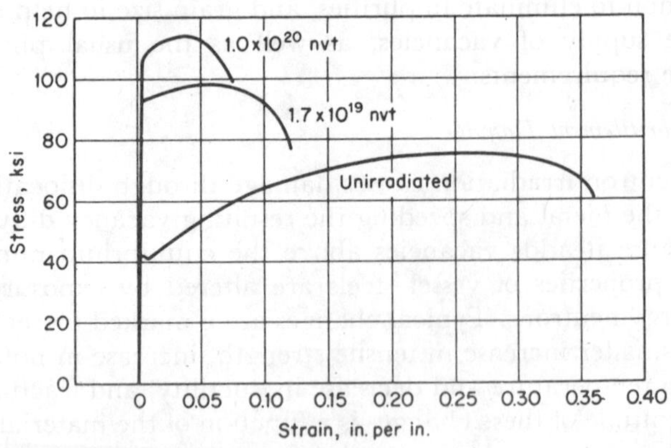

Fig. 5.29. Stress-strain Curves of Irradiated Carbon Steel, ASTM A-212B. Irradiation Temperature 200° F[25]

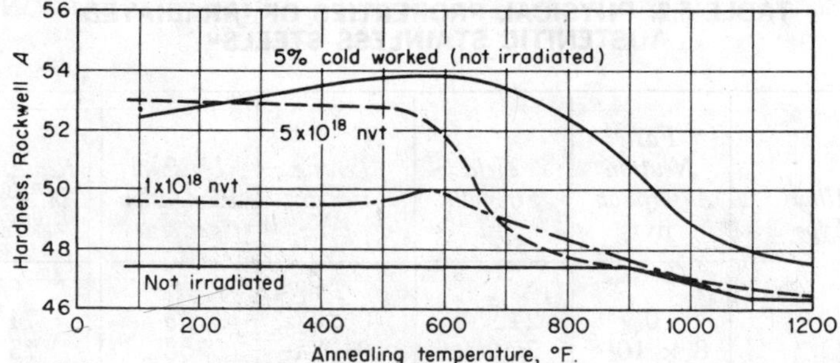

Fig. 5.30. Effect of Annealing Temperature on Irradiated ASTM A-212B Carbon Steel. (Annealing time = 1.5 Hr, Irradiation Temperature = 140°F)[25]

point to the ultimate tensile strength with accompanying gross loss of ductility resulting in a material condition prone to brittle fracture behavior, and of low fracture energy, as measured by the area under the stress-strain curve, Fig. 5.29.[25] Obviously, material in a brittle condition is unsuitable for pressure vessels subject to high alternating stresses of both pressure and thermal origin. A corrective restoration measure is annealing. Figure 5.30[25] shows this effect on annealing out the radiation-induced hardness, along with data for the same steel cold worked 5 percent before annealing. The greater effect of the higher neutron

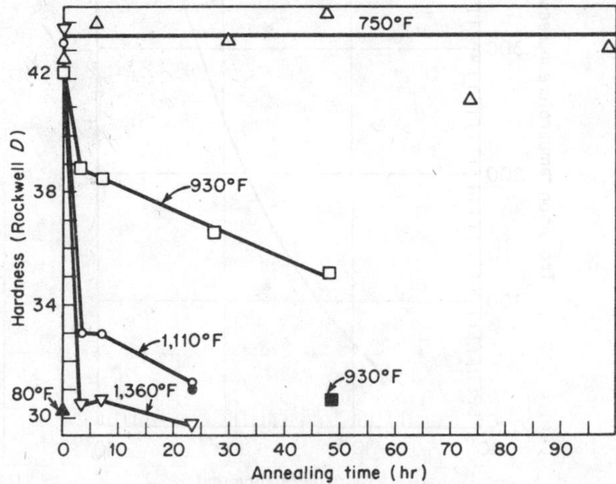

Fig. 5.31. Effect of Annealing Temperature and Time on the Hardness of Type 347 Stainless Steel Caused by Neutron Absorption of 2×10^{21} nvt (Solid Symbols Represent Unirradiated Control Samples)[40,42]

TABLE 5.2. PHYSICAL PROPERTIES OF IRRADIATED AUSTENITIC STAINLESS STEELS[41]

Alloy Type	Fast Neutron Absorption, nvt	Yield Strength psi	Tensile Strength, psi	Uniform Elongation per cent	Reduction in Area, per cent
304	0	24,350	86,300	63	74
	8×10^{19}	75,000	103,800	58	73
347	0	37,000	97,000	49	71
	4×10^{19}	96,500	114,800	25	62

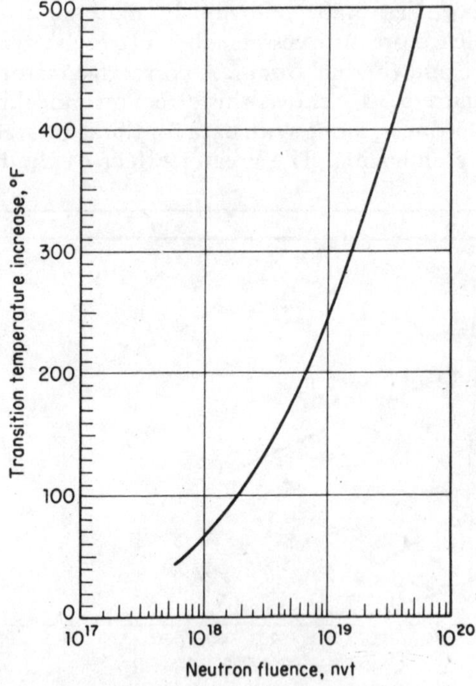

Fig. 5.32. Maximum Effect of Irradiation Embrittlement on the Transition Temperature Increase for Steels Irradiated below 450°F

absorption is apparent, and that recovery or decrease of hardness begins between 500° and 600°F with the recovery more rapid than that for 5 percent cold-worked material. These effects vary with individual materials. The irradiation effect on Type 347 austenitic stainless steel is shown in Table 5.2[41] and the annealing effect in Fig. 5.31.[40,42] These steels exhibit the usual increase in yield point and tensile strength, but retain a high degree of ductility.[41] The physical properties may be recovered by annealing at a temperature higher than that for carbon and low alloy steels, but as low as 930°F.

Characteristic brittleness of carbon and low alloy steels at low temperature has always been of concern in pressure vessels. Under irradiation these steels not only become brittle, but the temperature at which they remain brittle increases appreciably. This increase in material transition temperature is primarily a function of the fast neutron absorption and to a lesser extent the irradiation temperature, and is independent of whether the material is stressed or unstressed during the irradiation.[43,44,45] The effect of neutron absorption on ASTM A-212B and A-302B, two commonly used reactor vessel steels, is shown in Fig. 5.32 and is seen to raise the transition temperature[46,47] for

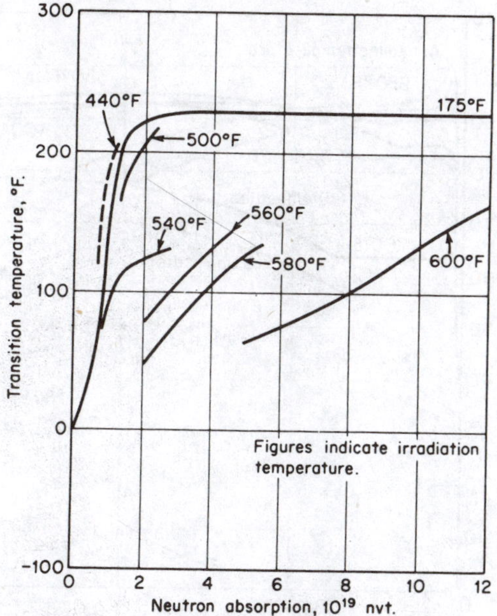

Fig. 5.33. Transition Temperature of Irradiated ASTM A-212B Carbon Steel Plate[25]

fluences above a threshold value normally taken as 1×10^{18} *nvt*. This value is based on observations that changes in properties are small and variations difficult to detect below a neutron absorption of 4.6×10^{18} *nvt*; hence, 1×10^{18} *nvt* represents an acceptable value below which irradiation embrittlement is considered inconsequential.

The temperature at which the irradiation takes place also affects the transition temperature of a material. This effect is illustrated in Fig. 5.33[25] for an ASTM-A-212B carbon steel, and shows irradiation temperatures below the order of 450°F to have little influence, while above this a mitigation countereffect takes place with indications that the effective recovery of the transition temperature is dependent not only on the irradiation temperature but also on the neutron absorption, nvt. This is of practical importance since power reactors and associated vessels operate at elevated temperatures, and this would seem to support the belief that elevated temperatures will anneal out irradiation-induced changes in physical properties.

The effect of annealing temperature[48] on ASTM A-212B steel following irradiation of 1.5×10^{19} *nvt* at 100°F is shown in Fig. 5.34.

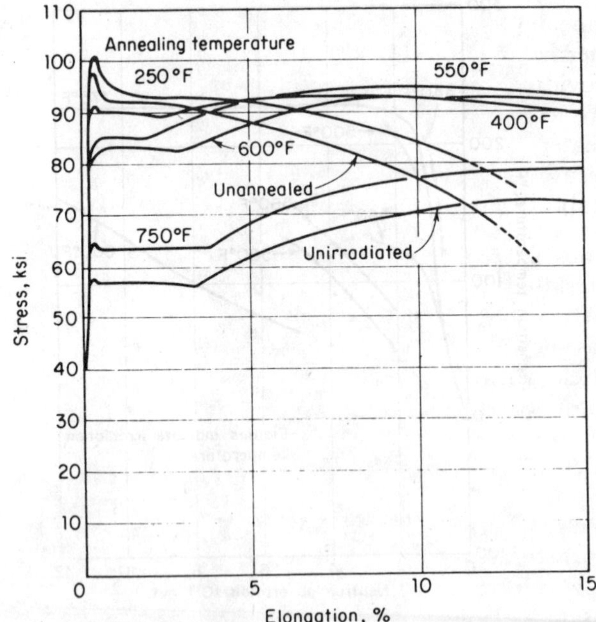

Fig. 5.34. Effect of Postirradiation Annealing on the Stress-Strain Curves of ASTM A-212B Steel[48]

Recovery of ductility and yield strength is substantial above 550–600°F and essentially complete recovery occurs at 750°F. While prediction of the combined effects of neurtron irradiation and in-service temperature poses a complex problem somewhat dependent upon each material composition, there are encouraging observations that for the commonly used carbon and manganese-molybdenum steels and temperatures in boiling water reactors (BWR) and pressurized water reactors (PWR) the neutron irradiation embrittlement effect may be largely self annealing.[49].[283] Although this trend is well established, evidence also indicates that irradiation effects for a given specification material are very susceptible to the residual elements of copper and phosphorous, which should be considered in appraising like specification material data, par. 5.10.2.

The effect of irradiation on fatigue life is discussed in Par. 5.18.5, on creep in Par. 5.20.4, and on hydrogen embrittlement in 5.21.

2. *Embrittlement Control*

The control of irradiation embrittlement damage of pressure vessel steels has focused on three basic methods:

1. Annealing of the vessel to restore the material to its original or intermediate material properties.
2. Providing initial ferritic grain size and metallurgical microstructure to give maximum high initial material toughness; hence, a greater leeway for increases in transition temperature.
3. Restricting residual chemical element content, which largely establishes the sensitivity of steels to irradiation embrittlement.

The most successful of these methods is that of control of the chemistry of residual elements. Studies of the irradiation embrittlement sensitivity, as measured by an increase in transition temperature, of steels has shown the prime importance of maintaining extremely low residual element content,[50–58] especially copper and phosphorus. The steels most widely used in nuclear reactor vessels, where this phenomenon occurs, are the carbon and manganese-molybdenum types. Figure 5.35 shows the effect of residual copper, while Fig. 5.36 shows that of phosphorus, on the ductile-brittle transition temperature of these steels.

The practical and economic significance of employing steels with initial low transition temperatures is that by so doing it is possible to compensate for operational restrictions in later vessel life and thereby avoid the expensive and time-consuming task of annealing the vessel to restore material toughness.

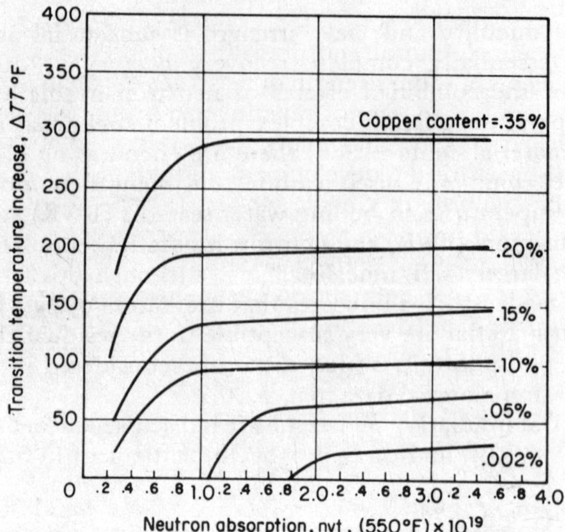

Fig. 5.35. Effect of Copper on the Transition Temperature of Manganese-Molybdenum Steel Plate

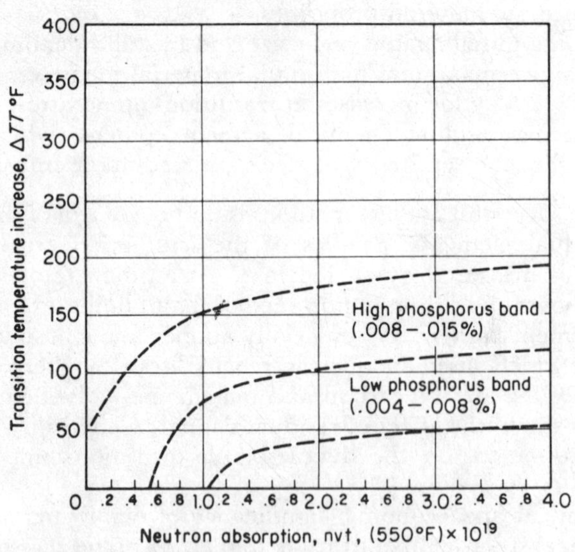

Fig. 5.36. Effect of Phosphorus on the Transition Temperature of Manganese-Molybdenum Steel Plate Having 0.16–0.22 Per cent Copper

5.11 Fatigue of Metals

A century ago structural and pressure vessel design was based entirely on concepts of static strength of materials. This proved adequate since there were few sources of repetitive stressing, in comparison to those which exist today, and many parts were designed with large factors of safety. With the development and use of power machinery, unexplainable failures of ductile materials well below the tensile strength, but exhibiting no gross plastic deformation, arose which were ascribed to "fatigue" since they usually occurred after a length of service. Later experiments showed the important factor is stress repetition, rather than duration of time, and that "fatigue failures" are "repeated stress failures." Improvements in design efficiency and economy require components to operate at high levels of both static and repeated stress; and accordingly, designing to resist fatigue failure is a major engineering concern, because it is the commonest cause of service failure.[285]

Fatigue failures are characterized by a fracture which involves little plastic flow, and is transgranular in nature as compared to intergranular which is characteristic of stress rupture failures. The fatigue process may be divided into three main stages: crack initiation, crack propagation to critical size, and unstable rupture of the remaining section. Figure 5.37 gives an example of a fatigue fracture showing the second and third stages. The crack initiated at a surface flaw and spread from this location during cycling until the section was sufficiently reduced for a final tensile fracture to occur. The second stage region has a "ground" or "rubbed" surface and frequently has "oyster shell" markings which focus on the origin or nucleus of failure, while the third stage region has a more jagged texture representing a tensile fracture. The presence of "oyster shell" markings is typical of a fatigue fracture. Many fatigue tests of metals have provided the following general observations of the behavior of metal structures that are useful in coping with the problem of design, construction and research to improve their fatigue resistance. These are elaborated upon in subsequent paragraphs.

1. Failure at much lower than the ultimate tensile stress occurs in most metals that exhibit some ductility in static tests, and the magnitude of the applied alternating stress range is the controlling fatigue life parameter.

2. Failure depends upon the number of repetitions of a given range of stress rather than the total time under load. The speed of loading is a factor of secondary importance except at elevated temperatures.

Fig. 5.37. Fatigue Fracture of a Shaft Showing Typical Oyster-Shell Marking Concentric with the Origin of Failure

3. Most metals have a safe range of stress, called the "endurance or fatigue limit," below which failure does not occur irrespective of the number of stress cycles.
4. Notches, grooves, or other discontinuities of section, including those associated with surface finishes, greatly decrease the stress range that can be substained for a given number of cycles.
5. The range of stress necessary to produce failure in a fixed number of cycles usually decreases as the mean tension stress of the loading cycle is increased.

5.12 Fatigue Crack Growth

It is now recognized that fatigue is a result of plastic deformation,[59,60] crack initiation and growth; and the principles of fracture mechanics, Par. 5.22, may be used to predict fatigue behavior. Stress produces slip lines in the crystals of metals[5,61] that develop into small cracks which subsequently grow, join others, and result in a fracture exhibiting no gross plastic deformation. Accordingly, they start at points of high stress concentration where sharp notches, material defects, etc., serve as points of nucleation. Once a crack has been initiated, it advances a finite amount with each loading cycle. At the start of the loading cycle the crack tip is sharp, but during extension and creation of an advancing plastic zone, it becomes blunted. The effect is a balance between the applied stress and amount of plastic extension at the crack tip which establishes the crack growth rate, Crack growth continues until the crack becomes large enough to trigger final instability. In brittle materials this means a fast running crack, Par. 5.22; while in ductile materials it means the remaining cross-sectional area can no longer support the applied load and a slow ductile shear type rupture occurs, Par. 5.8.

Applications of stress to a material containing a very sharp crack results in plastic deformation of the material about the crack tip. As the applied stress increases, the zone of plastically deformed material expands and the crack tip radius increases until a characteristic radius associated with the K_c material fracture toughness value for the material is reached, Par. 5.22.3. Once a plastic zone of critical size for the fracture has been developed at the crack tip, each succeeding application of stress in the crack-opening direction will cause extension of the crack and simultaneous motion of the plastic zone boundary in the direction of crack extension as evidenced by striations on the fracture surface. The distance the crack front advances each cycle is a function of the stress intensity factor range, ΔK.

The fatigue crack growth phenomenon is localized in the small volume of material about the advancing crack tip, and it is the local stress field range which determines crack growth rate. The local stress is defined in terms of the applied stress and the crack length, or the corresponding stress intensity factor, K. Hence, crack growth relates to the stress intensity factor range and because experimental data correlating these parameters plotted as straight lines on log-log coordinates, Paris[62] has suggested the simple power law

$$\frac{da}{dN} = C(\Delta K)^m \tag{5.12.1}$$

where

a = crack size, in. (depth of surface crack, one-half length of internal cracks, one-half the length of a through thickness crack).

a_i = initial crack size, in.

a_c = critical crack size, in.

N = fatigue life (number of cycles)

ΔK = range of stress intensity factor, ksi $\sqrt{\text{in.}}$

$\frac{da}{dN}$ = crack growth rate, in. per cycle

C = material intercept constant relating crack growth rate to stress-intensity range and crack size, and to be determined by tests. (The ordinate value of da/dN at $\Delta K = 1$ on a log-log plot, Fig. 5.38.)

m = material constant equal to the slope of the log da/dN versus logΔK curve determined by tests. (See Fig. 5.38.)

Q = crack shape factor

Since m was observed to be approximately 4 for all the ferrous and nonferrous metals commonly used in pressure vessel construction, Eq. 5.12.1 can be written

$$\frac{da}{dN} = C(\Delta K)^4 \tag{5.12.2}$$

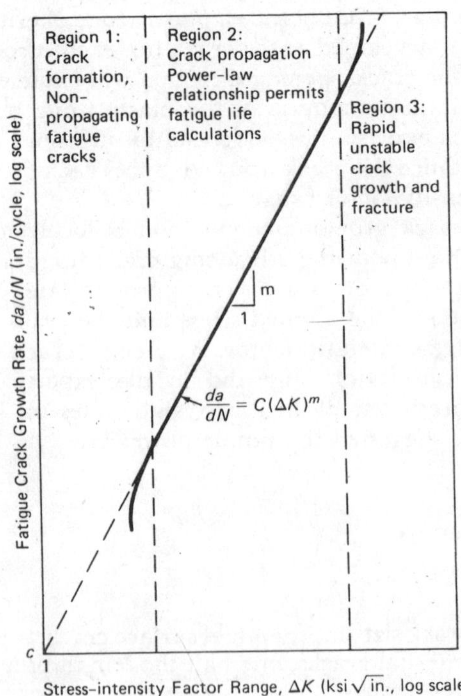

Fig. 5.38. Sigmoidal Fatigue Crack Propagation Curve[63]

and this equation is known as the fourth power law. The value of C varies with each material and its environment. The average value for all steels in air and room temperature is 4.0×10^{-24}. More recent research[63–68] has shown that the entire fatigue crack growth rate curve for most materials is of a sigmoidal form when plotted on log-log coordinates, as shown in Fig. 5.38. Pressure vessel steels are examples of this, Fig. 5.39. The power-law portion of the curve is limited by upper and lower inflection points. The lower inflection point indicates non-propagating cracks and occurs under very low stress intensities where the crack growth rates are of the magnitude of the atomic spacing of the crystal lattice (1×10^{-7} to 10^{-8} in./cycle). This threshold is approximately $K = 25$ ksi $\sqrt{\text{in.}}$ for ASTM-A533 steel which is widely used in pressure vessel construction. This is the endurance limit of the material. The upper inflection point is caused by the onset of rapid unstable crack extension prior to terminal fracture and places a critical limit on the fatigue resistance of the material.

Since pressure vessel service usually involves a limited number of pressure cycles, it is not necessary that the stress and crack size be kept below the threshold for propagation.[215,216] Accordingly, crack propagation[217,218,219] data can be used to determine the ΔK value that can be tolerated; or conversely, the number of cycles required to extend

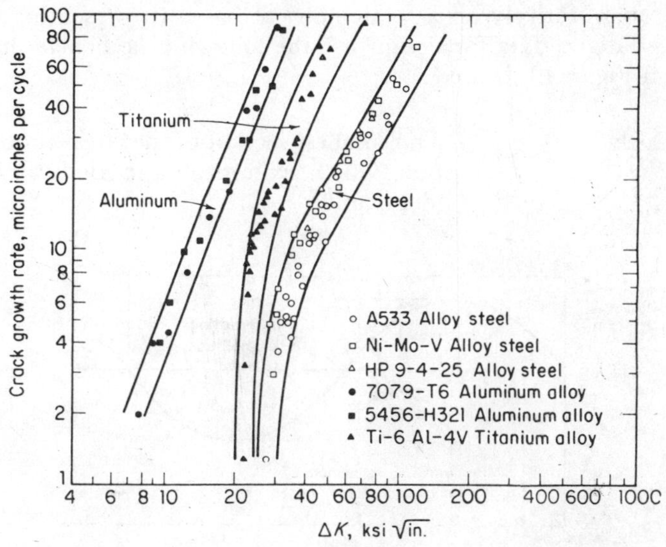

Fig. 5.39. Crack Growth Rate for Various Materials[63]

an initial crack size, a_i, to a critical size, a_c, can be established. Substituting Eq. 5.22.13 in Eq. 5.12.2 gives:

$$\int_{a_i}^{a_c} \frac{da}{a^2} = CQ^2\pi^2\Delta\sigma^4 \int_0^{N_F} dN \tag{5.12.3}$$

$$N_F = \frac{1}{CQ^2\pi^2\Delta\sigma^4}\left(\frac{1}{a_i} - \frac{1}{a_c}\right) \tag{5.12.4}$$

In applying this equation it is interesting to note that the typical flaws that occur in pressure vessels, such as surface flaws, are usually of similar proportions; hence, the constant Q is always approximately the same. However, the constant C that depends on the material and its environment varies widely. For instance, the presence of boiler water[220] will increase the crack growth rate of mild steel by a factor of 2.5; and with salt water this factor is 3.0.

5.13 Fatigue Life Prediction

Since fatigue failure involves the cumulative effect of numerous small-scale events taking place over many cycles of stress and strain and under various service environments, it is difficult to make predictions of the fatigue lifetime. However, certain aspects of fatigue can be treated quantitatively on a semiempirical basis. In presenting or organizing fatigue data for design use the following nomenclature, depicted in Fig. 5.40, is used:

Stress Cycle — The smallest section of the stress-time function which is repeated periodically and identically.

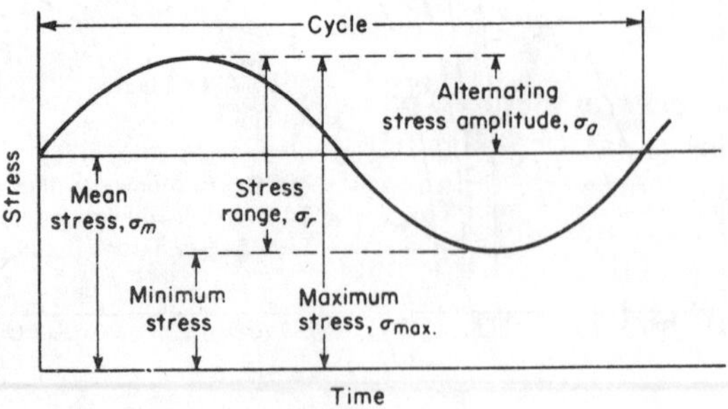

Fig. 5.40. Fatigue Cycle Nomenclature

Nominal Stress σ — The stress calculated by simple theory without taking into account variations in stress caused by geometrical discontinuities, such as holes, grooves, fillets, etc.

Maximum Stress, $\sigma_{max.}$ — The highest algebraic value of stress in the cycle.

Minimum Stress, $\sigma_{min.}$ — The lowest algebraic value of stress in the cycle.

Stress Range, σ_r — The algebraic difference between the maximum and minimum stress in one cycle, $\sigma_r = \sigma_{max.} - \sigma_{min.}$

Alternating Stress Amplitude, σ_a — One half the stress range, $\sigma_a = \sigma_r/2$.

Mean Stress, σ_m — The algebraic mean of the maximum and minimum stress in one cycle, $\sigma_m = (\sigma_{max.} + \sigma_{min.})/2$.

Stress Ratio, R — The algebraic ratio of the minimum stress to the maximum stress in one cycle, $R = \sigma_{min.}/\sigma_{max.}$

Stress Ratio, A — The algebraic ratio of the stress amplitude to the mean stress, $A = \sigma_a/\sigma_m$.

Stress Cycles Endured, n — The number of cycles endured at any stage of life.

Fatigue Life, N — The number of stress cycles which can be sustained for a given condition.

σ–n Diagram — A plot of Stress against number of cycles to failure.

Endurance Limit, σ_E — The value of stress below which a material can presumably endure an infinite number of cycles. This is the stress at which the σ–n diagram becomes horizontal or asymtotic thereto.

There are several methods of applying the load in endurance testing. The specimen may be subjected to direct tension and compression, to bending, to torsion or a combination of these. The simplest, and most frequently used method, is the rotating reversed bending test.[69] This consists of applying a load at the end of a standard cantilever fatigue test specimen which is rotated at constant speed, thereby creating full reversed bending stresses with each revolution. Data from such tests are usually reported as σ–n curves, and it is usual practice to plot $\sigma_{max.}$

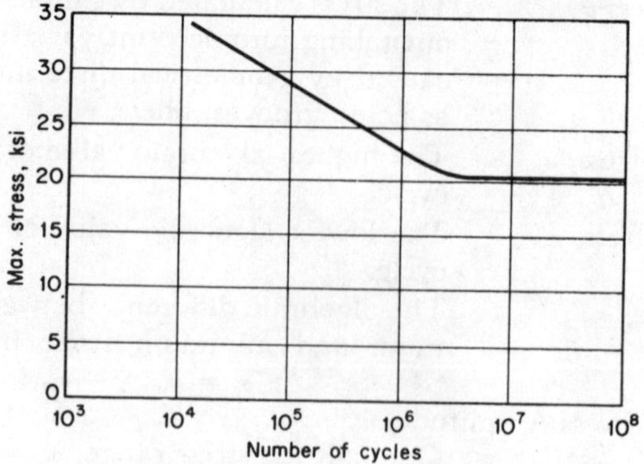

Fig. 5.41. Typical σ-n Diagram

against log n. In this manner the endurance limit is disclosed by a definite break in the curve. Figure 5.41 shows a typical σ–n diagram for mild steel. At the beginning $\sigma_{max.}$ decreases rapidly with increase in n, then the curve approaches asymptotically a stress value which shows no further decrease with increase in number of cycles; i.e., a value at which an unlimited number of cycles can be endured without failure. This is called the endurance limit of the material, σ_E. The endurance limit of ferritic steels tested in air at room temperature is reached at 10^6–10^7 cycles; whereas, for some other metals, and also for ferritic steels at elevated temperatures, Par. 5.18.1, the fracture stress continues to drop-off although the rate is small. Here, for most practical purposes, the fracture stress at 10^8 cycles can be used as the endurance limit. Investigations of steels[70] with tensile strengths to 350,000 psi have shown that endurance limit, Fig. 5.42, is best related to the product of tensile strength $\sigma_{ult.}$ and reduction in area d_a, Eq. 1.1.5, as

$$\sigma_E = .01\ d_a \sigma_{ult.} \tag{5.13.1}$$

For low and moderate strength steels this amounts to 40 to 55 percent of the ultimate tensile strength.[71] Fatigue tests show that in the region of the knee and to its right, Fig. 5.41 fatigue fracture correlates well with stress as the controlled variable. This is called high-cycle fatigue. But to the left of this region there is considerable scatter which is attributable to the fact that in this region the applied stress exceeds the yield strength of the material, thereby producing plastic instability in the test specimen. However, when strain is used as the controlled variable the test results in this region are consistently reliable and reproducible. Accordingly, in preparing fatigue curves the strains are multiplied by one-half the elastic modulus to give a pseudo stress

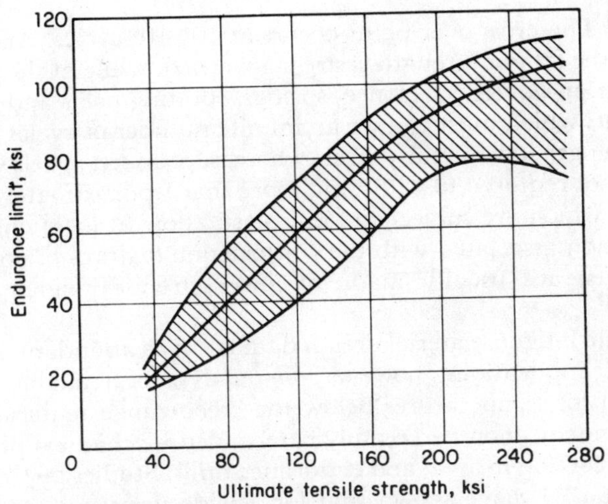

Fig. 5.42. Relationship Between Fatigue Endurance Limit and Tensile Strength[70]

amplitude. This is called low-cycle fatigue and is usually considered to encompass the region below 10^5 cycles.

Figure 5.43 gives typical results of three types of carbon and alloy steels, and shows that the comparative fatigue resistance of two materials in the low-cycle region can be the opposite in the high-cycle

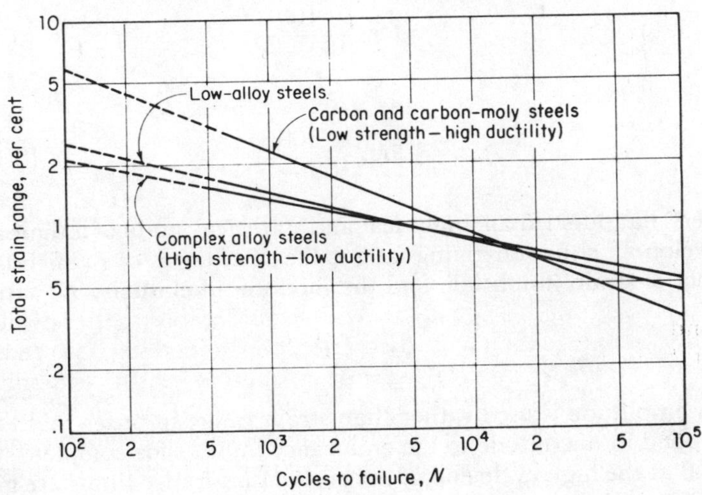

Fig. 5.43. Low-Cycle Fatigue Characteristics of Three Classes of Steel[72]

region.[72,73] The cross-over point occurs at 10^3–10^4 cycles. At the high-cycle end the fatigue strength is stress-governed, while at the low-cycle end it is strain-governed. Hence, springs, rotating parts and pulsating components which must endure an infinite number of cycles are made of high strength low ductility steel; whereas, pressure vessels which are normally not required to withstand more than approximately 10^5 load cycles but which are subject to local plastic flow at locations of stress concentration associated with the construction features of nozzles and attachments, are usually made of lower strength higher ductility material.[221]

Low cycle fatigue material design data is not as abundant as that for high cycle applications, however, an analysis of available data has shown that for temperatures below the creep range material fatigue life is dependent upon two readily determined mechanical properties; namely, reduction in area, and endurance limit. Studies by Coffin[74,75,76] and Manson [77,78] have shown that for a wide variety of materials at temperatures below the creep range the relationship

$$e_p\sqrt{N} = c \tag{5.13.2}$$

holds, where e_p is the plastic strain, and the constant c can be taken as one-half the fracture ductility (true strain at fracture).

Substituting the value of A from Eq. 1.1.5 in Eq. 5.2.19, the fracture ductility e^1 and material constant c are determined in terms of the percentage reduction in area d_a as

$$e^1 = \log_e \frac{100}{100 - d_a} \tag{5.13.3}$$

and

$$c = \tfrac{1}{2} \log_e \frac{100}{100 - d_a} \tag{5.13.4}$$

Langer[79] has pursued an analytical and statistical study of fatigue data to develop an equation giving fatigue life N in terms of σ which is the product of strain amplitude and the modulus of elasticity E

$$\sigma = \frac{e_t}{2}E \tag{5.13.5}$$

Strain amplitude is used rather than strain range in order that σ will correspond in magnitude to the endurance limit σ_E for complete stress reversal at the high cycle end of the curve. Endurance limits are given in this form. Designers work in terms of total strain e_t which is the sum

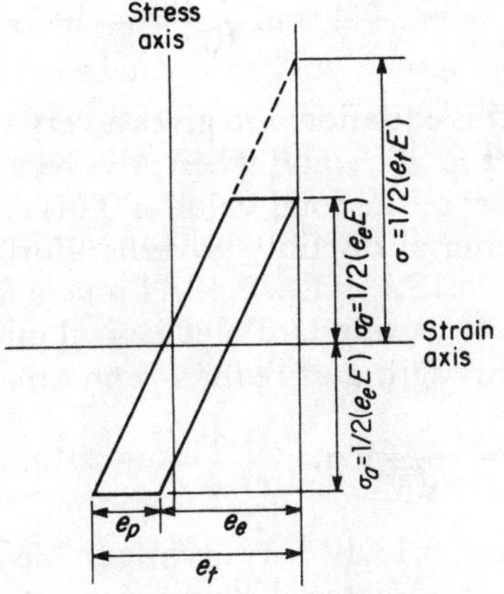

Fig. 5.44. Stress—Strain Cycle

of the elastic strain e_e and plastic strain e_p, Fig. 5.44,

$$e_t = e_e + e_p \tag{5.13.6}$$

but

$$e_e = 2\frac{\sigma_a}{E} \tag{5.13.7}$$

and substituting the value of e_t from Eq. 5.13.5 and e_e from Eq. 5.13.7 in Eq. 5.13.6 gives

$$\frac{2\sigma}{E} = \frac{2\sigma_a}{E} + e_p \tag{5.13.8}$$

$$\sigma = \sigma_a + \frac{e_p}{2}E \tag{5.13.9}$$

Substituting the value of e_p from Eq. 5.13.2 and the value of c from Eq. 5.13.4 in Eq. 5.13.9 gives

$$\sigma = \frac{E}{4\sqrt{N}} \log_e \frac{100}{100 - d_a} + \sigma_a \tag{5.13.10}$$

Since fatigue data shows that σ_a is a function of N, Eq. 5.13.10 cannot be used per se to establish a fatigue curve. However, if σ_a is replaced by the endurance limit σ_E a necessary requirement is satisfied; namely, that σ approaches σ_E as N approaches infinity. This is the high cycle end of the curve and Eq. 5.13.10 becomes

$$\sigma = \frac{E}{4\sqrt{N}} \log_e \frac{100}{100 - d_a} + \sigma_E \tag{5.13.11}$$

At low values of N this equation also gives a very satisfactory conservative correlation since $\sigma_a \geqq \sigma_E$ and when N is very small, less than 100, σ_a is a very small part of the total value σ. This is the low cycle end of the curve where material ductility governs. Further, substituting the value of σ_E from Eq. 5.13.1 in Eq. 5.13.11 give a fatigue life prediction equation in terms of the standard mechanical material test properties of ultimate tensile strength and reduction in area; namely

$$\sigma = \frac{E}{4\sqrt{N}} \log_e \frac{100}{100 - d_a} + .01\sigma_{ult}d_a \tag{5.13.12}$$

Equations 5.13.11 and 5.13.12 permit fatigue life σ–n curve prediction from readily obtainable material properties, and are in good agreement with low-cycle fatigue data. For design purposes, safety factors must be applied to give a design fatigue curve. These are customarily taken as a factor of safety of either 2.0 on stress amplitude, or 20 on cycles whichever is more conservative at each point, Fig. 5.45. The factor of 20 on cycles is to account for data scatter, size effect, surface finish, etc., and is made up of the product of the following subfactors:

Scatter of data (minimum to mean)	= 2.0
Size effect	= 2.5
Surface finish, environment, etc.	= 4.0

Endurance tests are usually carried out for completely reversed stresses, $\sigma_{max.} = -\sigma_{min.}$, whereas in many structures the stress variation is not a complete reversal. Such a cycle of fluctuating stress can be

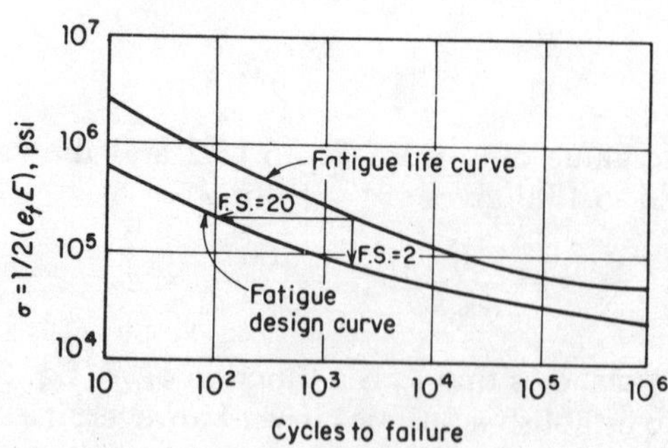

Fig. 5.45. Fatigue Life Curve and Fatigue Design Curve for Austenitic Stainless Steel

described by superposing a cycle of reversed stress, σ_r and a steady mean stress, σ_m, so that:

$$\sigma_{max.} = \sigma_m + \frac{\sigma_r}{2} = \sigma_m + \sigma_a \quad (5.13.13)$$

$$\sigma_{min.} = \sigma_m - \frac{\sigma_r}{2} = \sigma_m - \sigma_a \quad (5.13.14)$$

Very early studies and attempts to establish endurance limits under varying stresses showed the stress range σ_r necessary to produce fracture decreases as the mean stress σ_m increases and Gerber[80] proposed a parabolic law relating these, Fig. 5.46. Here the mean stress and stress range are expressed as a fraction of the ultimate tensile strength. The stress range is a maximum when the mean stress is zero (complete reversal of stress), and approaches zero when the mean stress approaches the ultimate strength (a static tensile test is equivalent to a quarter of a full reversed cycle). Presumably if the endurance limit for complete reversed stress and ultimate strength are known, the fatigue life for any fluctuating stress can be obtained from such curves. Further experiments showed that there is no general rule correlating mean stress and stress range, and that for many materials Goodman[81] has suggested their relation is best represented by a straight line, Fig. 5.46. Most schemes used for representing design data are modifications of this latter relationship.[82,83] For instance, one such method of representing the effect of steady and alternating stress on fatigue strength in design[84,85] analysis is shown in Fig. 5.47. Curves representing fatigue behavior are not always straight lines; however the expedient and conservative practice is to construct them in this manner. For instance, the design curves for 10^6 cycles would be represented as the broken line AB; hence, the use of this line is conservative and tells the designer immediately that all combinations of stress below this line are safe, and those above it will result in failure for this number of cycles.

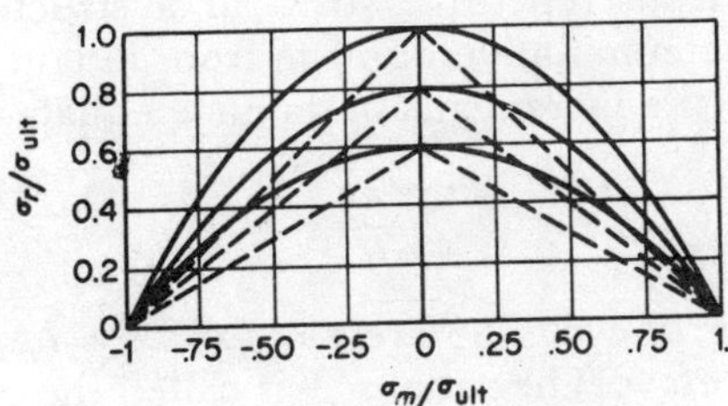

Fig. 5.46. Fatigue Diagram

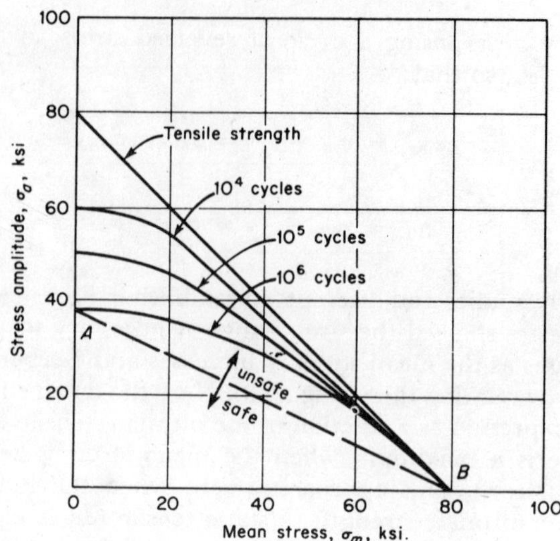

Fig. 5.47. Method of Representing the Combined Effect of Steady and Fluctuating Stress on Fatigue

Experiments have also shown that the stress range cannot be represented by summetrical curves as in Fig. 5.46 for both plus and minus values of mean stress, but that when this stress is compressive the material can withstand an appreciably higher stress range than when the stress is tension. This means, then, that curves endeavoring to depict the entire fatigue life pattern become nonsymmetrical; in fact, because the endurance limit in compression is so much higher than the limit in tension, for practical purposes, fatigue failures in compression need not be considered.

5.14 Cumulative Fatigue Damage

Practical service conditions often subject many structures to a number of cycles of stress of different magnitudes. One method of appraising the damage from repetitive stress to a structure suggested by Miner[86] is that the cumulative damage from fatigue will occur when the summation of the increments of damage equals unity; i.e.,

$$\sum \frac{n}{N} = 1 \tag{5.14.1}$$

where n = number of cycles at stress σ, and N = number of cycles to failure at same stress σ. The ratio n/N is called the cycle ratio since it represents this fraction of the total life which each stress value uses up.

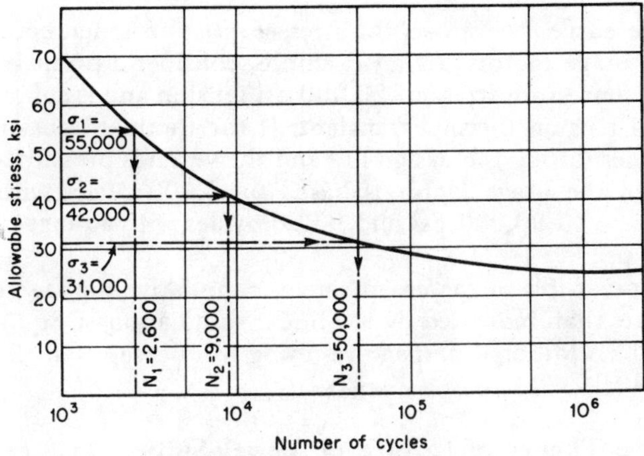

Fig. 5.48. Fatigue Evaluation

The value of N is determined from σ–n curves for the material. If the sum of these cycle ratios is less than unity, the structure is presumed safe. This is particularly important in designing an economic and safe structure which experiences only a relatively few cycles at a high stress level and the major number at a relatively low stress level. As an example, a vessel subjected to 500 cycles at a stress of 55,000 psi, 2000 cycles at 42,000 psi, and 10,000 cycles at 31,000 psi, and fabricated of material with allowable fatigue strength properties given by the σ–n curve of Fig. 5.48, would be considered safe fatigue-wise because the sum of the cycle ratios is less than 1.0.

$$\sum \frac{n}{N} = \frac{500}{2{,}600} + \frac{2{,}000}{9{,}000} + \frac{10{,}000}{50{,}000} = 0.61 \qquad (5.14.2)$$

Again, like most attempts to generalize fatigue behavior there are exceptions to this linear cumulative damage rule, Eq. 5.14.1. For instance, it does not take into account the order in which the stress cycles are encountered,[87–90] and it has been demonstrated that if the higher stresses are applied early in life, the cumulative usage factor at failure will be greater than unity.[91, 92, 286] However, for random time distribution of stress levels (and seldom is any other sequence known at the time of design of the structure) test results are in agreement with this linear cumulative damage rule and it is a recommended design method. When stress cycles of various magnitudes and frequencies occur it is important to correctly identify the range and repetition of each type of cycle because a small increase in stress range causes a large decrease in fatigue life, and this relationship varies for different portions of the fatigue curve. Therefore, the stresses must be added before calculating the usage factors. As an example, consider a point in a vessel which has a pressure stress of 20,000 psi tension and is subjected to a 70,000 psi tension thermal transient. If the thermal transient occurs 10,000 times during the design life and the vessel is pressurized 1,000 times; then the usage factor is based on 1,000 cycles with a stress range of zero to 90,000 psi and 9,000 cycles with a range of 20,000 to 90,000 psi.

At temperatures in the creep range, cumulative damage occurs at a faster rate than indicated by the linear summation of Eq. 5.14.1 because of the additional damage resulting from creep, Par. 5.20.2 and Eq. 5.20.10.

5.15 Stress Theory of Failure of Vessels Subject to Steady State and Fatigue Conditions

The mechanical properties of structural and pressure vessel materials are determined by simple uniaxial tension tests. When actual structural shapes are subject to simple tension or compression, the allowable stress upon which to base the design is taken as some fraction of the yield or ultimate stress obtained from these simple tension tests. However, in order to determine allowable design stresses for multiaxial stress conditions which occur in practice, several theories of failure have been developed. Their purpose is to predict when failure will occur under the action of combined stresses on the basis of data obtained from simple uniaxial tension or compression tests. Failure refers to either yielding or actual rupture of the material, whichever occurs first. In the case of ductile material, yielding occurs first and this is the basis of failure theories for these materials.

1. *Steady State Stress Condition*

The state of stress which can exist in a body can be determined by three principal stresses, σ_1, σ_2 and σ_3 acting on it as shown in Fig. 5.49. Tension is considered positive and compression negative, and in Fig. 5.49 the relation between the algebraic values of the principal stresses is $\sigma_1 > \sigma_2 > \sigma_3$. There are three stress theories of failure that are used for converting the uniaxial to combined stress data, and these have been widely used in pressure vessel design.

(a) "*Maximum Stress,*" *or Rankine theory.* This is the oldest theory of failure and is based on the maximum or minimum principal stress as a criteria of failure and postulates that failure occurs in a stressed body when one of the principal stresses reaches the yield point value in

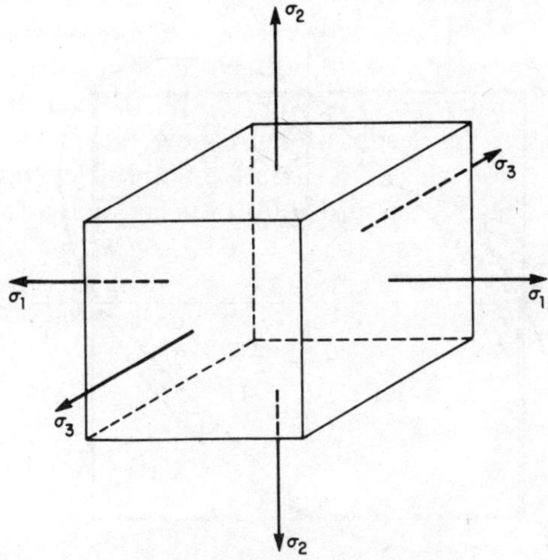

Fig. 5.49. Principal Stresses

simple tension $\sigma_{Y.P.}$, or compression $\sigma'_{Y.P.}$. The conditions of yielding of materials whose tension and compression properties are the same, such as mild steel, become as follows:

For $\sigma_1 > \sigma_2$ or σ_3, failure occurs when $\sigma_1 = \pm\sigma_{Y.P.}$
For $\sigma_2 > \sigma_1$ or σ_3, failure occurs when $\sigma_2 = \pm\sigma_{Y.P.}$
For $\sigma_3 > \sigma_1$ or σ_2, failure occurs when $\sigma_3 = \pm\sigma_{Y.P.}$

This theory is represented graphically in two dimensions in Fig. 5.50 for a material which has the same yield point in tension and compression. This plot is made by dividing the above expressions by $\sigma_{Y.P.}$,

$$x = \frac{\sigma_1}{\sigma_{Y.P.}} \tag{5.15.1}$$

$$y = \frac{\sigma_2}{\sigma_{Y.P.}} \tag{5.15.2}$$

and shows the locus of failure points is a square *ABCD*. According to this theory there is no yielding at stresses represented by points inside this square. While the maximum stress theory does best predict cohesive or brittle failure of materials, it does not always cover ductile material in which failure is a sliding or shearing action along planes inclined at 45° to the axis of the specimen, Par. 5.3. On these 45° planes neither the tensile or compressive stresses are a maximum but failure is caused by shearing.

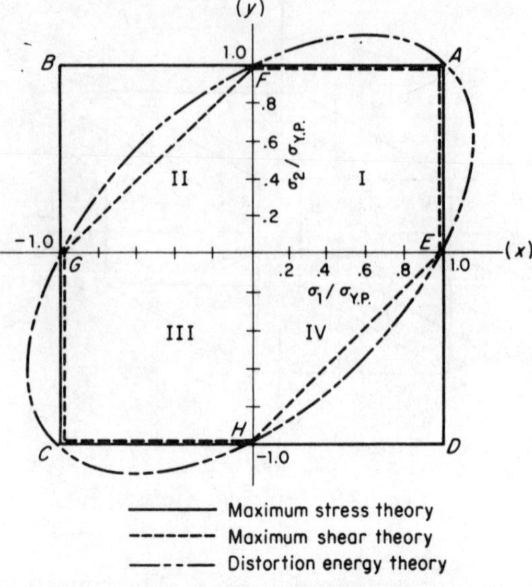

Fig. 5.50. Failure Theories

(b) "*Maximum Shear Stress*" *or Tresca theory*. This theory postulates that yielding in a body subject to combined stresses will occur when the maximum shear stress becomes equal to the maximum shear stress at yield point in a simple tension test. This theory is in better agreement with experimental results[1] for ductile materials whose tension and compression properties are the same than is the "maximum stress" theory, Fig. 5.51. The maximum shear stress is equal to half the difference of the maximum and minimum principal stresses; thus, for a member under combined stresses the shear stresses are:

$$\tau = \frac{\sigma_1 - \sigma_2}{2} \tag{5.15.3}$$

$$\tau = \frac{\sigma_2 - \sigma_3}{2} \tag{5.15.4}$$

$$\tau = \frac{\sigma_3 - \sigma_1}{2} \tag{5.15.5}$$

Likewise, the maximum shear stress in a tension test is equal to half the normal stress at yielding of the test specimen

$$\tau = \frac{\sigma_{Y.P.}}{2} \tag{5.15.6}$$

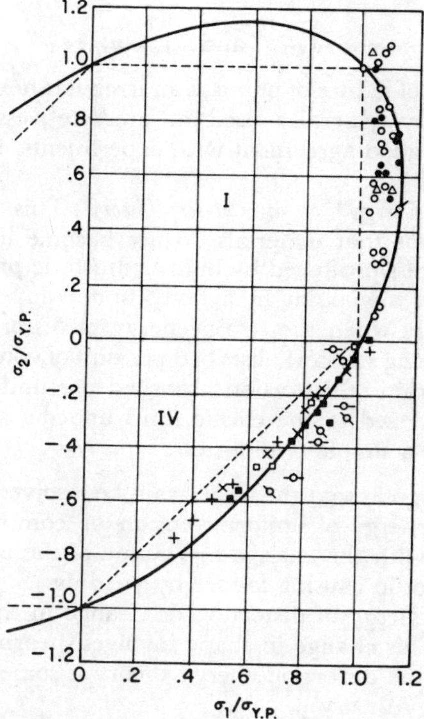

Fig. 5.51. Comparison of Biaxial Stress Test Data with Theories of Failure

Equating Eq. 5.15.6 successively to Eqs. 5.15.3, 5.15.4 and 5.15.5 gives the condition of yielding under combined stresses as:

$$\sigma_1 - \sigma_2 = \pm\sigma_{Y.P.} \tag{5.15.7}$$

$$\sigma_2 - \sigma_3 = \pm\sigma_{Y.P.} \tag{5.15.8}$$

$$\sigma_3 - \sigma_1 = \pm\sigma_{Y.P.} \tag{5.15.9}$$

The quantity on the left side of Eqs. 5.15.7, 5.15.8 and 5.15.9 is twice the shear stress and is called the shear stress intensity. In using these equations it is noted that σ_3 is the algebraically minimum principal stress.

A plot of this theory in two dimensions is also shown in Fig. 5.50. It was made by dividing Eqs. 5.15.7, 5.15.8 and 5.15.9 by $\sigma_{Y.P.}$ to give

$$x - y = \pm 1, \; x = \pm 1, \; y = \pm 1$$

where

$$x = \sigma_1/\sigma_{Y.P.} \quad \text{and} \quad y = \sigma_2/\sigma_{Y.P.}$$

It shows the locus of failure of points is an irregular hexagon *EAFGCH*. This is the theory generally used in pressure vessel design, par. 5.15.2(a). It is in good agreement with experiments, Fig. 5.51, and is simple to apply.

(c) "*Distortion Energy*" *or von Mises Theory.* This theory is based on test observations that materials do not become inelastic under a triaxial state of stress produced by high hydrostatic pressure, but that inelastic action at any point in a body under any combination of stresses begins only when the strain energy of distortion (change of shape due to shearing stresses) absorbed per unit of volume at the point is equal to the strain of distortion absorbed per unit volume at any point in a bar stressed to the elastic limit under a state of uniaxial stress as occurs in a simple tension test.

This theory proposes that the total strain be resolved into two parts:

(1) the strain energy of uniform tension or compression. (This is associated with the change in volume of the unit volume and has no effect in causing failure by yielding.)

(2) the strain energy of distortion or change in shape of the unit volume. (This change in shape involves shearing stresses; consequently, the distortion energy theory is sometimes called the shear energy theory).

and that only the part attributable to the strain energy of distortion be used to determine yielding or fracture of the material. This can be written as

$$U = U_v + U_d \tag{5.15.10}$$

where U_v is the energy of volume change per unit of volume, and U_d is the energy of distortion per unit of volume.

The total strain energy U produced in an element depends upon both the stress and strain on the element. If the stress σ gradually increases from zero and causes a strain e, the work done is $U = 1/2\sigma e$. For the three-dimensional case shown by the element in Fig. 5.49, and using the relation between stress and strain given by Eq. 2.3.3, the total energy is

$$U = \frac{\sigma_1 e_1}{2} + \frac{\sigma_2 e_2}{2} + \frac{\sigma_3 e_3}{2} \tag{5.15.11}$$

$$U = \frac{1}{2E}(\sigma_1^2 + \sigma_2^2 + \sigma_3^2) - \frac{\mu}{E}(\sigma_1\sigma_2 + \sigma_1\sigma_3 + \sigma_2\sigma_3) \tag{5.15.12}$$

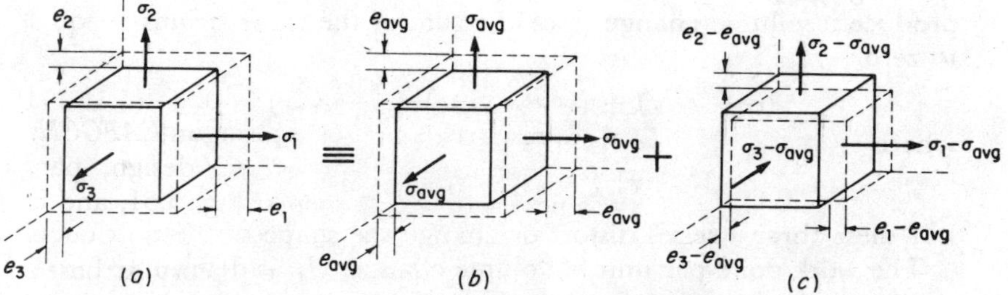

Fig. 5.52. General Representation of Distortion Energy Theory

in which tensile stress is taken as positive and compressive stress as negative.

This total strain can be divided into two parts by resolving the principal stresses Fig. 5.52*a* into two components states of stress as shown in Figs. 5.52*b* and *c*. The first component, Fig. 5.52*b*, is chosen as the average of the three principal stresses $\sigma_{\text{avg.}} = (\sigma_1 + \sigma_2 + \sigma_3)/3$ which produces an average strain $e_{\text{avg.}} = (e_1 + e_2 + e_3)/3$ in each direction equal to the average of the three principal strains. The strained unit volume is $(1 + e_1)(1 + e_2)(1 + e_3)$, and if e_v represents the change in unit volume due to straining

$$1 + e_v = (1 + e_1)(1 + e_2)(1 + e_3) = 1 + e_1 + e_2 + e_3 \tag{5.15.13}$$

$$e_v = e_1 + e_2 + e_3 \tag{5.15.14}$$

in which the products of strains are considered small and neglected.

Equation 5.15.14 states that the volume change is equal to the algebraic sum of the three principal strains. Likewise, the relation between the volume change and the principal stresses can be found by substituting in Eq. 5.15.14 the value of the principal strains from Eq. 2.3.3 as

$$e_v = e_1 + e_2 + e_3 = \frac{1 - 2\mu}{E}(\sigma_1 + \sigma_2 + \sigma_3) \tag{5.15.15}$$

or in terms of the average stress $\sigma_{\text{avg.}} = (\sigma_1 + \sigma_2 + \sigma_3)/3$

$$e_v = \frac{3(1 - 2\mu)}{E}\sigma_{\text{avg.}} \tag{5.15.16}$$

The second component state as shown in Fig. 5.52*c* consists of the remainder of each of the three principal stresses, and also each of the three principal strains. The average principal stress, Fig. 5.52*b*, produces the entire volume change, Eqs. 5.15.15 and 5.15.16. The remaining components of the three principal stresses, Fig. 5.52*c*, do not produce a volume change since the sum of the three strains is equal to zero.

$$(e_1 - e_{avg.}) + (e_2 - e_{avg.}) + (e_3 - e_{avg.}) = 0 \quad (5.15.17)$$

where

$$e_{avg.} = (e_1 + e_2 + e_3)/3$$

but these three stresses distort or change the shape of the unit cube.

The work done per unit of volume change, U_v, is determined from the stresses and strains shown in Fig. 5.52 as

$$U_v = 1/2\,\sigma_{avg.}e_{avg.} + 1/2\sigma_{avg.}e_{avg.} + 1/2\sigma_{avg.}e_{avg.} = 1/2\sigma_{avg.}e_v \quad (5.15.18)$$

and substituting the value of e_v from Eq. 5.15.16 in Eq. 5.15.18 gives

$$U_v = \frac{3}{2E}(1 - 2\mu)\sigma_{avg.}^2 \quad (5.15.19)$$

The energy of distortion U_d is then obtained by subtracting the energy of the volume change, Eq. 5.15.19, from the total strain energy U given in Eq. 5.15.12

$$U_d = \frac{1}{2E}(\sigma_1^2 + \sigma_2^2 + \sigma_3^2) - \frac{\mu}{E}(\sigma_1\sigma_2 + \sigma_1\sigma_3 + \sigma_2\sigma_3) - \frac{3}{2E}(1 - 2\mu)\sigma_{avg.}^2 \quad (5.15.20)$$

Substituting the value of $\sigma_{avg.} = (\sigma_1 + \sigma_2 + \sigma_3)/3$ in Eq. 5.15.20 gives

$$U_d = \frac{1 + \mu}{6E}[(\sigma_1 - \sigma_2)^2 + (\sigma_2 - \sigma_3)^2 + (\sigma_3 - \sigma_1)^2] \quad (5.15.21)$$

The foregoing is well illustrated by the uniaxial stress condition shown in Fig. 5.53*a*. It is seen that it is equivalent to a cube subjected to three equal principal stresses, each of which is equal to the average of the given set of principal stresses, Fig. 5.53*b*; and in Fig. 5.53*c* a set of

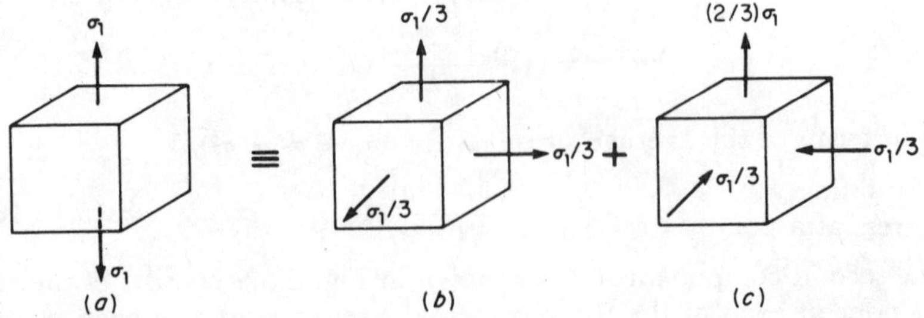

Fig. 5.53. Uniaxial Stress Representation of Distortion Energy Theory

principal stresses which when superimposed on those of Fig. 5.53*b* gives the original state of stress in Fig. 5.53*a*. The first condition, Fig. 5.53*b*, is one of uniform tension, and the second condition, Fig. 5.53*c*, is one of pure shear.

Equation 5.15.21 is the basis for determining the failure of ductile materials according to the distortion energy theory. According to this theory, yielding begins when the distortion energy, Eq. 5.15.21, reaches the value of the distortion energy at the yield point in a simple tension test. The latter is obtained by substituting $\sigma_1 = \sigma_{Y.P.}$, and $\sigma_2 = \sigma_3 = 0$ in Eq. 5.15.21 which gives

$$U_d = \frac{1+\mu}{3E}\sigma_{Y.P.}^2 \qquad (5.15.22)$$

Equating Eqs. 5.15.21 and 5.15.22 gives the general condition of yielding based on the distortion energy theory as

$$(\sigma_1 - \sigma_2)^2 + (\sigma_2 - \sigma_3)^2 + (\sigma_1 - \sigma_3)^2 = 2\sigma_{Y.P.}^2 \qquad (5.15.23)$$

In the case of two-dimensional stress, $\sigma_3 = 0$, and Eq. 5.15.23 gives the condition of yielding as

$$\sigma_1^2 - \sigma_1\sigma_2 + \sigma_2^2 = \sigma_{Y.P.}^2 \qquad (5.15.24)$$

Dividing Eq. 5.15.24 by $\sigma_{Y.P.}^2$ gives

$$\frac{\sigma_1^2}{\sigma_{Y.P.}^2} + \frac{\sigma_2^2}{\sigma_{Y.P.}^2} - \frac{\sigma_1\sigma_2}{\sigma_{Y.P.}^2} = 1 \qquad (5.15.25)$$

This is the equation of an ellipse and is shown plotted in Fig. 5.50.

In the case of pure shear, as occurs in a cylinder subject to torsion, the shearing stress is equal in magnitude to each of the two principal stresses occurring at 45° to the shear stress. Thus in Eq. 5.15.21, $\sigma_1 = \sigma$, $\sigma_2 = -\sigma$ and $\sigma_3 = 0$ and further $\sigma = \tau$, which gives

$$U_d = \frac{1+\mu}{6E}\{[\sigma - (-\sigma)]^2 + (\sigma)^2 + (-\sigma)^2\}$$

$$= \frac{1+\mu}{E}\sigma^2 = \frac{1+\mu}{E}\tau^2 \qquad (5.15.26)$$

Equating Eq. 5.15.22 to Eq. 5.15.26 gives the maximum shearing stress at a point when yielding starts in terms of the yield strength at the same point.

$$\frac{1+\mu}{E}\tau_{Y.P.}^2 = \frac{1+\mu}{3E}\sigma_{Y.P.}^2 \qquad (5.15.27)$$

$$\tau_{Y.P.} = 0.577\sigma_{Y.P.} \qquad (5.15.28)$$

2. *Fatigue Stress Theory, and Uniaxial and Biaxial Material Fatigue Strength*

(a) *Fatigue Stress Theory.* Internal pressure fatigue tests of cylinders conducted by Morrison, Crossland and Parry,[93] Fig. 5.54, also show that these failures subscribe to the "maximum shear stress theory." When these results are plotted against hoop stress, Fig. 5.54*a*, extremely wide deviations occur; hence, the maximum principal stress is not a criterion of failure. On the other hand, if these fatigue results are plotted against maximum shear stress, Fig. 5.54*b*, the spread is entirely reduced to one attributable only to experimental scatter, and is evidence that the "maximum shear stress theory" of failure is applicable to pressure vessels made of ductile materials and subject to fatigue. Accordingly, the good agreement of this theory of failure, under both steady state and fatigue stress conditions with experiments, have led to its adoption in the design of vessels.[222, 223, 224, 225, 287] It is the basis of many pressure vessel codes such as the American Society of Mechanical Engineers Boiler and Pressure Code for Nuclear Power Plant Components,[94] and Pressure Vessels, Division 2 and 3, Par. 7.4.

(b) *Uniaxial and Biaxial Material Fatigue Strength.* The study of elastic and plastic behavior of metals under uniaxial static and fatigue loading has been the path by which material properties are evaluated, specifications written and designs established. Most practical problems involve a biaxial stress field in which the stress is applied at a point in two orthogonal directions. These generally occur at free surfaces where the third stress is zero; however, this third stress component can be significant at the inner surface of a very high pressure vessel (where the pressure becomes the third component).

Just as with static theories of failure, Par. 5.15.1(b), the maximum shear stress theory has been found to predict the fatigue failure of a material under the action of multiaxial stress with good accuracy.[95–98] This makes it possible to correlate the calculated values of the three principal stresses into a single number, called the "stress intensity," which can be compared directly to the results of uniaxial tests. If σ_1, σ_2, and σ_3 are the three principal stresses at a point, and $\sigma_1 > \sigma_2 > \sigma_3$ algebraically, the stress intensity S, is equal to twice the maximum shear stress and is

$$S = 2\tau = \sigma_1 - \sigma_3 \qquad (5.15.29)$$

Fatigue curves used in the low cycle range are based on strain-cycling data. However, for ready comparison with elastically computed stresses, it is convenient to multiply the strains by the elastic modulus

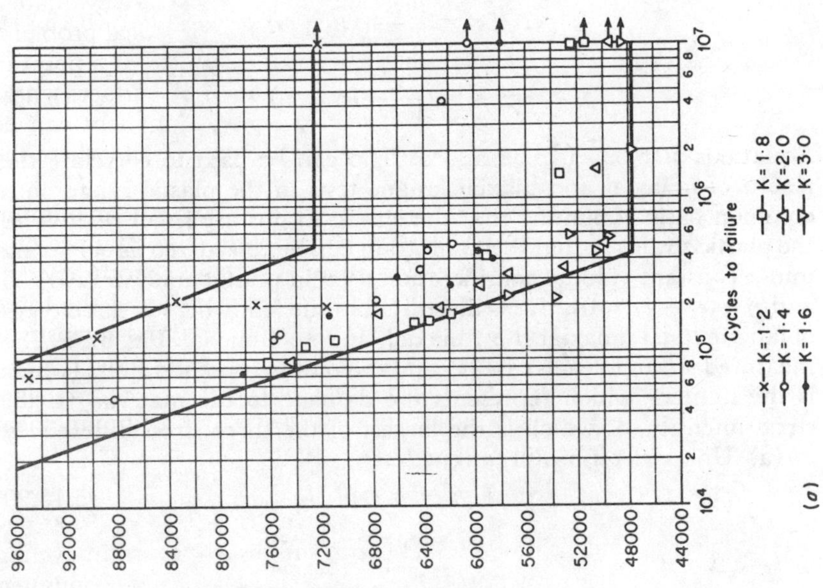

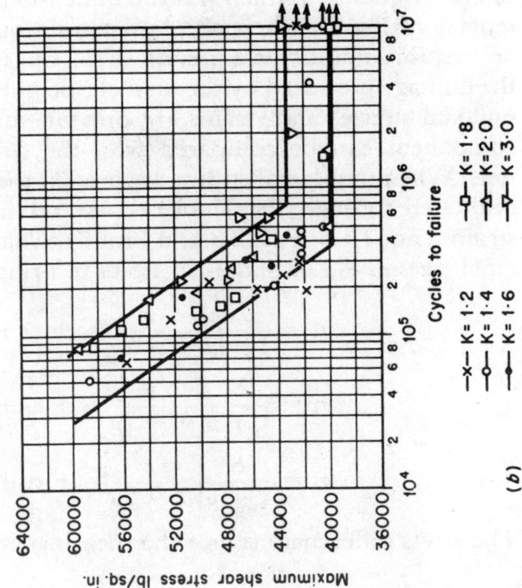

Fig. 5.54. Fatigue Life of Thick Cylinders for Various Ratios of K (Ratio of Outside Diameter to Inside Diameter) of 126,000 psi UTS Material, Plotted Against (a) Maximum Hoop Stress, and (b) Maximum Shear Stress[93]

to give a quantity which has the dimensions of stress, but only represents a real stress when no plastic strain occurs. When plastic strain occurs, this quantity is a pseudo-stress, but is nonetheless a measure of the damage produced by the total elastic plus plastic strain. When only uniaxial stresses and strains are present, the calculated values in the component can be compared with the strains measured in fatigue tests. When multiaxial strains are present their combined effect can be evaluated through the Hooke's Law relations between stresses and strains (Eq. 1.1.3, Par. 2.3 and Prob. 4, Chap. 2) which gives the principal stresses σ_1, σ_2, and σ_3 in terms of principal strains e_1, e_2 and e_3 as

$$\sigma_1 = \frac{E}{1+\mu}\left[\frac{\mu}{1-2\mu}(e_1 + e_2 + e_3) + e_1\right] \tag{5.15.30}$$

$$\sigma_2 = \frac{E}{1+\mu}\left[\frac{\mu}{1-2\mu}(e_1 + e_2 + e_3) + e_2\right] \tag{5.15.31}$$

$$\sigma_3 = \frac{E}{1+\mu}\left[\frac{\mu}{1-2\mu}(e_1 + e_2 + e_3) + e_3\right] \tag{5.15.32}$$

The stress differences (twice the shearing stress) are

$$\sigma_1 - \sigma_2 = \frac{E}{1+\mu}(e_1 - e_2) \tag{5.15.33}$$

$$\sigma_2 - \sigma_3 = \frac{E}{1+\mu}(e_2 - e_3) \tag{5.15.34}$$

$$\sigma_3 - \sigma_1 = \frac{E}{1+\mu}(e_3 - e_1) \tag{5.15.35}$$

Equations 5.15.33, 5.15.34 and 5.15.35 can be used to correlate the results of uniaxial and biaxial fatigue tests in the plastic range on a common basis. Assuming that the elastic strains are small relative to the plastic strains so that the total strain can be considered as occurring under constant volume consideration used in plastic analysis ($\mu = 0.5$ and $e_1 + e_2 + e_3 = 0$), this is illustrated in the following for three types of fatigue tests: uniaxial, 1 : 1 biaxial, and 2 : 1 biaxial, Fig. 5.55. The measured strain in each case is e and the required pseudo stress (which is the number which represents the damage to the material) is the stress intensity, S, which is the largest of the three stress differences.

(a) Uniaxial push-pull fatigue test.

$$\sigma_2 = \sigma_3 = 0$$

$$e_2 = e_3 = -\frac{e}{2}$$

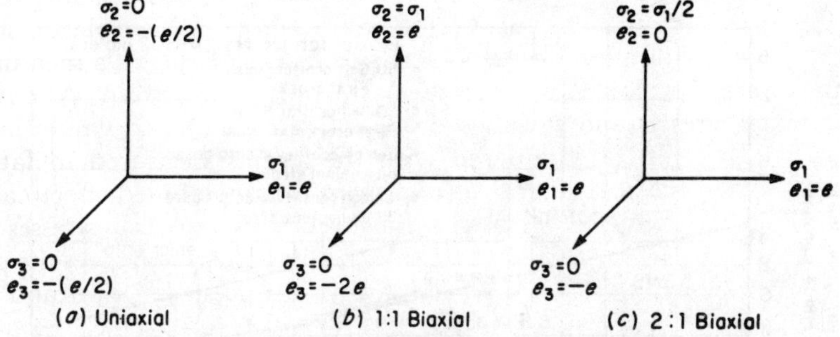

Fig. 5.55. Equivalent Strain Calculations for Uniaxial and Biaxial Fatigue Correlations

The stress intensity is from Eq. 5.15.33

$$S = \sigma_1 - \sigma_2 = \frac{E}{1 + 0.5}\left(e + \frac{e}{2}\right) = eE \tag{5.15.36}$$

(b) 1 : 1 Biaxial fatigue test

$$\sigma_3 = 0$$
$$\sigma_1 = \sigma_2$$
$$e_1 = e_2 = e$$
$$e_3 = -2e$$

The stress intensity is from Eq. 5.15.34

$$S = \sigma_2 - \sigma_3 = \frac{E}{1 + 0.5}(e + 2e) = 2eE \tag{5.15.37}$$

(c) 2 : 1 Biaxial fatigue test

$$\sigma_3 = 0$$
$$\sigma_2 = \sigma_1/2$$
$$e_1 = e$$
$$e_2 = \frac{\sigma_2}{E} - \frac{\mu\sigma_1}{E} = \frac{\sigma_1}{2E} - \frac{0.5\sigma_1}{E} = 0$$
$$e_3 = -e - e_2 = -e$$

The stress intensity from Eq. 5.15.35 is

$$S = \sigma_3 - \sigma_1 = \frac{E}{1 + \mu}(e_3 - e_1) = \frac{4}{3}eE \tag{5.15.38}$$

Figure 5.56 shows the correlation of uniaxial and biaxial fatigue data using this type of analysis.[96] The B&W tests of 1:1 biaxiality and the Lehigh tests of 2 : 1 biaxiality show large discrepancies when plotted

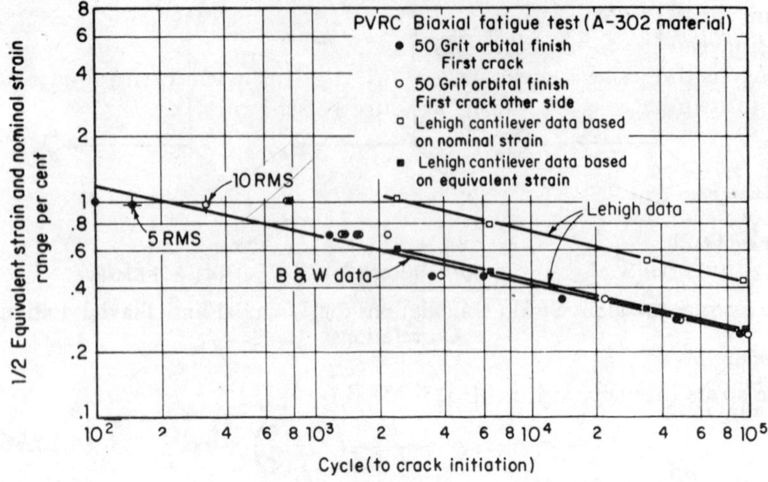

Fig. 5.56. Biaxial Fatigue Test Data for Manganese-Molybdenum, SA-302B, Steel

in terms of measured nominal strain, but when reduced to the common basis of stress intensity their agreement becomes excellent.

If only elastic strains are involved, such as occurs at the high cycle region of fatigue, say at the endurance limit, Eqs. 5.15.33, 34 and 35 may be used with a value of $\mu = 0.3$ instead of 0.5.

3. *Significance of Theories of Failure*

Figure 5.50 is a comparison of these three main stress theories of failure drawn for a material which has the same yield point in tension and compression, and subject to a two-dimensional stress pattern ($\sigma_3 = 0$). The maximum principal stress theory coincides with the maximum shear stress theory when both the principal stresses have the same sign, i.e., they are in the I and III quadrants, and the maximum deviation from the distortion energy theory is approximately 15 per cent. In an equal biaxial tension stress condition all three theories coincide, point A in Fig. 5.50. However, when the principal stresses have opposite signs, II and IV quadrants, there is considerable difference between the maximum stress theory and either the maximum shear stress or the distortion energy theory and only the latter two theories should be used because experimental results have shown them to be closer to actual failure tests. In fact, Fig. 5.51 giving a plot of tests of ductile material failures shows these to lie between the maximum shear stress and distortion energy theories with the former representing the lower boundary. Accordingly, the maximum shear stress theory is the more conservative. This, together with its simplicity, has led to its general adoption in pressure vessel stress analysis.

Example: The ASME Nuclear Power Code uses the maximum shear stress theory of failure and requires the minimum thickness, h, of a thin spherical vessel of inside radius, r, and subject to an internal pressure, p, to be:

$$h = \frac{pr}{4\tau - p}$$

Develop this formula. (τ is the allowable maximum shear stress.)

The maximum principal stress in a sphere is the hoop or longitudinal one as given by Eq. 2.2.7

$$\sigma_{max} = \frac{pr}{2h} \tag{5.15.39}$$

and the minimum stress is the average radial stress occurring over the wall thickness, or

$$\sigma_{min} = \frac{-p + 0}{2} = \frac{-p}{2} \tag{5.15.40}$$

Substituting Eqs. 5.15.39 and 5.15.40 in Eq. 5.15.4 gives

$$\tau = \frac{\sigma_{max} - \sigma_{min}}{2} = \frac{\frac{pr}{2h} - \left(-\frac{p}{2}\right)}{2} \tag{5.15.41}$$

$$h = \frac{pr}{4\tau - p} \tag{5.15.42}$$

In the practical design of pressure vessels this means that even the maximum stress theory is adequate as long as the state of stress is in the first and third quadrants, Fig. 5.50. In a thin cylindrical vessel under internal pressure the radial stress, even though it is negative, is small compared to the hoop and longitudinal stress and can be assumed equal to zero; hence, either the maximum principal stress theory or the maximum shear stress theory give approximately the same results. In a corresponding thick walled cylindrical vessel, however, the radial stress is not small in comparison with the hoop and longitudinal stress. Since this radial stress is compressive, and cannot be assumed equal to zero in Eqs. 5.15.4 and 5.15.5, these two theories no longer give approximately the same results. Accordingly, the maximum shear stress theory should be used. This same condition also applies to spherical or other vessel shapes.

5.16 Stress Concentrations

The magnitude of the applied stress is the predominate factor that determines fatigue crack growth and establishes fatigue life. Hence, anything that is done to a vessel or structure to create a "stress raiser or concentrator" directly influences its life since these are direct multipliers of the normal stress. Geometric discontinuities such as holes, grooves, notches, abrupt changes in cross section, etc., as well as thermal discontinuities, cause a local increase in stress, and proper allowance for the effects of these stress concentrations is the single most important factor in designing to resist failure by fatigue.[99,100] In view of this, Chapter 6 is devoted entirely to this item—its cause, effect and remedy.

5.17 Influence of Surface Effects on Fatigue

Most fatigue failures start at the surface of a material—hence the importance of choosing a surface finish compatible with the intended life of the structure. There are three ways by which fatigue life is influenced by surface effects: (1) the creation of stress risers due to surface roughness; (2) the establishment of actual strength differences between the outer shell and core of the material; and (3) the difference in stress levels obtained by the presence of residual stress.

1. *Surface Finish*

Critical changes in surface geometry may result from certain types of finishes, or even the direction in which the finish is applied.[101] For instance, scratches give rise to stress concentration effects, making it desirable to have the mechanical finishing performed in a direction parallel with the principal stress rather than normal to it. Table 5.3, based on the work of Horger and Neifert[102] shows the fatigue life of polished specimens to be some 10 percent higher than that of rough turned ones. This potential surface finish improvement varies somewhat with the material, being as high as 25 per cent for steel and 40 percent for aluminum alloys. It must be remembered that, although surface finishes are critical in obtaining the full endurance limit of a material, their effect is not so pronounced when a relatively few cycles[79,103] are involved and this transition is of most practical importance. For instance, in the design of vessels subjected to relatively

TABLE 5.3. EFFECT OF SURFACE ROUGHNESS ON ENDURANCE LIMIT (SAE 1045 STEEL, 69,700 UTS, 39,300 Y.P.)[102]

Specimen Diameter in.	*Type of Surface Finish*	*Max. depth of Finish Marks, Microinches*	*Endurance Limit, psi*
1.5	Rough turn	3,900	28,000
1.5	Smooth turn	140	28,000
1.5	0.000 Emery	80	31,000
1.5	Superfinish	35	31,000

few cycles, surface finishes beyond those inherent in the fabrication of the base metal, such as rolled plate, are not warranted. However, in the design of hydraulic cylinders for repeated ram operations, fine finishing is applicable and experiments have shown that honing[93] of the inside diameter (I.D.) has given an increase in fatigue life of 11 per cent with $K = 1.4$ where K = ratio of O.D./I.D., to 41 percent with $K = 1.8$. Although honing is a "smearing" finishing process which gives a combination of polish, cold work, and residual stress, the results are significant for this type of service.

Noll and Lipson[104] have demonstrated that the surface sensitivity of material increases with tensile strength, Fig. 5.57, which becomes particularly important when high tensile materials are used in design. The stronger the material the greater the influence of surface condi-

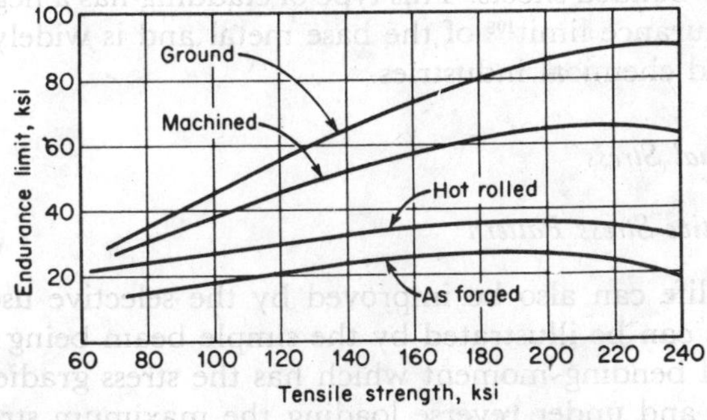

Fig. 5.57. Relation of Endurance Limit to Tensile Strength of Unnotched Specimens in Reverse Bending[104]

tioning because the sensitivity of a material to surface flaws and notches generally increases with tensile strength. The low fatigue life of as-received high-strength alloys is due to the presence of millscale which is easily cracked, thereby, creating stress concentrators.

2. *Surface Coatings*

It has been mentioned that the endurance limit of steels increase with their ultimate tensile strength; hence, a material which has a shell of tensile strength greater than its core shows a greater endurance limit than the base core material. Materials which have been processed to harden the surface (provided no deleterious metallurgical effects are embodied), such as by carburizing, nitriding, flame-hardening, and other surface hardening effects, exhibit this increase.[105,106] This effect is particularly pronounced and useful in members subjected to high stress gradients, such as shafts in bending and torsion where the high strength material can be located in the region of maximum stress.

Claddings for non strength purposes are frequently used in vessels to protect the vessel material proper from a corrosive media such as occurs in the chemical industry, or to protect the media from the corrosion contamination of the vessel material such as occurs in the food-processing industry, or with some nuclear reactor vessels. One of the commonest methods of metallic cladding is that of electroplating with a compatible corrosion resistant material (nickel, chromium, cadmium, zinc, copper, etc.). However, the effect of such coatings is to severely reduce the endurance limit[107] due to hydrogen embrittlement of the base metal by the electroplating process, Par. 5.21. Another method of accomplishing this is to clad the exposed surfaces with a corrosion resistant material, such as stainless steel, by deposited weld metal or by bonded sheets. This type of cladding has a negligible effect on the endurance limit[108] of the base metal and is widely used in the nuclear and chemical industries.

3. *Residual Stress*

(*a*) *Selective Stress Pattern*

Fatigue life can also be improved by the selective use of residual stress. This can be illustrated by the simple beam being subjected to an external bending moment which has the stress gradient shown in Fig. 5.58*a*, and under reverse loading the maximum stress will vary from equal values of tension and compression, σ_b. If, however, a thin outer layer is subjected to a residual compressive stress σ_r, Fig. 5.58*b*,

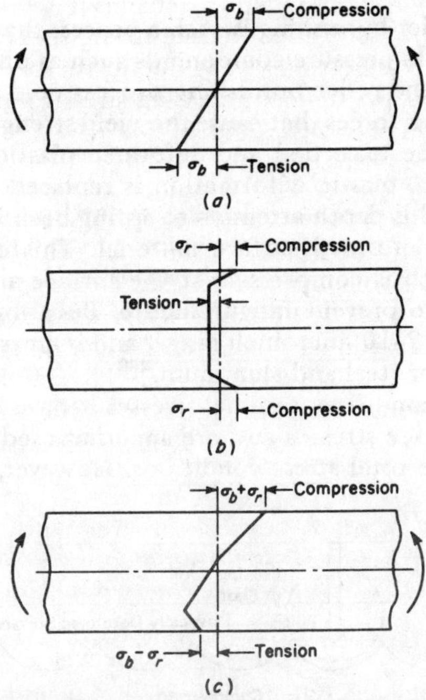

Fig. 5.58. Stress Gradient in a Simple Beam, (*a*) Bending Only, (*b*) Residual Stress Only, (*c*) Bending Plus Residual Stress

the net maximum stress on the surface will be $\sigma_b - \sigma_r$ on the tension side, and $\sigma_b + \sigma_r$ on the compression side, Fig. 5.58*c*. Under full reversed loading, the surface will be subjected to a maximum tension stress of $\sigma_b - \sigma_r$ which, for the surface, means longer life because a lower mean stress is applied. One of the most popular mechanical ways of obtaining this effect is by shot peening.[109] Peening is the mechanical working of metals by means of impact blows, Fig. 5.59*a*. This increases fatigue life by (1) increasing the tensile strength of the skin, and (2) reducing the applied tensile stress by the amount of compressive stress set up by the process. If peening is performed after the application of a preload, fatigue life is further increased. This is caused by a significant stress redistribution resulting from the greater depth over which the desirable compressive near-surface stresses can be developed,[288]

Likewise, coining or cold-rolling of the internal surface of cylinders can be used to obtain a comparable favorable residual stress distribution which is similar to that for a thick-walled cylinder with an interference fit.[110] Roller burnishing is such a process that is used on shafts and axles, as well as pressure components such as oil well pipe. Much as with peening, the roller burnishing process cold works the metal surface by applying forces that cause the yield strength of the contact surface metal to be exceeded and deformed plastically, Fig. 5.59*b*. At a certain depth, plastic deformation is replaced by elastic deformation. Metal at this depth attempts to spring back but is resisted by the permanently deformed surface material. This creates a residual stress pattern which is compressive at the surface and tensile below, and which helps to prevent fatigue failure. Burnishing improves the surface finish to a 2–10 microinch range, and increases fatigue life by a factor of three for steel and aluminum.[289]

In a stabilized condition, residual stresses behave in the same manner as static inservice stresses and are superimposed on applied ones to create favorable total stress conditions. However, residual stresses

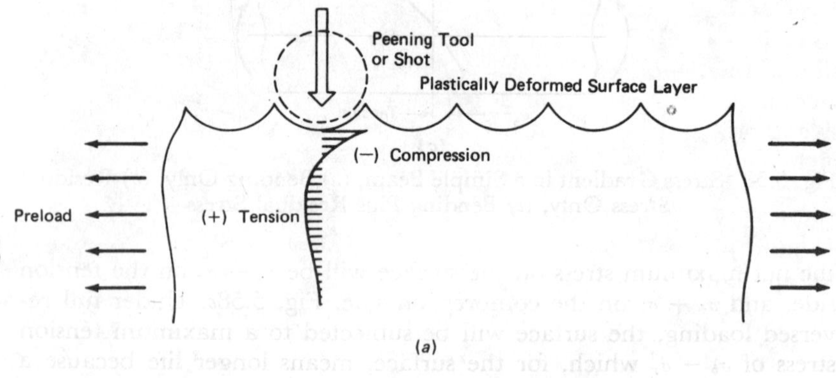

(*a*)

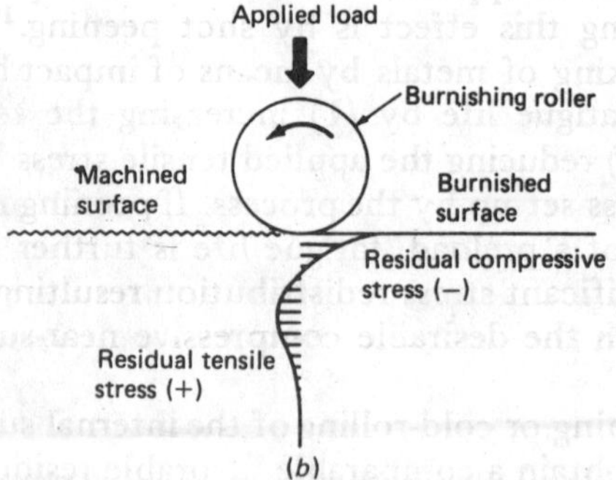

(*b*)

Fig. 5.59. (*a*) Peening. (*b*) Burnishing

can change during the service life of the component, particularly in an elevated temperature creep environment. Accordingly, it is difficult to take them into account when calculating fatigue life and σ-n tests remain the only firm evaluation method.

Cyclic stressing produces changes in the stress-strain properties of metals dependent upon the initial state of the metal and testing conditions. Metals that have had their strength increased by cold-working can be unstable under cyclic loading in the plastic range, and the result is a softening process rather than a hardening one.[111] Hence, a beneficial residual stress system in cold-worked metal may be rendered ineffective by such plastic cyclic loading. Accordingly, while the endurance limit is increased by peening, coining and cold-working, the low cycle large plastic strain fatigue strength may be little improved over that of the comparable annealed state.[112]

Grinding is a cutting operation that also produces very light peening which induces a highly localized compressive stress in a thin surface layer or skin, Fig. 5.60*a*. This condition prevails when the surface is cooled so as to prevent a significant temperature rise due to the friction of the grinding process.[290] When abusive grinding is done without a coolant, a rise in the surface metal temperature creates a compressive thermal stress which causes local yielding when a sufficiently high temperature is reached. When the locally heated yielded metal cools, it shrinks, thereby resulting in tensile thermal stresses, Fig. 5.60*b*. The residual stress distribution is reversed, and the beneficial peening residual compressive stress created by grinding is nullified. An approximation of the maximum value of the surface temperature

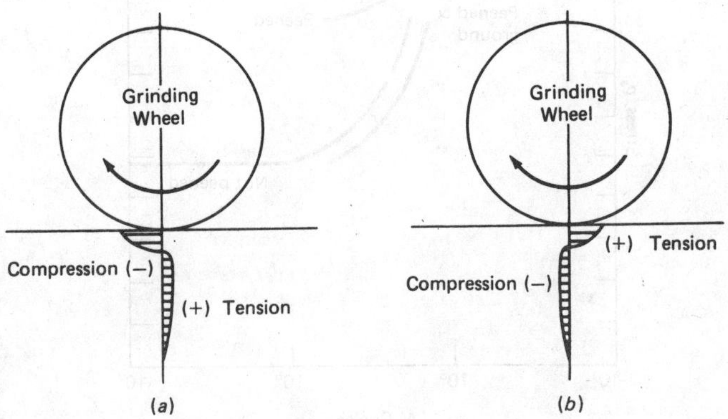

Fig. 5.60. Residual Stress Resulting from Grinding. (*a*) Cooled Surface. (*b*) Uncooled Surface

rise that can be tolerated without inducing a tensile residual surface skin stress can be found by assuming a condition of full two-dimensional thermal restraint, Eq. 2.11.3. For instance, with a carbon steel material at a temperature of 100°F, a yield strength of 30,000 psi, and a $E\alpha/(1-\mu)$ value of 280 psi/°F from Fig. 2.28; Eq. 2.11.3 gives:

$$\sigma_{Y.P.} = \frac{E\alpha\Delta T}{1-\mu} \tag{5.17.1}$$

$$\Delta T = \frac{30{,}000}{280} = 107°\text{F} \tag{5.17.2}$$

At a temperature rise above this, the surface metal is "upset" and upon cooling goes into a residual tensile stress state which is a condition more susceptible to stress corrosion, fatigue, or brittle failure. Peening creates a favorable initial surface compressive prestress which enhances the fatigue limit of the component to which it is applied. If this residual stress is removed by a grinding, polishing, or machining operation the fatigue life is accordingly reduced to essentially that of the original unpeened material,[291] Fig. 5.61. This means that peening or other forms of mechanically prestressing should be performed after finish grinding or machining to obtain geometric shape in order to assure the presence of the desirable prestress.

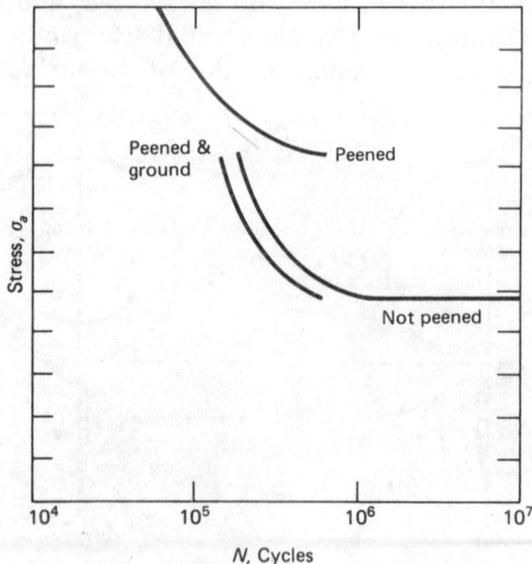

Fig. 5.61. Effect of Peening and Subsequent Grinding on Fatigue Life[291]

(*b*) *Nonselective Stress Pattern*

Fatigue life can be reduced by nonselective location and orientation of residual stresses by the same reason that the selective use of them improves it. Fabrication processes, such as welding, which indiscriminately introduce residual tensile stresses in nonselective locations,[113] magnitude and orientation can have an adverse effect. Pressure vessels, wherein the prime applied stress is a tensile stress acting over the full thickness of the metal, may well fall in this category. Hence, one of the reasons that vessel construction codes[94] require stress-relief of completed vessels prior to service operation is to minimize the potential adverse affect of randomly oriented residual stresses remaining from the fabrication process. Soviet investigations [114] have found that fatigue strength is impaired by residual tensile stresses, and an increase in the endurance limit of 10 per cent has been obtained by the elimination of these residual stresses through thermal stress relieving; while British investigations have shown improvements up to 40 per cent.[115,116]

Residual stresses associated with welding or weld repairs reach yield strength stresses in low strength materials, but do not reach this value in high strength materials which obtain their strength by quench and temper or mechanical working methods. This is due to the lower yield strength property of the deposited weld in its nonstrength enhanced state.[292, 293, 294]

5.18 Effect of the Environment and Other Factors on Fatigue Life

1. *Elevated Temperature Effect*

Many structures, such as vessels or piping for high-temperature steam, gases or liquids, as well as the rotating parts of turbines, etc., must also operate under cyclic conditions; hence, it is necessary to consider failure by fatigue as well as excessive distortion by creep.[226,227,228,229,230] Endurance limit tests on a variety of steels have shown no loss of fatigue strength for temperatures of from 0° to 650°F.[71] In fact, there is a slight improvement in these values for temperatures in the upper portion of this range. This may be attributable to the increase in tensile strength also noticed in this same temperature range, Par. 5.20, and the unimportance of creep, even for carbon steel, at these temperatures. Below[231,232] the creep range, temperature has little effect on the relationship between strain-range and fatigue life. Accordingly, within this temperature limitation, the effect of elevated temperature may be accounted for by multiplying the calculated stress by the ratio of the elastic modulus at the temperature for which the available fatigue curve was established to the elastic modulus at the operating service temperature. This modified value of stress is then used to enter the fatigue curve to find the corresponding number of cycles. At higher temperatures,[233] typical completely reversed stress (mean stress equal zero) σ–n curves indicate fatigue strength uncomplicated by creep, and exhibit the same general shape as room temperature curves with the possible exception that they do not approach their asymptotes as rapidly but with (1), a decrease in fatigue strength with increase in temperature and (2), a less rapid approach to their asymptotes, Fig. 5.62.

Fatigue is a cycle-dependent phenomenon, whereas creep is a time-dependent phenomenon; thus, the time to fracture by fatigue, or to failure by excessive creep may vary for the same combination of alternating and steady stress,[117,118] depending upon the frequency of the alternating stress. In the absence of actual data a method of combining fatigue and creep data for design purposes suggested by Tapsell[119] is shown in Fig. 5.63. The static creep strength at the design temperature is plotted on the abscissa. This may have several values depending upon the degree of distortion considered permissible. Its value for vessels or structures which can undergo some growth without interfering with their operation will be higher than that for rotating turbine parts which must maintain clearances. The fatigue strength for completely reversed stress at temperature is plotted on the ordinate and the two points connected by a straight line. Any combination of static and alternating stresses within the triangle so formed is considered

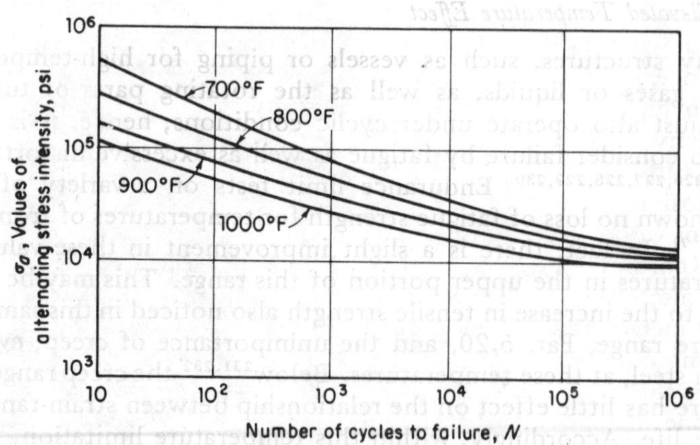

Fig. 5.62. Fatigue Design Curves at Elevated Temperatures

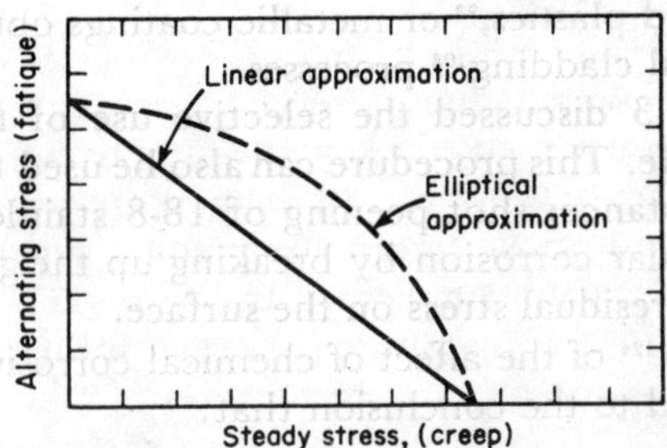

Fig. 5.63. Method of Constructing Curves for Combined Creep and Fatigue[119]

safe. Actually, indications are that the straight-line approximation is somewhat conservative, and an elliptical or power function curve is closer to the actual.[120, 121, 295]

2. *Corrosion Effect*

Corrosion can reduce fatigue life through the surface damage effect of roughening, and also by pitting which reduces the cross-sectional area, thereby increasing the magnitude of the applied stress. A more serious type of damage occurs when the corrosive and fatigue effects act simultaneously,[122] this is called "corrosion fatigue." Here, corrosion occurs both on the surface and in the crack after its formation with the corrosive media acting on the walls and base of the crack, and corroded particles, which separate and lodge in the crack, forming a wedge which forces the crack to deepen.[4] Tests with dry steam have showed no reduction in the endurance limit of carbon steels, whereas with wet steam or water an appreciable lowering of the endurance limit in the order of 20 per cent occurred.[123] This may be even higher with more corrosive media such as salt water, etc.; and many cases of failure in service of boiler and heat exchanger tubes, turbine blades, marine propeller shafts, etc., are attributed to corrosion fatigue.[234, 296, 297, 298, 329]

Means of combatting this are through the use of alloying elements, such as chromium in steel, to increase the ordinary corrosion resistance which also enhances corrosion-fatigue resistance. Likewise, the nonferrous metals, such as phosphor bronze and aluminum bronze, show excellent corrosion-fatigue resistance at low temperature, and the stainless steels at high temperatures. It must be remembered that there is no one material that is a cure-all, but it must be selected for the particular corrosive media and service environment. Another combatting method is through the use of corrosion protective coatings, such as rubber and plastics,[93] or metallic coatings obtained by nitriding, or other metal cladding[108] processes.

Paragraph 5.17.3 discussed the selective use of residual stress to increase fatigue life. This procedure can also be used to forestall stress corrosion. For instance: shot peening of 18-8 stainless steel protects it from intergranular corrosion by breaking up the grains and imposing a compressive residual stress on the surface.

Studies by Frost[124] of the affect of chemical corrosive environments on fatigue have led to the conclusion that:

1. No definite fatigue endurance limit exists when a corrosive environment is present.
2. The lowest fatigue strength for a given material and environment occurs when air, or oxygen, has contact with the material.

The effect of air or oxygen has been demonstrated by comparing fatigue tests conducted in air to those made in an air free environment such as in vacuum, dry inert atmospheres, air-tight protective coatings, or immersion in oil. The fatigue endurance limit of several materials tested in vacuum relative to that in air is shown in Table 5.4, and it has been deduced that any improvement in endurance limit is due to the absence or barrier to oxygen.[299]

Table 5.4 illustrates the large potential increase in the endurance limit of aluminum parts by machining these in vacuum and protecting them by air-tight coatings prior to their contact with air. This is of practical significance in the design of many instruments which must have extremely high fatigue endurance life. On the other hand, it also indicates the endurance limit of steel is not affected by an oxygen environment. At stresses higher than the endurance limit, however, fatigue

TABLE 5.4. EFFECT ON AIR OR OXYGEN ENVIRONMENT ON THE ENDURANCE LIMIT OF MATERIALS[124]

Material	Approximate Ratio, Fatigue Endurance Limit in Vacuum to Endurance Limit in Air
Mild Steel	1.0
Gold	1.0
Copper	1.2
Brass	1.3
Lead	2.5
Aluminum	5.0

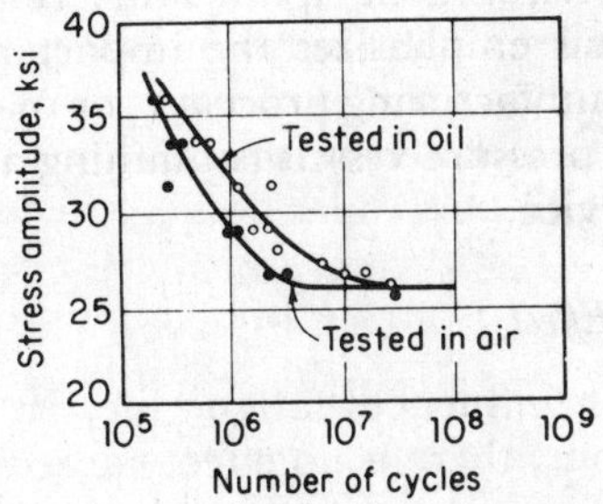

Fig. 5.64. Effect of an Air and an Air-Free Environment on the Endurance Limit of Mild Steel[124]

strength is affected. This is apparent in Fig. 5.64 showing the fatigue strength of mild steel in air, as compared to that submerged in oil. When specimens of the same steel are notched with varying radii grooves and fatigue tested in air, and in an air-free media; no difference in endurance limit due to the environment is noted unitl the notch assumes the sharpness of a crack. For instance, Fig. 5.65 shows the results of a fatigue test of a mild steel specimen with an 0.05 inch root radius notch and indicates no difference in endurance limit when tested in air or in an air-free media. However, when the notch is made as sharp as possible, root radius of 0.0001 inch, the endurance limit of the same material is considerably reduced in an air environment as compared to an air-free one, Fig. 5.66. Protecting mild steel from an

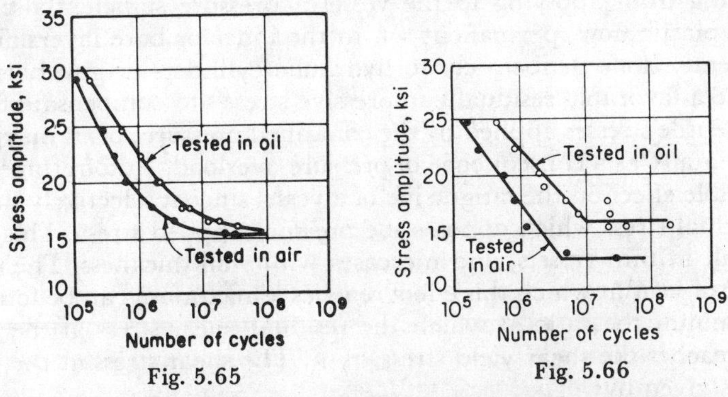

Fig. 5.65. Effect of an Air and an Air-Free Environment on a Mild Steel Specimen with a 0.05 inch Root Radius Notch[124]

Fig. 5.66. Effect of an Air and an Air-Free Environment on a Mild Steel Specimen with a 0.0001 inch Root Radius Notch[124]

oxygen containing environment significantly retards the propagation of fatigue cracks. This emphasizes the importance of avoiding construction features, manufacturing processes, or materials which present crack-like notches in pressure vessels containing a corrosive media and subject to fatigue service.

3. *Cycle Frequency Effect*

The rate of cycling influences fatigue life. In the range 12,000 to 420,000 cycles per hour, there is no effect caused by the varying rate. At rates less than this, recent tests reported on carbon and manganese-molybdenum steels commonly used in pressure vessels have shown that a change in cycling rate from 12,000 to 7 cycles per hour reduced the life about 15 percent.[71] Higher rates of cycling show some increase in endurance limits.[300]

Extremely high rates of 1,800,000 cycles per hour have shown an 8 percent increase in the endurance limit of aluminum and brass. Such variations are not regarded as of practical significance and need not be considered in normal vessel design practice at temperatures below the creep range. At elevated temperatures where the time-dependent[235,236,237] creep phenomenon occurs, a reduction in cycling rate allows more time for creep damage to accumulate and fatigue life is correspondingly reduced, par. 5.20.2.

4. *Autofrettage Effect*

Pressure vessels are often in an autofrettaged state, i.e., a condition resulting from applying to the vessel a pressure sufficiently high to cause plastic flow, permanent set, in the inner or bore layers. This is frequently done deliberately to hydraulic cylinders or gun barrels to induce a favorable residual compressive stress[18] to compensate for the high tensile stresses applied by the contained pressure; or, it may occur accidentally as a consequence of pressure overload. Autofrettage has a favorable effect on the fatigue life of a vessel since it selectively induces a residual stress which opposes the pressure applied stress. This effect is small in thin vessels, and increases with wall thickness. The radius ratio $b/a = K$ at which this effect reaches a maximum can be found by determining the ratio at which the residual shear stress at the inside wall reaches the shear yield stress, $\tau_{Y.P.}$. The shear stress at the inside wall is given by:

$$\tau_a = \frac{\sigma_{t_a} - \sigma_{r_a}}{2} \tag{5.18.1}$$

The residual stress remaining upon release of an autofrettage pressure is the difference between the stress created by the autofrettage pressure and that given by the elastic stress equations using the same autofrettage pressure, paragraph 2.10. In the case of a fully plastic autofrettaged cylinder, the residual tangential stress at the inside wall is found from Eqs. 2.10.10 and 2.8.20, letting the outside-to-inside radius ratio $K = b/a$ and $\log_e (1/K) = -\log_e K$, to be:

$$\underset{\text{residual}}{\sigma_{t_a}} = 2\tau_{\text{Y.P.}} (1 - \log_e K) - 2\tau_{\text{Y.P.}} \log_e K \left(\frac{1+K^2}{K^2-1}\right) \qquad (5.18.2)$$

Similarly, the residual radial stress at the inside wall is found from Eqs. 2.10.8 and 2.8.18 to be:

$$\underset{\text{residual}}{\sigma_{r_a}} = 0 \qquad (5.13.3)$$

Substituting the value of the residual tangential stress from Eq. 5.18.2, and the residual radial stress from Eq. 5.18.3 into Eq. 5.18.1 gives the residual shear stress at the inside wall of the cylinder:

$$\underset{\text{residual}}{\tau_a} = \tau_{\text{Y.P.}} \left[1 - \log_e K - \log_e K \left(\frac{1+K^2}{K^2-1}\right)\right] \qquad (5.18.4)$$

This reaches a value $-\tau_{\text{Y.P.}}$ at a diameter ratio $K = 2.2$, and at this value a fully autofrettaged cylinder offers maximum resistance to the applied pressure stress. The major effect of autofrettage is to reduce the mean stress in the inner part of the vessel wall thickness; hence, to increase fatigue life.

Fatigue tests on a variety of steels by Morrison, Crossland, and Parry[93] have shown an increase in endurance limit of 10 percent for a cylinder of radius ratio $K = 1.6$ to 26 percent for a very thick cylinder of ratio $K = 2.0$. The results of fatigue tests by Davidson, Eisenstadt and Reiner[130] on thick-wall fully autofrettaged cylinders subject to internal pressure also bear out this effect on fatigue life. These tests show an improvement in fatigue strength with increase in radius ratio K, reaching a maximum effect at a K value of approximately 2.0.

5. *Irradiation Damage Effect*

Since the basic effect of irradiation on steels is to increase the tensile and yield strength, and decrease the ductility mechanical properties, Par. 5.10, the fatigue properties are likewise affected and show an increase in the endurance limit and high cycle fatigue strength and a decrease in the low cycle fatigue strength as predicted Eqs. 5.13.11

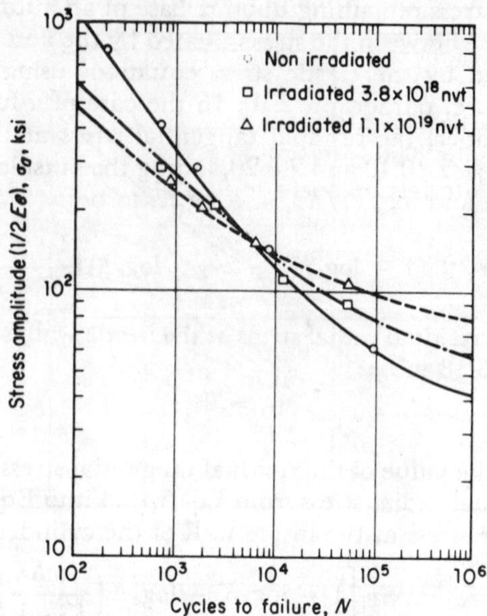

Fig. 5.67. Effect of Irradiation on the Fatigue Strength of ASTM A212-B Carbon Steel[131]

and 5.13.12. Figure 5.67 illustrates this effect for a carbon steel pressure vessel material, ASTM-A212-B, and shows this effect of reducing the low-cycle fatigue life and increasing the high-cycle fatigue life with increasing radiation damage.[131] It also shows a cross-over at approximately 10^4 cycles of the irradiated and non-irradiated material much as for non-irradiated materials of varying tensile strengths, Fig. 5.67. Neutron irradiation has a significant effect on fatigue properties and this affect varies with the type material.[132, 133]

The effect of increased temperature on fatigue properties of irradiated material is much the same as that on unirradiated material and is to reduce the fatigue strength particularly in the low-cycle region.[301]

5.19 Thermal Stress Fatigue

Thermal stresses are produced by restricting the natural growth of a material induced by temperature. The restriction can be an external constraint that prevents free expansion of the entire body, such as by the action of several materials which make up the structure having different coefficients of expansion. The more common case, however, is that of varying temperatures throughout the structure wherein the free thermal growth of each fiber is influenced by that of those surrounding it. Thus thermal stresses are dependent upon temperature distribution, with steep gradients giving rise to high stresses and pointing out the major way of increasing thermal fatigue life, especially when brittle materials or operation in the material brittle transition temperature range is involved. Failure under repetitive cycling is known as thermal stress fatigue. It is analogous to mechanical fatigue,[134] but because of continuously changing temperature during a thermal stress fatigue test, strain concentration, and other local effects, it is often difficult to predict thermal fatigue life from life in mechanical fatigue at constant temperature. It is fair to say, however, that of the several factors influencing fatigue, ductility is the predominate one in thermal fatigue. Metals deform under excessive stress and it is this plastic flow which prevents sudden failure and results in a more favorable stress redistribution, but this deformation also introduces the problem of thermal stress fatigue. Thermal stresses are basically self-limiting, and Coffin's[135,136] experiments have shown that while thermal stress cycling caused cracks to appear early in the life of the specimen, progress of the crack was quite slow in subsequent cycles. Under thermal shock the high induced thermal stress is confined to the region near the contact surface, Pars. 2.14 and 5.22.3. Hence, the high stress intensity, K, causing rapid crack growth is confined to this surface region and relatively slow growth prevails with distance into the thickness.[238] Figure 5.68 shows thermal fatigue or "alligator" cracking on the outside surface of a spherical pressure component subject to severe thermal cooling shocks. The approximately square or equi-distant crack pattern is indicative of an equal bi-axial thermal stress field such as occurs in cylinders, spheres and plates, Pars. 2.11, 2.12, and 3.4. This indicates the importance not only of thermal stress fatigue, per se, but also of the crack formed as a prime nucleus for the start of mechanical fatigue. The fracture path is predominately transgranular at low values of the maximum temperature, and intergranular at high values where creep is pronounced.

There is no one index, applicable to both steady-state and transient temperature conditions, for estimating the relative resistance of materials to thermal stress, but since these are directly proportional to the modulus E and the coefficient of expansion α, and the thermal gradient diminishes with the thermal conductivity k, a useful parameter is $E\alpha/k$. The greater its value, the more susceptible the material is to high thermal stresses. A plot of these thermal shock parameters is shown in Fig. 5.69 for several pressure vessel steels, and while they vary with

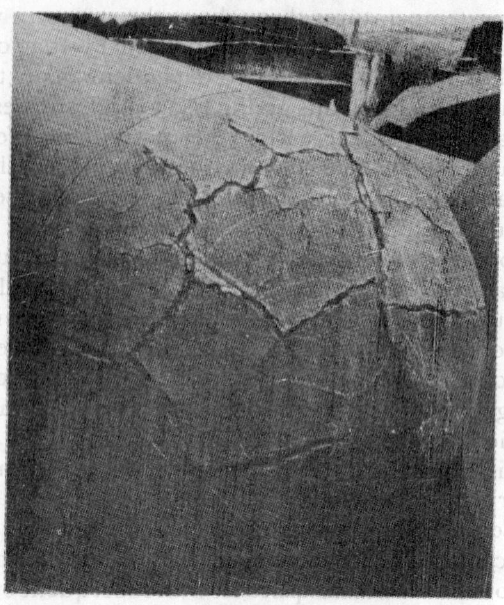

Fig. 5.68. Thermal Faigue, or Alligator, Crack Pattern on the Outside Surface of a Spherical Component.

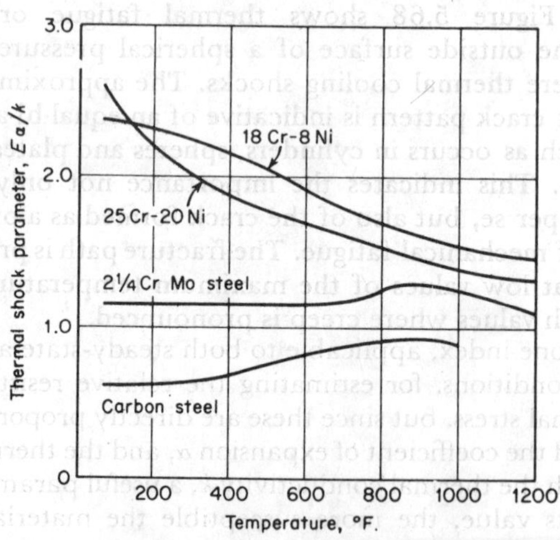

Fig. 5.69. Effect of Temperature on the Thermal Shock Parameter of Several Pressure Vessel Steels

temperature it is seen that the austenitic materials of the 18 Cr—8 Ni type give the highest values. As with mechanically induced fatigue, high thermal stresses can be substained if the number of cycles is limited. Thermal fatigue life can be expressed as a function of the total plastic strain amplitude per cycle by the same Coffin-Manson relationship[137, 138] applicable to low-cycle ambient temperature, Eq. 5.13.2; that is,

$$N^{\frac{1}{2}}\Delta e_p = c \qquad (5.19.1)$$

The constant c varies with the material, and Table 5.5 lists these for several structural and vessel materials. As an example of the use of this equation, determine the allowable number of cycles for the flange of a nuclear reactor vessel built of Type 347 stainless steel in which a stress analysis shows a maximum elastic stress σ_e = 120,000 psi is developed on its face during a temperature change from 600° to 200°F. Constructing the simplified stress-strain diagram for this point in the flange Fig. 5.70, the stress range $\Delta\sigma$ is the ordinate and the strain range Δe is the abscissa. The total strain range Δe is the sum of the elastic strain range Δe_e and the plastic strain range Δe_p, from which the plastic strain range Δe_p can be found and the number of cycles, N, computed as follows:

(1) Elastic thermal stress: σ_e = 120,000 psi (from elastic analysis assuming the material does not yield even though the stress so calculated exceeds the yield point).

(2) Equivalent total strain range: $\Delta e = \sigma_e/E = 120{,}000/27 \times 10^6 = 0.0044$ in./in.

TABLE 5.5. MATERIAL CONSTANT VS. MEAN TEMPERATURE[137]

Material	*Mean Temperature* (°F)					
	70	400	570	660	930	1110
18 Cr—8 Ni, Type 347 . .	0.96	0.78		0.68	0.59	0.44
Carbon Steel, Type 1020 . .	0.46					
1.7 Mn Steel	0.48					
2½ Cr—Mo Steel	0.42					
3 Cr—0.4 Mo Steel	0.60		0.56		0.43	
13 Cr Steel	0.62		0.60		0.57	
2011 Aluminum	0.88					
2024 Aluminum	0.35					
5.0 Mg Aluminum	0.30					

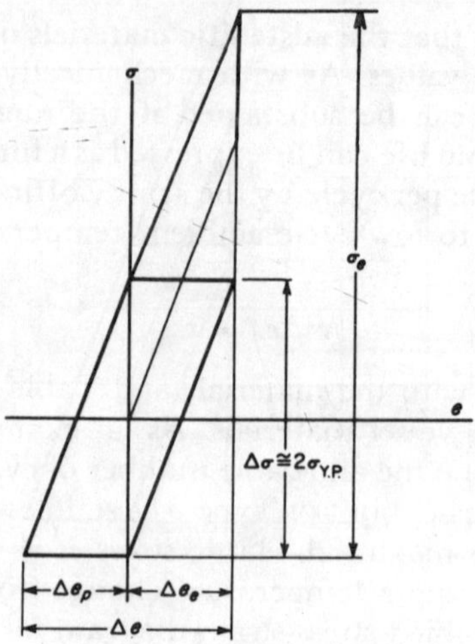

Fig. 5.70. Stress-Strain Diagram

(3) Assuming yield point (30,000 psi) is not exceeded, elastic strain range: $\Delta e_e = 2(30{,}000/27 \times 10^6) = 0.0022$ in./in.
(4) Plastic strain range: $\Delta e_p = \Delta e - \Delta e_e = 0.0044 - 0.0022 = 0.0022$ in./in.
(5) Number of cycles to failure from Eq. 5.19.1: $N^{\frac{1}{2}}\, \Delta e_p = c = 0.78$ (see Table 5.4 for value of c at mean temp. $= \dfrac{600 + 200}{2} =$ 400°F), $N^{\frac{1}{2}} = 0.78/0.0022 = 355$, $N = 126{,}000$ cycles.

Although tests show that mean stresses decrease and shake-down under cyclic straining, they do have a pronounced effect on the total creep strain.[139] As the mean stress level rises, the combined effects of creep and thermal cycling considerably increase the plastic strain per cycle; hence, comparably shorten the total lifetime.[140, 141, 239, 302]

Thermal fatigue and mechanical fatigue frequently occur together in pressure vessels since each is generally a function of the operating service or power demand with time. Just as with mechanical and pressure loadings, cumulative thermal fatigue loadings must also be grouped so as to produce the greatest effect. For instance, consider a vessel subject to the following variations, Fig. 5.71

20 variations: $\Delta T_1 = 250$ heating (Oabc)

10 variations: $\Delta T_2 = 150$ cooling (Odef)

100 variations: $\Delta T_3 = 100$ cooling (dghi)

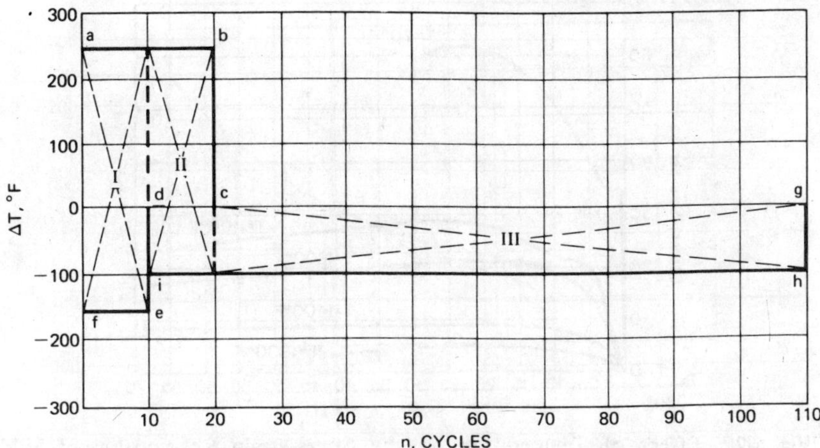

Fig. 5.71. Method of Grouping Thermal Loadings to Produce the Maximum Cumulative Fatigue Effect

In order to establish fatigue life, these ranges must be lumped so as to produce the greatest effects as follows:

$$\text{I, 10 cycles, } \Delta T_1 = 250 + 150 = 400$$

$$\text{II, 10 cycles, } \Delta T_2 = 250 + 100 = 350$$

$$\text{III, 90 cycles, } \Delta T_3 = 100$$

5.20 Creep and Rupture of Metals at Elevated Temperatures

Many pressure vessels and other engineering structures are subjected simultaneously to the action of stress and high temperature. This is the case for vessels and piping used in nuclear power plants, boilers, and the chemical and other process industries. The continual increase in the temperatures of operation has placed great practical importance on the strength of material at elevated temperatures, and the development of materials to cope with this trend.[240, 303]

In general, the strength properties (yield point and ultimate strength) decrease with high temperature while the ductile properties (elongation and reduction in area) increase.[241] The tensile test diagrams for a medium carbon vessel steel for several very high temperatures are shown in Fig. 5.72. At these temperatures the ultimate strength falls rapidly, the yield point becomes less pronounced and above 1100°F loses its characteristics, and the modulus of elasticity

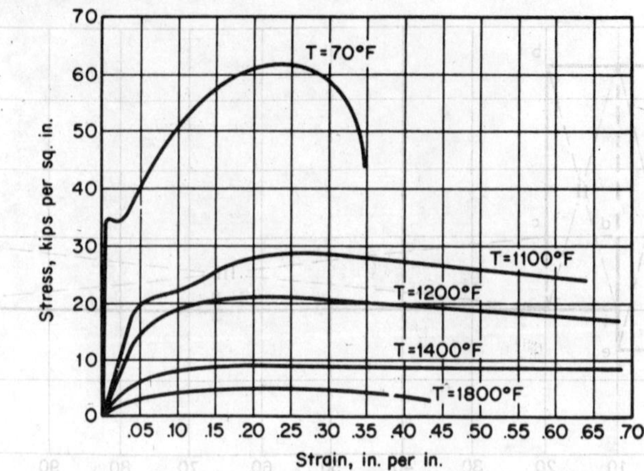

Fig. 5.72. Effect of Temperature on the Stress-Strain Relationship of Mild Carbon Steel

represented by the slope of the straight portion of the curve likewise decreases. The strength properties of this material are not suitable for vessel design at these temperatures, but are important in the vessel forming and forging fabrication process. At these elevated temperatures the ultimate elongation does not always continue to increase with a rise in temperature, but reaches a peak and falls rapidly thereafter to final rupture. Figure 5.73 shows the nature of this loss of ductility. Optimum metal forming temperature can be established by determining this peak, Fig. 5.77. A summary of test results of mild carbon steel properties are contained in Fig. 5.74. These show some increase in ultimate tensile strength, and decrease in the elongation and reduction of area for temperatures up to about 500° F, but reversal of this trend with further increase in temperature.

In hot metal forming, the elapsed time required to complete the operation within the optimum temperature range is critical. Figure 5.75 shows the drop in temperature of various thickness plates upon removal from the heating furnace. The rapid drop in temperature within the first few minutes applies to all thicknesses, but is particularly pronounced for the thinner ones because of their limited heat storage. In actual practice, the available elapsed forming operation time is further reduced by the accelerated cooling effect from cold dies, lubricants and air currents.

Example: **It is desirable to hot form flat plate of mild carbon steel having the characteristics given in Fig. 5.72 into a spherical shape in a press, Fig. 5.76. What is the approximate optimum metal forming temperature to minimize cracking of the metal during the forming operation?**

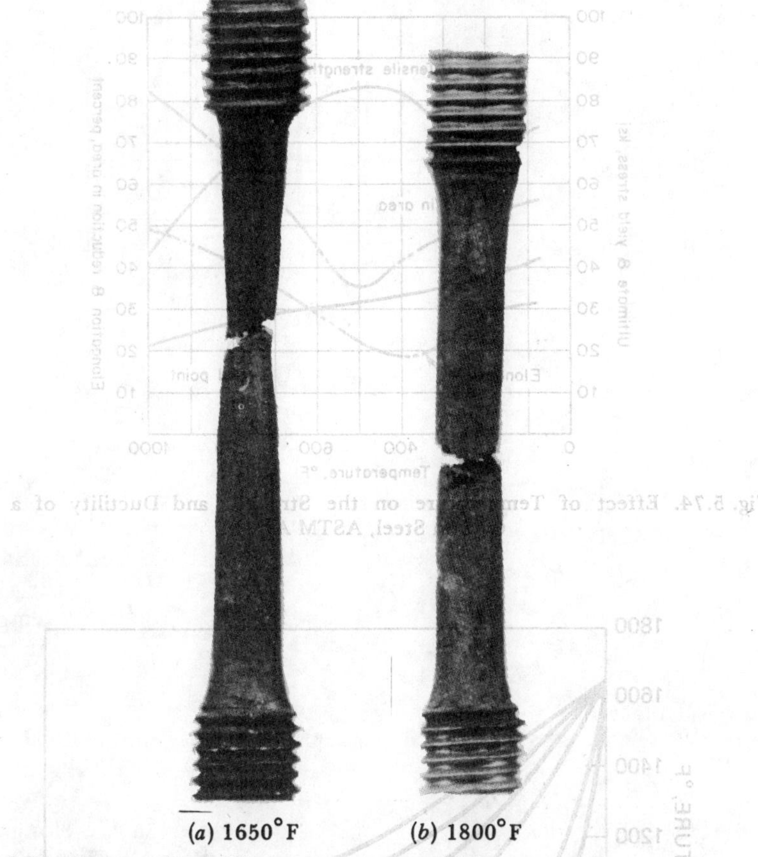

(*a*) 1650°F (*b*) 1800°F

Fig. 5.73. Nature of Ductility Loss at High Elevated Temperatures, Mild Carbon Steel. Specimens Were of the Same Material and Initially the Same Length.

Solution: Using elongation as a measure of the forming property of the metal, and replotting the data of Fig. 5.72 to show the ultimate elongation, e_u, versus temperature, T, and establishing $de_u/dT = 0$ gives an optimum forming temperature of approximately 1600°F, Fig. 5.77. It is significant to note that merely increasing the metal temperature does not always improve the hot forming operation. Also, by similarly plotting the utlimate tensile strength versus temperature, the variation in the pressing force required can be estimated. For instance, the ratio of the press forming force required at the optimum forming temperature of 1600°F to that at another metal forming temperature, say 1100°F, is approximately 7/30 = 23 percent. In the hot metal forming practice, it is frequently necessary to compromise the optimum metal temperature with that for the forming press force capability. Hence, the ordinary standard tensile test mechanical properties can be used as a practical means of determining the optimum metal forming temperature and press capacity.

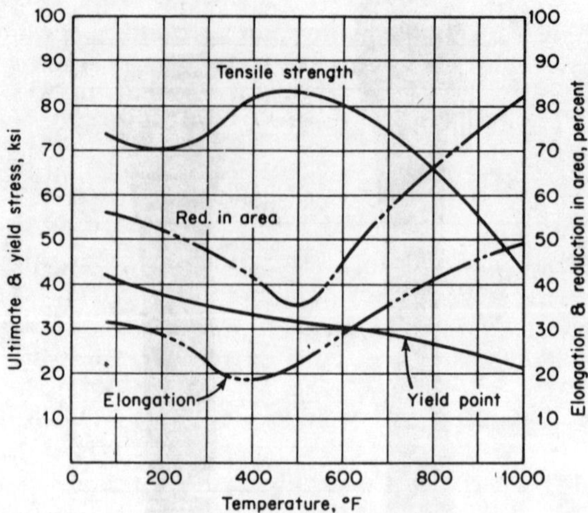

Fig. 5.74. Effect of Temperature on the Strength and Ductility of a Mild Carbon Steel, ASTM A-212B

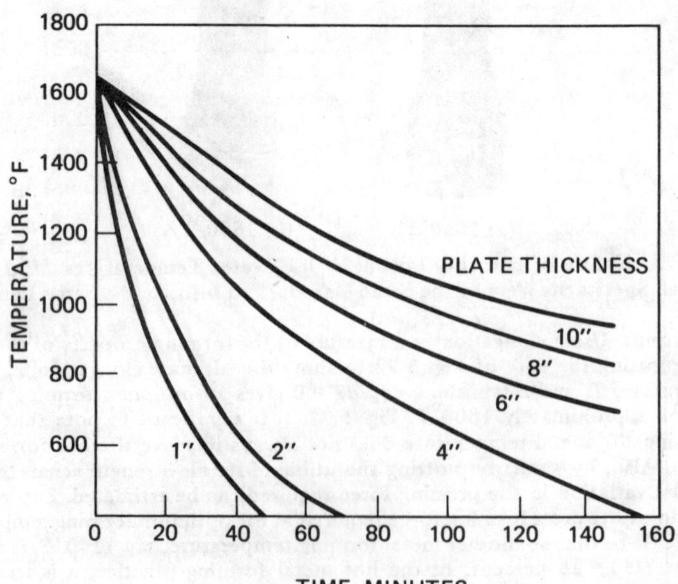

Fig. 5.75. Rate of Air Cooling of Steel Plate Measured at One-Quarter Thickness Below Surface

Fig. 5.76. Heated Circular Steel Plate Being Positioned in Press to Form a Spherical Head (Courtesy The Babcock & Wilcox Co.)

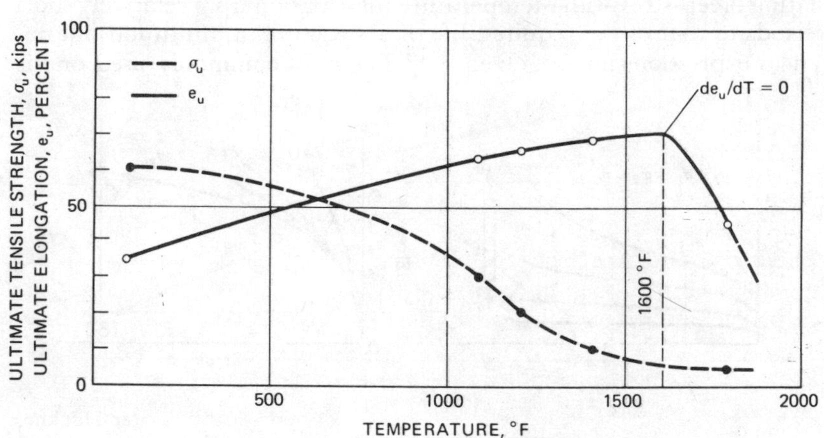

Fig. 5.77. Optimum Forming Temperature

1. *Creep*

At elevated temperatures the deformation of metals continue with no increase in stress. This is called "creep" and is defined as the time-dependent inelastic deformation of materials. Creep properties are obtained by subjecting tensile specimens to a constant load at a constant temperature and observing the axial strains at selected time intervals.[242,243] From this data a series of constant stress creep-strain curves may be plotted, Fig. 5.78*a* or alternatively constant time creep-stress curves, called isochronous creep curves, may be plotted, Fig. 578*b*.

Creep curves for metals exhibit three characteristic behavior regions. In Fig. 5.79*a OA* is the instantaneous deformation that occurs immediately upon application of the load and may contain both elastic and plastic deformation. The portion *AB* is the primary stage in which the creep is changing at a decreasing rate as a result of strain hardening. The deformation is mainly plastic. The portion *BC* is the secondary steady state stage in which the deformation is plastic. In this stage the creep rate reaches a minimum and remains constant as the effect of strain-hardening is counterbalanced by an annealing influence. Here the creep rate is a function of stress level and temperature. The portion *CD* is the tertiary stage in which the creep continues to increase and is also accompanied by a reduction in cross-sectional area and the onset of necking, hence increase in acting stress; thereby, resulting in fracture.[304, 305, 306]

In order to use creep data in design and provide a means of extrapolating creep-stress-time-temperature information from relatively short periods to that of the required life of a structure, a multitude of analytical expressions have evolved.[142,143] The most commonly used one is

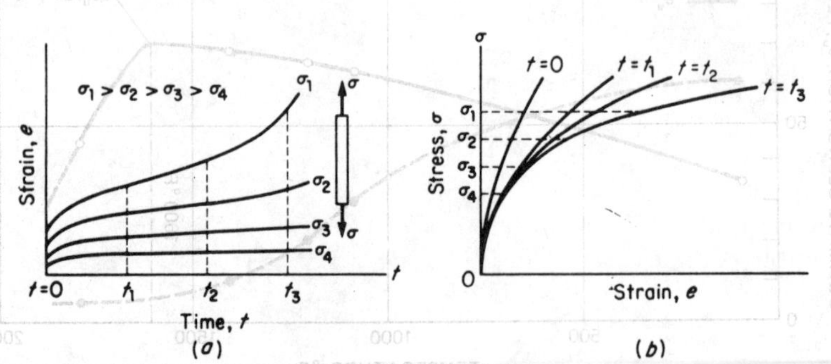

Fig. 5.78. Creep Curves. (*a*) Constant Stress Creep-time Curves, (*b*) Isochronous Stress-Strain Curves

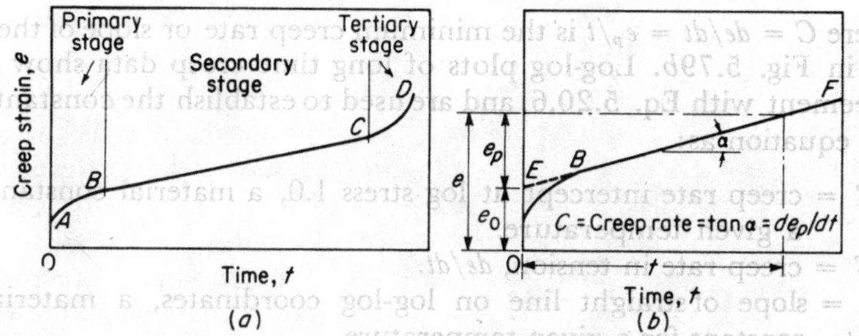

Fig. 5.79. (*a*) Typical Creep Curve, (*b*) Creep Rate from Secondary Stage

the log-log method which considers the total creep strain, e, to be composed of an initial intercept strain e_0 and a time dependent creep strain, e_p,

$$e = e_0 + e_p \tag{5.20.1}$$

That is, in Fig. 5.79*b* the creep-time curve OBF is replaced by the two straight lines *OE* and *EF*. Tests of metals show that e_0 is related to stress by the relation,

$$e_0 = A\sigma^m \tag{5.20.2}$$

where A and m are temperature dependent material constants. Likewise, tests show that the time dependent creep strain, e_p, may also be expressed by the power function,

$$e_p = tB\sigma^n \tag{5.20.3}$$

where B and n are material constants which are dependent upon the material and temperature. Substituting Eqs. 5.20.2 and 5.20.3 in Eq. 5.20.1 gives an expression for the creep strain in terms of stress and time that can be used in selecting a design stress for an estimated life.

$$e = A\sigma^m + tB\sigma^n \tag{5.20.4}$$

For applications involving long time periods, such as steam and gas turbines, boiler superheater tubes, etc., the initial strain e_0 becomes negligible compared to e_p, so Eq. 5.20.4 reduces to,

$$e = e_p = tB\sigma^n \tag{5.20.5}$$

$$C = \frac{e_p}{t} = B\sigma^n \tag{5.20.6}$$

where $C = de/dt = e_p/t$ is the minimum creep rate or slope of the line *EF* in Fig. 5.79*b*. Log-log plots of long time creep data show good agreement with Eq. 5.20.6, and are used to establish the constants for this equation as:

B = creep rate intercept at log stress 1.0, a material constant for a given temperature.
C = creep rate in tension, de/dt.
n = slope of straight line on log-log coordinates, a material constant for a given temperature.

This log-log straight-line relationship is shown in Fig. 5.80 for a $2\frac{1}{4}$ percent chromium 1 percent molybdenum steel.[244]

Equation 5.20.5 is the usual design basis for structures that must undergo longtime use and/or retain prescribed deformations. One design method frequently used is based on that stress to give a maximum permissible arbitrary amount of creep, usually 0.01 or 0.10 per cent per 1000 hours, corresponding by extrapolation to 1 per cent extension in 10,000 and 100,000 hours. Such typical creep stress curves are shown in Fig. 5.81 for several carbon, low alloy, and austenitic pressure vessel materials.[144] Chromium, molybdenum, and nickel are major alloying elements for high-temperature service metals. In using

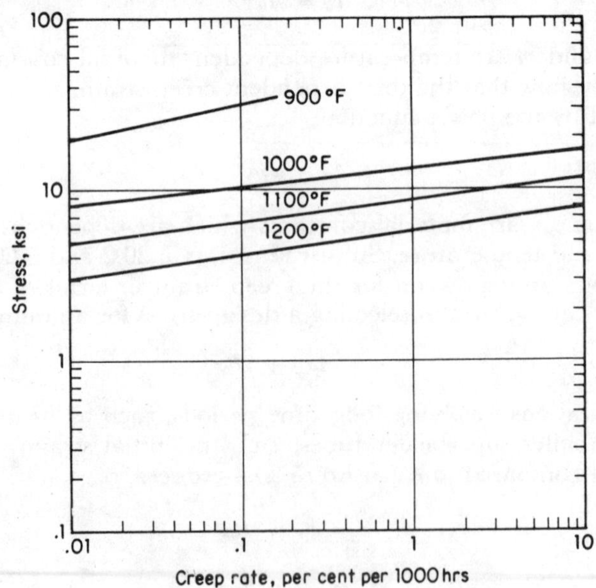

Fig. 5.80. Creep Rate Curve for 2¼% chromium 1% Molybdenum Steel (*Data Courtesy The Babcock & Wilcox Company*)[144]

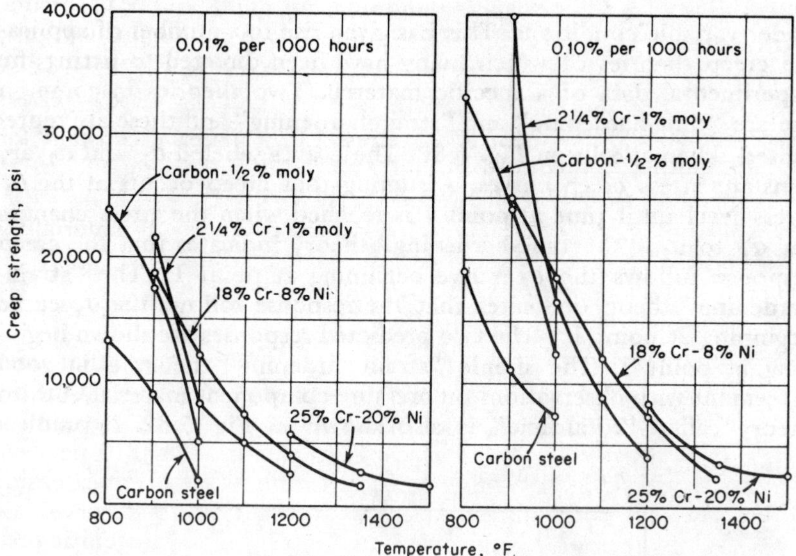

Fig. 5.81. Creep Strength of Several Pressure Vessel Materials[144]

creep data, the designer must establish the expected service life and corresponding amount of permissible permanent deformation; and accordingly, choose the stress that satisfies these conditions.[145] As an example, structures that involve closely fitting moving parts, such as turbines, are designed for a low value of permissible creep; whereas, vessels, heat exchanger tubes, etc., are designed for a higher creep since small deformations do not influence their operation.[146,147]

The creep behavior of materials is not only sensitive to stress and time, but also to their environment (atmosphere, neutron irradiation etc.,) physical properties, past strain history, etc. For instance, carburizing, oxidizing and nitriding atmospheres increase creep life, while neutron irradiation exposure may reduce it. Accordingly, while it is necessary to have laws that aid the extrapolation of creep data because industrial progress cannot wait for the results of lifespan tests of the structure; it is essential to use creep data obtained from maximum available time tests, conducted in environments simulating the actual service one. Extrapolation of short time data to long times are not always reliable and must be used with caution.

Equation 5.20.5 describes only the creep strain during a constant stress and does not cover the case of a variable stress. Since pressure components are subject to varying stresses, methods are required for using the constant stress creep response to predict their behavior under variable conditions. This has given rise to a number of applicable creep theories of which many have been directed to fitting the experimental data of a specific material. Two theories in common use are "time-hardening" and "strain-hardening" and these are represented schematically in Fig. 5.82. The curves labeled σ_1 and σ_2 are constant stress creep curves. Assuming that creep occurs at the σ_1 stress level until time at point 1 is reached when the stress changes for σ_1 to σ_2; "the time-hardening" theory indicates that the creep response follows the σ_2 curve beginning at point 1′. The "strain-hardening" theory indicates that the response follows the σ_2 curve beginning at point 1″. The two predicted responses are shown beginning at point 1. The simple "strain-hardening" theory is in good agreement with observations on pressure component material. A third theory called "total-time" is also shown in Fig. 5.82. It predicts

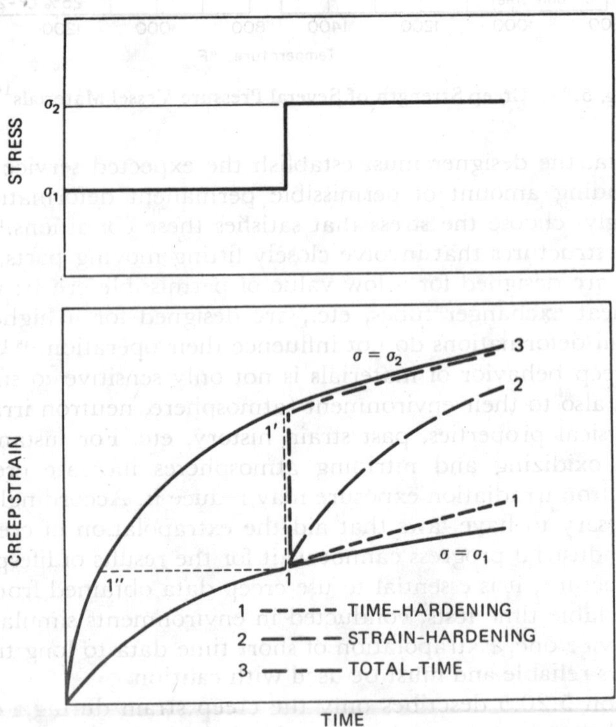

Fig. 5.82. Schematic Representation of Time-Hardening, Strain-Hardening and Total-Time Theories of Creep Response Under a Stepwise Varying Load

creep behavior greatly different from observations and is not advocated as a method of predicting creep behavior. However, it is useful in the approximate analysis of pressure components, and together with "time-hardening" theory forms the bounds for all other theories for predicting creep response to variable stress. This is a useful tool when the actual creep law is unknown.

When creep strengths at time and temperature are not covered by existing data, the Larson-Miller parameter determination method may be used for estimating this information. This is

$$P = T(C + \log_{10} t)$$

where

t = time at T, hours
C = a constant, 20 for steel
P = a parameter
T = absolute temperature, ^0R

This equation is based on a parametric study of creep strengths developed at different times and temperatures; and hence, it is a method of determining an equivalent time at some temperature from a known time at another temperature. This is useful not only in estimating temperature material properties, but also in appraising the qualification acceptability of uncontemplated temperatures. The following examples are illustrative:

Example 1: A design established on a creep rupture basis of 1000°F at 100,000 hours is now required to operate at 1100°F. What is the estimated service life?

Putting $T = 460 + 1000$, and $t = 100{,}000$ in this parameter equation gives

$$P = (460 + 1000)\ (20 + \log 100{,}000)$$

$$P = 36{,}500$$

Now using this equation with $P = 36{,}500$, $T = 460 + 1100$ and solving for t gives

$$36{,}500 = 1560(20 + \log t)$$

$$\log t = 3.4$$

$$t = 2{,}500 \text{ hours}$$

or the estimated service life has been reduced by 97.5 percent.

Example 2: The properties of a material have been qualified at a stress-relieving temperature of 1150°F for 50 hours. A vessel made from this material undergoes a two-hour temperature excursion to 1180°F during the stress-relieving operation. What is the equivalent of this excursion in terms of hours at the material qualification temperature of 1150°F?

Putting $T = 460 + 1180$ and $t = 2$ in the parameter equation gives

$$P = (460 + 1180)\ (20 + 0.303) = 33{,}300$$

Now placing $P = 33{,}300$ and $T = 460 + 1150$ in this equation and solving for t gives

$$33{,}300 = 1610\ (20 + \log t)$$

$$\log t = .683$$

$$t = 4.8 \text{ hours}$$

or 4.8 hours must be added to the normal time at 1150°F to obtain the total time. If this does not exceed the material qualification time, the material is acceptable.

2. *Creep Rupture*

Failure due to creep rupture is an important design consideration.[245,246,247,248] Under constant stress and temperature conditions expected service life can be established from standard creep rupture data.[148,149] However, since most members are not subject to either constant stress or constant temperature, creep-rupture damage criteria which will predict time to rupture in such members having multi-axial states of stress using time to rupture data obtained from tension tests have evolved.[150,249] Two of these are the "life-fraction" rule and the "strain-fraction" rule.

The "life-fraction" rule is based on the premise that the expenditure of each individual rupture life-fraction of the total life at elevated temperature is independent of all other fractions of the life to rupture, and that when the fractional life used up at different stress levels and temperatures is added up, it will equal unity; namely,

$$\sum \frac{t_i}{t_{ri}} = 1 \tag{5.20.7}$$

where t_i is the time at temperature i and t_{ri} is the creep rupture time at temperature i. As an example, a cylindrical tube made of $2\frac{1}{4}$ percent chromium material with the creep-rupture properties shown in Fig.

5.83 is initially designed for a life of 100,000 hours at 1000°F with a stress of 12,000 psi. After 10,000 hours (1/10 of the design life), a malfunction subjects the tube to a stress of 15,000 psi at 1100°F. The predicted remaining life would then be 0.9 of 500 hours, or about three weeks. If the malfunction is corrected after two weeks and operation is returned to the 12,000 psi, 1000°F condition the life remaining in the tube is 100,000 (1-1/10–2/3) = 23,000 hours. Thus, the malfunction lowered the original life expectancy from 100,000 to 33,000 hours or a reduction of 67 percent.

The "strain-fraction" rule is expressed by the relation

$$\sum \frac{e_i}{e_{Ii}} = 1 \tag{5.20.8}$$

where e_i is the strain for a given stress and temperature and e_{Ii} is the strain at rupture under the same stress and temperature. The methods of Eqs. 5.20.7 and 5.20.8 are simple, give results in good agreement with experiment, and are well adapted to design analysis. The strain-fraction rule best fits those materials which exhibit appreciable cracking throughout their life while the life-fraction rule is in better agreement with those materials which show little cracking until final rupture is approached.[151,152] Some studies have shown that a geometric mean of

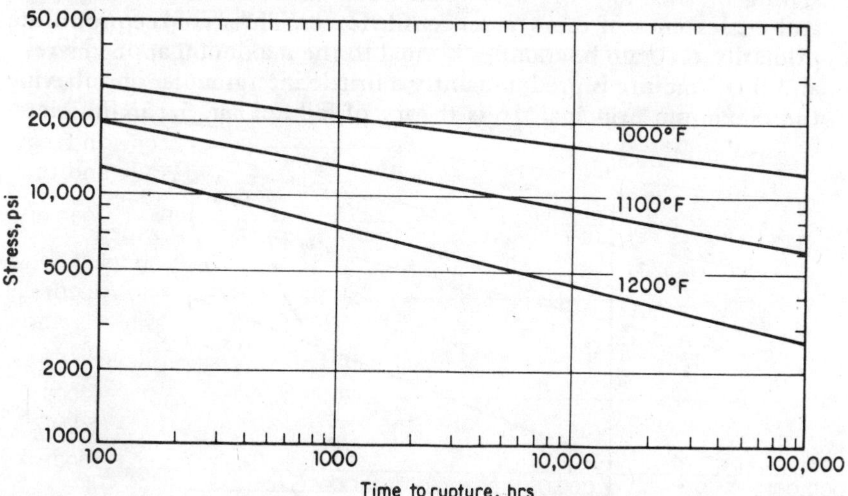

Fig. 5.83. Creep Rupture Curve for 2¼% Chromium 1% Molybdenum Steel (*Data Courtesy The Babcock & Wilcox Company*)[144]

these two approaches represents a good over-all data fit; that is,

$$\sum\left(\frac{t_i}{t_{Ii}}\cdot\frac{e_i}{e_{Ii}}\right)^{\frac{1}{2}} = 1 \tag{5.20.9}$$

Creep-rupture and fatigue subscribe to the same linear cummulative damage concept, and these effects may be combined by

$$\sum\frac{n}{N} + \sum\frac{t_i}{t_{Ii}} = 1 \tag{5.20.10}$$

It is recognized that this simple rule cannot fully account for large amounts of strain hardening, metallurgical changes and order of loading; accordingly, a damage factor less than unity (0.6–1.0) is frequently employed.[153,330]

Figure 5.84 shows the nature of the creep-rupture strain at constant temperature. Curve A is the typical one based on nominal or "engineering" stress, Eq. 5.2.2. If this strain based on constant "true" stress, Eq. 5.2.1, curve B would describe its behavior, i.e., it would be linear until rupture. In practice, however, rupture is described by curve C, which indicates that geometric instability (necking, Par. 5.2) is not the principal cause of rupture, but rather it is deterioration or damage of the material itself.

The phenomenon of creep rupture is one of formation of voids by sliding or shearing forces at grain boundaries, and the enlargement and coalescence of these under tensile forces. These voids concentrate primarily on grain boundaries normal to the maximum applied stress, and final fracture is predominantly a brittle intergranular one obeying the maximum principal stress theory of failure, Par. 5.15.1(a). Mate-

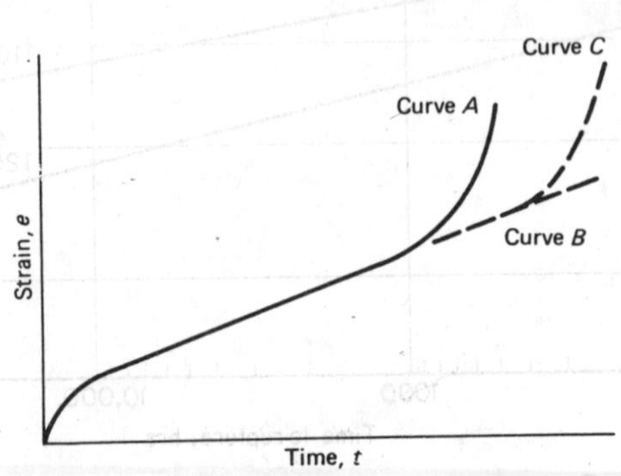

Fig. 5.84. Variation of Strain with Time.

rial damage accumulates with time with the result that the proportion of the material available to carry load is reduced, thereby increasing the applied stress until fracture occurs. This damage (voids, fissures, cracks, etc.) occurs essentially uniformly over a relatively large cross-sectional area, or volume, of the material and the resulting stress is an average one. Accordingly, in describing this stress condition the ligament efficiency analogy, Par. 1.5, is applicable. In this case it is called a "continuity damage factor," ψ, and is the proportion of the material continuous effective area, A_{eff}, remaining after a period of time, t, expressed as a percent of the original cross-sectional area, A_0:

$$\psi = \frac{A_{eff}}{A_0} \tag{5.20.11}$$

If a specimen of initial cross-sectional area, A_0 is subject to a force, F, the initial stress, σ_0, is

$$\sigma_0 = \frac{F}{A_0} \tag{5.20.12}$$

After a period of time, t, the effective stress, σ_t, increases to

$$\sigma_t = \frac{F}{A_{eff}} = \frac{\sigma_0}{\psi} \tag{5.20.13}$$

Assuming the rate of decrease in material damage continuity is a function of stress, and representing this as a power function

$$\frac{d\psi}{dt} = - C\sigma_t^n \tag{5.20.14}$$

and substituting the value of σ_t form Eq. 5.20.13 into Eq. 5.20.14 gives

$$\frac{d\psi}{dt} = - C\left(\frac{\sigma_0}{\psi}\right)^n \tag{5.20.15}$$

At time $t = 0$ there is 100 percent material continuity, $\psi = 1$, whereas at time of rupture, $t = t_r$, the continuity is reduced to zero, $\psi = 0$. Integrating Eq. 5.20.15 for constant stress, σ_0, gives

$$\int_1^0 \psi^n \, d\psi = - \int_0^{t_{r0}} C\,\sigma_0^n \, dt \tag{5.20.16}$$

$$- \frac{1}{n+1} = - C\,\sigma_0^n \, t_{r0} \tag{5.20.16a}$$

$$t_{r0} = \frac{1}{C(n+1)\,\sigma_0^n} \tag{5.20.17}$$

where t_{r0} is the time to rupture under stress σ_0.

Under a variable nominal stress condition, $\sigma_{0,t}$, the integration of Eq. 5.20.15 gives the time to rupture t_r under stress $\sigma_{0,t}$ as

$$\int_1^0 \psi^n\, d\psi = -\int_0^{t_r} C\,\sigma_{0,t}^n\, dt \tag{5.20.18}$$

$$1 = \int_0^{t_r} C(n+1)\,\sigma_{0,t}^n\, \mathrm{dt} \tag{5.20.19}$$

From Eq. 5.20.17 the time to rupture, t_{r0}, for a constant stress of $\sigma_{0,t}$ maintained throughout the life of the specimen is

$$t_{r0} = \frac{1}{C(n+1)\,\sigma_{0,t}^n} \tag{5.20.20}$$

and substituting from Eq. 5.20.20 into Eq. 5.20.19 gives

$$\int_0^{t_r} \frac{dt}{t_{r0}} = 1 \tag{5.20.21}$$

Equation 5.20.21 is the continuous integral form of the "life fraction rule," Eq. 5.20.7, which is widely used to evaluate the effect of step-block changes in temperature on the service and operation of high temperature components.

The main components of a fossil-fuel fired boiler which operate in the creep range are the superheater and reheater tubes, and their junction headers and associated piping. In these components, the pressure is internal and the heat source is external. They are designed for the operational life of the boiler using methods as described above. Corollarily, if metal temperatures are monitored in service, life of these components can be predicted by employing creep data for the particular material. When this is less than the planned operational life, changes in boiler operation, tube arrangement or tube material may be altered to achieve the required life. Creep rupture failures are the most frequent type occuring in superheater and reheat boiler tubes. Figure 5.85 shows the nature of the creep rupture of such a tube in which the fracture path is predominantly intergranular and the fracture face is perpendicular to the direction of maximum stress (hoop stress). Gross swelling distortion and rupture edge thinning are char-

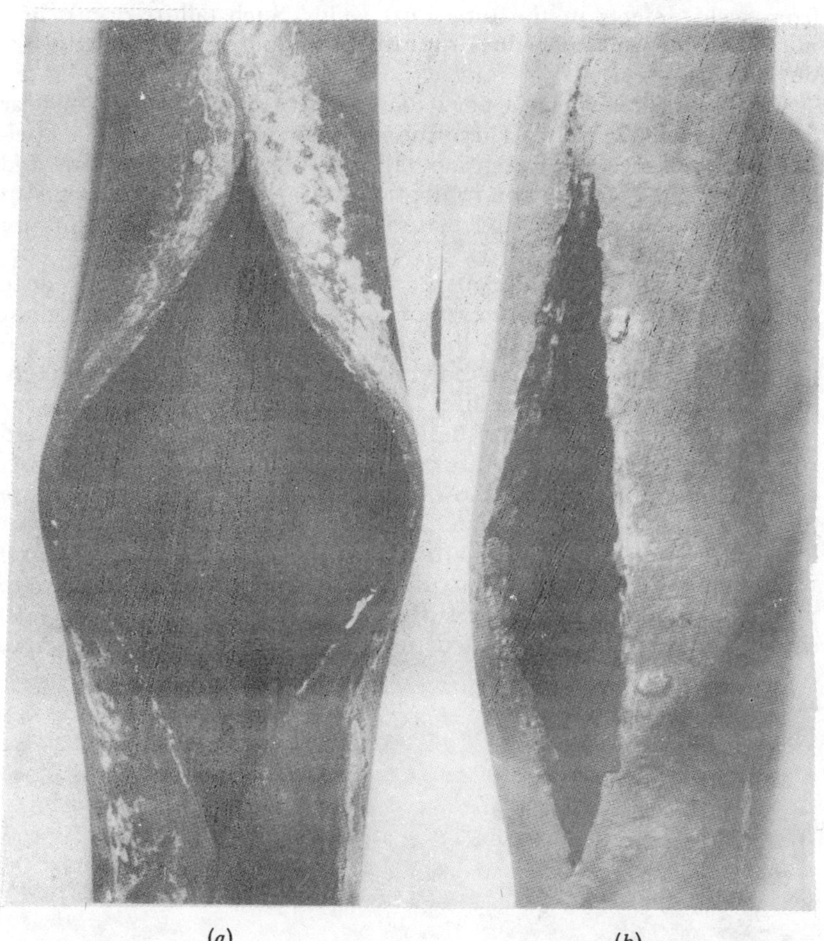

(*a*) (*b*)

Fig. 5.85. Tube Creep Rupture, Showing: (*a*) Gross Swelling and Rupture Edge Thinning, Characteristic of Short-Time Creep; and (*b*) Limited Swelling and Little Rupture Edge Thinning, Characteristic of Long-Time Creep.

acteristic of short-time creep failure, Fig. 5.85*a*, whereas limited swelling and little rupture edge thinning is characteristic of long-time creep rupture, Fig. 5.85*b*. Here, design emphasis must be placed on correct estimation of maximum metal temperature, since small increases cause very large reductions in time to rupture at a given stress. A major contributing factor to these failures is creep cracking starting from weld defects in the tube butt welds. Such failure occurs in a much shorter time than does rupture of the parent tube metal without defects.

Creep rupture is also a particular concern in the rods of nuclear fuel elements, Fig. 5.86. These tubular rods encase uranium fuel pellets and their swelling during operation applies internal pressure and heat. Here the high neutron irradiation damage environment poses an accelerated reduction in the material creep strength with exposure time, Pars. 5.18.5 and 5.20.1.

In high temperature design the basic approaches are: (a) reduce the number, size and complexity of components subject to the highest temperatures, and (b) reduce the loads (mechanical stress) and thermal gradients (thermal stress) on individual components. When these techniques longer give tolerable life, it becomes necessary to circumvent the creep problem by novel design features. Internal insulation and water jackets, walls, or coils are two methods frequently used in vessel construction. Figure 5.87 shows a high pressure, high temperature (4,000°F) pebble bed air heater. In order to preclude the danger of local refractory insulation failure causing a "hot spot" on the vessel wall, Par. 6.14.4, multiple circuit helical water cooling coils, were placed adjacent to the inside surface of the vessel wall. Figure 7.30 shows a similar scheme adapted to a prestressed reinforced concrete

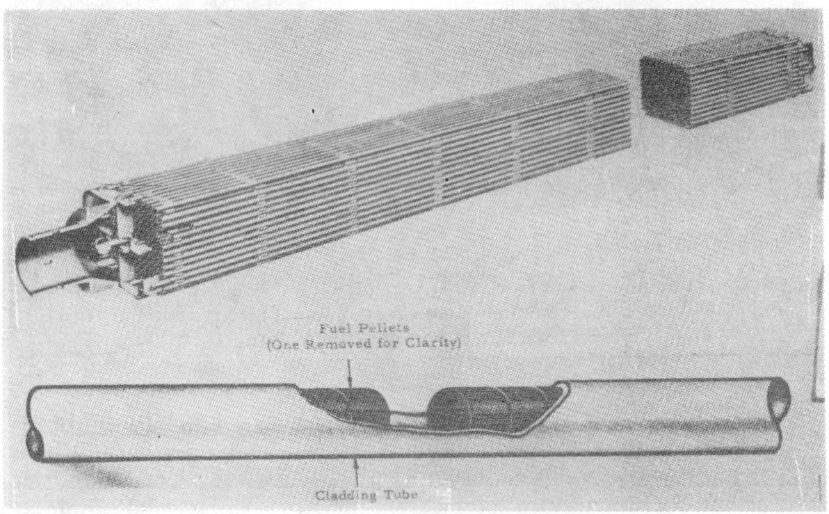

Fig. 5.86. Nuclear Fuel Element Consisting of Tubular Fuel Rods Encasing Uranium Pellets (*Courtesy of The Babcock & Wilcox Company*)

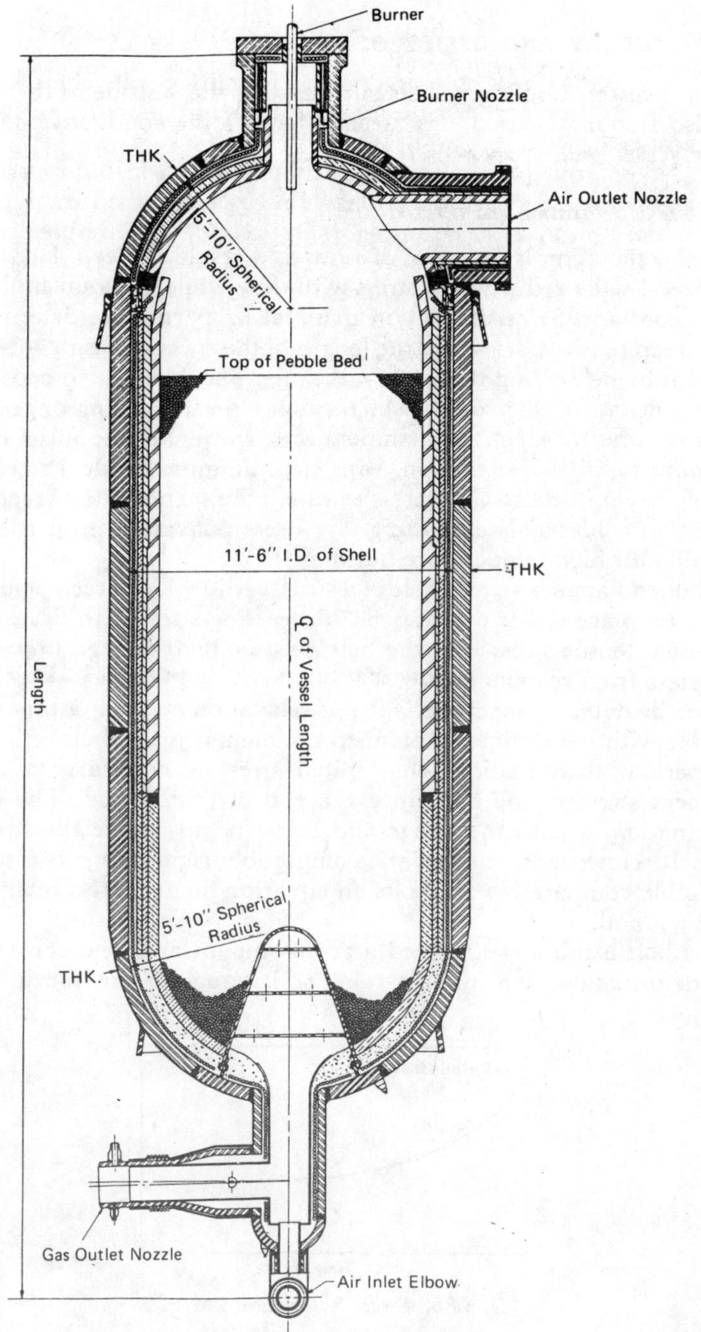

Fig. 5.87. Internally Insulated and Internally Cooled Pressure Vessel.

pressure vessel. Cooling jackets surrounding the outside of the vessel are also frequently used as a means of providing conductive cooling of the vessel wall.

3. *Stress Relaxation and Stress Relief*

Stress relaxation is the relief of stress as a result of creep. It is characterized by the reduction of stress with time while the total strain remains constant. Stress relaxation material properties are determined from creep tensile tests where the length of the specimen is maintained constant by decreasing the stress with time. The result is a creep stress relaxation curve, Fig. 5.88, which relates the remaining or residual stress to time for a constant temperature. The higher the initial stress, the more rapid the relaxation, with the minimum residual stress becoming asymptotic to that stress at which the second stage creep rate is nil.[250, 251] Materials exhibiting low stress relaxation properties are desirable for high temperature bolting.[331, 332]

A bolted flange is an example of a structure in which creep deformation takes place under conditions of decreasing stress with time. Here the initial tensile stress σ_0 in the bolt must be high enough to prevent the stress from relaxing below that necessary to prevent leakage. This is done by entering the material stress relaxation curve at a time comparable with the ordinarily planned maintenance and bolt retightening period, then finding what initial stress is necessary to retain sufficient stress in the bolt until the end of that period. The creep deformation of both the flange and bolt tend to reduce the stress in the bolt. However, if that of the comparably rigid flange is assumed negligible compared to the bolts an equation for the stress relaxation can be found.

If a bolt is initially stretched a given amount and held constant at this deformation, the initial strain, e_0, is equal to the elastic time-

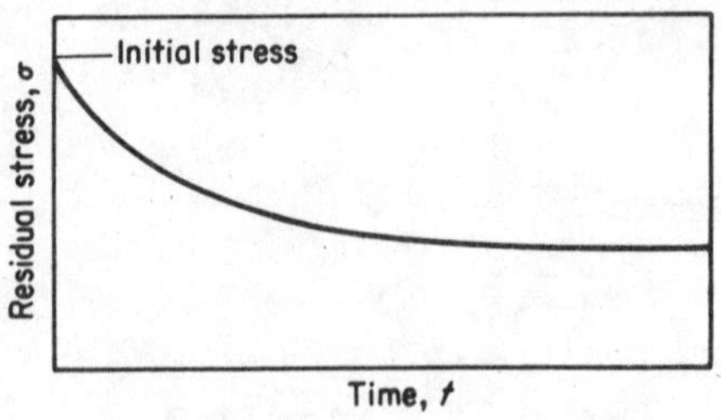

Fig. 5.88. Stress Relaxation Curve

dependent strain, e, plus the strain due to creep, e'.

$$e_0 = e + e' \tag{5.20.11}$$

substituting the values $e_0 = \sigma_0/E$ and $e = \sigma/E$ in Eq. 5.20.11 and differentiating with respect to time gives

$$\frac{1}{E}\frac{d\sigma_0}{dt} + \frac{de'}{dt} = 0 \tag{5.20.12}$$

Assuming the creep law of Eq. 5.20.5 and noting

$$\frac{de'}{dt} = C = B\sigma^n \tag{5.20.13}$$

and combining Eqs. 5.20.12 and 5.20.13 gives

$$dt = -\frac{1}{EB}\frac{d\sigma}{\sigma^n} \tag{5.20.14}$$

and integrating between the limits σ and σ_0 gives

$$t = \left[\frac{1}{(n-1)BE}\right]\left[\frac{1}{\sigma^{n-1}}\right]\left[1 - \left(\frac{\sigma}{\sigma_0}\right)^{n-1}\right] \tag{5.20.15}$$

where t is the time required for the initial stress σ_0 to decrease to σ.

Equation 5.20.15 considers only the elasticity attributable to the bolt shank proper; whereas, in practice there is additional elasticity in the bolt end fastening. When this is considered, an expression similar to Eq. 5.20.15 can be developed which shows that as a result of this additional bolt end fastening elasticity, the relaxation of the bolt stress proceeds more slowly than in the case of assumed fixed ends; that is, a higher percentage of the total elongation is now elastic. Such additional elasticity can be augmented by the use of tall sleeve nuts or a combination of long bolts with collars under the nuts. This requires large creep deformations before a lower stress level is reached.

High temperature effects also have their desirable attributes. The rate at which the effect of strain-hardening is removed depends upon the temperature. In practice it is frequently desirable to eliminate these effects resulting from fabrication procedures. This can be done in a short time by annealing the metal at certain specific high temperatures dependent upon the type of metal, or the same effect can be obtained at a much lower temperature acting over a longer period of time.[154] The same relaxation phenomenon is also used to reduce the residual stresses induced in weldments by their differential cooling during the fabrication process.[155] This is commonly called "stress-relieving" and consists of uniformily raising the temperature of the structure to a high temperature for a brief period of high relaxation followed by a slow cool which minimizes the thermal gradients throughout, thereby, avoiding the reinstatement of high thermal stresses.[156,157] Knowing the creep properties of the material, the value of the residual stress can be reduced to any desired level. It is likewise a factor that should be considered in the detail design of parts embodying stress concentrations and subjected to the simultaneous action of stresses and high temperature.[158,159] At points of high stress concentration, the rate of creep is larger; hence, creep will result in a more favorable stress redistribution.[160] Low residual stresses enhance fatigue life and reduce susceptibility to brittle fracture.

4. *Effect of Irradiation on Creep*

Radiation affects the micromechanism related to creep by disrupting the internal structure of the material. Its manner is unpredictable, therefore, it is desirable to determine creep parameters under operating conditions. In general, tests have shown that larger creep occurs under irradiation than under equivalent thermal tests only thus indicating the significance of irradiation which is believed attributable to atomic dislocation climb by the absorption of irradiation induced vacancies.[161]

5.21 Hydrogen Embrittlement of Pressure Vessel Steels

Many media which pressure vessels must contain produce critical changes to the physical properties of the vessel material during service One of these that is often encountered is hydrogen, which under the action of high pressure and/or high temperature produces two effects: (1) a diffusion into the metal as atomic hydrogen and a process of recombining to its molecular form within the metal, thereby creating extremely high pressures with resulting surface bulging or blistering, and (2) a metallurgical decarburizing, and reducing effect on sulfides or oxides present in the steel. Hydrogen attack results in a loss of ductility, fracture strength and a resultant cracking or delayed brittle type failure.[162,163] This phenomenon is called "hydrogen embrittlement." The effects of hydrogen become more pronounced as the strength levels of steels are increased.[164,252,253] Hence, the selection of high strength steels to obtain an advantage for high pressure large diameter minimum weight vessels deserves particular economic and engineering consideration. Factors such as increased strength, the presence of notches or other sources of stress concentration, or cold working make the effects of hydrogen more pronounced. This action is retarded by the introduction of stable carbide forming alloy elements, and the chromium-molybdenum steels are frequently used for this service. These steels become increasingly hydrogen embrittlement resistant with increase in the alloy content,[162, 165, 307] Fig. 5.89.

The mechanism of hydrogen damage in steel is one of the propagation of microcracks under high stress. When a combination of hydrogen and stress becomes critical at these stress risers, crack propagation proceeds away from the point of high hydrogen concentration and is arrested in a region of low concentration. The hydrogen then moves through the material along the stress gradient to the crack tip to establish a new stress concentration tip—and the process repeats itself to failure. Since the diffusion of hydrogen from the old to the new crack tip requires a period of time ranging from minutes to days, depending on the hydrogen available, delayed fracture occurs. Embrittlement is accelerated by slow strain rates which allow time for the hydrogen to accumulate at the crack tip, and at elevated temperatures where diffusion rates are high.[166,167] However, if hydrogen is present in the material and the stress, residual plus acting, is greater than that value known as the lower critical stress a brittle fracture will occur. The time may vary from minutes to days, and the stress may be as low as 20 per cent of the yield strength. High strength materials, and particularly the heat treated steels, are very susceptible to this phenomenon. This is shown in Fig. 5.90 for a steel which has been heat treated to three

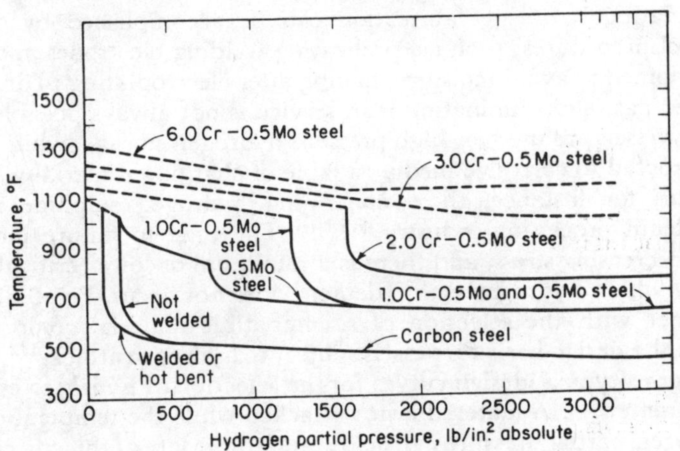

Fig. 5.89. Recommended Maximum Operating Limits for Steels in Hydrogen Service[162,165]

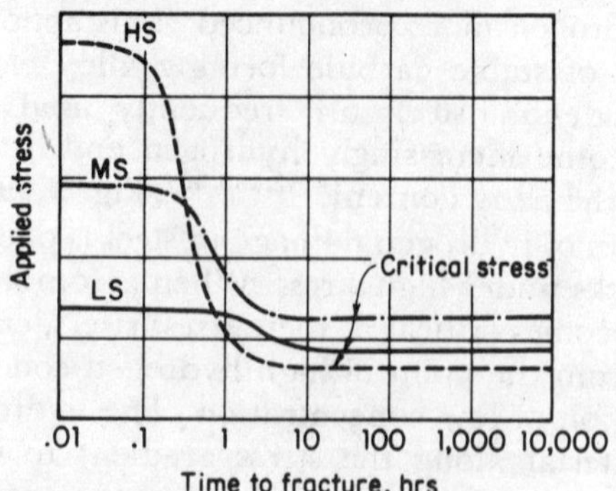

Fig. 5.90. Hydrogen Embrittlement Delayed Failure for a Steel Heat Treated to Low (LS), Medium (MS), and high Strength (HS) Levels[168]

different strength levels and which indicates that as the strength is increased so also is the susceptibility to delayed fracture, and the ratio of critical stress to strength decreases.[168] This is due to the fact that the most significant parameter is a high stress, either applied or residual. Hence, the heated treated steels, which are generally characterized by high residual stresses, will have a total stress much closer to the true fracture stress than will low strength steels.

Hydrogen embrittlement must be forestalled if the full strength of a material is to be used. This requires preventing hydrogen from entering the material during (1) fabrication, and (2) during service. Eliminating it during fabrication can be accomplished by quality control procedures; such as, preheating welding electrodes and parts to be joined to avoid moisture, baking after electroplating to drive out infused gas, etc. Eliminating it in service is not always possible since many vessels are used for high pressure hydrogen storage, while others are exposed to corrosive media capable of charging material with hydrogen; for instance, those containing hydrogen sulphide in the petroleum processing industry.[169,170,171] The time to failure increases with decreasing stress, and there is a minimum or lower critical stress below which hydrogen embrittlement does not occur, Fig. 5.89. This together with the selection of a compatible material composition, forms the design basis for vessels subject to hydrogen attack.[172] Figure 5.89 gives typical design curves for the selection of hydrogen environment materials. A material is not attacked while the temperature and hydrogen partial pressure fall below and to the left of the corresponding material curve.

The effect of hydrogen embrittlement on fatigue life is largely dependent on the mean stress because a stress gradient must be maintained long enough for hydrogen to reach the crack tip. Hydrogen embrittlement does shorten fatigue life if the mean stress is tensile, with the endurance limit of high strength steels reduced approximately 40 per cent.[173]

Multiple-layer vessels, Fig. 2.5, intended to avoid hydrogen attack through the venting of each layer to prevent the build-up of hydrogen pressure, have also been used in this environment.[174]

Nuclear reactors, which use water as a coolant and hence evolve hydrogen by radiolysis from the nuclear core, handle this problem by a continual hydrogen scavenging system. Storage vessels for missiles, on the other hand, which must operate at extremely high stresses to minimize the weight carried aloft, have experienced failures due to hydrogen in suspension in the material traceable to the electrolytic action of water used in the hydraulic proof test.[175] These vessels were constructed of a 5 per cent chrome tool steel heat treated to give a yield strength of 240,000 psi and an ultimate tensile strength of 270,000 psi with a working stress approaching the yield strength. This points out the fundamental importance of both minimizing stress concentrations in vessels designed to low factors of safety and considering the various media to which these vessels are to be subjected throughout their life.

The effect of neutron irradiation, paragraph 5.10, on hydrogen embrittlement is to augment this phenomenon when the conditions conducive to the latter are favorable. However; with the carbon and low alloys steels used in water-cooled nuclear reactors the concentration of hydrogen, which results from radiolysis and the corrosion reaction between the coolant and the steel in the pressure vessels, is not of sufficient concentration in combination with the radiation-increased tensile strength to create a problem. For instance, the notch tensile strength of a highly hydrogen charged A-212B carbon steel irradiated to 3×10^{19} showed little increase in susceptibility toward hydrogen embrittlement.[176] Accordingly, for these materials irradiation assisted hydrogen embrittlement does not present a condition prone to delayed hydrogen failure in water-cooled nuclear reactors.

5.22 Brittle Fracture

Brittle materials are those which have little ductility. Cast iron, glass and concrete at room temperature are examples of such materials. They crack normal to the tensile stress; and accordingly, for designs involving combined stresses the "maximum stress" theory of failure is applicable, paragraph 5.15.1. These materials have low toughness since they undergo little deformation prior to fracture. They substantially follow Hooke's law up to fracture so that the tensile test diagram is represented by a straight line *OA* in Fig. 5.91. In static tensile tests the energy is represented by the area under the tensile test curve from which it can be concluded that a high toughness material must have both high strength and large ductility. In brittle materials the high stress concentrations associated with notches, flaws, etc., remain right up to the breaking point since the material has little ductility or ability to stretch and hence redistribute the local high stress more uniformly.[308, 309] Accordingly, points of stress concentration have a great weakening effect and must be minimized by the judicious addition of properly disposed reinforcing material at these locations. Figure 5.92 shows the typical nature of a brittle fracture in a pressure vessel.

Materials that exhibit a ductile behavior in a normal environment can also act brittlely with a change in environment, such as a lower temperature. For instance, this was widely encountered in the merchant ships of World War II which operated satisfactorily in tropical and temperate waters, but experienced numerous failures when operating in frigid waters. Likewise, pressure vessels that have been satisfactorily pressurized have experienced brittle fracture when subjected to the same pressure at a lower temperature. Brittle fracture is the most dangerous of the types of failure since it can occur at stresses considerably below the yield point, and without prior noticeable deformation,[177, 178] Fig. 5.93. This type of failure is characterized by a very rapid crack propagation of speeds up to six thousand feet per second. The action can be described as a spontaneous one in which the small amount of required driving energy is entirely derived from the release of elastic strain energy. The fractures occur normal to the wall surface and are of a crystalline texture indicating that the individual grains of the steel have failed by cleavage of crystal planes.

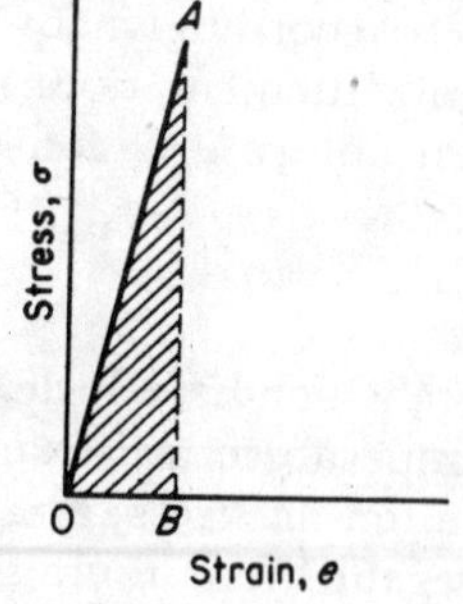

Fig. 5.91. Stress-Strain Diagram for a Brittle Material

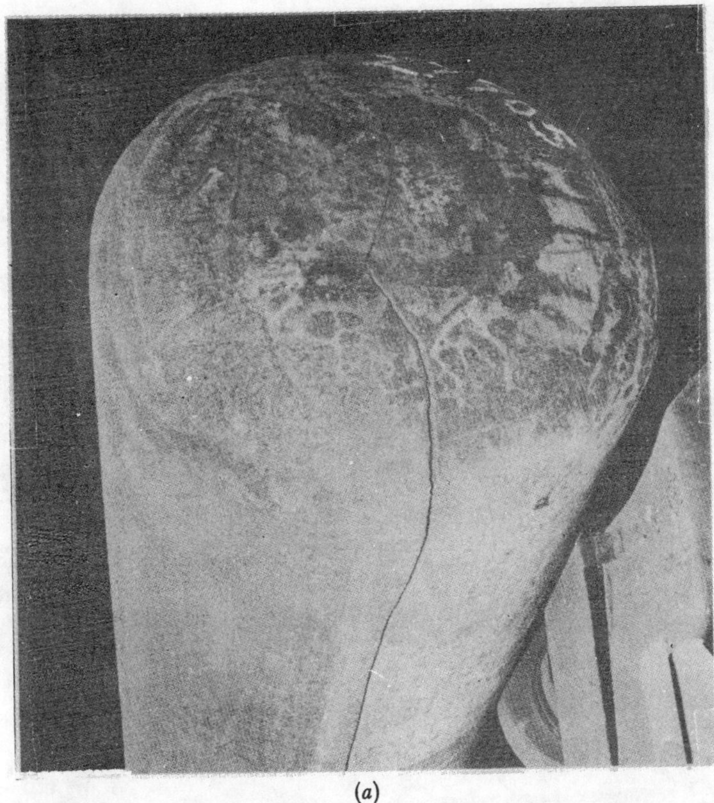

(a)

Fig. 5.92. Brittle Fracture in a Pressure Cylinder and Torospherical Head Showing (a) it to Run the Length of the Cylinder and Branch Across the Head and, (b) the Fracture Surface, as Shown by a Cut-Through Section, to be Perpendicular to the Wall and to the Direction of the Maximum Principal Stress in Compliance with this Theory of Failure, Par. 5.15.1(a)

There is little evidence of plastic flow except for very thin shear lips at the free edges. This is in contrast with ductile fracture with its 45° tear involving large plastic deformation and accompany high energy absorption, Fig. 5.94. The point of initiation of a brittle fracture can often be determined by an examination of the markings on the surfaces created by the crack.[179,333] These markings form a "chevron" or "herringbone" pattern and point in the direction from which the crack started. This is usually found to be a point of severe stress concentration such as a design geometry, fabrication flaw, or material defect, Fig. 5.95.

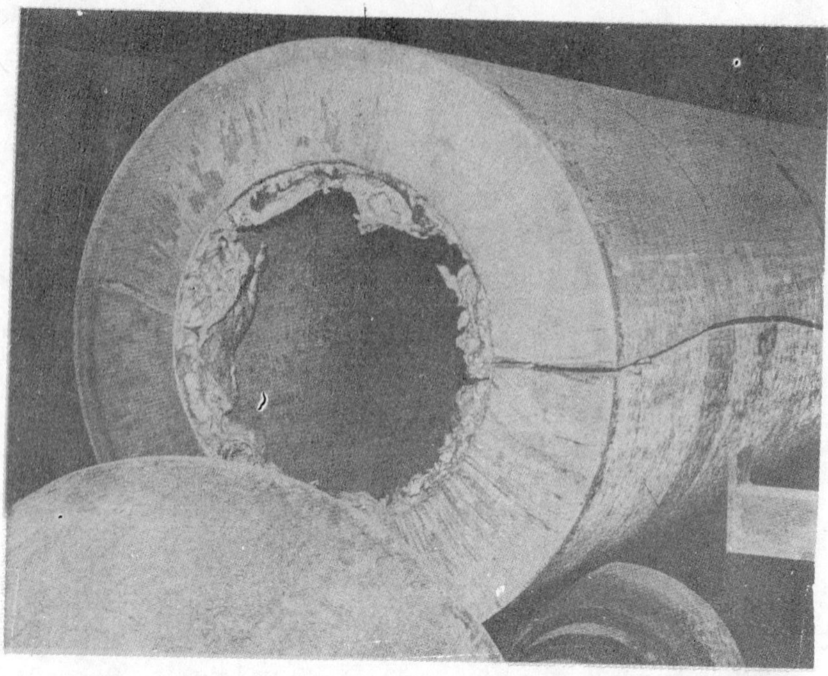

(*b*)

Figure 5.92. (continued)

Fig. 5.93. Brittle Failure of a High Pressure Thick-Walled Chemical Reactor Vessel Which Occurred During Hydrostatic Test in the Manufacturer's Shop[178]

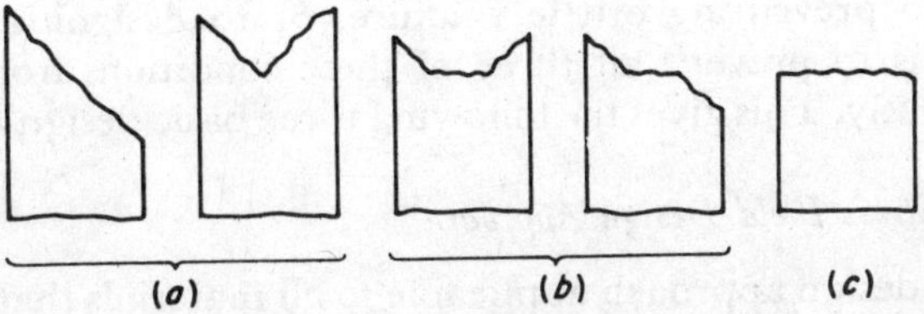

Fig. 5.94. Common Types of Fracture Appearance; (*a*) Shear, (*b*) Mixed Shear and Flat, (*c*) Flat or Brittle

Brittle fracture requires the simultaneous occurrence of three conditions:

1. a high stress field,
2. a material prone to brittle behavior by virtue of its environment, such as operating at a temperature below its nil ductility temperature (NDT is the transition temperature at which the basic mode of failure changes from ductile to brittle, Fig. 5.27).
3. a trigger such as notch or flaw.

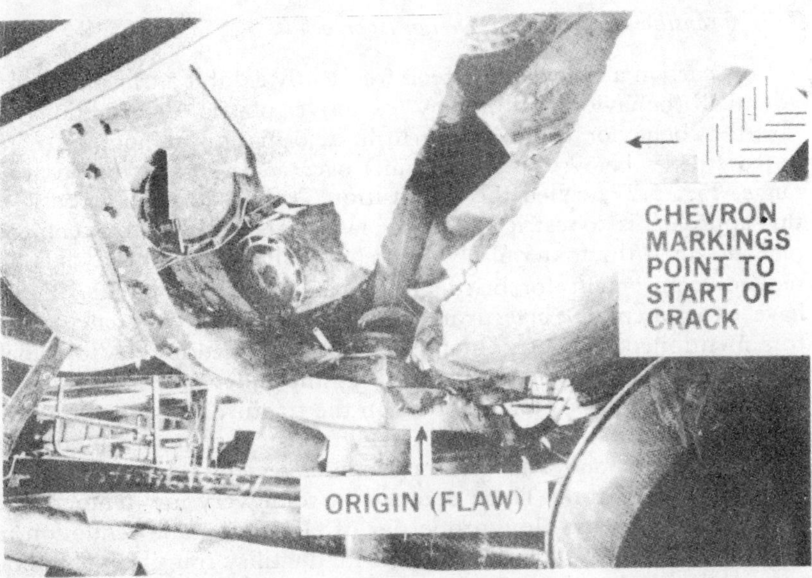

Fig. 5.95. Brittle Fracture of a Steam Drum During a Cold Hydrostatic Test. The Chevron Markings on the Flat Fracture Surface Point to the Origin of the Crack, a Large Flaw in a Welded Nozzle[179]

The key to preventing brittle fracture, or to designing with brittle materials, is to prevent all three of these conditions from occurring simultaneously. This gives the following three basic design approaches:

1. *Low Stress Field Design Approach*

This is a design approach applicable to all materials that act brittley in their operational environment and contain notches, flaws, etc. It consists of employing a low allowable stress (high factor of safety) thereby avoiding the first condition for brittle fracture. As an example, a glass beverage bottle is a very serviceable pressure vessel even though it is made of a material that has little toughness and has many stress concentration notches and flaws. It works satisfactorily because the stress is extremely low.

Pressure vessels for the nuclear, power, or chemical industry on the other hand are subject to high stresses because of the very high pressures these vessels must contain; hence, this approach of permitting only very low stresses cannot be used. Accordingly, it is necessary to resort to the following two design approaches to prevent brittle fracture.

2. *Transition Temperature Design Approach*

Low carbon and low alloy steels frequently exhibit a transition from "ductile" behavior (failure by excessive plastic deformation) to "brittle" behavior (failure with little deformation at nominal computed stresses below the yield point) over a very small temperature range, Fig. 5.27, called the "Transition Temperature." The basis of this approach is to restrict the temperature of operation to a temperature at which the material always behaves ductilely. This avoids the second requirement for brittle fracture. Pellini[180, 181, 310] and Puzak have experimentally investigated and correlated the influencing factors mentioned above to obtain a practical engineering fracture analysis diagram based on temperature as the independent parameter. The general effects of temperature on the fracture stress transition of steels are shown in Fig. 5.96. A flaw-free steel shows an increasing tensile strength and still greater increasing yield strength with decreasing temperature. These coincide at some very low temperature at which the plastic flow properties of elongation and reduction in areas are nil. This temperature is the nil ductility transition temperature in the absence of a flaw (NDT no-flaw). If the vessel material is flaw-free there is no need to consider the brittle fracture properties of steel. Likewise, the closer a material approaches this condition the

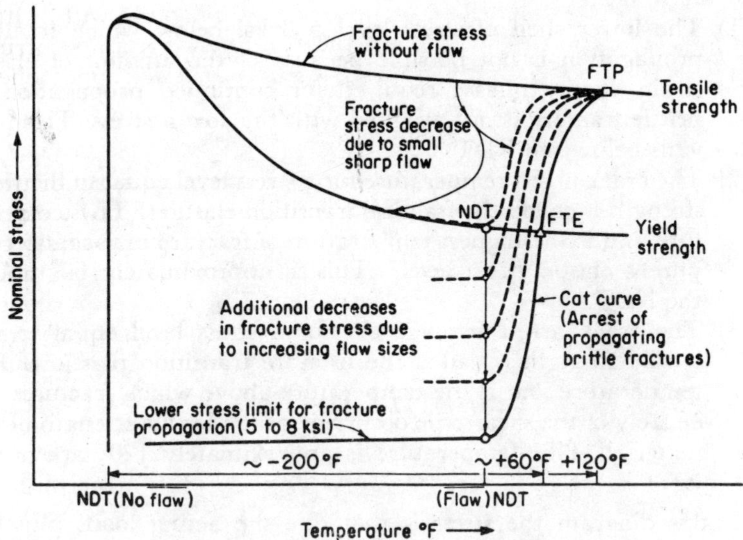

Fig. 5.96. Transition Temperature Features of Steel[180,181]

less susceptible it is to brittle type failure; hence, the importance of quality control of base material and manufacturing processes. Commercial materials do have flaws or inherent minute defects to which may be attributed the fracture stress decrease in the region of the material transition temperature. The highest temperature at which the decreasing fracture stress for fracture initiation due to these small flaws becomes contiguous with the yield strength curve of the steel is called the nil ductility transition temperature (NDT). Below the NDT temperature the fracture stress curve for these small inherent flaws follows the course of the yield strength curve to lower temperatures as indicated by the dashed curve, Fig. 5.96. At the NDT, increases in flaw size result in a progressive lowering of the fracture stress curve to lower values of nominal stress. This decrease in the fracture stress is approximately inversely proportional to the square root of the flaw size. The result is a family of fracture stress curves showing a marked increase in the stress required for fracture as the temperature is increased above the NDT temperature. The lowest boundary curve of this family is called the crack arrest temperature (CAT) curve. This curve establishes the temperature of arrest of a propagating brittle fracture for various levels of applied nominal stress. This curve is the key to vessel design and operation and has three significant features:

(1) The lower shelf of stress level 5–8 ksi below which fracture propagation is not possible because of the amount of elastic strain energy release required for continued propagation of brittle fracture is not attained with this low a stress. This prevails below the NDT.

(2) The crack arrest temperature for a stress level equal to the yield strength is called the fracture transition elastic (FTE) temperature and is the highest temperature of fracture propagation for purely elastic stress levels. This is approximately 60° above the NDT.

(3) The crack arrest temperature for a stress level equal to the tensile strength is called the fracture transition plastic (FTP) temperature and is the temperature above which fractures are entirely of the shear type occurring at the tensile strength of the material. This temperature is approximately 120° above the NDT.

In this diagram the stress is that due the acting loads plus the residual stresses resulting from the fabricating process (such as welding). The load stresses and residual stresses are additive if superposition does not cause yielding or partial fracture. Welded vessels contain residual stresses which are not entirely eliminated by the final stress relieving heat treatment they receive. The amount remaining depends upon the creep strength of the material at the stress relieving temperature, and this may conservatively be taken as the short-time yield strength of the material. The fracture analysis diagram has four primary reference points which may be used as design criteria for the engineering selection of fracture-safe steels, design stresses, or minimum service temperatures. These are illustrated in Fig. 5.97 and defined as follows:

(1) NDT Temperature Design Criterion. This is based on the existence of an acting yield point stress level, and coexistence of only small flaws less than one inch in size. Restricting the service temperature to above this value assures ductile behavior. It provides fracture protection by preventing crack initiation; however, it does not provide for the arrest of a propagating crack.

(2) NDT Plus 30°F Design Criterion. This criterion is based on a stress level of the order of 1/2 yield strength, commonly used for commercial vessel design, and its relation to the CAT curve. Restricting service temperature to above this value obviates the flaw-size evaluation problem. That is, fractures cannot initiate, or can not propagate in this stress field.

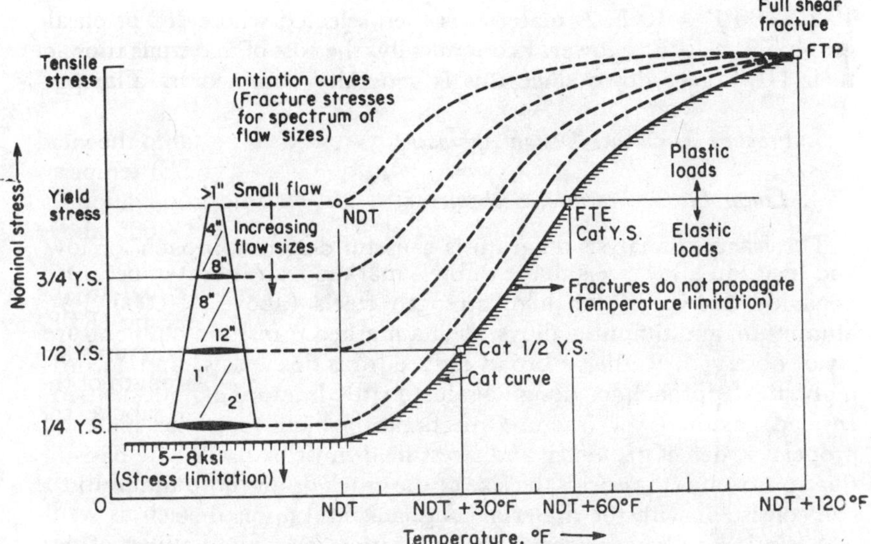

Fig. 5.97. Fracture Analysis Diagram for the Engineering Selection of Fracture-Safe Steels Based on the NDT Temperature.[180,181]

(3) NDT Plus 60°F Design Criterion. This criterion is based on the same considerations as design criterion 2, except the level of stress is the yield strength of the material. It is applicable to vessels designed to high stress levels, or which are subject to high test pressures, or to severe thermal stress conditions. This criterion is in frequent engineering use, and is the basis of the American Society of Mechanical Engineer's Nuclear Code requirements.

(4) NDT Plus 120°F Design Criterion. This criterion is based on the premise that the vessel will be subject to a stress level above the yield point of the material. This restricts service to full shear fracture temperatures in order to develop maximum fracture resistance.

These criteria relate to the fabrication, design, flaw-size and stress levels consequent to the mode of operation, all of which require engineering judgment. To illustrate the use of this diagram, assume that engineering judgment dictates the use of the second design criterion, NDT Plus 30°F, and that 40°F if the lowest expected service temperature. The highest permissible NDT temperature of the material to be used is then determined from the temperature scale of the diagram as

40°F − 30°F = 10°F. A material is then selected whose NDT, paragraph 5.9, is 10°F or lower. Economically, the cost of materials rises as their NDT goes down since this is generally obtained by alloying.

3. *Fracture Mechanics Design Approach*

A. *Linear Elastic Fracture Mechanics*

The fracture analysis diagram is a useful design approach for low and medium alloy steels that exhibit a marked transition temperature behavior. However, for high strength steels (above 150,000 TS), aluminum, and titanium alloys, such a marked transition temperature is not observed. Rather a broad diffuse transition exists; and accordingly, this approach for design against brittle fracture cannot be used. Instead, methods of fracture mechanics which reconcile material properties, defect geometry and stress field must be used. The basis of this approach is to restrict the size of the crack or defect to that critical size consistent with the material toughness and applied stress.

A crack may be considered as the limiting form of an ellipse of decreasing minor axis. Equation 6.4.1 gives the maximum stress occurring at the ends of an elliptical hole in a plate subject to uniform tens.on perpendicular to the direction of the major axis as

$$\sigma_1 = \sigma(1 + 2a/b) \qquad (5.22.1)$$

This stress may be expressed in terms of the radius of curvature ρ at the end of the crack. Since for an ellipse $\rho = b^2/a$, Eq. 5.22.1 can be written

$$\sigma_1 = \sigma\left(1 + 2\sqrt{\frac{a}{\rho}}\right) \qquad (5.22.2)$$

Denoting the half crack length, a, equal to the semi-major axis of the ellipse, Eq. 5.22.2 for a sharp crack-like ellipse where $a \gg \rho$ becomes

$$\sigma_1 \cong 2\sigma\sqrt{\frac{a}{\rho}} \qquad (5.22.3)$$

so that the stress concentration factor for the infinitely sharp crack is

$$K_t = \frac{\sigma_1}{\sigma} = 2\sqrt{\frac{a}{\rho}} \qquad (5.22.4)$$

Since a crack may be considered as the limiting form of an ellipse of decreasing minor axis it can be concluded that the stress concentration factor should be proportional to $(a/\rho)^{\frac{1}{2}}$. Thus the low fracture strengths of brittle solids can be explained, at least qualitatively, by assuming that the stress concentrating flaw gives rise to local stresses exceeding the theoretical fracture strengths. The mathematical treatment of stability conditions in localized cracks or other sources of stress concentration is called "linear elastic fracture mechanics," and is based on elastic analysis which assumes that stress is proportional to strain. It is based on the work of Griffith[182] (see paragraph 5.22.3B) and Irwin.[183,184] In the early 1920's Griffith proposed an energy criteria for brittle failure and performed a series of experiments on glass spheres to verify his theory. Irwin extended this principle to include plastic work and solved the elasticity problem for the stresses and strains around the tip of the crack and the rate of release of strain energy per increment of crack length. At points (r, θ) in the area of a crack tip in an infinite plate subject to tension perpendicular to a through the thickness crack of length $2a$, Fig. 5.98, the elastic stresses[183] are:

$$\sigma_x = \frac{\sigma\sqrt{\pi a}}{\sqrt{2\pi r}} \cos\frac{\theta}{2}\left(1 - \sin\frac{\theta}{2}\sin\frac{3\theta}{2}\right) \qquad (5.22.5)$$

$$\sigma_y = \frac{\sigma\sqrt{\pi a}}{\sqrt{2\pi r}} \cos\frac{\theta}{2}\left(1 + \sin\frac{\theta}{2}\sin\frac{3\theta}{2}\right) \qquad (5.22.6)$$

$$\tau_{xy} = \frac{\sigma\sqrt{\pi a}}{\sqrt{2\pi r}} \sin\frac{\theta}{2}\cos\frac{\theta}{2}\cos\frac{3\theta}{2} \qquad (5.22.7)$$

Directly ahead of the crack, $\theta = 0$, $\sigma_x = \sigma_y = \sigma(a/2r)^{\frac{1}{2}}$ and $\tau_{xy} = 0$.

Equations 5.22.5, 5.22.6 and 5.22.7 indicate that the local stresses near a flaw depend on the product of the nominal stress and the square root of the flaw size a. This product is called "stress intensity factor" to emphasize this fundamental relationship and has the dimension (lb/sq in.) $\sqrt{\text{in.}}$ The stress intensity factor K for a sharp through crack in an infinitely wide plate is defined as

$$K = \sigma\sqrt{\pi a} \qquad (5.22.8)$$

and the local stresses near the crack are then written

$$\sigma_x = \frac{K}{\sqrt{2\pi r}} \cos\frac{\theta}{2}\left(1 - \sin\frac{\theta}{2}\sin\frac{3\theta}{2}\right) \qquad (5.22.9)$$

$$\sigma_y = \frac{K}{\sqrt{2\pi r}} \cos\frac{\theta}{2}\left(1 + \sin\frac{\theta}{2}\sin\frac{3\theta}{2}\right) \qquad (5.22.10)$$

$$\tau_{xy} = \frac{K}{\sqrt{2\pi r}} \sin\frac{\theta}{2}\cos\frac{\theta}{2}\cos\frac{3\theta}{2} \qquad (5.22.11)$$

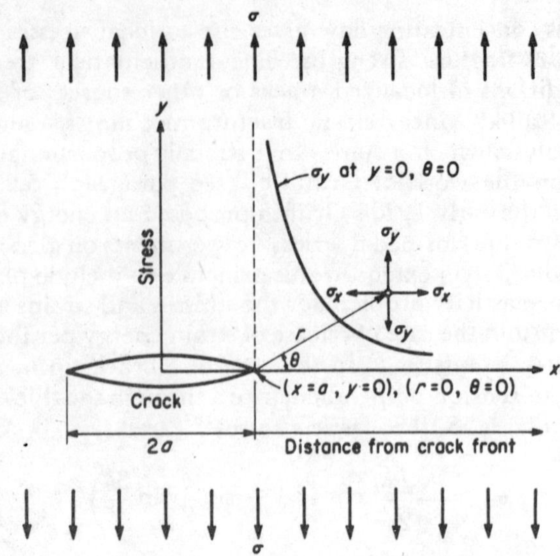

Fig. 5.98. Wide Plate with a Central Through Crack Perpendicular to the Applied Stress

It is seen that all three stresses depend on the same constant K, which is a one parameter representation of the stresses in the area of a crack tip. It is a purely numerical quantity which, if known, provides complete knowledge of the stress field at the crack tip.

A comparison of Eqs. 5.22.3 and 5.22.8 shows that K is proportional to the limiting value of the elastic stress concentration factor K_t as the root radius approaches zero. This relation[185] is

$$K = \lim_{\rho \to 0} \tfrac{1}{2}\sigma K_t \sqrt{\pi\rho} \tag{5.22.12}$$

Thus K values for variously shaped flaws in structures can be determined when values of the elastic stress concentration factor are known.[186] In all cases the stress intensity factor has the form

$$K = \sigma\sqrt{Q\pi a} \tag{5.22.13}$$

where Q is a function of the crack geometry. Table 5.6 gives values of Q for several crack shapes where the body containing the crack is sufficiently large that its bounding surfaces do not influence the stress field at the crack tip.[254,255]

Figure 5.99 shows the three failure modes treated by fracture mechanics. The first mode is the opening mode, the second is the edge sliding mode, and the third is a tearing mode. The stress intensity factors associated with these modes are K_{I}, K_{II}, and K_{III}, respectively. Since these K factors are one parameter representations of the stress field at the tip of the crack, it is assumed that they should also be a good theory of failure for materials. Experiments have verified this for average net section stresses below 0.8 of the yield stress. The basic assumption in fracture mechanics is that unstable fracture (where the crack continues to extend with no increase in applied stress) occurs when K reaches a critical value designated K_c, commonly called fracture toughness. K_c is determined experimentally for the point at which rapid crack propagation occurs in fracture-toughness tests, similar to the way tensile strength is determined in a tensile test.[187] Such tests are

TABLE 5.6. STRESS INTENSITY FACTORS FOR TENSILE LOADING

Type of Crack	Geometry	Value of Q in Basic Form of Stress Intensity Factor $K = \sigma\sqrt{Q\pi a}$	Stress Intensity Factor, K_I (Plane Strain)
Case 1 Infinite Plate with Central Through Crack.	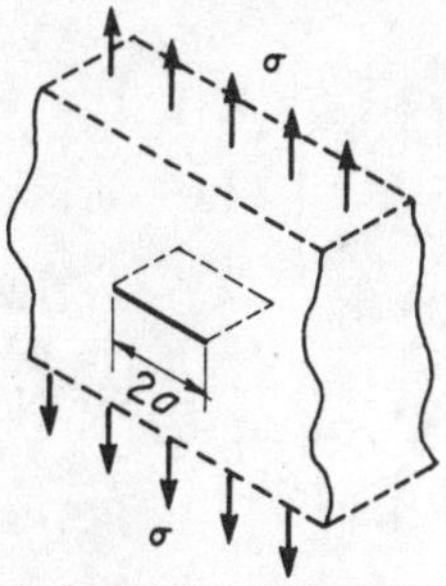 σ, $2a$	1.0	$K_I = \sigma\sqrt{\pi a}$
Case 2 Semi-Infinite Plate with Through Edge Crack. (Long Shallow Surface Crack)	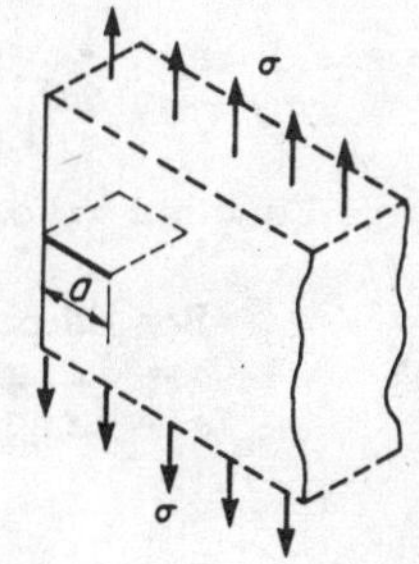 σ, a	1.2	$K_I = 1.1\sigma\sqrt{\pi a}$

TABLE 5.6 (continued)

Case 3 Internal Circular Crack in an Infinite Body.	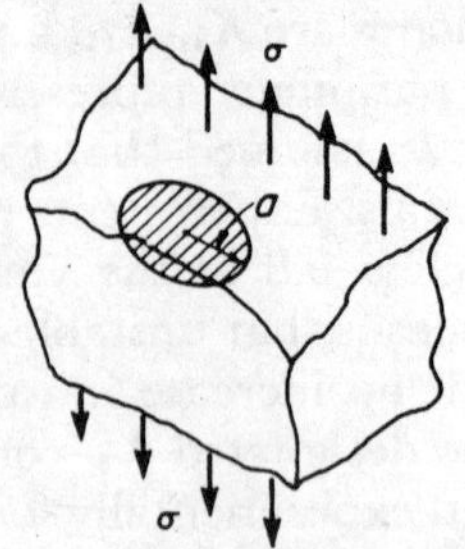	$\frac{4}{\pi^2}$	$K_I = 2\sigma\sqrt{\frac{a}{\pi}}$
Case 4 Internal Elliptical Crack in an Infinite Body.	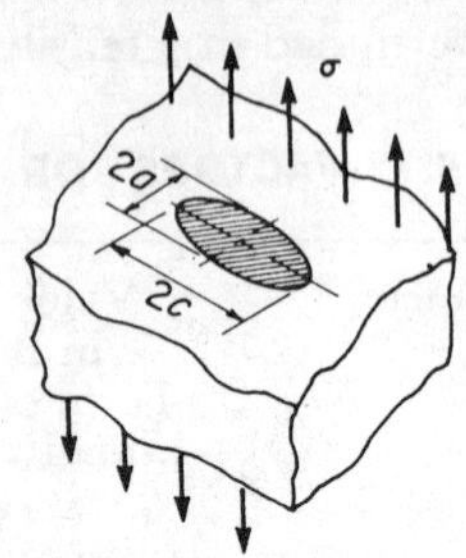	$\frac{1}{\Phi^2}$ (See Note 1)	$K_I = \frac{\sigma}{\Phi}\sqrt{\pi a}$
Case 5 Semi-Infinite Body with Elliptical Surface Crack.	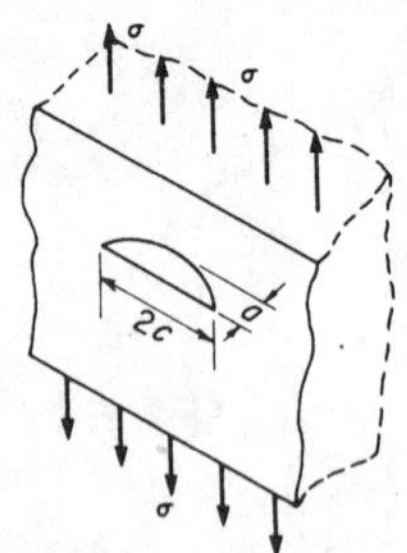	$\frac{1.2}{\Phi^2}$ (See Note 1)	$K_I = \frac{1.1\sigma}{\Phi}\sqrt{\pi a}$

Note 1

$$\Phi = \int_0^{\pi/2} \left(1 - \frac{c^2 - a^2}{c^2} \sin^2 \theta d\theta\right)^{1/2}$$

Φ Is the Complete Elliptic Integral Function Whose Value is:

a/c	Φ	a/c	Φ
0.0	1.000	0.6	1.277
0.1	1.016	0.7	1.345
0.2	1.051	0.8	1.418
0.3	1.097	0.9	1.493
0.4	1.151	1.0	$\pi/2$
0.5	1.211		

For Elliptical Cracks:
a = Length of Semi-Minor Axis.
c = Length of Semi-Major Axis.

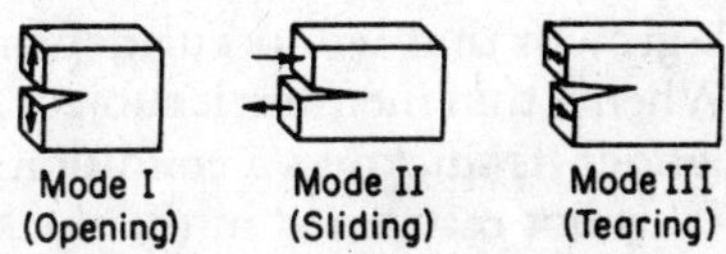

Fig. 5.99. Three Basic Modes of Crack Surface Displacement

performed to give static, dynamic, and arrest toughness properties designated, respectively, K_{Ic}, K_{Id}, and K_{Ia}. Figure 5.100 shows these relative toughness properties for a heavy section manganese-molybdenum steel plate material commonly used for nuclear reactor vessels,[188] and also compares these to the same material which has been highly irradiated in order to illustrate the service problem that can be encountered in such vessels unless the irradiation damage is mitigated by the annealing effect of elevated temperature, paragraph 5.10.

It is important to appreciate the difference between K and K_c. The stress-intensity factor K is simply a coefficient in an equation describing the elastic stresses in the vicinity of a crack tip. Fracture toughness K_c is a particular value of K corresponding to unstable propagation of the crack. As far as the crack is concerned, any configuration and load distribution that produces a given value K is equivalent to any other such configuration and load distribution. Thus K_c is independent of loading conditions and may be considered a material constant. It can be used in a way very similar to the way ultimate strength is used in designing a structure.

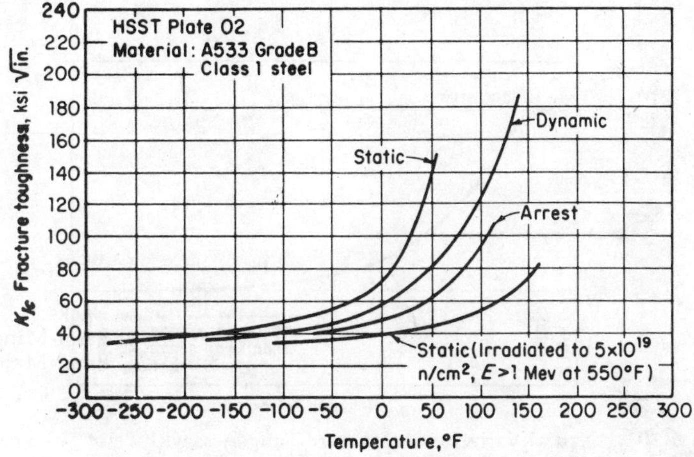

Fig. 5.100. Relative Fracture Toughness Properties of Heavy Section Manganese-Molybdenum Steel Plate[188]

The mode of crack growth that will occur depends upon the thickness of the material. When a thin member is subject to in-plane loading that is uniform throughout its thickness a condition of plane stress prevails since the material is not restrained in the thickness direction and considerable plastic flow accompanies the cracking process. In this case most metals fail in a 45° shear fracture appearance, Fig. 5.94*a* and *b*, which is a mixture of Mode I and Mode II, crack surface displacement, Fig. 5.99. If a thick member is similarly subject to in-plane loading that is uniform throughout its thickness, a condition of plane strain prevails since strain in the thickness direction is suppressed by the very thickness of the material and noticeably less plastic flow is associated with the cracking process. The associated fracture has a flat or square appearance, Fig. 5.94*c*, which is primarily a Mode I crack surface displacement. In elasticity theory a state of plane strain occurs when the strain in the z direction is zero, Eq. 3.2.4. This condition occurs in the interior of very thick plates. Therefore, in a thick plate a condition of plane stress occurs on the surface, and a condition of plane strain on the interior, Fig. 5.101. Accordingly, the fracture toughness K_c of a material will vary with the thickness of the material. Figure 5.102 shows the effect of plate thickness on fracture toughness and the fracture appearance in terms of percentage of square fracture. A condition of plane stress associated small thicknesses gives high values of fracture toughness. These approach a limiting minimum value under a condition of plane strain as thickness increases. The plane strain value, K_{Ic}, is used in engineering design since this is the minimum value for the material. Plasticity associated with fracture

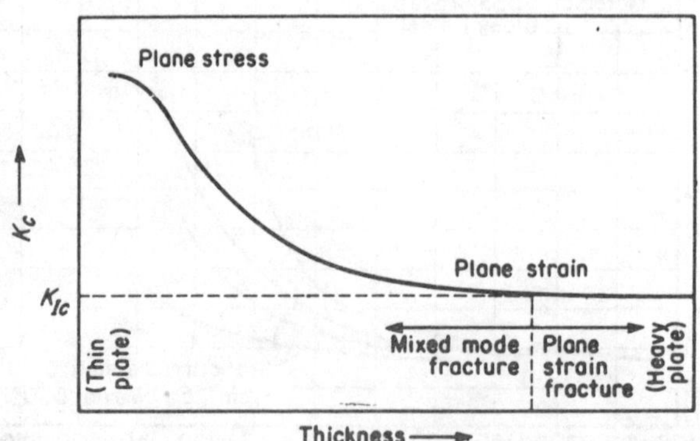

Fig. 5.101. Typical Variation in Fracture Toughness with Plate Thickness

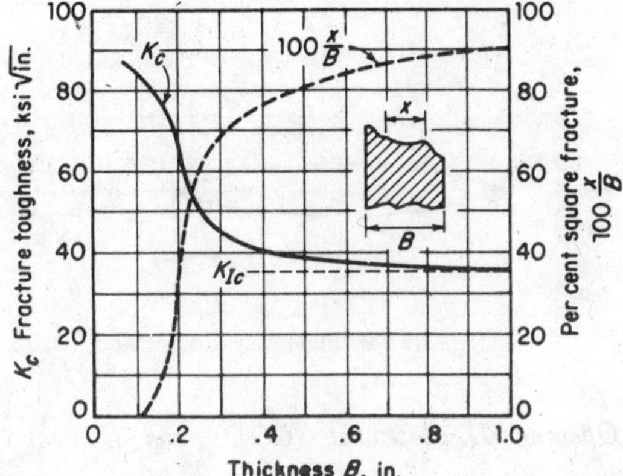

Fig. 5.102. Variation in Fracture Toughness and Fracture Appearance with Thickness for 7075-T6 Aluminum. This Variation is also Qualitatively Typical of Steels

decreases with increasing material thickness (increasing percentage of square fracture $100x/B$).

Figure 5.103 depicts schematically the shape of the plastic zone at the edge of a crack. It is large at the surface and rapidly diminishes in size as the center is approached; therefore, if the plate is thick the dominating portion is the small plastic zone in the center region. The effect of this small plastic zone on the linear elastic analysis around the tip of the crack can be corrected by increasing the crack length to include the radius of the plastic zone,[189] Fig. 5.104

$$r_p \approx \frac{1}{2\pi}\left(\frac{K_{Ic}}{\sigma_{Y.P.}}\right)^2 \tag{5.22.14}$$

The nature of the stress around a crack tip in tension is illustrated by the photoelastic isochromatic fringe pattern[311] of Fig. 5.105.

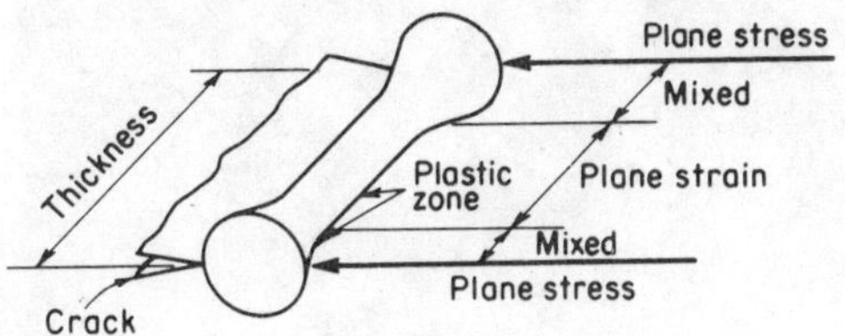

Fig. 5.103. Plastically Deformed Zone at a Crack Front

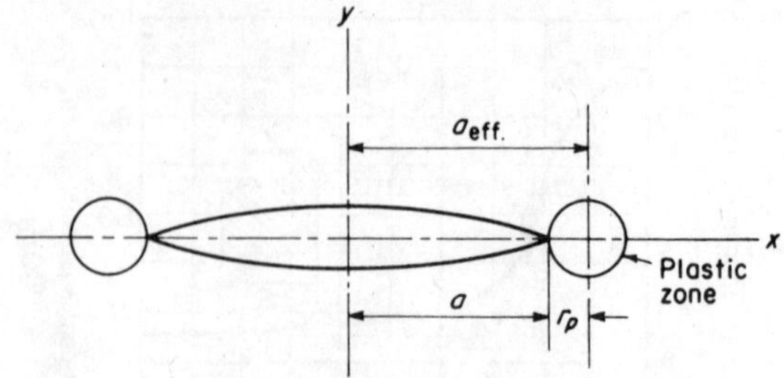

Fig. 5.104. Plasticity Correction

B. Crack Opening Displacement (COD) and General Yield Fracture Mechanics

Linear elastic fracture mechanics is most applicable to materials in which little local plasticity is associated with fracture. In tough materials where a considerable plastic zone is created ahead of the crack, Wells [190,191] has proposed that unstable extension of a crack occurs at a critical value of local displacement near the tip of the crack, crack opening displacement or COD, and assumes that this critical value is the same in actual structures as it is in a test specimen of similar thickness. COD is a measure of this critical plastic zone. This approach and

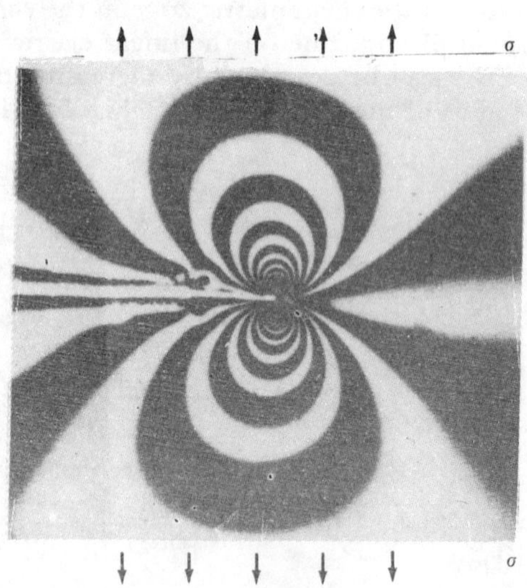

Fig. 5.105. Isochromatic Fringe Pattern Around a Crack Tip in Tension

that of linear fracture mechanics become similar when the plastic zone is small, and the critical crack extension force, G_c, is

$$G_c = \sigma_{Y.P.}\delta_c \tag{5.22.15}$$

where $\sigma_{Y.P.}$ is the yield stress and δ_c is the critical COD.

The critical value of COD occurring at fracture[256] is dependent upon the same factors as K_{Ic}; i.e., material, temperature, strain rate, etc. These values are likewise established experimentally by extensometer measurements on notched test specimens for a material in its environment (temperature, etc.) for comparison to these values calculated in the actual structure. Figure 5.106 shows a typical load-displacement curve.[192] Point A is that of deviation from linearity, or secant modulus line, point B is known as "pop-in" which indicates abrupt instability and is accompanied by a sudden small load drop and/or increase in displacement, and C is the maximum load. Pop-in is identified with plane strain toughness, K_{Ic}; however, since it is not always clearly defined the COD, δ_c, is usually taken at the maximum load. The COD to fracture is less under dynamic conditions than under static conditions just as are the corresponding stress intensities, K_{Ic}. Accordingly, the relevant loading rate determines the parameter to be employed and should match that of the actual structure; i.e., static for those which experience only slowly applied loads and dynamic for those subject to the rapidly applied loadings of impacts and explosions.

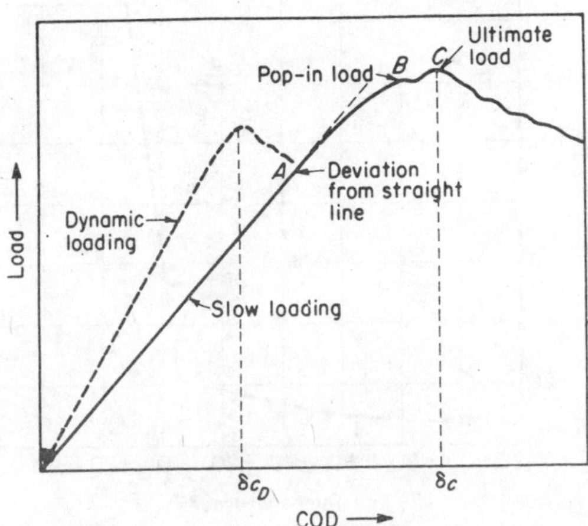

Fig. 5.106. Crack Opening Displacement Curve Showing Fracture Parameters for Static and Dynamic Initiation[192]

The critical value of COD exhibits a transition with temperature[193, 312, 334] that can be used as a failure criterion or structural design basis for the prevention of plain strain crack propagation, Fig. 5.107.

As the plastic zone spreads from the tip of the crack, the crack opening displacement, COD, at the tip of the crack increases. This displacement is related to the plastic zone size by

$$\delta = \frac{8e_{Y.P.}a}{\pi} \log_e \sec\left(\frac{\sigma}{\sigma_{Y.P.}} \frac{\pi}{2}\right) \tag{5.22.16}$$

where

δ = crack tip COD
a = half crack length
σ = applied stress remote from and normal to the crack plane
$\sigma_{Y.P.}$ and $e_{Y.P.}$ = yield properties of the material in the yield zone at the crack tip

If the applied stress is a function of the yield stress, Eq. 5.22.16 reduces to

$$\delta = (\text{Constant})\, e_{Y.P.}a \tag{5.22.17}$$

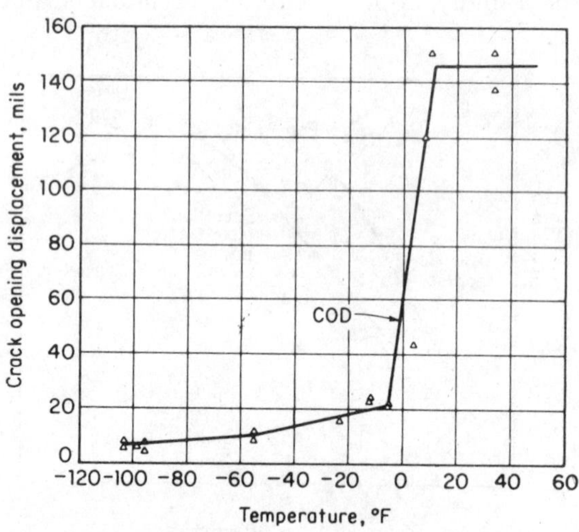

Fig. 5.107. Variation in Statically Initiated Critical COD with Temperature for a Surface-Cracked Specimen in Tension, Manganese-Molybdenum ASTM A533 Grade B Steel[193]

As an illustration, Eq. 5.22.16 for a typical design stress of $\sigma = (2/3)\sigma_{Y.P.}$ shows the COD developed at crack length "a" to be

$$\delta = 1.7 e_{Y.P.} a \tag{5.22.18}$$

Hence, in order to avoid failure the material at the crack tip must be able to tolerate this level of applied COD,

$$\delta_c > 1.7\, e_{Y.P.} a \tag{5.22.19}$$

or

$$a_{max.} < 0.6 \frac{\delta_c}{e_{Y.P.}} \tag{5.22.20}$$

Figure 7.3 shows the behavior of a crack in a very ductile material subject to a bending moment. This rubber model provides a qualitative visualization of the high localized strains in the immediate vicinity of the crack tip. Figure 7.16*b* shows the ductile shearing nature of the failure of a thick-walled steel cylinder under internal pressure in the presence of a longitudinal internal surface crack. In contrast, Fig. 7.16*a* shows the brittle fracture of a similar thick-walled cylinder of a less tough material that is an elastic-plastic brittle one obeying general yield fracture mechanics.

C. *Energy Consideration of Brittle Failure*

The sudden onset of unstable crack propagation can also be explained by considering the balance between the elastic energy, δE_e, released and the energy absorbed by the fracture process during unit extension of the crack. The absorbed energy is accounted for by two principal mechanisms; namely, (1) the surface free energy created per increment of crack extension, δS, and (2) the energy of plastic deformation, δW, per increment of crack extension

$$\delta E_e = \delta S + \delta W \tag{5.22.21}$$

Griffith[182] considered this energy balance for a purely brittle material (glass) wherein $\delta W = 0$ in Eq. 5.22.21 for the condition of a crack subject to a tensile stress normal to the plane of the defect and established the critical crack size as

$$a = \frac{2ES}{\pi\sigma^2} \tag{5.22.22}$$

Equation 5.22.16 may be written as

$$a\sigma^2 = \text{constant} \tag{5.22.23}$$

or

$$\sigma = \frac{\text{constant}}{\sqrt{a}} \tag{5.22.24}$$

which show the basic relationship between the stress to cause fracture and the critical dimensions of a crack in a brittle solid when the applied tensile stress is normal to the plane of the defect. Many materials such as metals fracture with concurrent plastic deformation and it has been observed that the amount of energy required to produce even the small amounts of deformations observed in brittle metallic fractures is several orders of magnitude greater than the surface free energy ($\delta W \gg \delta S \approx 0$). Such a condition applie to steel for which Eq. 5.22.21 can be written

$$\delta E_e = \delta W \tag{5.22.25}$$

Irwin[183] has evaluated this change in elastic energy with the introduction of a through crack in an infinite plate subject to a uniaxial stress normal to the crack plane as

$$G = \frac{\pi\sigma^2 a}{E} \tag{5.22.26}$$

where G is the strain energy release rate, in lb/in.². Equation 5.22.26 is of the same form as Eq. 5.22.22 and likewise can be written

$$a\sigma^2 = \text{constant} \tag{5.22.27}$$

or

$$\sigma = \frac{\text{constant}}{\sqrt{a}} \tag{5.22.28}$$

The value of G at the onset of unstable crack propagation is called the critical strain energy rate, G_c, and is a measure of the material fracture toughness. As long as the inequality $G > G_c$ exists, the fracture process continues.

The strain energy release rate, G, and stress intensity factor, K, are directly related. The relationships for mode I are:

$$G_{\text{I}} = \frac{(1 - \mu^2)}{E} K_{\text{I}}^2 \text{ for plane strain} \tag{5.22.29}$$

$$G_{\text{I}} = \frac{K_{\text{I}}^2}{E} \text{ for plane stress} \tag{5.22.30}$$

Either G_{Ic} or K_{Ic} are referred to as "fracture toughness," and values determined under one set of conditions are applicable to other conditions such as geometries, flaw size and type of loading.

Nuclear reactor vessels employ tough carbon and low alloy steels that are highly temperature sensitive which requires impractically thick specimens, in the order of twelve inches, to measure a valid value of K_{Ic} at room temperature. These vessels are also subject to neutron irradiation, and to monitor this effect on the K_{Ic} fracture toughness of the vessel material, small size surveillance specimens that will not restrict fluid coolant flow are used. Thus for the range of interest it is not feasible to obtain valid K_{Ic} data; and as a result, other methods of establishing the K_{Ic} fracture toughness data or bounding values are being pursued. These include the equivalent energy method of Witt,[194,195] the J-integral strain energy method of Rice,[196,197] the stress-concentration method of Irvine and Quirk,[198] and the near tip strain method of Liu.[199]

D. J-Integral Evaluation of Fracture Toughness

The stress intensity parameter, K, is widely used for correlation of fatigue crack growth rates and brittle fracture evaluations. But, since it is based on linear elastic analysis, its validity is restricted to situations where the size of the plastic zone occurring at the crack tip is small compared to the other significant dimensions. The fracture toughness approach, however, may be extended to situations involving large scale plasticity by the use of the J-integral concept. The J-integral test procedure overcomes the size limitation in K_{Ic} measurements and allows consistent measurements of fracture toughness in much smaller specimens than is possible with the K_{Ic} procedure. A definition of J which provides a good physical understanding of the concept is[257]

$$J = -\frac{\partial U}{\partial a} \tag{5.22.31}$$

where ∂U is the change in strain energy per unit of thickness corresponding to an infinitesimal increase in crack length of a loaded body. This change in strain energy can be visualized as the decrease in area under a load-displacement curve corresponding to an increase in crack length at constant displacement Fig. 5.108. The change in strain energy is negative, so J is a positive quantity. The critical J value corresponding to a critical amount of Mode I crack growth, Fig. 5.99, in a loaded body is the quantity of interest, J_{Ic}. When plastic deformation is small, J is equivalent to G, Eqs: 5.22.29 and 5.22.30, the strain energy release rate. Thus

$$J = J_{Ic} = G_{Ic} = \frac{K_{Ic}^2}{E} \text{ for plane stress} \tag{5.22.32}$$

$$K_{Ic}^2 = \frac{E}{1-\mu^2} J_{Ic} \text{ for plane strain} \tag{5.22.33}$$

This corresponds to the initial straight-line portions of the load-displacement curves in Fig. 5.108. For elastic-plastic materials, J losesits interpretation in terms of the potential energy available for crack extension, but retains physical significance as a measure of the intensity of the characteristic crack tip strain field and it is this strain intensity interpretation of J that is important; i.e., it applies as well to geometries with large plastic deformation.[258, 259] Also, J_{Ic} values measured from small specimens with significant plastic deformation are the same as the linear elastic G_{Ic} values measured for large specimens of the same material. Simple approximations allow J values to be estimated from a single compact small size specimen experimental load versus load-point displacement curve, Fig. 5.109. A solution[260, 261] of Eq. 5.22.31 for this specimen in which yielding is confined to the uncracked ligament, b, is

$$J = \frac{2}{bB} \int_0^\delta P d\delta = \frac{2A}{bB} \tag{5.22.34}$$

where A is the total area under a load-displacement curve and the displacement is due to the introduction of the crack only. The latter

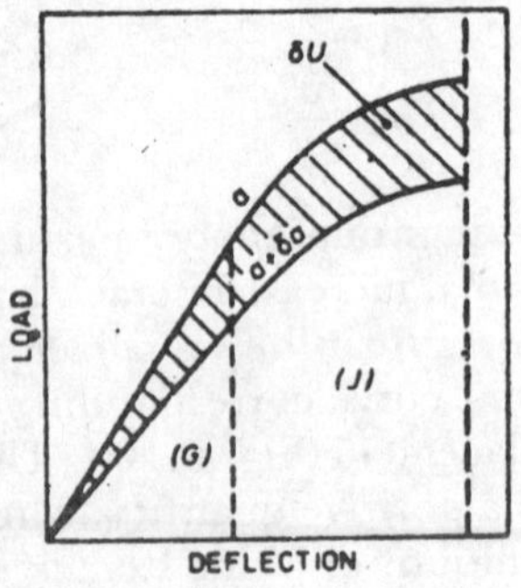

Fig. 5.108. Load-Displacement Curve.[257]

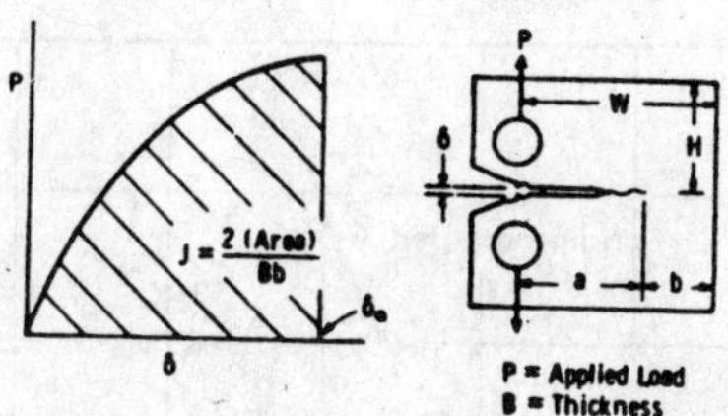

Fig. 5.109. Determination of J from a Load-Displacement Curve for a Compact Specimen.[257]

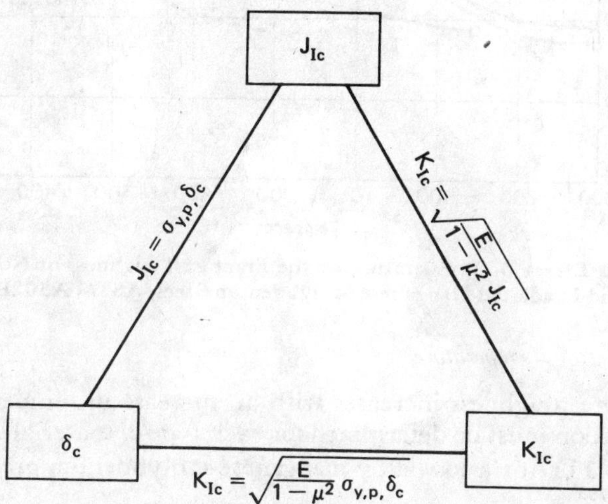

Fig. 5.110. Basic Relationship of Major Fracture Toughness Parameters

is satisfied when $a/W > 0.6$; i.e. there is primarily bending in the ligament and yielding is confined to the ligament.

The theoretical interrelationship between the major fracture toughness parameters[262] is shown in Fig. 5.110.

5.23 Effect of Environment on Fracture Toughness

Since fracture toughness, K_c or G_c, is a material property it is affected by environment as are other material properties. Temperature affects all materials; hence, it is a major design concern for all vessels and structures. Other adverse environments such as chemicals, moisture, radiation, loading rate, etc., are more characteristic of individual materials but nonetheless important.

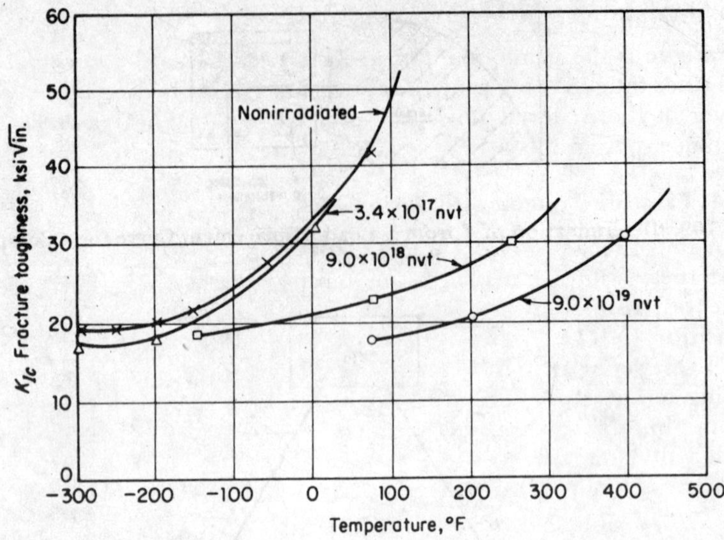

Fig. 5.111. Effect of Temperature on the Fracture Toughness on Non-irradiated, and Irradiated Manganese-Molybdenum Steel, ASTM A302B[200]

1. *Effect of Temperature*

Fracture toughness increases with an increase in temperature and this variation must be determined for each type of material. It is shown in Fig. 5.111 for a low alloy manganese-molybdenum pressure vessel steel.[200, 263]

2. *Effect of Radiation*

The material of nuclear reactor vessels is also subject to irradiation damage. Its effect on fracture toughness is a decrease in this property with an increase in neutron absorption. This is similar to the static properties of elongation and reduction in area in the same environment, paragraph 5.10. Neutron absorption below 3.4×10^{17} nvt has a negligible effect, while in the range of 3.4×10^{17} to 4.8×10^{18} nvt the observed changes are small and variation difficult to detect.[39,200] Accordingly, a value of 1×10^{18} nvt is generally accepted as that above which the effect of irradiation must be considered. This variation is shown in Fig. 5.111 for a low alloy manganese-molybdenum pressure vessel steel. This effect varies with chemical composition and metallurgical treatment of the material, with the ferritic steels being much more susceptible to this damage than the austenitic steels.[201, 202, 264]

3. *Effect of Fatigue*

Fatigue is the prime manner in which cracks grow to critical size; and since fatigue crack growth rate varies as approximately the fourth power of the material stress intensity factor, it is of paramount importance, paragraph 5.12.

5.24 Fracture Toughness Relationships

In the absence of direct fracture toughness measurements, K_{Ic} and K_{Id}, these values must be estimated by correlations with the small scale test results from Charpy V-notch (CVN) impact, nil-ductility transition (NDT) temperature, or dynamic tear (DT) testing.[313] The CVN test reports fracture energy and appearance with respect to the testing temperature, Fig. 5.112. The NDT temperature test provides the minimum temperature at which a small flaw will not grow when loaded dynamically to the yield strength. The DT test is similar to the Charpy test and utilizes impact testing to measure fracture energy as a function of temperature. The CVN, NDT and DT tests were not designed to predict failure load, or critical flaw size, but can provide this information through correlation with the fracture toughness. In addition to these correlations, crack opening displacement (COD), *J*-integral, and other methods have been correlated with fracture toughness.

Of these, the CVN is the most widely used because it requires little material and is economical to perform. This correlation falls into two categories based on the regions of the Charpy curve they describe—the

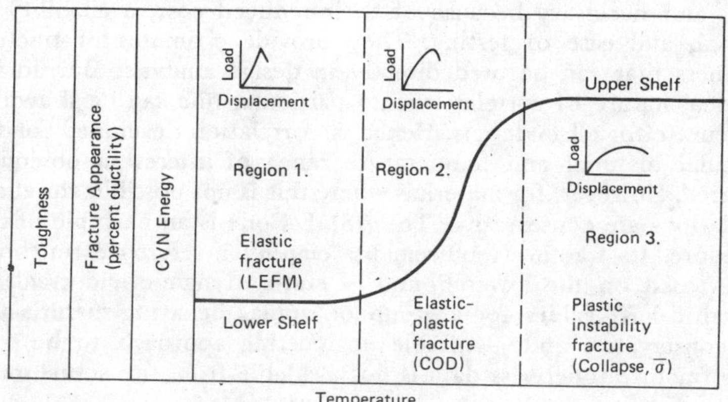

Fig. 5.112. Schematic Transition Curve Illustrating Various Failure Modes.

upper shelf and the transition region, Fig. 5.112. In the upper shelf region the rate of loading has little effect and the Rolfe-Novak-Barsom correlation[314] is representative of steels with yield strengths in the range of 110 to 246 ksi.

$$\left(\frac{K_{Ic}}{\sigma_{Y.P.}}\right)^2 = 5\left(\frac{CVN}{\sigma_{Y.P.}} - 0.05\right) \tag{5.24.1}$$

where

K_{Ic} = static fracture toughness, ksi $\sqrt{\text{in.}}$
K_{Id} = dynamic fracture toughness, ksi $\sqrt{\text{in.}}$
$\sigma_{Y.P.}$ = yield strength, ksi
CVN = energy, ft. lb.

Likewise, correlations between fracture toughness and CVN results in the transition temperature region can be expressed by the relationship

$$K_{Ic}, K_{Id} = A(CVN)^n \tag{5.24.2}$$

where the values of A and n are found from a suitable curve fit to the data.[315] A conservative estimate of K_{Ic} and K_{Id} is provided by the relationship[204, 265]

$$K_{Ic} = 9.35(CVN)^{0.63} \tag{5.24.3}$$

and

$$K_{Id} = 21.6(CVN)^{0.17} \tag{5.24.4}$$

Correlations between fracture toughness and small scale tests are useful and necessary because of their reduced cost, availability of material, and ease of testing. They provide estimates of fracture toughness that can be used directly in design analysis. Due to the empirical nature of correlations, no particular one can be shown to be accurate for all materials. Hence, a correlation developed for the particular material and temperature range of interest is obviously preferred. However, for materials where this is not possible, the above correlations are conservative. The ASME Code is an example of this procedure. Its fracture requirements employ a reference toughness curve based on the lower bound of static, dynamic and crack arrest critical K_I values for a group of similar behaving materials. It is a conservative and justifiable engineering approach to be used when fracture toughness data is not available from the actual material specification.

5.25 Criteria for Design with Defects

1. *Leak-before-break*

One criterion for selecting the material or operating limits of a vessel is called "leak-before-break." It is based on the fracture mechanics concept that a detectable leak can be supported by the material without resulting in a catastrophic failure.[205,266] Experience with thin wall vessels containing through cracks perpendicular to the hoop stress and of length twice the thickness of the vessel has shown that they will deform sufficiently under pressure to permit leakage. Accordingly, substituting the value for the total crack length, $2a$, equal to twice the vessel thickness, $2h$, in Eq. 5.22.13 gives

$$K = \sigma\sqrt{\pi h} \tag{5.25.1}$$

Example: The minimum environmental temperature of an unirradiated cylindrical pressure vessel constructed of manganese-molybdenum steel, ASTM 302B, is 0°F, and it is subjected to an operating hoop stress of 20,000 psi. Assuming a residual stress value of 10,000 psi. (equal to the short-time yield strength at the stress relieving temperature of the vessel), what is the maximum thickness for which the "leak-before-break" criterion applies?

Answer: Substituting the following values

$$K_{Ic} = 33{,}000 \text{ psi} \sqrt{\text{in.}} \text{ from Fig. 5.111}$$
$$\sigma = 20{,}000 + 10{,}000 = 30{,}000 \text{ psi}$$

in Eq. 5.25.1 gives

$$h = \frac{K_{Ic}^2}{\sigma^2\pi} = \frac{33{,}000^2}{30{,}000^2 \times 3.14} = 0.38 \text{ in.}$$

This is a severe design criterion that permits the presence of a large unfavorable oriented defect by requiring a compensating high toughness property in the material employed. The conservatism in the leak-before-break criterion lies entirely in the assumed defect size and location. It is recommended that this criterion be used only when a realistic limit based on the actual size of the defect cannot be established.

2. *Defect Size Evaluation*

Another more realistic criterion is to establish the relationship between the failure stress and the actual defect size as determined by nondestructive examination methods, and maintain operating stresses within these limits. Figure 5.113 shows the relationship between applied tensile stress,[206] defect size and K_I for the case of a long semi-

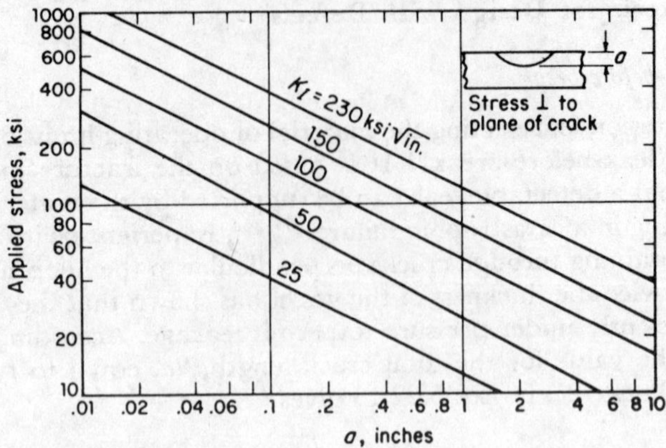

Fig. 5.113. Stress vs. Crack Depth for Long Shallow Semi-elliptical Surface Cracks[206]

elliptical surface crack in a section much thicker than the depth of the crack, Table 5.6. High strength steels have both characteristically low material toughness and are basically subject to high stresses, and this figure shows the decreasing defect size that can be tolerated under these conditions. It also emphasizes the importance of the available nondestructive examination methods employed prior to service and during service.[267, 268, 269, 270, 328]

Example: Consider a solid shaft of radius b on which a cylinder of outside radius c is shrunk on with a radial interference δ, Fig. 5.114. The interface pressure created by the interference is given by Problem 9, Chapter 2, which when substituted in Eq. 2.8.20 gives the maximum hoop stress developed in the cylinder as

$$\sigma_{max} = \frac{E}{2}\left(\frac{\delta}{b}\right)\left[\left(\frac{b}{c}\right)^2 + 1\right] \tag{5.25.2}$$

If the cylinder contains a small crack of length "a" radiating from its inner surface, Fig. 5.114, the allowable radial interference that can be tolerated without causing the cylinder to fracture can be found from Case 2, Table 5.6 as

$$K_I = 1.1\, \sigma_{max} \sqrt{\pi a} \tag{5.25.3}$$

(The crack length "a" is assumed to be small in comparison with "b".) Substituting the value of σ_{max} from Eq. 5.25.2 in Eq. 5.25.3 gives the critical radial interference as

$$\delta_c \cong \frac{bK_{Ic}}{E\sqrt{a}} \frac{1}{1+\left(\frac{b}{c}\right)^2} \tag{5.25.4}$$

If the cylinder is made of a high tensile steel, AISI 4340, with an ultimate tensile strength of σ_{ult} = 240 ksi, fracture toughness of K_{Ic} = 60 ksi $\sqrt{\text{in.}}$, μ = 0.3 and E = 30,000 ksi, and b = 0.5 in., Eq. 5.25.4 becomes

$$\delta_c = \frac{10^{-3}}{\left[1+\left(\frac{b}{c}\right)^2\right]\sqrt{a}} \tag{5.25.5}$$

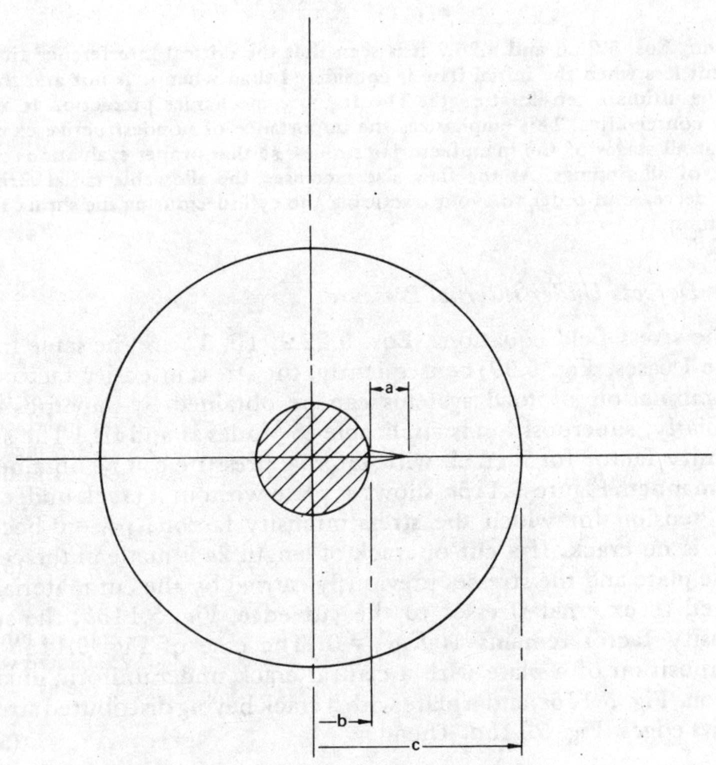

Fig. 5.114. Flaw in a Shrink-Fit Cylinder

For a ratio c/b = 3.5 and an initial flaw size a = 0.02 in.,

$$\delta_c = \frac{10^{-3}}{[1 + (0.285)^2] \sqrt{0.02}} = 0.0065 \text{ in} \qquad (5.25.6)$$

The conventional maximum stress criteria, which does not account for initial flaws, predicts failure when σ_{max} in Eq. 5.25.2 reaches σ_{ult} or

$$\delta_c = \frac{2b\,\sigma_{ult.}}{E} \frac{1}{\left[\left(\frac{b}{c}\right)^2 + 1\right]}$$

$$= \frac{2(0.5)\ (240)}{30{,}000\ [1 + (0.285)^2]} = 0.0074 \text{ in.} \qquad (5.25.7)$$

From Eqs. 5.25.6 and 5.25.7 it is seen that the critical interference fit is 12 percent less when the initial flaw is considered than when it is not and the criteria is ultimate tensile strength. The fracture mechanics prediction is always more conservative. This emphasizes the importance of nondestructive examination at all stages of the manufacturing process so that proper evaluations can be made of all findings. As the flaw size increases, the allowable radial shrink-fit must decrease in order to avoid fracturing the cylinder during the shrink-fitting operation.

3. *Defects Under Internal Pressure*

The stress field equations, Eqs. 5.22.9, 10, 11 are the same for all mode I cases, Fig. 5.99; consequently, the stress intensity factors for a combination of load systems can be obtained by superposition. (Similarly, superposition is applicable to Modes II and III.) The stress intensity factor for a crack with internal pressure can be obtained in this manner. Figure 5.115*a* shows a plate without a crack under uniaxial tension for which the stress intensity facror $K_{I(a)} = 0$ because there is no crack. If a cut or crack of length $2a$ is made in the center of the plate and the stresses previously carried by the cut material are applied as external stresses to the cut edge, Fig. 5.115*b*, the stress intensity factor remains as $K_{I(b)} = 0$. The case of Fig. 5,115*b* is a superposition of a plate with a central crack under uniform uniaxial tension, Fig. 5.115*c* and a plate with a crack having distributed stresses σ at its edges, Fig. 5.115*d*. Then

$$K_{I(c)} + K_{I(d)} = K_{I(b)} = 0 \qquad (5.25.8)$$

$$K_{I(d)} = -K_{I(c)} \qquad (5.25.9)$$

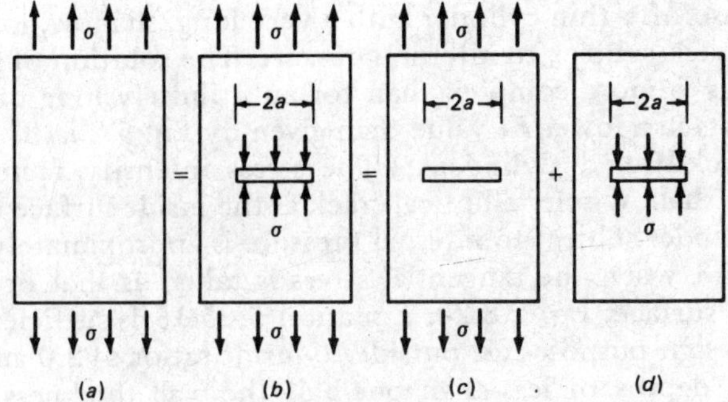

Fig. 5.115. Superposition, Crack Under Internal Pressure

and from Eq. 5.22.8 or Case I, Table 5.6,

$$K_{\mathrm{I}(d)} = -\sigma\sqrt{\pi a} \tag{5.25.10}$$

The case of a crack with internal pressure p_i is equivalent to that of Fig. 5.115*d* but the pressure acts in a direction opposite to the stress σ. Accordingly, the sign of $K_{\mathrm{I}(d)}$ is reversed in Eq. 5.25.10 to give

$$K_{\mathrm{I}} = p_i\sqrt{\pi a} \tag{5.25.11}$$

as the stress intensity factor for a crack with internal pressure.

(a) Thin-Walled Cylinders. Cracks occurring in practice often initiate at exposed surfaces and tend to grow inward to assume semi-elliptical shapes. When such a defect is subject to both direct pressure and a field stress, their individual effects must be added together (superpositioned) to obtain the resultant stress intensity factor.[316] For instance, the stress intensity factor K_{I} for a semi-elliptical crack at the inside surface of a thin cylinder subject to internal pressure is then:

$$K_{\mathrm{I}} = \frac{1.1}{\Phi}(\sigma_t + p_i)\sqrt{\pi a} \tag{5.25.12}$$

where a is the crack depth (one-half the minor axis of the elliptical crack, Case 5, Table 5.6), σ_t is the tangential stress, Eq. 2.2.4, and p_i is the internal pressure. When the elliptical crack shape is long and slender, which is the frequent case in vessels, a/ℓ approaches zero and the value of Φ becomes one, so Eq. 5.25.12 becomes

$$K_{\mathrm{I}} = 1.1(\sigma_t + p_i)\sqrt{\pi a} \tag{5.25.13}$$

for the case of a thin cylinder with a very long, shallow, longitudinal internal crack subject to internal pressure. The solution of semi-elliptical cracks is more complex than for an infinitely long one, but always results in a lower K_I value than given by Eq. 5.25.13.

(b) Thick-Walled Cylinders. The stress intensity factor for the same long shallow semi-elliptical crack at the inside surface of a thick-walled cylinder subject to internal pressure is approximately given by Eq. 5.25.13 when the tangential stress is taken as that occurring at the inside surface, Eq. 2.8.20. Equation 5.25.13 is sufficiently accurate for design purposes for outside-to-inside ratios of 2.0 and greater, and crack depths of less than one-half the wall thickness.[317,318] In fracture mechanics evaluations the stresses generated from all conditions must be considered (pressure, thermal, discontinuity, external loading, and residual).

Example: The thick-walled cylinder shown in Fig. 5.116 is 24 in. OD and 12 in. ID and contains a semi-elliptical longitudinal defect 0.4 in. deep on the inside surface of aspect ratio $a/\ell = 0.1$. The material has a fracture toughness $K_{Ic} = 164$

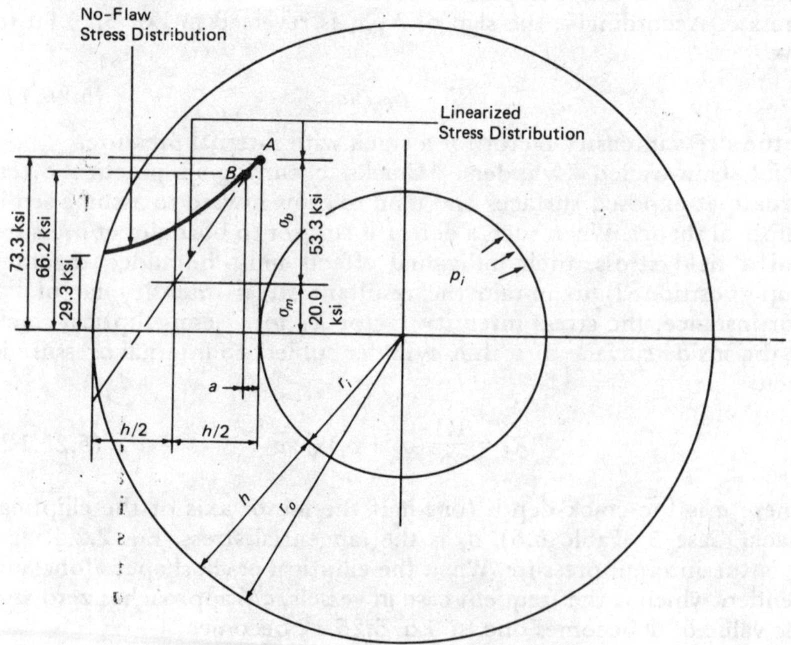

Fig. 5.116. Linearized Representation of Stresses

ksi $\sqrt{\text{in.}}$ at the service temperatue, and vessel operates at 44,000 psi. Determine if the vessel is satisfactory from a brittle fracture viewpoint under those service conditions.

This can be established by using Eq. 5.25.13 to calculate the acting stress intensity factor K_I. If this value is less than the critical material value, K_{Ic}, the vessel is satisfactory. The tangential stress at the inside surface is, per Eq. 2.8.20,

$$\sigma_t = p_i \frac{(r_i^2 + r_o^2)}{(r_o^2 - r_i^2)} = 73.3 \text{ ksi}$$

Substituting this value of the inside tangential stress, and the value of $a = 0.4$ in. and $p_i = 44{,}000$ psi, in Eq. 5.25.13 gives

$$K_I = 1.1(73.3 + 44.0)\sqrt{3.1416 \times 0.4} = 140 \text{ ksi} \sqrt{\text{in.}}$$

Accordingly, a defect of this size will not result in brittle fracture because the acting stress intensity factor is less than the critical one for the material. However, defect size grows with fatigue, Par. 5.21; hence, this conclusion is time dependent and inspections during service must be a part of all operating procedures.[335,336,337]

Another method of solution that is frequently used when the stress varies throughout the thickness, and is employed by the ASME Code, Section XI, is to resolve the acting stress at the defect location into membrane and bending stress components with respect to the wall thickness by converting the actual nonlinear stress distribution into a linear representation of the actual stress distribution. It is an approximate one that gives conservative results, and consists of first determining the tangential stresses in the absence of the flaw at the location corresponding to the two extremities of the flaw depth, points A and B, Fig. 5.116. A pseudo-linear stress distribution is then obtained by drawing a straight line through points A and B. This gives the requisite stress distribution at these locations. (It is not a linearization of the entire stress gradient through the wall thickness.) The components of this linear distribution, σ_m and σ_b, are those occurring at the inside surface, point A, and are taken as the acting field or global stress. The resulting stress intensity factor is then given by

$$K_I = \sigma_m M_m \sqrt{\frac{\pi a}{Q}} + \sigma_b M_b \sqrt{\frac{\pi a}{Q}} \qquad (5.25.14)$$

and adding to this the effect of internal pressure on the flaw from Eqs. 5.25.11 and 5.25.13

$$K_I = \sigma_m M_m \sqrt{\frac{\pi a}{Q}} + \sigma_b M_b \sqrt{\frac{\pi a}{Q}} + 1.1\, p_i \sqrt{\pi a} \qquad (5.25.15)$$

where:

σ_m = membrane stress, psi
σ_b = bending stress, psi
$\sigma_{Y.P.}$ = specified minimum yield strength, psi
a = flaw depth, inches
ℓ = major axis of semi-elliptical flaw, inches
Q = flaw or defect shape parameter to be determined from Fig. 5.117 using $(\sigma_m + \sigma_b)/\sigma_{Y.P.}$ and the defect geometry ellipse aspect ratio, a/ℓ
M_m = correction factor for the membrane stress that adjusts the basic stress intensity factor to account for finite thickness and flaw location, Fig. 5.118
M_b = correction factor for the bending stress that adjusts the basic stress intensity factor to account for finite thickness and flaw location, Fig. 5.119

Similarly, methods are availabe for establishing the stress intensity factors for various shape and oriented defects located on the surface or below the surface.[319]

Example: The thick-walled cylinder in the previous example is made of a 130,000 psi specified minimum yield strength material. The material fracture toughness and flaw size remain the same. Determine if the vessel is satisfactory from a brittle fracture viewpoint using the linearized stress distribution method.

The stress gradient through the wall is given by Eq. 2.8.19 and the linearized

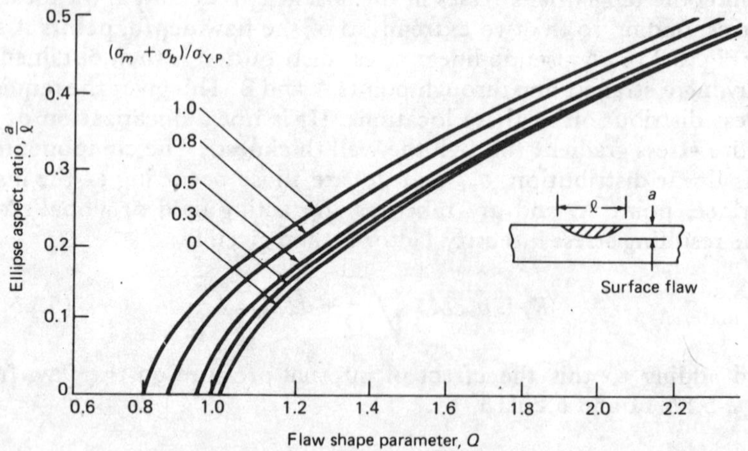

Fig. 5.117. Shape Factors (*Courtesy The American Society of Mechanial Engineers*)

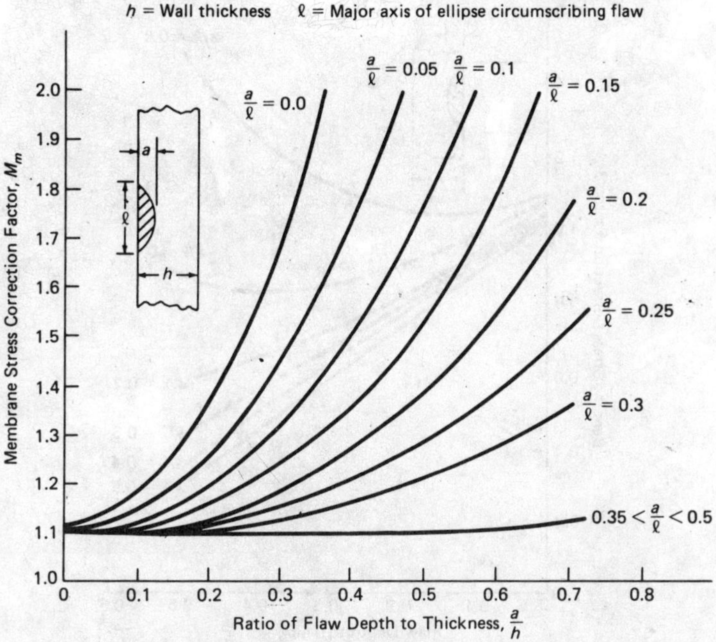

Fig. 5.118. Membrane Stress Correction Factors for Surface Flaws (*Courtesy The American Society of Mechanical Engineers*)

stress gradient is shown in Fig. 5.116, from which σ_m = 20.0 ksi and σ_b = 53.3 ksi. For a value of $(\sigma_m + \sigma_b)/\sigma_{Y.P.} = (20.00 + 53.3)/130 = 0.56$, $a/h = 0.4/6 = 0.07$ and $a/\ell = 0.1$, the memberane stress correction factor M_m is obtained from Fig. 5.118 as 1.0 and the bending stress correction factor is obtained from Fig. 5.119 as 1.1. Also from Fig. 5.117, $Q = 1$, so Eq. 5.25.15 reduces to

$$K_I = 1.1(\sigma_m + \sigma_b + p_i)\sqrt{\pi a}$$
$$= 1.1(20.0 + 53.3 + 44.0)\sqrt{3.1416 \times 0.4} = 140 \text{ ksi} \sqrt{\text{in.}}$$

This is in agreement with the previous maximum stress at defect location method, and likewise indicates that a defect of this size will not result in brittle fracture because the acting stress intensity factor is less than the fracture toughness of the material, namely, 164 ksi $\sqrt{\text{in.}}$

5.26 Integrity Assessment Diagram

A simple engineering procedure to assess the failure of vessels is that shown in Fig. 5.120. This diagram is a safety-failure space one defined by the stress intensity/fracture toughness ratio as the ordinate

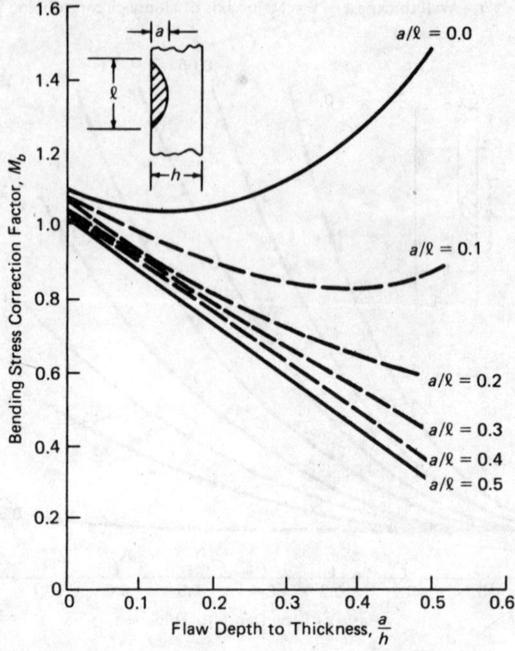

Fig. 5.119. Bending Stress Correction Factor for Surface Flaws (*Courtesy The American Society of Mechanical Engneers*)

and the applied stress/plastic collapse stress (ultimate tensile strength) ratio as the abscissa. It is based on the premise that the vessel will fail by either a brittle fracture or a plastic collapse mechanism, and that these two mechanisms are connected by a transition curve from linear elastic fracture mechanics, LEFM, to plastic instability or collapse, ultimate tensile strength. This forms a safety-failure space, the ordinates of which are given by [320, 321, 338]

$$K_r = \frac{S_r}{\sqrt{\frac{8}{\pi^2} \log_e \sec\left(\frac{\pi}{2} S_r\right)}} \tag{5.26.1}$$

where

$$K_r = K_I / K_{Ic} \tag{5.26.2}$$

$$S_r = \sigma / \sigma_{ult.} \tag{5.26.3}$$

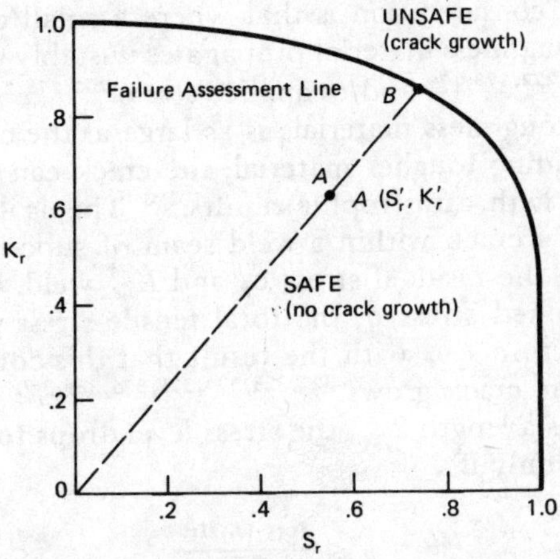

Fig. 5.120. Failure Assessment Diagram

and

K_I = stress intensity factor for the vessel to be assessed, ksi $\sqrt{\text{in.}}$
K_{Ic} = material fracture toughness (critical stress intensity), ksi $\sqrt{\text{in.}}$
$K_r' = K_I/K_{Ic}$ for a particular assessment point
$S_r' = \sigma/\sigma_{\text{ult.}}$ for a particular assessment point
σ = applied stress, psi
$\sigma_{\text{ult.}}$ = ultimate tensile stress, psi

In order to use Fig. 5.120 two parameters must be calculated for the acting or postulated conditions, namely, a plastic collapse parameter (S_r') and a linear elastic fracture mechanics parameter (K_r'). If the point lies inside the assessment curve the vessel is safe from failure. Further, since both K_r' and S_r' are directly proportional to the applied load or pressure, the closeness of the point K_r', S_r' to the failure assessment curve, as measured radially from the origin of the diagram, determines how near the vessel is to failure; or, referring to Fig. 5.120, the factor of safety is OB/OA.

5.27 Significance of Fracture Mechanics Evaluations

All vessels have defects or flaws that are inherent in their construction process. The critical size may have been built-in during construction, initiated in service, or grown slowly from a subcritical size by service conditions such as fatigue or creep.[271]

A significant consideration is that where a small defect in a local region of low toughness material propagates unstably under the action of high residual[272,273,274] and/or applied stress.[207,275,276,277] If the local region of low toughness materials is as large as the critical crack size for the surrounding tougher material, the crack can run through the tough material with catastrophic results.[278] This is illustrated in Fig. 5.121 showing a crack within a weld seam of subcritical length $2a_1$ with respect to the residual stress σ_r and K_{Ic} weld. Upon superposition of the applied stress σ, the total tensile stress surrounding the crack increases to $\sigma + \sigma_r$ with the result that this condition becomes unstable and the crack grows.[322,323,324,325,326,339,340] However, when the crack attains a length $2a_2$, the stress level drops to σ but the crack will be arrested only if

$$\sigma \leqq \frac{K_{Ic} \text{ plate}}{\sqrt{\pi a_2}} \tag{5.26.1}$$

This emphasizes the importance of minimizing the residual stress present in vessels because the fracture stress calculated by the methods of fracture mechanics represent the total stress occurring at the instant of fracture. Surface cracks are more injurious than internal cracks. For instance, transverse cracks in pressure vessels of larger sizes than those

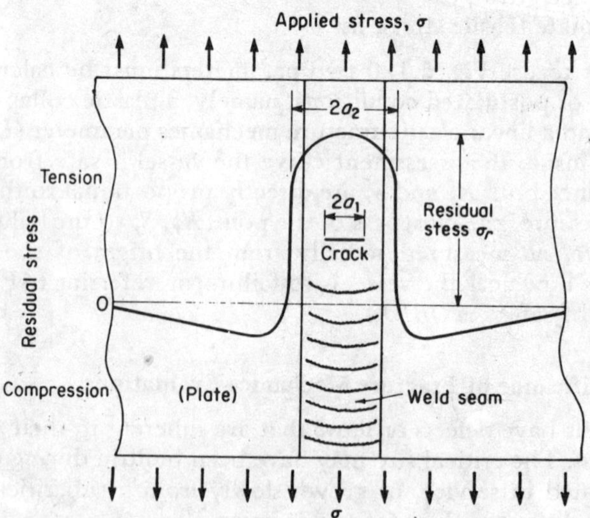

Fig. 5.121. Residual Stress Distribution in a Weld Seam Containing a Crack

given by the employment of plane strain K_{Ic} values can be tolerated if they are located near the midpoint of the wall thickness because:

1. Exposure of a crack to a free surface essentially doubles its effective size relative to the stress level near the crack tip, Table 5.6, Case 1 vs Case 2.
2. The midthickness crack is subject to little stress due to superimposed bending usually present due to vessel non-circularity or local imposed bending moments.
3. Extension of the midthickness crack does not give access for or contact with the pressure contained inside or outside the vessel which would directly increase the pressure stress and/or expose the crack face to a potentially harmful environment.

5.28 Effect of Warm Prestressing on the Ambient Temperature Toughness of Pressure Vessel Steels

The toughness of pressure vessel steels containing notches and cracks can be improved by warm prestressing or prestraining; thereby, improving the brittle fracture characteristics of vessels and structures fabricated from these materials. Prestressing a material with a flaw or notch (and all materials or structures have natural flaws or fabricated notches) in tension, increases the subsequent fracture strength at a lower temperature. This effect increases with the amount of prestress.[208]

Prestressing the material at a temperature above the intended service temperature, and at which temperature a plastic zone at the crack tip can be established without creating excessive crack growth and triggering a brittle fracture, results in an improvement in brittle fracture resistance which comes primarily from the mechanical effects of the plastic zone created at the crack tip.[209,210] Here the crack is blunted, the material is work hardened, and upon removal of the applied stress the crack tip is left in a favorable state of compressive residual stress.[279,280,327]

Most pressure vessel construction codes, such as those of the American Society of Mechanical Engineers, ASME, require a hydrostatic test at a pressure in excess of the service or design pressure and at a temperature above the NDT; hence, this test provides an inherent asset beyond establishing initial pressure and leak integrity.

It is important, however, to also appraise the subsequent mitigating metallurgical effects to which the material may be subjected.[211,212,213] Degradation of material properties occurs when metallurgical effects interact and offset mechanical ones. For instance, strain aging of the highly strained material near the crack tip promoted by elevated temperature service can greatly reduce the toughness of this region and render the structure more susceptible to brittle fracture with prestressing than without prestressing.[214] As an example, 15 per cent plastic prestraining at 200°F of a commonly used carbon steel pressure vessel material, ASTM 516 Grade 70, gave a 20 percent increase in fracture resistance at 75° F. Whereas, subsequent aging at 650° F for 1,000 hours reduced it again to below this value by 30 percent.

REFERENCES

1. J. R. Low, "Behavior of Metals Under Direct or Non-reversed Loading," American Society of Metals Congress, October 23, 1948.

2. W. E. Cooper, "The Significance of the Tensile Test to Pressure Vessel Design," Welding Research Supplement, pp. 49–56, January, 1957.

3. T. C. Hsu, S. R. Davies and R. Royles, "A study of Stress-Strain Relationship in the Work-Hardening Range," *ASME Transactions*, September, 1967.

4. J. J. Gelman, "Fracture in Solids," *Scientific American*, Vol. 202, No. 2 pp. 94–104, February, 1960.

5. A. S. Tetelman and A. J. McEvily, Jr., *Fracture of Structural Materials*, John Wiley & Sons, New York, 1967.

6. V. F. Zackay, "The Strength of Steel," *Scientific American*, pp. 72–81, August, 1963.

7. A. Nadai, *Theory of Flow and Fracture of Solids*, McGraw-Hill Book Co. Inc., New York, 1950.

8. M. S. Gregory and I. F. Jones, "Yield Under Plane-Stress Conditions," *Experimental Mechanics*, March, 1971.

9. D. Radaj, "Detection of Lueders' Lines by Means of Brittle Coatings," *Experimental Mechanics*, March, 1969.

10. R. I. Mair and E. E. Banks, "Spread of Yield With Mild Steel," *Experimental Mechanics*, Vol. 13, No. 2, February, 1973.

11. J. F. Harvey, "The State of Stress in Riveted and Welded Joints as Determined by Plastic Flow Observations," Cornell University, Ithaca, N. Y., 1937.

12. M. J. Manjoine, "Influence of Rate of Strain and Temperature on Yield Stresses in Mild Steel," *ASME Journal of Applied Mechanics*, Vol. 11, pp. 211–218, December, 1944.

13. D. P. Kendall and T. E. Davidson, "The Effect of Strain Rate on Yielding in High Strength Steels," *ASME Publication Paper No.* 65-WA/Met-8, 1965.

14. R. J. Goodenow, "Yielding and Flow Characteristics of Low-Carbon Steel Between Ambient and Liquid Nitrogen Temperature," *ASME Publication Paper No.* 69-Met-17, 1969.

15. Thelsch and Helmut, "Strain Aging of Pressure Vessel Steels," *The Welding Journal*, *Research Supplement*, pp. 283–290, 1951.

16. J. D. Baird, "Strain Aging of Steel—A Critical Review," *Iron and Steel*, May, 1963.

17. C. J. Osburn, A. F. Scotchbrook, R. D. Stout and B. G. Johnson, "Effect of Plastic Strain and Heat Treatment," *The Welding Journal*, *Research Supplement*, pp. 337–353, 1949.

18. R. V. Milligan, "Influence of Bauschinger Effect on Reverse Yielding in Thick-Walled Cylinders," *ASME Publication Paper No.* 69-SESA-4, 1969.

19. J. H. Holloman and C. Zenes, "Conditions of Fracture of Steel," *ASME Transactions*, pp. 283–296, 1944.

20. W. T. Lankford, "Effect of Cold Work on the Mechanical Properties of Pressure Vessel Steel," *The Welding Journal*, April, 1956.

21. S. S. Toz, R. D. Stout, and B. G. Johnson, "Further Tests on Effects of Plastic Strain and Heat Treatment," *The Welding Journal, Research Supplement*, pp. 576–584, 1951.

22. J. H. Gross and R. D. Stout, "The Performance of High-Strength Pressure-Vessel Steels," *The Welding Journal, Research Supplement*, pp. 115–119, March, 1956.

23. J. F. Gormley, "Effect of Prestrain on the Fatigue Properties of Steel," *ASME Publication Paper No.* 61-SA-17, 1961.

24. P. W. Bridgeman, *Studies in Large Plastic Flow and Fracture*, McGraw-Hill Book Co., Inc., New York, 1952.

25. R. G. Berggren, "Radiation Effects in Ferritic Steels," U.S. Atomic Energy Commission Report, Office TID-7588, 1960.

26. W. S. Pellini and R. W. Judy, "Significance of Fracture Extension Resistance (R Curve) Factors in Fracture-Safe Design for Nonfrangible Metals," *Welding Research Council Bulletin* 157, December, 1970.

27. F. J. Freely, M. S. Northup, R. R. Kleippe and M. Gensamer, "Studies on the Brittle Fracture of Tankage Steel Plates," *The Welding Journal, Research Supplement*, Vol. 12, pp. 596–607, December, 1955.

28. A. L. Brown and J. B. Smith, "Failure of Spherical Hydrogen Storage Tank," *Mechanical Engineering*, Vol. 66, pp. 392–397, 1944.

29. A. J. Babecki and P. P. Puzak, "Fabrication and Service Factors Involved in Failure of Welded Steam Receivers," *The Welding Journal, Research Supplement*, pp. 320–324, July, 1958.

30. W. R. Apblett "Investigation—Brittle Failure of Feedwater Heater Forging," Foster Wheeler Corporation, Bulletin No. SP65-6, 1965.

31. Report on the Brittle Fracture of a High-Pressure Drum at Cockenzie Power Station, by the South of Scotland Electricity Board, January 1967, Welding Research Abroad, Welding Research Council, October, 1967, New York.

32. D. L. Cheever and R. E. Monroe, "Failure of a Welded Medium Carbon Steel Heat Exchanger," *ASME Publication Paper No.* 70-PVP-3, 1970.

33. F. B. Harnel, "An Investigation of the Impact Properties of Vessel Steels," Reports of Progress of Welding Research Council, No. 9, p. 31, September, 1960.

34. K. Masubuchi, R. E. Monroe and D. C. Martin, "Interpretive Report on Weld-Metal Toughness," Welding Research Council Bulletin 111, 1966.

35. R. Rote and J. H. Proctor, "A-300 Alloy Steels," *Chemical Engineering*, pp. 119–122, June 27, 1960.

36. C. P. Sullivan, "A Review of Some Microstructural Aspects of Fracture in Crystalline Materials," *Welding Research Council Bulletin 122*, May, 1967.

37. L. Porse, "Reactor Vessel Design Considering Radiation Effects," *ASME Publication Paper No.* 63-WA-100, 1963.

38. E. T. Wesser and W. H. Pryle, "Brittle Fracture Characteristics of a Reactor Pressure Vessel Steel," *ASME Transactions*, Paper 60-MET-9, 1960.

39. R. H. Sterne and L. E. Steele, "Steels for Commercial Nuclear Power Reactor Pressure Vessels," *Nuclear Engineering and Design*, Vol. 10, pp. 259–307, 1959, North-Holland Publishing Co., Amsterdam.

40. D. S. Billington, "Relaxing Reliance on Emperical Data," *Nucleonics*, Vol. 18, No. 9, September, 1960.

41. R. G. Berggren, W. E. Brundage, W. W. Davis, N. E. Hinkle, and J. C. Zukas, "Tensile and Stress-Rupture Properties of Irradiated Stainless Steels," U. S. Atomic Energy Commission Report TID-7588, October, 1959.

42. R. E. Bailey and M. A. Filliman, ASTM-STP-233, p. 84, 1958.

43. J. R. Hawthorne and F. J. Loss, "The Effects of Coupling Nuclear Radiation with Static and Cyclic Service Stresses and of Periodic Proof Testing on Pressure Vessel Behavior," Naval Research Report 6620, August, 1, 1967.

44. R. W. Nichols and D. R. Harris, "Brittle Fracture and Irradiation Effects in Ferritic Pressure Steels," Symposium on Radiation Effects on Metals and Neutron Dosimetry, ASTM STP 341, 1963.

45. M. B. Reynolds, "The Effect of Stress on the Radiation Stability of ASTM A 302B Pressure Vessel Steel," Materials Research Standards 3, pp. 644-645, 1963.

46. W. S. Pellini, L. E. Steele and J. R. Hawthorne, "Analysis of Engineering and Basic Research Aspects of Neutron Embrittlement of Steels," Symposium on Radiation Damage in Solids and Reactor Materials, Venice, Italy, May, 1962.

47. L. E. Steele and J. R. Hawthorne, "New Information on Neutron Embrittlement and Embrittlement Relief of Reactor Pressure Vessel Steels," Naval Research Laboratory Report 6160, October 6, 1964.

48. G. D. Whitman, G. C. Robinson and A. W. Savolainen, "Technology of Steel Pressure Vessels for Water-Cooled Nuclear Reactors," Oak Ridge National Laboratory, ORNL-21, December, 1967.

49. C. Z. Serpan, L. E. Steele and J. R. Hawthorne, "Features and Results of Several Programs of Radiation-Damage Surveillance of Power Reactor Pressure Vessels," *ASME Journal of Basic Engineering*, March, 1967.

50. U. Potapovs, and J. R. Hawthorne, "The Effect of Residual Elements on the Response of Selected Pressure Vessel Steels and Weldments to Irradiation at 550°F *Nuclear Applications, NVAPA*, Vol. 6, No. 1, January, 1969.

51. J. R. Hawthorne, "Demonstration of Improved Radiation Embrittlement Resistance of A-533-B Steel Through Control of Selected Residual Elements," ASTM STP 484, American Society Testing Materials, 1970, pp. 96–127.

52. J. R. Hawthorne and U. Potapovs, "Initial Assessments Notch Ductility Behavior of A-533 Pressure Vessel Steel with Neutron Irradiation," Naval Research Laboratory, NRL Report 6772, November 22, 1968.

53. L. E. Steele, J. R. Hawthorne, and R. A. Gray, "Neutron Irradiation Embrittlement of Several High Strength Steels," Naval Research Laboratory, NRL Report 6419, September 7, 1966.

54. L. E. Steele, Irradiation Effects on Reactor Structural Materials, Naval Research Laboratory, NRL Report 1593.

55. C. Z. Serpan, "Neutron Radiation Embrittlement of La Crosse Reactor Pressure Vessel Steel and Weldments," Nuclear Engineering and Design, 1968, p. 95–107.

56. R. G. Berggren, W. J. Stelzman, and T. N. Jones, "Radiation Hardening and Embrittlement in ASTM A-533-B," Oak Ridge National Laboratory, ORNL 4590, February, 1970, p. 127–133.

57. J. R. Hawthorne, "Postirradiation Dynamic Tear Performance of 12 inch A-533-B Submerged Arc Weldment," Naval Research Laboratory WHAN-FR-401, October, 1970.

58. J. R. Hawthorne, "A Radiation Resistance Weld Metal for Fabricating A533-B Reactor Vessels," Naval Research Laboratory, NRL Report 2328 August 15, 1971.

59. E. Orowan, "Theory of Fatigue of Metals," *Proceedings of London Royal Society*, Vol. 171A, pp. 79–105, 1939.

60. E. S. Machlin, "Dislocation Theory of the Fatigue of Metals," NACA TN 1489, 1948.

61. A. M. Freudenthal, "Fatigue Mechanisms, Fatigue Performance, and Structural Integrity," Proceedings: Air Force Conference on Fatigue and Fracture of Aircraft Structures and Materials, Miami Beach, Florida, December 1969, AFFDL TR 70-144 (1970).

62. P. C. Paris, "The Fracture Mechanics Approach to Fatigue," Syracuse University Press, 1964.

63. Mowbray, Andrews and Brothers, "Fatigue-Crack Growth Rate Studies of Low Alloy Pressure Vessel Steels," ASME Paper 68-PVP-23, 1968.

64. T. W. Crooker and E. A. Lange, "The Influence of Yield Strength and Fracture Toughness on Fatigue Design Procedures for Pressure Vessel Steels." ASME Paper 70-PVP-19, 1970.

65. W. G. Clark, "Fracture Mechanics and Nondestructive Testing of Brittle Materials," ASME Paper 71-PVP-4, 1971.

66. M. R. Gross, "Fatigue—Crack Growth in Perspective," Naval Engineers Journal, pp. 44–48, 1970, Washington, D. C.

67. W. G. Clark, "Fatigue Crack Growth Characteristics of Heavy Section ASTM A533 Grade B, Class 1 Steel Weldments," ASME Paper 70-PVP-24, 1970.

68. T. W. Crooker, "Designing Against Structural Failure Caused by Fatigue Crack Propagation," *Naval Engineers Journal*, Vol. 84, No. 6, December, 1972, Washington, D. C.

69. "Manual on Fatigue Testing," *ASTM Special Technical Publication No.* 91, 1949.

70. W. E. Duckworth, "The Achievement of High Fatigue Strength in Steel," *Metallurgia*, Vol. 69, No. 412, February, 1964.

71. L. F. Kooistra, "Effects of Plastic Fatigue on Pressure Vessel Materials and Design," *The Welding Journal*, March ,1957.

72. J. H. Gross, "PVRC Interpretive Report of Pressure Vessel Research, Section 2—Materials Considerations," *Welding Research Council Bulletin 101*, November, 1964.

73. R. D. Stout and A. W. Pense, "Effect of Composition and Microstructure on the Low Cycle Fatigue Strength of Structural Steels," *ASME Publication Paper No.* 64-Met-9, 1964.

74. L. F. Coffin, "A Study of the Effects of Cyclic Thermal Stresses on a Ductile Metal," *ASME Transactions* 79, 1954.

75. L. F. Coffin, "The Resistance of Material to Cyclic Strains," *ASME Transactions*, Paper 57-A-286, 1957.

76. L. F. Coffin, "The Stability of Metals Under Cyclic Plastic Strain," *ASME Transactions*, Paper 59-A-100, 1959.

77. S. S. Manson, "Behavior of Materials Under Conditions of Thermal Stress," NACA TN 2933, July, 1953.

78. S. S. Manson, "Fatigue: A Complex Subject—Some Simple Approximations," *Experimental Mechanics*, July, 1965.

79. B. F. Langer, "Design of Pressure Vessels for Low-Cycle Fatigue," *ASME Publication Paper No.* 61-WA-18, 1961.

80. S. Timoshenko, *Strength of Materials*, Part II, D. Van Nostrand Company Inc., Princeton, N. J., 1956.

81. H. J. Grover, S. A. Gordon, and L. R. Jackson, "Fatigue of Metals and Structures," U.S. Government Printing Office, Washington, D. C., 1960.

82. J. G. Sessler and V. Weiss, "Low Cycle Fatigue Damage in Pressure Vessel Steels," *ASME Publication Paper No.* 62-WA-233, 1962.

83. H. O. Fuchs, "A Set of Fatigue Failure Criterion for Fatigue," *ASME Publication Paper No.* 64-Met-1, 1964.

84. J. Dubuc, J. R. Vanasse, B. Biron, and A. Bazergui, "Effect of Mean Stress and Mean Strain in Low-Cycle Fatigue of A-517 and A-201 Steels," *ASME Publication Paper No.* 69-PVP-1, 1969.

85. R. W. Nichols, *Pressure Vessel Engineering Technology*, Elsevier Publishing Company, 1971.

86. M. A. Miner, "Cumulative Damage in Fatigue," *Journal of Applied Mechanics*, Vol. 12, p. A-159, September, 1945.

87. R. R. Gatts, "Application of a Cumulative Damage Concept to Fatigue," *ASME Publication Paper* 60-WA-144, 1960.

88. R. R. Gatts, "Cumulative Fatigue Damage with Progressive Loading," *ASME Publication Paper* 62-WA-292, 1962.

89. S. Tanaka, "On Cumulative Damage Impulse Fatigue Tests," *ASME Publication Paper* 62-WA-17, 1962.

90. T. D. Scharton and S. H. Crandall, "Fatigue Failure Under Complex Stress Histories," *ASME Publication Paper* No. 65-Met-3, 1965.

91. E. Z. Stowell, "A Study of the Energy Criterion for Fatigue," *Nuclear Engineering and Design*," Vol. 3 pp. 32–40, 1966, North-Holland Publishing Company, Amsterdam.

92. J. Dubuc, B. Q. Thang, A. Bazergui, and A. Biron, "Unified Theory of Cumulative Damage in Metal Fatigue," *Welding Research Council Bulletin* 162, June, 1971.

93. J. L. M. Morrison, B. Crossland, and J. S. C. Parry, "The Strength of Thick Cylinders Subjected to Repeated Pressure," *ASME Transactions*, Paper 59-A-167, 1959.

94. ASME Section III, Nuclear Power Plant Components, and Section VIII, Division 2, Pressure Vessels.

95. G. Welter and J. A. Choquet, "Triaxial Tensile Stress Fatigue Testing," *Welding Research Supplement*, pp. 565–570, December, 1963.

96. K. D. Ives, L. F. Kooistra, and J. T. Tucker, "Equibiaxial Low-Cycle Fatigue Properties of Typical Pressure-Vessel Steels," *ASME Publication Paper No.*, 65-Met-19, 1965.

97. K. J. Pascoe and J. W. R. deVilliers, "Low Cycle Fatigue of Steels Under Biaxial Straining," *Journal of Strain Analysis*, Vol. 2, No. 2, 1967.

98. J. J. Kibler and R. Roberts, "The Effect of Biaxial Stresses on Fatigue and Fracture," *ASME Publication Paper No.* 70-PVP-17, 1970.

99. T. J. Dolan, "Significance of Fatigue Data in Design of Pressure Vessels," *Welding Research Supplement*, pp. 255–260, May 1956.

100. E. Z. Stowell, "The Calculation of Fatigue Life in the Presence of Stress Concentrations," *Nuclear Engineering and Design*, Vol. 8, pp. 313–316, 1968, North-Holland Publishing Company, Amsterdam.

101. B. C. Hanley and T. J. Dolan, "Surface Finish," *Metals Engineering Design, ASME Handbook*, 1953.

102. O. J. Horger and H. R. Neifert, "Effect of Surface Condition on Fatigue Properties, Surface Treatment of Metals," ASM Symposium, 1941.

103. A. C. Low, "Short Endurance Fatigue," International Conference on Fatigue of Metals, New York, November 28–30, 1956.

104. G. C. Noll and C. Lipson, "Allowable Working Stresses," *SESA Proceedings*, Vol. III, No. 11, pp. 89–109, 1946.

105. J. O. Almen and A. L. Boegehold, "Rear Axle Gears, Factors which Influence Their Life," *ASTM Proceedings*, Vol. 35, 1935.

106. J. H. Zimmerman, "Flame Strengthening," *Iron Age*, February, 1940.

107. R. R. Moore, "Metallic Coatings," *Metals Engineering Design, ASME Handbook*, 1953.

108. C. E. Bowman and T. J. Dolan, "Biaxial Fatigue Studies of High Strength Steels Clad with Stainless Steel," University of Illinois Report T&AM, No. 164, May, 1960.

109. H. F. Moore, "Shot Peening and the Fatigue of Metals," *Metals Engineering Design, ASME Handbook*, 1953.

110. P. V. Kasgard, E. E. Day and A. S. Kobayashi, "Exploratory Study of Optimum Coining for Improvement of Fatigue Life," *Experimental Mechanics*, October, 1964.

111. J. Morrow and F. R. Tuler, "Cyclic Produced Changes in Stress-Strain Properties of Metals," *Report of Progress, Welding Research Council*, Vol. XIX, April 1964.

112. H. V. Cordiano, "Effect of Residual Stresses in the Low Cycle Fatigue Life of Large Scale Weldments in High Strength Steel," *ASME Publishing Paper No.* 69-SESA-6, 1969.

113. L. Tall, "Residual Stresses in Welded Plates—A Theoretical Study," Welding Research Council, January, 1964.

114. M. B. Shapiro, "Effect of Isothermal Heat Treatment on Fatigue Strength and Wear Resistance of Steel," Brutcher Translation from Metallovedenie i Obrabolka Metallov, March, 1957.

115. T. R. Gurney, "Effect of Stress Relief on Fatigue Strength of ½ in. Thick Transverse Butt Welds Containing Slag Inclusions," British Welding Research Report, February, 1965.

116. E. J. Mills, T. J. Atterbury, L. M. Cassidy, R. J. Eiber, A. R. Duffy, A. G. Imgram, and K. Masubuchi, "Design, Performance, Fabrication, and Material Considerations for High-Pressure Vessels," March, 1964, Redstone Arsenal, Alabama.

117. T. C. Yen, "Thermal Fatigue—A Critical Review," *Welding Research Council Bulletin No.* 72, October, 1961.

118. J. Dubuc and A. Biron, "Effect of Creep in Low-Cycle Fatigue of Pressure Vessel Steel," *ASME Publication Paper No.* 69-PVP-2, 1969.

119. H. J. Tapsell, "Fatigue at High Temperatures, Symposium on High Temperature Steels for Gas Turbines," Iron and Steel Institute, London, pp. 169–174, 1950.

120. A. E. Carden, "Thermal Fatigue of a Nickel-Base Alloy," *ASME Publication Paper No.* 64-Met-2, 1964.

121. A. E. Carden, "Thermal-Fatigue Resistance: Material, Geometric, and Temperature Field Considerations," *ASME Publication Paper No.* 65-GTP-5, 1965.

122. H. R. Copson, "Stress Corrosion, Corrosion Fatigue, and Erosion Corrosion," *Metals Engineering Design, ASME Handbook*, 1953.

123. D. J. McAdam, Jr., "Influence of Cyclic Stress on Corrosion Pitting of Steels in Fresh Water," and "Influence of Stress Corrosion on Fatigue Life," *Journal of Research, National Bureau of Standards*, Vol. 24, 1940.

124. N. E. Frost, "The Effect of Environment in the Propagation of Fatigue Cracks in Mild Steel," *Applied Materials Research*, Vol. 3, No. 3, July, 1964.

125. G. N. Krouse, *ASTM Proceedings*, Vol. 34, 1934.

126. J. B. Conway and J. T. Berling, "A New Approach to the Prediction of Low-Cycle Fatigue Data," *Transactions of the Metallurgical Society of AIME*, May, 1969.

127. J. B. Conway, and J. T. Berling, "A New Correlation of Low-Cycle Fatigue Data Involving Hold Periods," General Electric Co., Nuclear Systems Programs, Cincinnati, Ohio, U. S. AEC Contract AT(40–1)-2847, June 1969.

128. C. R. Brinkman, "The Influence of Hold Time on the Elevated Temperature Fatigue Properties of Steels and Alloys—A Literature Survey," Idaho Nuclear Corporation USAEL Research and Development Report, Contract AT(10-1)-1230, June 1969.

129. L. A. James, "Hold-Time Effects on Elevated Temperature Fatigue-Crack Propagation of Type 304 Stainless Steel," *Nuclear Technology*, Vol. 16, No. 3, December 1972.

130. T. E. Davidson, R. Eisenstadt, and A. N. Reiner, "Fatigue Characteristics of Open-end Thick Wall Cylinders Under Cyclic Internal Pressure," *ASME Publication Paper No.* 62-WA-164, 1962.

131. J. M. Steichen, T. T. Claudson and R. A. Moen, "Irradiation Effects on Fatigue Properties of Reactor Cladding and Structural Materials," *ASME*, December, 1968.

132. E. B. Norris and R. D. Wylie, "Post-Irradiation Fatigue Properties of Base Metals and Weldments," *ASME Publication Paper No.* 69-PVP-3, 1969.

133. L. A. James and J. A. Williams, "The Effect of Temperature and Neutron Irradiation Upon the Fatigue-Crack Propagation Behavior of ASTM A533 Grade B, Class 1 Steel," Heavy Section Steel Technology Program Technical Report No. 21, Hanford Engineering Development Laboratory, HEDL-TME 72–132, September, 1972.

134. S. S. Manson, "Thermal Stresses in Design," *Machine Design*, pp. 126–133, September 4, 1958.

135. L. F. Coffin, "A Study of the Effects of Cyclic-Thermal Stresses in Ductile Metal," *ASME Transactions*, Vol. 76, pp. 931–950, 1954.

136. L. F. Coffin, "An Investigation of Thermal-Stress Fatigue as Related to High-Temperature Piping Flexibility," *ASME Publication Paper No.* 56-A-178, 1956.

137. L. F. Coffin, "Thermal Stress Fatigue," *Product Engineering*, June, 1957.

138. S. S. Manson, *Thermal Stress and Low-Cycle Fatigue*, McGraw-Hill Book Co., New York, 1966.

139. L. F. Coffin, "Introduction to High-Temperature Low-Cycle Fatigue," *Experimental Mechanics*, May, 1968.

140. C. W. Lawton, "High-Temperature Low-Cycle Fatigue: A Summary of Industry and Code Work," *Experimental Mechanics*, June, 1968.

141. E. Z. Stowell, "Theory of Metal Fatigue at Elevated Temperatures," *Nuclear Engineering and Design*, pp. 239–257, 1969, North-Holland Publishing Co., Amsterdam.

142. L. Glen, "The Shape of Creep Curves," *ASME Publication Paper No.* 62-WA-133, 1962.

143. M. M. Abo El Ata, and I. Finnie, "A Study of Creep Damage Rules," *ASME Publication Paper No.* 71-WA/Met-1, 1971.

144. "Properties of Carbon Alloy Steel for High Temperature and High Pressure Service," *Technical Bulletin*, The Babcock and Wilcox Co., New York, N.Y.

145. J. A. Gulya and R. A. Swift, "Petroleum Pressure Vessels," *ASME Publication Paper No.* 70-Pet-26, 1970.

146. F. P. J. Rimrott, E. J. Mills, and J. Marin, "Prediction of Creep Failure Time for Pressure Vessels," *ASME Transactions*, Paper 60-APM-7, 1960.

147. D. R. Miller, "Thermal-Stress Ratchet Mechanism in Pressure Vessels," *ASME Transactions*, Paper 58-A-129, 1958.

148. E. L. Robinson, "Effect of Temperature Variation in the Long-Time Rupture Strength of Steels," *ASME Transactions*, Vol. 74, 1952.

149. P. N. Randall, "Cumulative Damage in Creep-Rupture Tests of a Carbon Steel," *ASME Transactions*, Vol. 84, 1962.

150. R. M. Goldhoff, "Uniaxial Creep-Rupture Behavior of Low Alloy Steel Under Variable Loading Conditions," *ASME Transactions*, Vol. 87, 1965.

151. K. G. Brickner, G. A. Ratz, and R. F. Domagala, "Creep-Rupture Properties of Stainless Steel at 1600, 1800 and 2000 F," *ASTM Technical Publication No.* 369, 1965.

152. D. J. Wilson, "Sensitivity of the Creep-Rupture Properties of Waspaloy Sheet to Sharp-Edged Notches in the Temperature Range 1000–1400 Degree F." *ASME Publication No.* 71-WA/Met-3, 1971.

153. C. E. Jaske and others, "Low Cycle Fatigue and Creep Fatigue of Incoloy Alloy 800," *Scientific and Technical Aerospace Reports*, Vol. 10, No. 21, Nov. 8, 1972.

154. W. D. Doty, "Postweld Heat Treatment," *Welding Research Council Bulletin*, 1974.

155. A. E. Asnis and G. A. Ivashchenko, "The Tempering Temperature for Relieving the Residual Stresses in Welded Structures," *Automatic Welding*, Vol. 25, No. 6, June, 1972.

156. T. W. Greene and C. R. McKinsey, "Controlled Low-Temperature Stress Relieving of Pressure Vessels," *Welding Research Council Supplement*, March, 1956.

157. A. Vetters and S. A. Ambrose, "The Heat Treatment of Weldments," *Australian Welding Journal*, March, 1971.

158. E. L. Robinson, "Steam-Piping Design to Minimize Creep Concentrations," *ASME Publication Paper No.* 54-A-186, 1954.

159. S. B. Kim, "Creep and Relaxation Analysis of a Piping System," *ASME Publication Paper No.* 71-WA/PVP-3, 1971.

160. H. Poritsky and F. A. Fend, "Relief of Thermal Stresses Through Creep," *ASME Transactions*, Paper 58-A-41, 1958.

161. A. P. Boresi and O. M. Sidebottom, "Creep of Metals Under Multiaxial States of Stress," *Nuclear Engineering and Design*, Vol. 18, 1972, North-Holland Publishing Company.

162. C. G. Interrante, G. A. Nelson, C. M. Hudgins, "Interpretive Report on Effect of Hydrogen in Pressure-Vessel Steels," *Welding Research Council Bulletin* 145, October, 1969.

163. R. L. Klueh, "Bubbles in Solids," *Science & Technology*, No. 93, October, 1969.

164. W. Beck and others, "Hydrogen Stress Cracking of High Strength Steels," *Scientific Technical Aerospace Reports*, Vol. 10, No. 20, October 23, 1972.

165. C. H. Samans, "Which Steel for Refinery Service," *Hydrocarbon Processing and Petroleum Refiner*, Vol. 42. No. 11, November, 1963.

166. J. W. Coombs, R. E. Allen and F. H. Vitovec, "Creep and Rupture Behavior of Low Alloy Steels in High Pressure Hydrogen Environment," *ASME Publication Paper No.* 64-Met-5, 1964.

167. J. B. Greer and others, "Effect of Temperature and State of Stress on Hydrogen Embrittlement of High Strength Steel," *Corrosion*, Vol. 28, No. 10, October, 1972.

168. R. P. McNitt, "Unmasking Hydrogen Embrittlement," *Machine Design*, Vol. 44, No. 29, November 30, 1972.

169. C. M. Hudgins, R. L. McGlasson, P. Mehdizadeh, and W. M. Rosborough, "Hydrogen Sulphide Cracking of Carbon and Alloy Steels," *Corrosion-National Association of Corrosion Engineers*, Vol. 22, August, 1966.

170. E. E. Galloway, "Hydrogen Damage in a 1250 psi Boiler," *ASME Publication Paper No.* 63-PWR-6, 1963.

171. I. Le May, "Hydrogen Evolution and Steam Dissociation in Steam Generators," *ASME Publication No.* 64-WA/BFS-2, 1964.

172. R. C. DeHart and A. G. Pickett, "The Evaluation of High-Strength Materials for Use in Severe Environments," *ASME Publication Paper No.* 66-PET-25, 1966.

173. E. B. Norris and R. D. Wylie, "Effect of Hydrogen on the Strength of Austenitic and Nickel-Base Alloys," *ASME Publication Paper No.* 69-PVP-4, 1969.

174. D. W. McDowell, "Designing to Prevent High Temperature Hydrogen Attack of Pressure Vessels," *ASME Publication Paper No.* 70-Pet-20, 1970.

175. M. E. Shank, C. E. Spaeth, C. W. Cooke, and J. E. Coyne, "Solid-Fuel Rocket Chambers for Operation at 240,000 psi and Above," *Metal Progress*, Vol. 76, No. 5, November, 1959.

176. A. D. Rossin, T. H. Blewitt, and A. R. Troiano, "Hydrogen Embrittlement in Irradiated Steels," *Nuclear Engineering and Design*, Vol. 4, pp. 446–458, 1966, North-Holland Publishing Company, Amsterdam.

177. E. V. Bravenc, "Analysis of Brittle Fractures During Fabrication and Testing," *ASME Paper* 71-PVP-53, 1971.

178. R. Weck, "Brittle Fracture of a Thick Walled Pressure Vessel," *British Welding Research Assoc. Bull.* 7(6), June 1966.

179. Report on the Brittle Fracture of a High-Pressure Boiler Drum at Cockenzie Power Station, by The South of Scotland Electricity Board, January 1967, Welding Research Abroad, Welding Research Council, Vol. XIII No. 8, October, 1967, New York.

180. W. S. Pellini and P. P. Puzak, "Fracture Analysis Diagram Procedures for the Fracture-Safe Engineering Design of Steel Structures," *Welding Research Council Bulletin* 88, May, 1963.

181. W. S. Pellini, "Principles of Fracture Safe Design," 1971 Adams Lecture, *Supplement to the Welding Journal*, March 1971, pp. 91–s to 109–s, and April 1971, pp. 147–s to 161–s.

182. A. A. Griffith, "The Phenomenon of Rupture and Flow in Solids," *Phil. Trans. Roy. Soc.* (*London*), Vol. 221, 1921.

183. G. R. Irwin, "Analysis of Stresses and Strains Near the End of a Crack Traversing a Plate," *ASME Journal of Applied Mechanics*, Vol. 24, 1957.

184. G. R. Irwin, "Crack-propagation Behaviors," *Journal for the Society of Experimental Mechanics*, June, 1966.

185. Whitman, Robinson and Savolainen, "Technology of Steel Pressure Vessels for Water-Cooled Nuclear Reactors," Oak Ridge National Laboratory Report ORNL-NSIC-21, 1967.

186. P. C. Paris and G. C. Sih, "Stress Analysis of Cracks in Fracture Toughness Testing," *ASTM Special Technical Publication* 381, 1965.

187. "Fracture Toughness Testing and its Application," ASTM, STP 381, American Society of Testing and Materials, April 1965.

188. F. J. Witt, "HSST Fracture Investigations — An Overview," Heavy Section Steel Technology Program, Paper No. 26, 1972, Oak Ridge National Laboratory, Oak Ridge, Tennessee.

189. A. S. Tetelman and A. J. McEvily, *Fracture of Structural Materials*, John Wiley & Sons, New York, 1967.

190. A. A. Wells, "Unstable Crack Propagation in Metals — Clevage and Fast Fracture," Cranfield Crack Propagation Symposium, September, 1961, Vol. 1, Royal College of Aeronautics, Cranfield.

191. G. R. Egan, "Weld Defects and Fitness for Purpose — Brittle Fracture," *Welding Research Abroad, Welding Research Council*, Vol. XVIII, No. 3, November, 1972.

192. E. T. Wessel, "Fracture Mechanics, State of the Art of the WOL Specimen for K_{Ic} Fracture Toughness Testing," *Engineering Fracture Mechanics*," Vol. 1, pp. 77–103, 1968, Pergamon Press.

193. P. N. Randall, "Gross Strain Measure of Fracture Toughness of Steels," Paper No. 3, Heavy Section Steel Technology Program, November 1, 1969, Oak Ridge National Laboratory.

194. F. J. Witt, "Equivalent Energy Procedures for Predicting Gross Fracture, USAEC Report ORNL-TM-3172, Oak Ridge National Laboratory.

195. F. J. Witt and T. R. Mager, "A Procedure for Determining Bounding Values on Fracture Toughness K_{Ic} at Any Temperature, October, 1972, USAEC Report ORNL-TM-3894, Oak Ridge National Laboratory.

196. J. F. Rice, "A Path Independent Integral and the Approximate Analysis of Strain Concentration by Notches and Cracks," *ASME Journal of Applied Mechanics*, June, 1968.

197. J. G. Merkle, "Analytical Applications of the J-Integral," Paper No. 6, Heavy Section Steel Technology Program, April 25–26, 1972, Oak Ridge National Laboratory.

198. W. H. Irvine and A. Quirk, "The Application of the Stress-Concentration Theory of Fracture Mechanics to the Assessment of the Fast Structure Characteristics of Thick Walled Nuclear Reactor Pressure Vessels," U.K.A.E.A., Risley, 1971.

199. J. S. Ke and H. W. Liu, "The Measurements of Fracture Toughness of Ductile Materials," American Iron and Steel Institute Symposium on Fracture and Fatigue, May 3–5, 1972, Washington, D. C.

200. M. L. Parrish and D. J. Seman, "Fracture Toughness of Nonirradiated and Irradiated Reactor Vessel Steels," WAPD-TM-565, UC-80: Reactor Technology, U. S. Department of Commerce, March 1968.

201. F. J. Loos, J. R. Hawthorne, C. Z. Serpan, Jr., and P. P. Puzak, "Analysis of Radiation-Induced Embrittlement Gradients on Fracture Characteristics of Thick-Walled Pressure Vessel Steels," ASME Paper 71-PVP-7, 1971.

202. J. G. Merkle, "Fracture Safety Analysis Concepts for Nuclear Pressure Vessels, Considering the Effects of Irradiation," *ASME Journal of Basic Engineering*, June, 1971.

203. G. T. Hahn and A. R. Rosenfield, "Sources of Fracture Toughness: The Relationship Between K_{Ic} and the Ordinary Tensile Properties of Metals," Battelle Memorial Institute, 1967.

204. R. H. Sailors and H. T. Corten, "Relationship Between Material Fracture Toughness Using Fracture Mechanics and Transition Temperature Tests," Heavy Section Steel Technology Program, March 25–26, 1971, Oak Ridge National Laboratory, USA.

205. R. E. Johnson, "Fracture Mechanics: A Basis for Brittle Fracture Prevention," WAPD-TM-505, UC-80: Reactor Technology TID-4500, U.S. Department of Commerce, 1965.

206. B. F. Langer, "Design-Stress Basis for Pressure Vessels," *Experimental Mechanics*, January, 1971.

207. K. Ikeda and H. Kikara, "Brittle Fracture Strength of Welded Joints," *Supplement to The Welding Journal*, March, 1970, pp. 106–s to 114–s.

208. E. A. Steigerwald, "Influence of Warm Prestressing on the Notch Properties of Several High Strength Alloys," *Transactions ASME*, Vol. 54, No. 3, September, 1961.

209. A. J. Brothers and S. Yukawa, "The Effect of Warm Prestressing on Notch Fracture Strength," *Transactions ASME*, Vol. 85, No. 1, March, 1963.

210. R. W. Nichols, "Overstressing as a Means of Reducing the Risk of Subsequent Brittle Failure," *British Welding Journal*, October, 1968.

211. C. Mylonas and K. C. Rockley, "Exhaustion of Ductility by Hot Straining—An Explanation of Fracture Initiation Close to Welds," *Welding Journal Welding Research Supplement*, July, 1961.

212. W. R. Andrews, "Effect of Loading Sequence on Notch Strength of Warm Prestressed Alloy Steel," *ASME Publication Paper No.* 70-PVP-10, 1970.

213. L. N. Svccop, A. W. Pense, and R. D. Stout, "The Effects of Warm Overstressing on Pressure Vessel Steel Properties," The Welding Journal Research Supplement Vol. 49, No. 8, August, 1970, and "Extension Study of the Effects of Warm Prestressing on Ambient Temperature Toughness of Pressure Vessel Steels," Pressure Vessel Research Committee Report September 20, 1971.

214. T. C. Harrison and G. D. Fearnehough, "The Influence of Warm Prestressing on the Brittle Fracture of Structures," *ASME Publication Paper No.* 72-MAT-D, 1972.

215. L. A. James, "Estimation of Crack Extension in a Piping Elbow Using Fracture Mechanics Techniques,' *ASME Publication Paper No.* 74-PVP-14, 1974.

216. T. W. Crooker, "The Role of Fracture Mechanics in Fatigue Design," *ASME Publication Paper No.* 76-DE-5, 1976.

217. A. M. Sullivan and T. W. Crooker, "Analysis of Fatigue-Crack Growth in a High-Strength Steel," *ASME Publication Paper No.* 75-WA/PVP-22, 1975.

218. J. C. Radon and L. E. Culver, "Fatigue-Crack Propagation in Metals," *Experimental Mechanics,* March, 1976.

219. J. D. Burk and F. V. Lawrence, "Influence of Bending Stresses on Fatigue Crack Propagation Life in Butt Joint Welds," *Welding Research Journal*, February, 1977.

220. T. L. Gerber, J. D. Heald and E. Kiss, "Fatigue Crack Growth in SA 508-CL2 Steel in a High Temperature, High Purity Water Environment," *ASME Publication Paper No.* 74-Mat-2, 1974.

221. B. R. Ellingwood, "Probabilistic Assessment of Low-Cycle Fatigue Behavior of Structural Welds," *ASME Publication Paper No.* 75-PVP-29, 1975.

222. Y. Hara, J. Uedo, T. Okamoto, M. Uedo and Y. Terado, "Hydrostatic and Cylic Pressure Tests for a Heavy Walled Pressure Vessel," Third International Conference on Pressure Vessel Technology, Tokyo, Japan, 1977.

223. N. Nishihara, Y. Yamaguchi and S. Hattori, "Fatigue Strength of Thick-Walled Very High Pressure Cylinders Composed of High Strength Steels," Third International Conference on Pressure Vessel Technology, Tokyo, Japan, 1977.

224. N. J. I. Adams, "An Analysis and Prediction of Failure in Tubes," Third International Conference on Pressure Vessel Technology, Tokyo, Japan, 1977.

225. R. P. Harrison, B. J. L. Darlaston and C. H. A. Townley, "Failure Assessment of Pressure Vessels Under Yielding Conditions," Third International Conference on Pressure Vessel Technology, Tokyo, Japan, 1977.

226. A. W. Pense and R. D. Stout, "Elevated Temperature Fatigue of Pressure Vessel Steels," *Welding Journal,* August, 1975.

227. H. M. Minami, "Pressure Vessel Wall Thickness Limits Based on Thermal Stress Ratchetting and Creep Fatigue Ineraction," *ASME Publication Paper No.* 74-WA/PVP-11, 1974.

228. G. T. Yahr and W. K. Sartory, "Development of Simplified Analysis Methods for Ratchetting and Creep-Fatigue—A Status Report," *ASME Publication Paper No.* 76-WA/PVP-19, 1976.

229. J. T. Fong, "Energy Approach for Creep-Fatigue Interactions in Metals at High Temperatures," *ASME Publication Paper No.* 75-PVP-30, 1975.

230. P. Shahinian, "Creep-Fatigue Crack Propagation in Austenitic Stainless Steel," *ASME Publication Paper No.* 75-WA/PVP-1, 1975.

231. M. Doner, "An Analysis of Elevated Temperature Fatigue and Creep Crack Growth," *ASME Publication Paper No.* 75-WA/GT-16, 1975.

232. J. M. Corum and C. E. Pugh, "High-Temperature Structural Design Program Semiannual Progress Report for Period Ending December 31, 1976," Oak Ridge National Laboratory Report ORNL-5281, November, 1977.

233. M. Sakane and M. Ohnami, "Effect of Multiaxiality of Stress on Metallic Creep-Fatigue Interaction at Elevated Temperatures," Third International Conference on Pressure Vessel Technology, Tokyo, Japan, 1977.

234. J. V. Jolliff and A. Thiruvengadam, "Effect of Hydrostatic Pressure on Corrosion-Fatigue at High Frequency," *ASME Publication Paper No.* 73-WA/Oct-1, 1973.

235. T. Bui-Quoc, R. E. Saheb and A. Biron, "Evaluation of Hold-Time Effect in Fatigue at Elevated Temperature on a Stainless Steel," *ASME Publication Paper No.* 75-WA/PVP-17, 1975.

236. L. F. Coffin, A. E. Carden, S. S. Manson, L. K. Severud and W. L. Greenstreet, "Time-Dependent Fatigue of Structural Alloys," Oak Ridge National Laboratory, ORNL-5073, January, 1977.

237. T. Udoguchi, Y. Asada, S. Mitsuhashi and Morimoto, "An Approach to Evaluate Strain-Rate and Hold-Time Effects on High Temperature Low-Cycle Fatigue," Third International Conference on Pressure Vessel Technology, Tokyo, Japan, 1977.

238. J. G. Blauel, J. F. Kalthoff and D. Stahn, "Model Experiments for Thermal Shock Fracture Behavior," *ASME Publication Paper No.* 74-Mat-11, 1974.

239. J. R. Corum and W. K. Sartory, "Elastic-Plastic-Creep Analysis of Thermal Ratchetting in Straight Pipe and Comparisons with Test Results," *ASME Publication Paper* No. 73-WA/PVP-4, 1973.

240. G. F. Barcikowski, S. S. Blackburn, G. P. Hammer, D. A. Harris, D. G. Milligan, and W. R. Sylvester, "Locate Problems in Radiant and Convective Sections Early to Improve Availability," *Power,* Vol. 122, No. 3, March, 1978.

241. H. J. McQueen and W. J. M. Tegart, "The Deformation of Metals at High Temperatures," *Scientific American,* Vol. 232, April, 1975.

242. B. D. Clay and H. E. Evans, "The Computation of Constant-Load Creep Curves" *Nuclear Engineering and Design,* Vol. 39, November 1976.

243. P. R. Paslay and C. H. Wells, "Uniaxial Creep Behavior of Metals Under Cyclic Temperature and Stress or Strain Variations," *ASME Publication Paper No.* 76-WA/APM-6, 1976.

244. R. L. Klueh, "The Relationship Between Rupture Life and Creep Properties for 2-1/4 Cr-1 Mo Steel," *ASME Publication Paper No.* 74-PVP-20, 1974.

245. I. W. Goodall, and R. A. Ainsworth, "Failure of Structures by Creep," Third International Conference on Pressure Vessel Technology, Tokyo, Japan, 1977.

246. I. Berman, J. M. Chern and G. D. Gupta, "A Parametic Study of Elastic-Plastic-Creep Buckling of a Thin Cylindrical Shell," *ASME Publication Paper No.* 74-PVP-37, 1974.

247. R. P. Goel, "On the Creep Rupture of a Tube and a Sphere," *ASME Publication Paper No.* 75-WA/APM-4, 1975.

248. N. Jones, "Creep Buckling of a Complete Spherical Shell," *ASME Publication Paper No.* 76/WA/APM-8, 1976.

249. E. M. Smith, "Creep Testing at Variable Temperature and Load, Experimental and Computational Methods," *ASME Publication Paper No.* 75-WA/Mat-2, 1975.

250. M. Sabbaghian and M. R. Eslami, "Creep Relaxation of Axisymmetric Thermal Stresses in Thick Wall Cylindrical Vessels," *ASME Publication Paper No.* 74-PVP-9, 1974.

251. L. K. Severud, "Background to the Elastic Creep-Fatigue Rules of the ASME Boiler & Pressure Vessel Code Case 1592," *Nuclear Engineering and Design*, February, 1978.

252. S. Havel, M. Tvrdý, L. Hyspecká and K. Maxanec, "Evaluation of the Effect of Hydrogen on the Properties of Selected Pressure Vessel Steels," Third International Conference on Pressure Vessel Technology, Tokyo, Japan, 1977.

253. E. W. Johnson, "Hydrogen Embrittlement of Austenitic Steel Weld Metal with Special Consideration Given to the Effects of Sigma Phase," Welding Research Council Bulletin 240, August, 1978.

254. J. P. Gyekenyesi, and A. Mendelson, "Stress Analysis and Stress Intensity Factors for Finite Geometry Solids Containing Rectangular Surface Cracks," *ASME Publication No.* 77-WA/APM-5, 1977.

255. H. C. Wu, R. F. Yao, and M. C. Yip, "Experimental Investigation of the Angled Elliptic Notch Problem in Tension," *ASME Publication Paper No.* 77-WA/APM-6, 1977.

256. R. J. Podlasek and R. J. Eiber, "Predicting the Fracture Initiation Transition Temperature in High Toughness, Low Transition Temperature Line Pipe with the COD Test," *ASME Publication Paper No.* 74-Mat-14, 1974.

257. J. R. Rice, P. C. Paris and J. G. Merkle, "Some Further Results of J-Integral Analysis and Estimates," ASTM STP 536, American Society for Testing and Materials, 1973.

258. N. E. Dowling, "Geometry Effects and the J-Integral Approach to Elastic-Plastic Crack Growth," *Cracks and Fracture*, STP 601, American Society for Testing and Materials, 1976.

259. T. Kanazawa, S. Machida, S. Kaneda and M. Onozuka, "A Study on Practical Aspects of the J-Integral Fracture Criterion," Third International Conference on Pressure Vessel Technology, Tokyo, Japan, 1977.

260. J. G. Merkle and H. T. Corten, "A J-Integral Analysis for the Compact Specimen, Considering Axial Force As Well As Bending Effects," *ASME Publication Paper No.* 74-PVP-33, 1974.

261. S. Y. Zamrik and F. M. Bahgat, "A J-Integral Analysis to Fracture Toughness of Plates Containing Surface Cracks," *ASME Publication Paper No.* 75-PVP-16, 1975.

262. G. R. Egan, "Some Relationships Between Fracture Toughness, Applied Stress or Strain and Flaw Size," The Welding Institute, Cambridge, U.K., September, 1972.

263. C. B. Buchalet and W. H. Bamford, "Method for Fracture Mechanics Analysis of Nuclear Reactor Vessels under Severe Thermal Transients," *ASME Publication Paper No.* 75-WA/PVP-3, 1975.

264. D. M. Parks, "Interpretation of Irradiation Effects on Fracture Toughness of a Pressure Vessel Steel in Terms of Crack Tip Stress Analysis," *ASME Publication Paper No.* 75-Mat-9, 1975.

265. E. A. Lange, "Dynamic Fracture-Resistance Testing and Methods for Structural Analysis," Welding Research Council Bulletin 229, August 1977, Welding Research Council, New York, New York.

266. C. W. Smith, M. Jolles and W. H. Peters, "Stress Intensities in Flawed Pressure Vessels," Third International Conference on Pressure Vessel Technology, Tokyo, Japan, 1977.

267. G. C. Sih, "The Role of Fracture Mechanics in Design Technology," *ASME Publication Paper No.* 76-DET-58, 1976.

268. R. Erdol, and F. Erdogan, "A Thick-Walled Cylinder with an Axisymmetric Internal or Edge Crack," *ASME Publication Paper No.* 77-WA/APM-8, 1977.

269. G. M. Wilkowski and R. J. Eiber, "Review of Fracture Mechanics Approaches to Defining Critical Size Girth Weld Discontinuities," Welding Research Council Bulletin 239, July 1978.

270. R. G. Hoagland, P. C. Gehlen, A. R. Rosenfield and G. T. Hahn, "Analysis of Crack Arrest in Reactor Pressure Vessels," *ASME Publication Paper No.* 78-Mat-16, 1978.

271. W. J. Lambertin and F. H. Vaughan, "Evaluation of the Risk of Catastrophic Brittle Fracture in Existing Equipment," *ASME Publication Paper No.* 74-Pet-25, 1974.

272. T. Nishimura, "Analysis of Axially Symmetrical Residual Stresses in Bars and Tubes," *Experimental Stress Analysis*, May, 1978.

273. E. F. Rybicki and R. B. Stonesifer, "Computation of Residual Stresses Due to Multipass Welds in Piping Systems," *ASME Publication Paper No.* 78-PVP-104, 1978.

274. A. M. Nawwar and J. Shewchuk, "On the Measurement of Residual-Stress Gradients in Aluminum-Alloy Specimens," *Experimental Mechanics*, July, 1978.

275. J. F. Knott, *Fundamental of Fracture Mechanics*, John Wiley & Sons, New York, 1973.

276. M. C. Shaw, "Brittle Fracture Under a Complex Sate of Stress," *ASME Publication Paper No.* 75-WA/Prod - 14, 1975.

277. R. H. Hawley and D. C. Drucker, "Brittle Fracture of Precompressed Steel as Affected by Hydrostatic Pressure, Temperature and Strain Concentration," Experimental Mechanics, January, 1973.

278. P. B. Crosley and E. J. Ripling, "Plane Strain Crack Arrest Characterization of Steels," *ASME Publication Paper No.* 75-PVP-32, 1975.

279. K. Satoh, M. Toyoda, Y. Kawaguchi, and K. Arimochi, "Influence of Hot Straining Embrittlement on Brittle Fracture in Welded Steel Plates," Transactions of Japan Welding Society, Vol. 8, No. 1, April, 1977.

280. F. J. Loss, R. A. Gray, Jr., and J. R. Hawthorne, "Significance of Warm Prestress to Crack Initiation During Thermal Shock," *Nuclear Engineering and Design*, North-Holland Publishing Co., April, 1978.

281. N. L. Peterson and S. D. Harkness, "Radiation Damage in Materials," American Society for Metals, 1976.

282. S. E. Yanichko and T. R. Mager, "Effects of Irradiation on A508 Class 2 and 3 Forging Grade Steels," *ASME Publication Paper No.* 78-Mat-21, 1978.

283. J. N. Kass, A. J. Giannuzzi and D. A. Huges, "Radiation Effects in Boiling Water Reactor Pressure Vessel Steels," *ASME Publication Paper No.* 78-Mat-11, 1978.

284. B. Kony and P. C. Paris, "On the Cup and Cone Fracture of Tensile Bars," *Fracture Mechanics*, ASTM STP 677, American Society for Testing and Materials, 1979.

285. S. Kocanda, *Fatigue Failure of Metals*, Sijthoff & Noordhoff International Publishers, The Netherlands, 1978.

286. D. F. Bryan and J. M. Potter "Effect of Load Spectrum Variables on Fatigue Crack Initiation and Propagation," American Society for Testing and Materials, STP-714, May 1980.

287. K. Zander and J. Liljebald, "Fatigue Strength Calculations of Prestressed Pressure Vessels for Isostatic Presses," ASME Publication Paper No. 79-PVP-117, 1979.

288. A. Niku-Lari, "Residual Stresses and Surface Finish in Shot-Peened Components and Materials," *Experimental Techniques*, February and March 1983.

289. W. J. Westerman "Industry Rediscovers Roller Burnishing", *Machine Design*, August 25, 1983.

290. K. Masubuchi, "Nondestructive Measurement of Residual Stresses in Metals and Metal Structures", RSIC-410 Redstone Information Center, April 1965.

291. I. J. Polmear, "Effects of Peening on the Fatigue Performance of Aluminum Alloy Fillet Welds," AWRA Report P3-18-79, Welding Research Abroad, Welding Research Council, April 1981.

292. W. A. Ellingson and W. J. Shack, "Residual-stress Measurements on Multipass Weldments of Stainless-steel Piping," *Experimental Mechanics*, September, 1979.

293. E. F. Rybicki and R. B. Stonesifer, "An Analysis Procedure for Predicting Weld Repair Residual Stresses in Thick Walled Vessels," ASME Publication Paper No. 79-PVP-31, 1979.

294. Grzegorz Glinka, "Effect of Residual Stresses on Fatigue Crack Growth in Steel Weldments Under Constant and Variable Amplitude Loads," *Fracture Mechanics*, ASTM STP 677, American Society for Testing and Materials, 1979.

295. R. N. Miller and R. L. Smith, "The Effect of Various Environments on the Fatigue Cracking of High Strength Aluminum Alloys," ASME Publication Paper No. 80-C2/PVP-11, 1980.

296. R. Wei, G. W. Simmons, and R. Brazill, "Fatigue Crack Growth in $2\frac{1}{4}$ Cr–1 Mo Steel Exposed to Hydrogen Containing Gases," ASME Publication Paper No. 79-PVP-102, 1979.

297. W. J. Mills and L. A. James, "The Fatigue-Crack Propagation Response of Two Nickel-Base Alloys in a Sodium Environment," ASME Publication Paper No. 79-PVP-83, 1979.

298. J. L. Yuen and J. F. Copeland, "Fatigue Crack Growth Behavior of Stainless Steel Type 316 Plate and 16-8-2 Weldments in Air and High Carbon Liquid Sodium," ASME Publication Paper No. 79-PVP-105, 1979.

299. D. A. Hale, C. W. Jewett and J. N. Kass, "Fatigue Crack Growth Behavior of Four Structural Alloys in High Temperature High Purity Oxygenated Water," ASME Publication Paper No. 79-PVP-104, 1979.

300. P. Shahinian and K. Sadananda, "Effects of Stress Ratio and Hold-Time on Fatigue Crack Growth in Alloy 718," ASME Publication Paper No. 79-PVP-84, 1979.

301. B. M. Ma, "Elevated Temperature, Cyclic Loadings and Irradiation Effects on Fatigue Crack of LMFBR Pressure Vessel," ASME Publication Paper No. 79-PVP-59, 1979.

302. K. M. Lal and I. LeMay, "The Effect of Overloads on Fatigue in Structural Components," ASME Publication Paper No. 80-C2/PVP-9, 1980.

303. C. C. Schultz, "Consistent Creep and Rupture Properties for Creep-Fatigue Evaluation," ASME Publication Paper No. 79-PVP-119, 1979.

304. A. I. Smith and A. M. Nicolsen, *Advances in Creep Design*, John Wiley & Sons, Inc., New York, 1971.

305. R. K. Penny and D. L. Marriott, *Design for Creep*, McGraw-Hill Book Company (UK) Limited, London, 1971.

306. C. D. Pomeroy, *Creep of Engineering Materials*, Mechanical Engineering Publications, Limited, London, 1978.

307. "Embrittlement of Pressure Vessel Steels in High Temperature, High Pressure Hydrogen Environment," Final Report, January 1983, The Japan Pressure Vessel Research Council.

308. D. M. Fryer, "Failure Analysis of a 414 MPa (60,000 psi) Isostatic Press," ASME Publication Paper No. 79-PVP-18, 1979.

309. R. L. Cloud and G. A. Spiering, "Failure of a High Pressure Vessel in Ammonia Service," ASME Publication Paper No. 81-PVP-8, 1981.

310. P. P. Puzak and F. J. Loss, "Metallurgical and Mechanical Considerations in Selection of a Fracture-Safe Explosives Containment Vessel," ASME Publication Paper No. 79-PVP-24, 1979.

311. G. D. Whitman and R. H. Bryan," Heavy-Section Steel Technology Program Quarterly Progress Report for July-September 1980, NUREG/CR-1806, ORNL/NUREG/TM-419, Oak Ridge National Laboratory.

312. J. D. Harrison, "The 'State-of-the-Art' in Crack Opening Displacement (COD) Testing and Analysis", *Welding Research Abroad*, April 1981.

313. R. Roberts and C. Newton, "Interpretive Report on Small Scale Test Correlations with K_{Ic} Data," Welding Research Council Bulletin 265, February 1981.

314. S. T. Rolfe and J. M. Barsom, *Fracture and Fatigue Control: Applications of Fracture Mechanics*, Prentice-Hall, Inc., Englewood Cliffs, New Jersey, 1977.

315. J. R. Dougan, "Relationships Between Charpy V-Notch Impact Energy and Fracture Toughness," Report NUREG/CR-2362, ORNL/TM-7921, March 1982, Oak Ridge National Laboratory.

316. J. H. Underwood, "Stress Intensity Factors for Internally Pressurized Thick Wall Cylinders," ASTM STP 513, 1972.

317. J. A. Kapp and R. Eisenstadt, "Crack Growth in Externally Flawed, Autofrettaged Thick-Walled Cylinders and Rings," Fracture Mechanics, ASTM STP 677, American Society for Testing and Materials, 1979.

318. M. Nishihara, Y. Yamaguchi and S. Hattori, "Fatigue Design of Thick-Walled Cylinders for Very High Pressure", *R&D Kobe Steel Engineering Reports*, Vol. 28, No. 2, April 1978.

319. D. Broek, *Elementary Engineering Fracture Mechanics*, Noordhoff Internatioanl Publishing Leyden, The Netherlands, 1974.

320. R. P. Harrison, K. Loosemore, and I. Milne, "Assessment of the Integrity of Structures Containing Defects," CEGB Report No. R/H/6, Central Electricity Generating Board, United Kingdom, 1976.

321. J. M. Bloom and S. N. Malik, "Procedure for Assessment of the Integrity of Nuclear Pressure Vessels and Piping Containing Defects," EPRI Topical Report NP-2431, June 1982.

322. S. I. Pu and M. A. Hussain, "Residual Stress Redistribution Caused by

Notches and Cracks in a Partially Autofrettaged Tube," ASME Publication Paper No. 81-PVP-6, 1981.

323. R. H. Bryan, P. P. Holtz, S. K. Iskander, J. Merkle, and G. D. Whitman, "Test of Thick Vessel with a Flaw in Residual Stress Field," ASME Publication Paper No. 79-PVP-29, 1979.

324. J. M. Bloom, "An Analytical Assessment of the Effects of Residual Stresses and Fracture Properties of Service Performance of Various Weld Repair Processes," ASME Publication Paper No. 80-C2/PVP-140, 1980.

325. E. D. Walker, "Some Aspects of Residual Stress in Parts Heat Treated by the Induction Method," *Residual Stress for Deisgners and Metallurgists*, Publication of the American Society for Metals, 1981.

326. J. R. Shadley, N. Merah, and E. F. Rybicki, "The Effect of Induction Heating Stress Remedies on Existing Flaws in Pipes," ASME Publication Paper 82-PVP-45, 1982.

327. F. J. Loss, R. A. Gray, Jr., and J. R. Hawthorne, "Investigation of Warm Prestress for the Case of Small ΔT During a Reator Loss-of-Coolant Accident," ASME Publication Paper No. 79-PVP-62, 1979.

328. P. O. Metz, "Facture Control of Pressure Vessels Up to $2\frac{1}{2}$ Inches Thick," Welding Research Council Bulletin Bulletin 291, January 1984.

329. D. G. Chakrapani, "Stress-Corrosion Affects a Variety of Alloys in Pulping Digester Pressure Vessels," ASME Publication Paper 84-MAT-1, 1984.

330. R. Gomuc and T. Bui-Quoc, "An Analysis of the Fatigue/Creep Behavior of 304 Stainless Steel Using a Continuous Damage Approach," ASME Publication Paper 84-PVP-38, 1984.

331. P. S. Chen, W. A. Herman and A. W. Pense, "Relaxation Stresses in Pressure Vessels," Pressure Vessel Research Committee, May 1984.

332. H. Kraus and W. Rosenkrans, "Creep of Bolted Flanged Connections," Welding Research Council Bulletin 294, May 1984.

333. D. R. McIntyre, "Analyzing Explosions and Pressure Vessel Ruptures," ASME Publication Paper 84-PVP-76, 1984.

334. G. W. Wellman, S. T. Rolfe, and R. H. Dodds, "Failure Prediction of Notched Pressure Vessels Using CTOD Approach," ASME Publication Paper 84-MAT-13, 1984.

335. D. M. Fryer, "Protection From Failure of Medium and Large High Pressure Vessels," ASME Pressure Vessel and Piping Division Conference, San Antonio, Texas, 1984.

336. E. P. Dahlberg and W. L. Bradley, "Failure Analysis of An Ammonia Pressure Vessel," ASME Publication Paper 84-MAT-14, 1984.

337. J. C. Tafuri, G. V. Krishna and R. Roberts, "Engineering Applications of Fracture Mechanics to Pressure Vessel Safety Assurance," ASME Publication Paper 84-MAT-9, 1984.

338. K. K. Yoon, H. W. Slager, W. A. Pavinich, and J. M. Bloom, "Application of the Failure Assessment Diagram to the Establishment of Pressure-Temperature Limits for a Pressurized Water Reactor," ASME Publication Paper 84-MAT-11, 1984.

339. D. G. Hooton, "Residual Stress Effects in LMFBR Fracture Assessment—Procedures," ASME Publication Paper 84-PVP-21, 1984.

340. R. Zhou, A. W. Pense, M. L. Basehore, and D. H. Lyons, "A Study of Residual Stress in Pressure Vessel Steels," Department of Metallurgy and Materials Engineering, Lehigh University, Bethlehem, Pennsylvania and Battelle-Columbus Laboratories, Columbus, Ohio, May 1984.

6

Design-Construction Features

6.1 Localized Stresses and Their Significance

The ordinary formulas that have been developed for the stresses in vessels, Chapters 2, 3, and 4, are based on the assumption that there is continuous elastic action throughout the member and the stress is distributed on any cross section of the member by a mathematical law. For instance, in the case of simple tension and compression it was assumed that the stress was uniformly distributed over the entire cross section. When these basic assumptions are voided by abrupt changes in section, great irregularities in stress distribution occur which, in general, are developed only in a small portion of the member hence, they are called localized stresses or stress concentrations. In pressure vessels they occur at transitions between thick and thin portions of the shell, and about openings, nozzles, or other attachments.

In approaching the study of localized stresses it is well to note that their significance does not depend solely on their absolute value. It also depends on:

1. The general physical properties of the material, such as its ductility.
2. The relative proportion of the member highly stressed to that understressed which affects the reserve strength it can develop in resisting excessive loads.
3. The kind of loading to which the member is subjected; i.e., whether it is static or repeated loading.

For example, stress concentrations in a pressure vessel subjected to only a steady pressure, or one repeated for only a few times, are of little importance provided the vessel is made of a ductile material, such as mild steel, which yields at these highly stressed locations allowing the stress to be transferred from overstressed fibers to adjacent understressed ones. Riveted boiler drums are a good example of this. Localized stresses in the region of the rivet holes reach yield point values, yet failure does not occur since they are made of a ductile material and operated at substantially steady-state conditions.

On the other hand, when the loading is a repetitive one, localized stresses become significant even though the material is ductile and has a large measure of static reserve strength, paragraph 5.11. Accordingly, the stresses given by the ordinary formulas which are based on average stress conditions, and which do not account for such local effects, must be multiplied by a theoretical "stress concentration factor, K_t," defined as the ratio of the maximum stress to the average stress, to obtain the maximum stress value. As pointed out in paragraph 1.3, a mathematical analysis of these localized stresses is frequently impossible or impracticable; hence, experimental methods of stress analysis are frequently used.[1]

6.2 Stress Concentration at a Variable Thickness Transition Section in a Cylindrical Vessel

Cylindrical vessels are often made with a thick and thin section in the shell connected by a tapered transition section, Fig. 6.1. This gives rise to an irregular stress distribution in this region that can be ap-

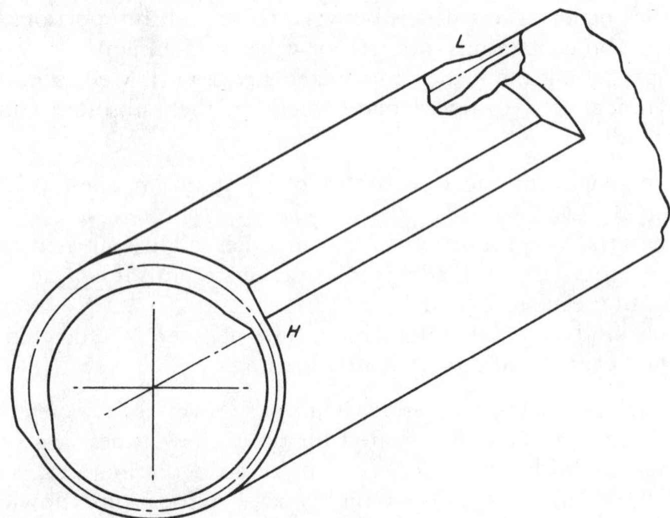

Fig. 6.1. Cylindrical Vessel with Thick and Thin Shell Courses and Tapered Transition Section

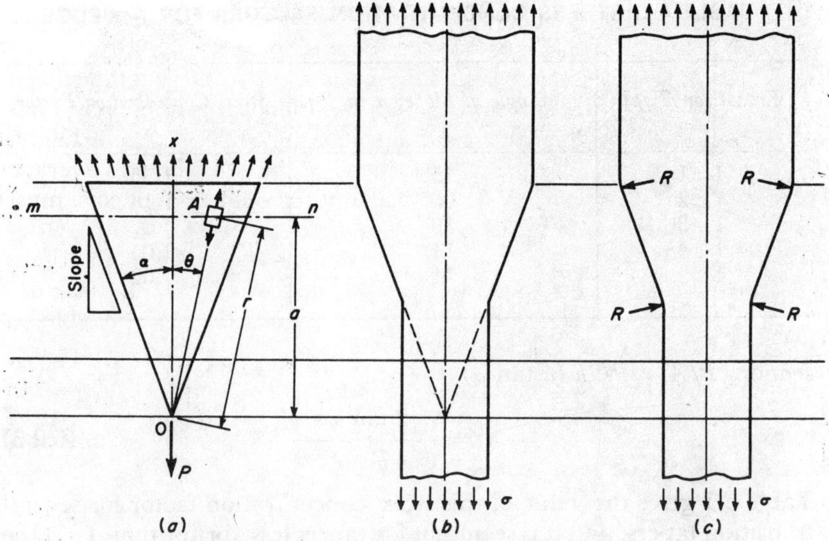

Fig. 6.2. Tapered Transition Section

proximated by comparing it to a symmetrical wedge of thickness h when loaded as shown in Fig. 6.2*a* and 6.2*b*. An exact solution[2] shows an Element A is in simple radial tension equal to:

$$\sigma_r = \left(\frac{1}{\alpha + \frac{1}{2}\sin 2\alpha}\right)\frac{P\cos\theta}{hr} \tag{6.2.1}$$

where θ is the angle a radius OA makes with the wedge axis, and r is the radius of the point A from the wedge apex 0. The normal stress σ_x on a cross section mn perpendicular to the axis of symmetry of the wedge is not uniform but varies as $\cos^2\theta$, and substituting $r = a/\cos\theta$ in Eq. 6.2.1.

$$\sigma_x = \sigma_r \cos^2\theta = \left(\frac{1}{\alpha + \frac{1}{2}\sin 2\alpha}\right)\frac{P\cos^4\theta}{ah} \tag{6.2.2}$$

This shows the normal stress is a maximum at the center of the cross section ($\theta = 0$) and a minimum at the faces $\theta = \alpha$. The stress concentration factor K_t can be found by dividing this maximum value ($\theta = 0$) obtained from Eq. 6.2.2. by the average stress at this cross

TABLE 6.1. STRESS CONCENTRATION FACTORS FOR A WEDGE

Transition Taper	*Angle of Slope, α in deg.*	*Stress Concentration Factor, K_t*
1 : 1	45°	1.55
1 : 2	26°	1.16
1 : 3	18°	1.07
1 : 4	14°	1.04
1 : 5	11°	1.02

section, $P/A = P/2h\,(a \tan \alpha)$,

$$K_t = \frac{2 \tan \alpha}{(\alpha + \frac{1}{2} \sin 2\alpha)} \tag{6.2.3}$$

Table 6.1 gives the value of this stress concentration factor for several transition tapers, and it is seen that for tapers less abrupt than 1 : 3 the maximum stress varies less than 10 per cent from the average. The assumption of uniform distribution of normal stresses over a cross section gives satisfactory results if the variation in cross section along the transition is gradual. In order to forestall high values of such transition stresses in vessel design, construction specifications[3] frequently state the allowable degree of taper. In actual practice the peaks and valleys at the juncture of the transition with the straight are also rounded to minimize these stress concentration effects, Fig. 6.2*c*. This latter refinement is not susceptible to analytical evaluation, but has been investigated photoelastically.[4,5]

Circular fillets are also used to make transitions from thick to thin sections. Stress concentration factors for this condition have been determined photoelastically by Frocht,[6] from which the data shown in Fig. 6.3 for the case of tension, and Fig. 6.4 for the case of bending, have been taken. These figures point out the importance of, first, avoiding great differences in thicknesses and, second, using as large a fillet radius as possible in order to reduce the stress concentration factor. This approach applies whether the transition is between vessel shell courses, opening reinforcement, or structural attachments. The stress concentration factors established for simple uniaxial tensile configurations, Fig. 6.3, may also be used with conservatism, 6 to 26 per cent, in the biaxial case of circumferential fillets between axially symmetric components of pressure vessels; such as, shell-flange junctures, shell and flat head intersections, and transitions between different

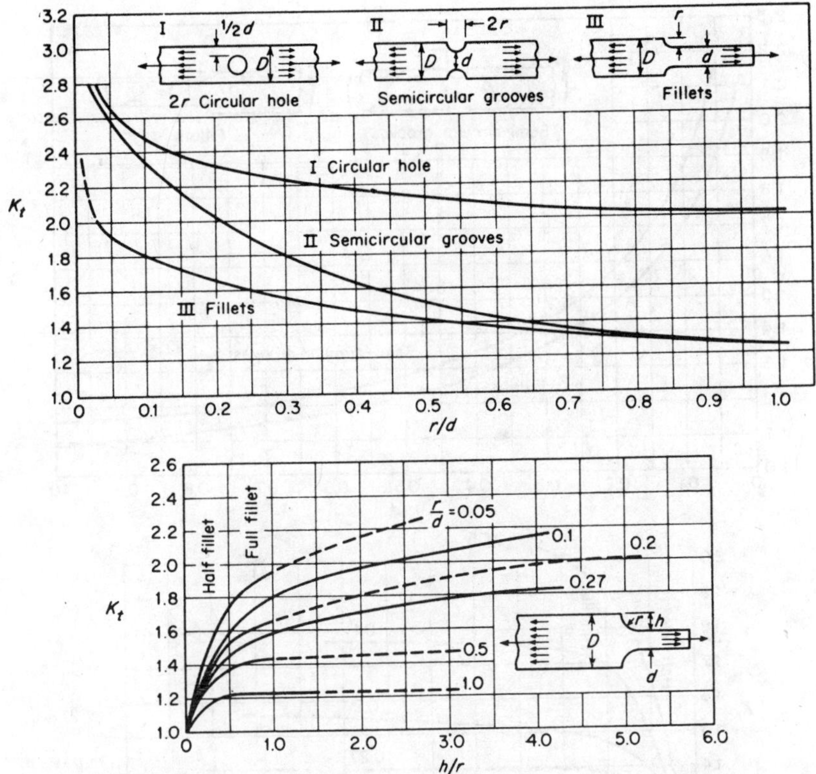

Fig. 6.3. Stress Concentration Factors for Circular Fillet Transitions Between Thick and Thin Sections in Tension[6]

thicknesses.[7,8] Various other noncircular fillet profiles such as elliptical, hyperbolic, etc., have been investigated[9,10,11] and, although these may give small improvements, they are impracticable and unwarranted in pressure vessel design.

Stress concentrations are also applicable to the case of bending of variable cross-sectional members. The maximum stress is greater in proportion to the abruptness of the discontinuity than that given by the simple bending formula for constant cross-sectional members, as shown by

$$\sigma_{max.} = K_t \sigma \tag{6.2.4}$$

where σ is the stress at the point under consideration obtained from the simple beam formula, and K_t is the stress concentration factor. In

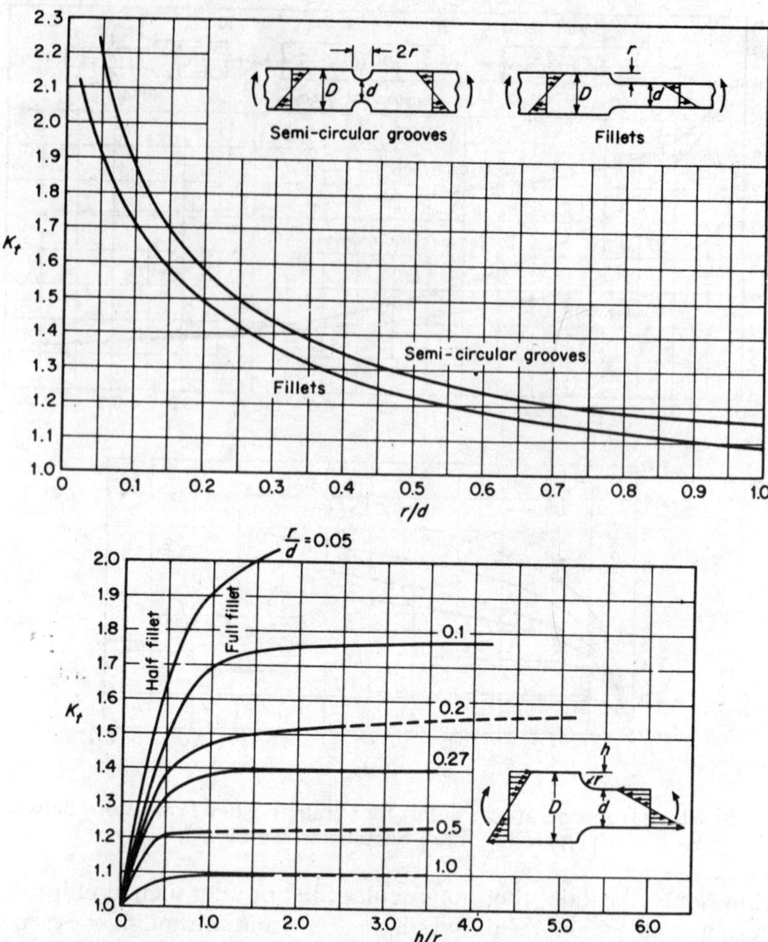

Fig. 6.4. Stress Concentration Factors for Circular Fillet Transitions Between Thick and Thin Sections in Bending[6]

only a very limited number of cases can the factor K_t be obtained analytically, but recourse must be made to experimental results.

6.3 Stress Concentration About a Circular Hole in a Plate Subjected to Tension

1. *Analytical*

When a small circular hole is made in a plate subjected to uniform tension σ, Fig. 6.5, a high stress concentration occurs near the hole as

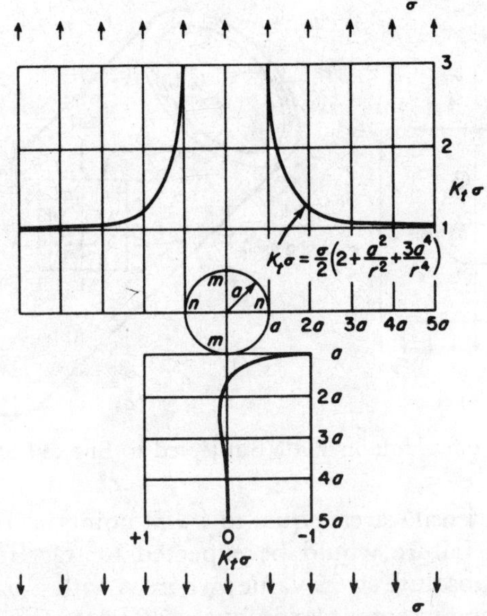

Fig. 6.5. Variation in Stress in a Plate Containing a Circular Hole and Subjected to Uniform Tension

given by Eq. 5.5.2. This has its maximum value at the edge of the hole, $r = a$, on a diametrical section *nn* perpendicular to the tension axis of value:

$$K_t\sigma = \frac{\sigma}{2}\left(2 + \frac{a^2}{r^2} + \frac{3a^4}{r^4}\right) = 3\sigma \qquad (6.3.1)$$

Although the exact theory is based on a small hole in an infinite plate, Fig. 6.5 shows that the effect of the small hole is extremely limited and damps out rapidly; hence, for practical purposes the formula can be used for plates of widths more than five times the hole diameter. For plate widths smaller than this, special analytical and photoelastic analyses have been performed.[12,13,118]

Figure 6.5 also shows the variation in the stress concentration, plotted from Eq. 5.5.2, along a diametrical section *mm* parallel to the direction of tension where it is a minimum. At the edge of the hole it produces a compressive stress, tangent to the hole, equal to the tensile stress σ applied at the ends of plate, and damps out very rapidly with distance from the edge of the hole.

If the direction of the applied stress is changed so as to be compression instead of tension, it is necessary only to change the sign of the stresses discussed above so that a compressive stress equal to 3 σ occurs

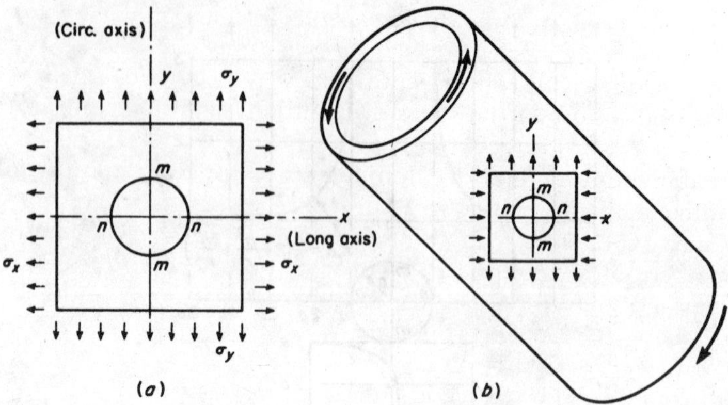

Fig. 6.6. Hole in Plate Subjected to Bi-axial Stress

at points n and a tensile stress equal to 1 σ at points m. In the case of a ductile material, failure would be expected to start from the point with the highest absolute stress value; whereas with a brittle material, which is very strong in compression but weak in tension (such as glass, cast iron, concrete, etc.), the cracks usually start at the point of maximum tensile stress (algebraically maximum stress value).

The stress concentrations about a circular hole in a cylinder or sphere with stresses applied by internal or external pressure can be obtained from the cases of simple tension or compression by using the method of superposition. In the case of a cylinder stressed by pressure, the longitudinal stress is half the hoop stress, Fig. 6.6a; therefore the maximum stress at point n on the longitudinal axis is $3\sigma_y - \sigma_x = 3\sigma_y - \frac{1}{2}\sigma_y = 2.5\sigma_y$, and at point m on the circumferential axis is $3\sigma_x - \sigma_y = \frac{3}{2}\sigma_y - \sigma_y = \frac{1}{2}\sigma_y$. In the case of a sphere, the two principal stresses are equal, $\sigma_x = \sigma_y = \sigma$, and the maximum stress concentration is $3\sigma_y - \sigma_x = 3\sigma - \sigma = 2\sigma$. In the special case of pure shear, $\sigma_y = -\sigma_x = \sigma$, the stress at point n is 4σ and at point m is -4σ, or four times the stress applied at the edge of the plate. This condition of high stress concentration is present in a thin cylinder with a small circular hole subjected to torsion, Fig. 6.6b.

2. *Model*

Theoretical solutions for the stress distribution at a section of discontinuity exist for only a few cases, such as a circular or elliptical hole, so this information is generally determined experimentally. The investigation may be conducted on a scale model or a prototype using any suitable material, and measuring strains at the section of discontinuity with sensitive short gage length extensometers. The latter is critical and necessary to obtain satisfactory results, because of the highly localized nature of the stress distribution at the place in question.

One fundamental method of obtaining stress concentration factors is by loading a model of the structure until yielding occurs at the points of maximum stress and comparing this with the normal stress in the member. This yielding can be observed on steel models having polished surfaces by the occurrence of Lueders' lines, paragraph 5.4. Figure 6.7 shows the Lueders' lines at the edge of a circular hole in a mild carbon steel plate subjected to tension in the vertical direction. The lines accurately showed the points of maximum stress concentration and gave a value of 2.1. This is less than the value of 3.0 given by the mathematical solution, Eq. 6.3.1, which is explainable by two facts: (1) at first the yield point producing stress is restricted to two points at which it is difficult to observe any visible flow until the load is increased to make these flow lines spread and become easily visible; and (2) this small region of plastically deformed material is surrounded by portions where the stress does not exceed the yield point, hence it

Fig. 6.7. Lueders' Lines at the Edge of a Hole in a Plate Subject to Tension in the Vertical Direction (Mild Carbon Steel)

tends to retard the formation of Lueders' lines which are produced by sliding at 45° to the direction of the tension.

The use of Lueders' lines to determine stress concentration factors has been used on a variety of structures, such as boiler heads, structural members, etc., and until recently was one of the few methods applicable to three-dimensional problems. The development of three-dimensional photoelastic techniques, however, has now made this relatively quick and inexpensive method the prime means of determining theoretical stress concentration factors.[1,14,15]

6.4 Elliptical Openings, Stress Concentration

Round openings are used most frequently in pressure vessels because this shape is readily fabricated; but occasionally an elliptical opening is used for a special purpose, such as a manway or handhole opening and is chosen so as to permit the insertion through its own opening of a pressure sealing cover plate. It may also be extended to the shape of a nozzle, so chosen for favorable stress considerations, since the stress concentration factor can be increased or decreased, depending upon the stress field and orientation of the major axis of the ellipse.[16,17,18]

As in the case of a circular hole, the problem of an elliptical hole in a plate, Fig. 6.8, subject to a uniform tensile stress applied at the ends of the plate, has been solved mathematically.[9,177] When the major axis is perpendicular to the direction of tension, the maximum stress occurs at the ends of the major axis and is given by the equation

$$\sigma_1 = \sigma(1 + 2\,a/b) \tag{6.4.1}$$

and at the ends of the minor axis is

$$\sigma_2 = -\sigma \tag{6.4.2}$$

The stress increases with the ratio a/b, Eq. 6.4.1, making it apparent that a very narrow hole perpendicular to the direction of tension produces a very high stress concentration. Because of this, cracks perpendicular to the direction of applied force tend to spread. Conversely, crack spreading can be halted by drilling holes at the ends of the crack, thereby removing the sharp corners which promoted this condition.

When the tension σ is parallel to the major axis, the stress at the ends of the minor axis is

$$\sigma_2 = \sigma(1 + 2\,b/a) \tag{6.4.3}$$

and at the end of the major axis is

$$\sigma_1 = -\sigma \tag{6.4.4}$$

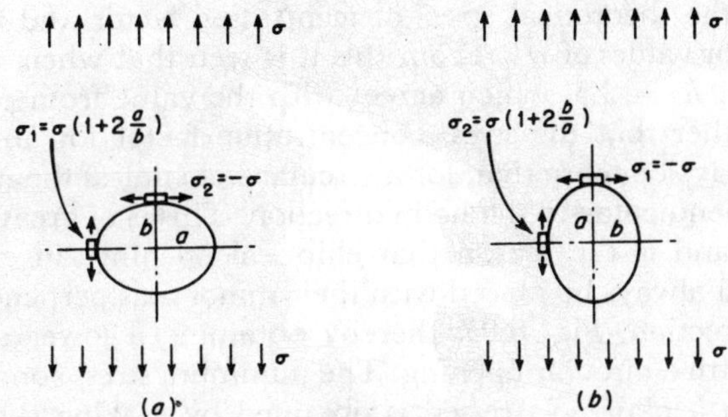

Fig. 6.8. Stress Condition at the Edge of an Elliptical Hole in a Plate in Tension

The maximum stress value occurs at the end of the minor axis, Eq. 6.4.3, and approaches σ when the ellipse is very slender; thus cracks parallel to the direction of tension are less prone to propagate than those perpendicular thereto.

1. *Elliptical Openings in a Cylinder*

The stress concentration factor for an elliptical hole in a cylinder with major axis perpendicular to the direction of hoop stress, Fig. 6.9*a*, can be found by superposition of the individual hoop stress from Eq. 6.4.1 and 6.4.2 and the longitudinal stress from Eqs. 6.4.3 and 6.4.4, if one remembers that the longitudinal stress in a cylinder is one half the hoop stress. Thus, the stress at the end of the major axis, where σ is the hoop stress, is

$$\sigma_1 = \sigma(1+2\,a/b) - \sigma/2 = \sigma(\tfrac{1}{2}+2\,a/b) \qquad (6.4.5)$$

$$\sigma_1 = K_t\sigma \qquad (6.4.6)$$

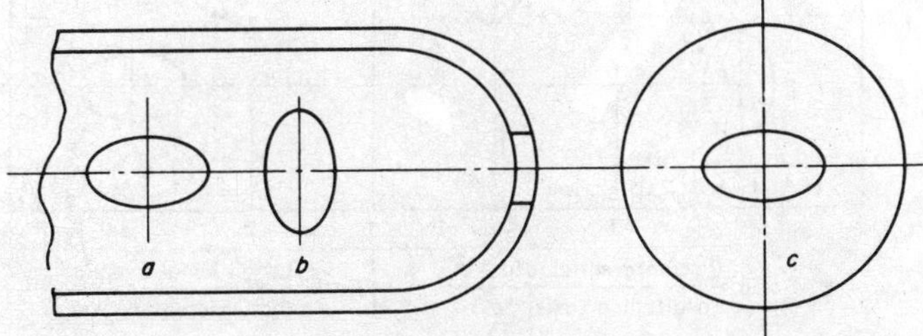

Fig. 6.9. Use of Elliptical Openings in Cylindrical and Spherical Parts of a Pressure Vessel

where K_t is the theoretical stress concentration factor and is plotted in Fig. 6.10 for values of a/b. From this it is seen that when $a/b = 1$, a circular hole, $K_t = 2.5$, which agrees with the value from paragraph 6.3, and further that the stress concentration factor for an elliptical hole will always be more than for a circular hole unless the minor axis is made perpendicular to the hoop direction. This is of great practical importance and is the reason that elliptical openings in cylindrical vessels should always be placed with their minor axis perpendicular to the hoop direction, Fig. 6.9*b*, thereby obtaining a lower maximum stress than with a circular opening. The minimum stress concentration for a hole in a plate so stressed is obtained by making the opening elliptical with the lengths of the axes inversely proportional to the applied stresses in the same direction; hence, for a cylindrical pressure vessel this means an ellipse with axis ratio 1:2, giving a stress concentration factor $K_t = 1.5$, Fig. 6.10.

This type of selective orientation can also be extended to reinforced openings or nozzles.[19,119] Figure 6.11 shows the measured stress con-

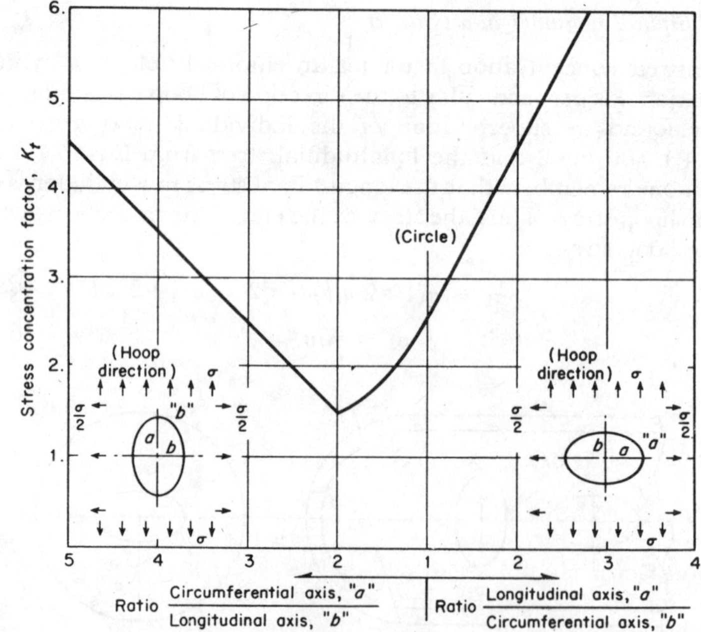

Fig. 6.10. Maximum Theoretical Stress Concentration Factor for an Elliptical Hole in a Cylindrical Vessel

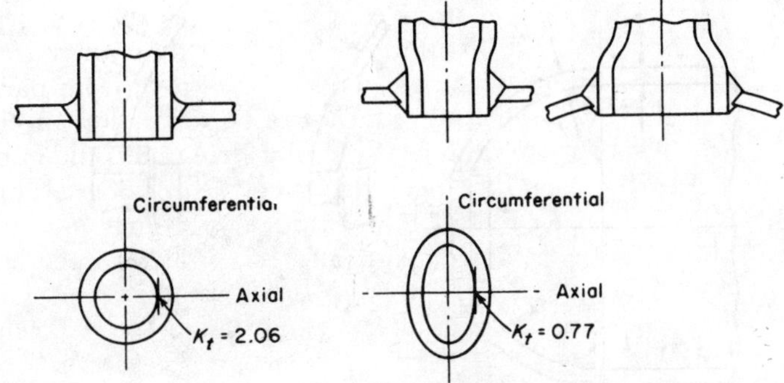

Fig. 6.11. Comparison of Stress Concentration Factors for Similarly Reinforced Circular and Elliptical Openings (Ratio Axis = 1:2) in a Cylindrical Pressure Vessel[20]

centration factor for an elliptical shaped nozzle and a similarly constructed and reinforced round nozzle. A reduction from $K_t = 2.06$ for the circular nozzle to $K_t = 0.77$ for the elliptical nozzle was obtained.[20] Although elliptical shape nozzles are desirable stresswise for the vessel, and are used for special applications, they present an economic problem, because the change in cross section from an ellipse to a circle (for connecting to piping) is costly to manufacture.

2. *Elliptical Openings in a Sphere*

The stress concentration factor for an elliptical opening in a sphere, Fig. 6.9*c*, can likewise be found from superposition, noting that the two principal stresses in a sphere are equal. Therefore, from Eq. 6.4.1 and 6.4.2, the maximum stress at the edge of the major axis is

$$\sigma_1 = \sigma(1+2\,a/b) - \sigma \tag{6.4.7}$$

$$\sigma_1 = \sigma(2\,a/b) \tag{6.4.8}$$

$$\sigma_1 = K_t\sigma \tag{6.4.9}$$

When $a/b = 1$, a circular hole, Eq. 6.4.8 gives a factor of $K_t = 2.0$, which agrees with paragraph 6.3. This is the minimum stress concentration factor obtainable in a spherical vessel, and it is seen that the stress for an elliptical opening will always be higher than that for a circular opening in the ratio of a/b. In practice, stress concentrations for either round or elliptical openings in cylindrical vessels with hemispherical heads are frequently handled by making the head

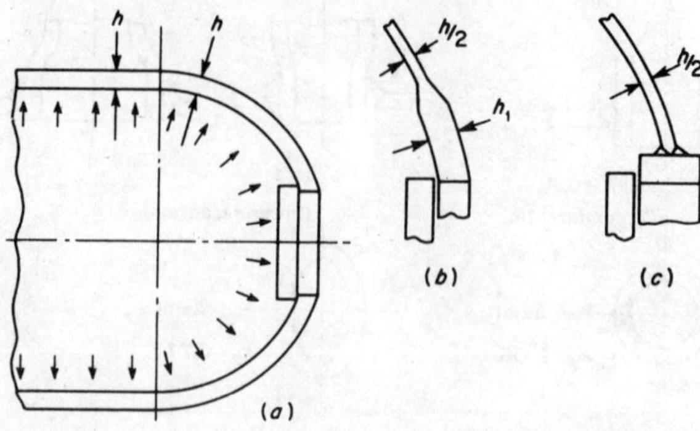

Fig. 6.12. Methods of Reinforcing Openings in Hemispherical Vessel Heads

thickness the same as that required for the cylindrical part, Fig. 6.12*a*. Since this is twice the thickness required for a sphere, the maximum stress is reduced 50 per cent. Openings of this type are generally referred to as "inherently reinforced openings," since the stress level in the entire structure has been reduced and no local additions of material in the region of the opening have been added. When this design approach is not used, but the entire thickness is reduced in accordance with the normal stress in the hemisphere, it is necessary to reinforce the opening with additional material to compensate for this high local stress, Fig. 6.12*b* and *c*.

6.5 Stress Concentration Factors for Superposition, Dynamic and Thermal-transient Conditions

1. *Superposed Notch Stress Concentration*

When notches are far apart from each other, the stress concentration factor of each notch is not influenced by the geometrical parameters of each other, Fig. 6.13*a*. However, when notches are located near each other the maximum stress depends both on the geometrical parameters of the notches and their proximity.[21]

(a) *Notches near to each other.*

When notches are placed next to each other, Fig. 6.13*b*, the resulting stress concentration factor is lower than that for the single most severe notch; in fact, adding adjacent notches is one means that is frequently used to reduce the effect of a single severe notch, paragraph 6.18.

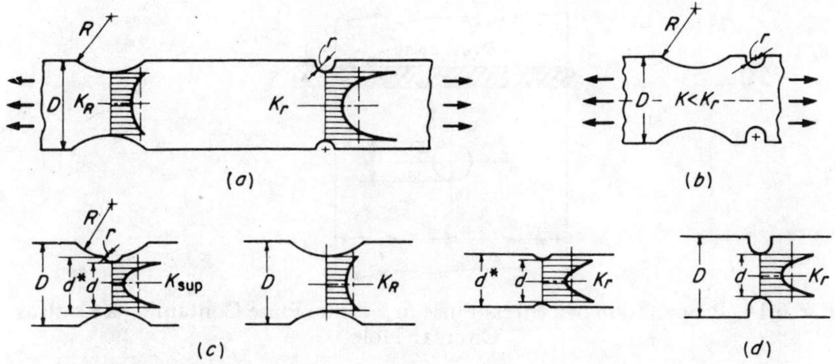

Fig. 6.13. Effect of Notch Proximity on Stress Concentration Factors[21]

(b) *Notches superposed.*

When the notches are so close as to become superposed, Fig. 6.13*c*, the resulting maximum stress concentration factor is the product of the two individual ones using the notch dimensions for each notch as shown in Fig. 6.13*c*.

$$K_{\text{sup.}} \leqq K_R \times K_r \tag{6.5.1}$$

The stress concentration factor for a superposed notch is less than that for the single smaller radius notch of same total depth, Fig. 6.13*d*. In fact, the superposed notch is one of the means that is used to reduce the effect of a single severe one, paragraph 6.18.

2. *Dynamic Stress Concentration*

The stress concentration factors thus far discussed have been derived from analytical and experimental studies under the condition of a static stress. The design procedure for obtaining the maximum stress associated with such geometric discontinuities is to apply the applicable stress concentration factor to the nominal stresses obtained from the simple strength of materials formulas. This design method is satisfactory for static loading conditions. When dynamic loading is involved, the usual recourse is to follow the same procedure; i.e., the statically determined stress concentration factor is applied to the nominal dynamic stress due to impact or shock.[22,23]

Dynamic loading can originate from mechanical impact or from the shock of explosions which give stress pulses of short duration.

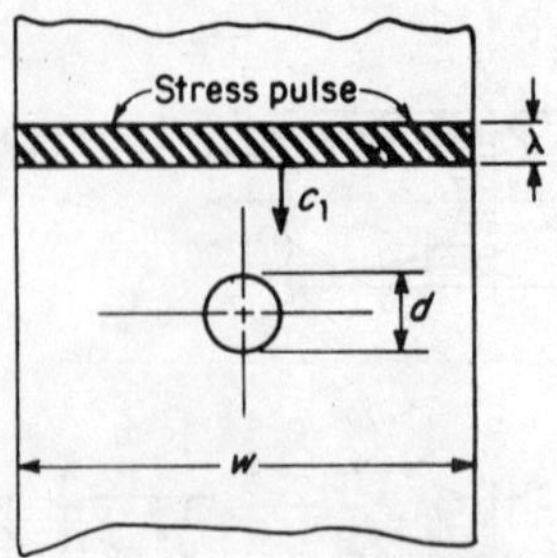

Fig. 6.14. Propagation of a Stress Pulse in a Finite Plate Containing a Central Circular Hole

Figure 6.14 shows a central circular hole in a finite plate subject to a short stress pulse. The pulse wave length λ depends on the type material and the duration of the applied load and is given by

$$\lambda = c_1 t \tag{6.5.2}$$

where c_1 is the dilatational wave velocity of the material and t is the loading duration in seconds. The value of c_1 is given by

$$c_1 = \sqrt{\frac{E}{\rho(1 - \mu^2)}} \tag{6.5.3}$$

where ρ is the mass density of the material. The resulting stress concentration factor at the edge of this hole, determined experimentally, is given in Fig. 6.15. This shows that there is a pronounced variation in the stress concentration factor with the pulse wave length, particularly when it is of the same length as the diameter of the hole. This is due to the fact that when the band of stressed material is about the same length as the discontinuity, diameter of the hole in this case, there is insufficient time, or space, for a secondary pulse to be reflected from the hole boundary and reinforce the main pulse. As the band is further diminished relative to the hole diameter, there is even less opportunity for reinforcement to occur and it is seen from Fig. 6.15 that for values of d/λ greater than one, the stress concentration factor approaches 1.0. The time duration t of the applied load to obtain such a low stress concentration factor is extremely short; for instance, for a 1 inch diameter hole in steel it is in the order of 2 microseconds (1 microsecond = 0.000001 second). Conversely, for long duration pulses the band of stressed material becomes large, approaching a static loading condition, and there is sufficient wave reinforcement to bring about a high stress concentration factor.

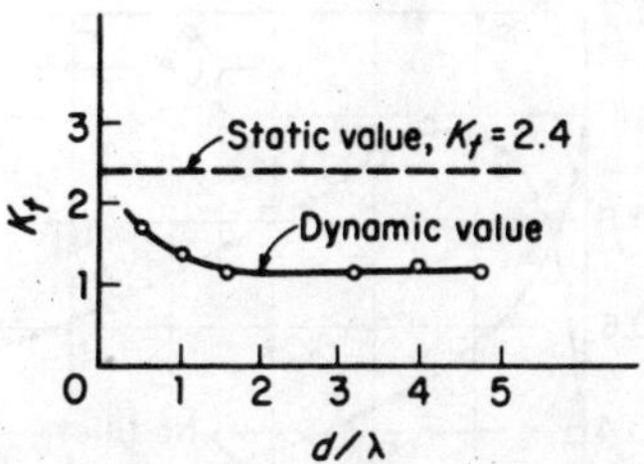

Fig. 6.15. Comparison of the Static and Dynamic Stress Concentration Factor at the Edge of a Circular Hole in a Finite Plate,[22] Fig. 6.14

While the usual design procedure mentioned above for dynamic conditions gives satisfactory results, it may under some conditions (very short pulses) give very conservative results. It is not always possible to take advantage of the latter unless the pulse is firmly established and not likely to be associated with random ones of greater duration and their associated higher stress concentration factors. Accordingly, this phenomenon is best viewed as another inherent material safety factor, much like that associated with rapid loading conditions wherein many steels exhibit an increase in their yield and ultimate strength properties, paragraph 5.6.

3. *Thermal Stress Concentrations*

The theoretical maximum thermal stress associated with a thermal transient is from Eq. 2.11.2 equal to $\alpha E\Delta T$. This occurs at time zero and is called thermal shock.[24,25,120] It is confined to a boundary or surface layer; hence, the notch shape does not influence its magnitude and the thermal stress concentration factor is unity with the upper bound for the maximum thermal stress

$$\sigma_{\text{max. therm.}} = (1)\ \alpha E\Delta T_{t=0} \tag{6.5.4}$$

However, after the start of the transient the shape of the notch does have an effect on the heat flow which in turn establishes the thermal gradient and thermal stress, paragraph 2.14. Accordingly, these thermal stresses are greater than those in a un-notched member subject to the same thermal shock, $\alpha E\Delta T$. This is illustrated in Fig. 6.16 which shows the thermal stress concentration factor at the root of a semi-circular fillet and that at the corresponding edge of an un-notched plate versus thermal transient time. As the temperature effects penetrate into the member with the passage of time the fillet begins to function as a stress raiser. The late time of maximum stress is due to the interaction of the falling stress levels in the member and the in-

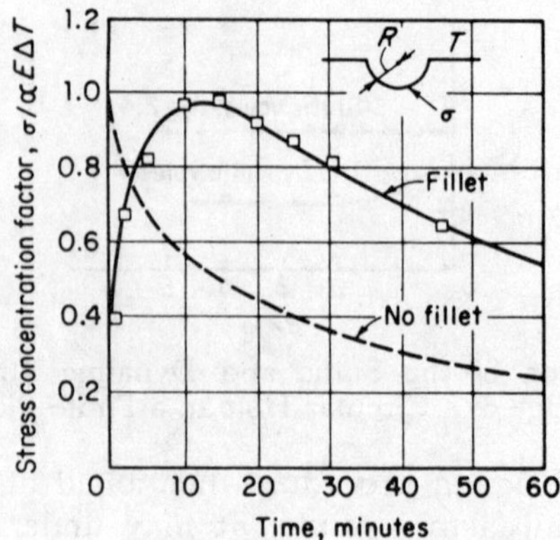

Fig. 6.16. Stress Concentration Factor at the Fillet Root, and the Face of an Unnotched Member Subject to a Thermal Transient[25]

creasing concentrating effect of the fillet. The persistence of high stresses for the fillet indicates the importance of accurate evaluation of the temperature field and accompanying thermal stress. This difficulty has led to the common design practice of accepting maximum thermal stress concentration caused by a structural discontinuity as equal to the thermal shock stress, Eq. 2.11.2; namely,

$$\sigma_{\text{max. thermal}} = \alpha E \Delta T_{t=0} \tag{6.5.5}$$

or the maximum thermal stress is simply assumed equal to the mechanical stress concentration factor times the maximum thermal stress which would occur in a corresponding member without the geometric discontinuity,

$$\sigma_{\text{max. thermal}} = K_{\text{mech.}} \alpha E \Delta T_{t=x} \tag{6.5.6}$$

6.6 Theory of Reinforced Openings

The most commonly used openings in vessels that require reinforcement are those for nozzles. The reinforcement material should be integral with the vessel or nozzle, such as obtained with weldments or forgings as compared to riveted or bolted constructions, so as to offer no additional thermal discontinuities under conditions of thermal transients, other than those presented by geometric discontinuities. The two basic requirements for reinforcement are:

A. Sufficient metal be added to compensate for the weakening effect of the opening, yet preserving the general dilation or strain pattern predominating in the vessel.

B. The reinforcing material be placed immediately adjacent to the opening, but suitably disposed in profile and contour so as not to introduce an over-riding stress concentration itself.

1. Reinforcement of an opening cannot be unquestionably obtained by adding huge amounts of material because this has the reverse effect, but similar end result, of under-reinforcing, and creates a "hard spot" on the vessel which does not allow its natural growth under pressure or general strain pattern occurring throughout the vessel to take place at this over-reinforced location. The result is a local overstressing, which may be visualized as similar to 'pinching a balloon." Determining the correct amount of reinforcement is difficult. For instance, if $A_1 = 2rh$ is the decrease in cross-sectional area due to the hole, and the total cross-sectional area of the reinforcing is A_2, photoelastic investigations[26] of a number of nozzles in cylindrical and spherical vessels with various percentages of reinforcement, A_2/A_1, have shown negligible improvement in maximum stress when this ratio is increased from 65 to 115 percent. The same experiments have shown the stress to be less sensitive to the amount of reinforcing material than to the distribution of this material.

2. The boundaries for the addition of effective reinforcing material can be obtained by examining the stress gradient along the cross section *mm* with distance from the edge of the edge of the hole, Fig. 6.17. By exact theory, Eq. 5.5.2 for a cylindrical vessel subject to internal pressure, wherein the longitudinal stress is one half the hoop stress, this stress is

$$\sigma_1 = \frac{\sigma}{2}\left(1 + \frac{a^2}{r^2}\right) - \frac{\sigma}{2}\left(1 + 3\frac{a^4}{r^4}\right)\cos 2\theta_{(\theta=\pi/2)} + \frac{\sigma}{4}\left(1 + \frac{a^2}{r^2}\right) - \frac{\sigma}{4}\left(1 + 3\frac{a^4}{r^4}\right)\cos 2\theta_{(\theta=0)} \tag{6.6.1}$$

$$\sigma_1 = \frac{\sigma}{4}\left(4 + 3\frac{a^2}{r^2} + 3\frac{a^4}{r^4}\right) \tag{6.6.2}$$

The stress decreases rapidly with distance from the edge of the hole, as shown by the shaded area in Fig. 6.17*a*. At the edge of the hole, $r = a$, and from Eq. 6.6.2 the maximum stress is 2.5σ. At a distance from the edge of the hole equal to the radius, $r = 2a$, the stress has fallen to 1.23σ.

In like manner, the variation in stress in the region of a circular hole in a spherical vessel subject to internal pressure, wherein the two principal stress are equal, is:

$$\sigma_1 = \sigma(1 + a^2/r^2) \tag{6.6.3}$$

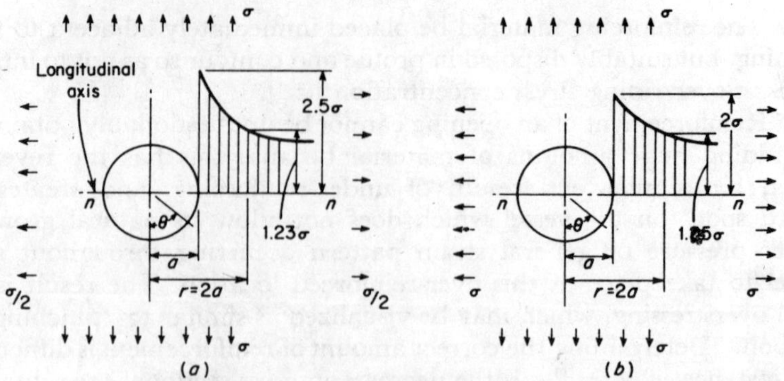

Fig. 6.17. Variation in Stress in Region of a Circular Hole in (*a*) Cylinder, (*b*) Sphere Subjected to Internal Pressure

A corresponding radical decrease with distance from the edge of the hole also occurs, reaching a maximum at the edge of the hole equal to 2σ and falling to a value of 1.25σ at a distance from the edge of the hole equal to the radius, Fig. 6.17*b*.

Accordingly, at a distance from the hole edge equal to the radius, the effect of the opening on the stress is negligible, and this distance is usually accepted as the boundary limit for effective reinforcement to the vessel surface, Fig. 6.18. The boundary limit in the direction perpendicular to the plate surface can be approximated from the deflection characteristics of the nozzle or ring which is doing the re-

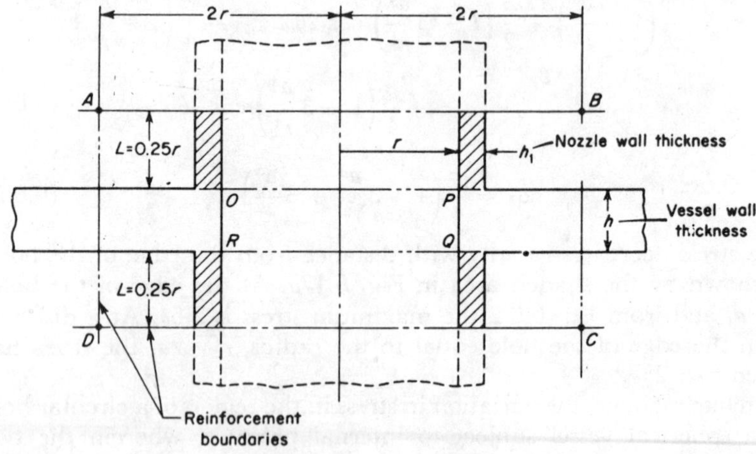

Fig. 6.18. Reinforcement Boundaries for Circular Openings in Cylindrical and Spherical Vessels

inforcing. In the case of a cylindrical nozzle this is a distance L beyond the surface of the vessel equal to $1/\beta$, and if an average nozzle wall thickness is taken as one tenth the nozzle radius, this gives

$$L = \frac{1}{\beta} = \frac{\sqrt{rh_1}}{1.285} = \frac{\sqrt{0.1r^2}}{1.285} = 0.25r \tag{6.6.4}$$

Thus, the reinforcement boundary limits, $ABCD$, can be completely established by the nozzle radius, as shown in Fig. 6.18. If it is required to replace the area $OPQR$ removed from the vessel plate by the opening, the shaded nozzle area may be counted as compensating reinforcing material. Should additional area be required, this must be located within the boundary $ABCD$ for full effectiveness. This is the basic area replacement method of reinforcement used in pressure vessel design. Recent analytical and experimental analyses have verified this simple method, while refining its details and accuracy.[27,28,29,121,122]

6.7 Nozzle Reinforcement, Placement, and Shape

1. *Single Nozzles*

Nozzle opening reinforcement must be placed within the boundary limits given in paragraph 6.6 to be effective. Further, experiments[30] have shown that the placement or location of this reinforcement within the boundary limits, all other factors constant, has a pronounced effect on the stress concentration factor as shown in Fig. 6.19.

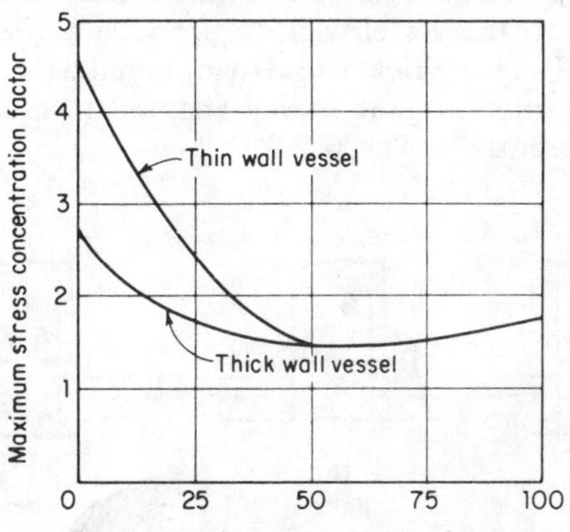

Fig. 6.19. Effect of Location of Reinforcement on the Stress at a Nozzle in a Spherical Vessel (Ratio Nozzle to Vessel Diameter = 0.10)

This shows that the minimum stress concentration factor is obtained with "balanced" reinforcement. This is explainable by the fact that reinforcement material evenly disposed both inside and outside of the vessel surface introduces no eccentricity or unbalance to create local bending moments and stresses, Fig. 6.20*c*. It is not always possible or feasible to use balanced reinforcement; in fact, outside reinforced nozzles are the most prevalent type and in which case the stress concentration factor is some 20 per cent higher than for a balanced one with the same amount of reinforcement, Fig. 6.20*b*. Inside reinforcement is the least desirable, with the stress concentration factor being higher in thin wall vessels than in thick wall vessels, Fig. 6.20*a*.

Maximum reinforcing action with minimum material is obtained when the material is kept within the effective boundary limits. It should be mentioned, however, that stress concentration factors can be further reduced, although at the expense of additional material, by extending the reinforcement beyond the effective boundary limits so that it acts as a basic increase in vessel thickness thereby reducing the maximum stress associated with the opening directly in proportion with the thickness, Fig. 6.12. As an example, if the basic stress concentration factor for an unreinforced opening in a plate of thickness h is 2.5, the relative stress concentration factor for the same hole reinforced by increasing the plate thickness in the vicinity of the hole to h_1 is 2.5 (h/h_1), Fig. 6.12*b*. This concept applies to both balanced and unbalanced reinforcements. It is basically the most direct, although not necessarily the most inexpensive, way to reinforce an opening; namely, increase the vessel wall thickness in the vicinity of the opening. Figure 6.21 *d* and *e* shows its use, in combination with an internal piping attachment to the vessel wall, to produce a nozzle of low stress concentration. Here a thick ring is employed as a shell course, and provides a construction that is well adapted to multiple nozzles arranged circumferentially, Par. 6.7.2.

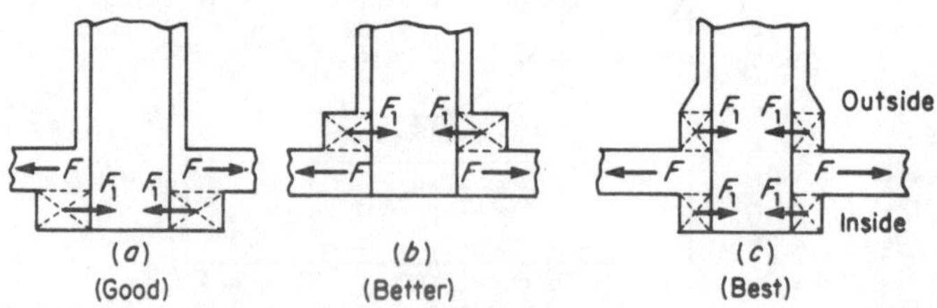

Fig. 6.20. Diagrammatic Location of Nozzle Opening Reinforcement, (*a*) Unbalanced Inside, (*b*) Unbalanced Outside, (*c*) Balanced

So far, the basic principles of the amount of reinforcing material, the boundary limits, and the general distribution have been considered. Equally important is the detail shape of the transition, fillets, etc., connecting the reinforcing material to the parent vessel material. Experiments have located the points of maximum stress at two locations:[122,123,124] the inside corner A, and the outside corner B of the nozzle. Fig. 6.21. They have also shown that stresses can be minimized at the inside[38] corner A by rounding this corner to a radius equal to 25 to 50 percent of the thickness of the vessel wall, and at the outside corner by a generous radius of the same proportion or a taper with transition radii, Figs. 6.21*b* and *c*. These peak stresses are only of a very local nature,[39,40] as shown in Fig. 6.22, but are none the less significant in establishing the fatigue life of a vessel. Nozzles may well be a "weak link in the chain" under repetitive stress condition, and the designer should give particular attention to the minimizing of these peak stresses. Figure 6.23 shows the fatigue failure of a vessel. The typical oyster-shell markings focus the origin on a material defect in a high local stress region at the inside corner of the nozzle from whence it propagated throughout the nozzle and vessel.[41, 125, 126, 127, 178]

For practical purposes, constant radii transition fillets, grooves, etc., are continually referred to as a means of reducing stress concentrations. It is not inferred that a radii is the only transition curve or by any means the optimum from a stress viewpoint (though it is most frequently used from the practical fabricating viewpoint of drilling, reaming, grinding, etc.). Just as in a highway or railroad line, a spiral is used to form the transition from a tangent to the final curve radius,

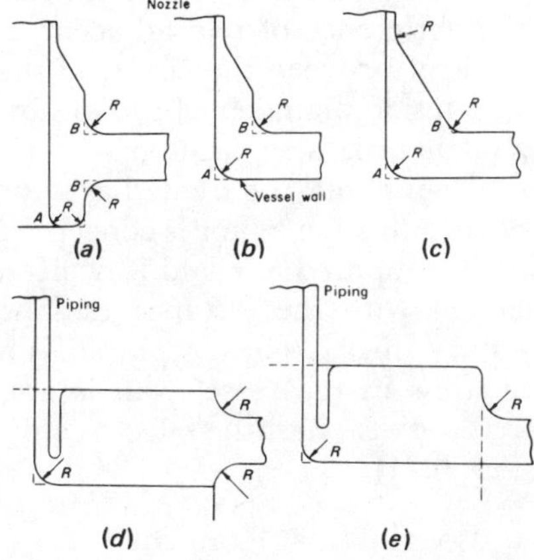

Fig. 6.21. Details of Nozzle Reinforcement to Reduce Stress Concentration

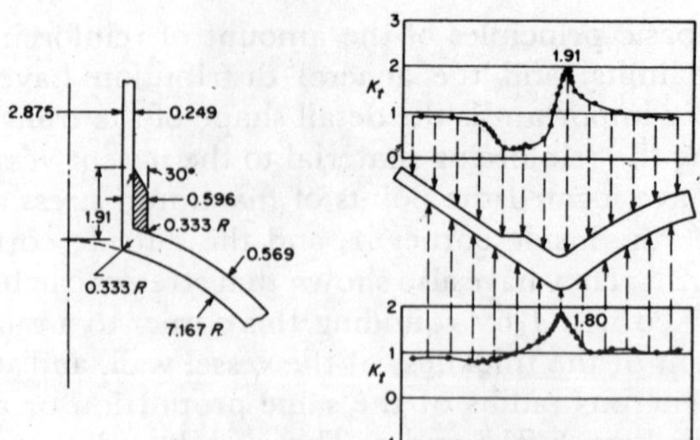

Fig. 6.22. Variation in Stress Concentration About a Circular Reinforced Nozzle in a Sphere Resulting from Internal Pressure[26]

a transition introducing a more gradual change of radii, such as an ellipse,[42] etc., will also show an improvement in stress concentration factor over a single radius connecting two tangents. Although such a refinement is seldom required in vessel design, and single radii fillets are ample, it is cited to emphasize the fundamental principle of "gradual change in geometric cross section" to minimize unbalance in stress patterns.

2. *Multiple Nozzle Arrangements*

Multiple reinforced nozzle arrangements require special consideration when they are very closely spaced because their individual effects become overlapping.[43] Experimental investigations have shown that the zone of mutual interference of two adjacent nozzles is no wider than $2(r + h_1)$. That is, if between the circles of the intersection with the shell of the nozzle inside diameter of radial nozzles or the ellipses of non-radial nozzles there is a separation of at least $2(r + h_1)$, the maximum stress in the vessel may be treated as that for a single nozzle insofar as the stress concentration factor is concerned. When it is noted that h_1 is usually small compared to r and is neglected, this reduces to $2r$ and is in agreement with the distance established by the basic reinforcement boundary theory, paragraph 6.6. This means that the average membrane stress in the vessel wall is not increased by the presence of reinforced nozzles; i.e.; the effect is only one of local stress concentration, if (Fig. 6.24)

$$\text{Ratio } \frac{\text{Longitudinal ligament}}{\text{Opening diameter}} = \frac{L}{d_i} \geqq 1.0 \tag{6.7.1}$$

(a)

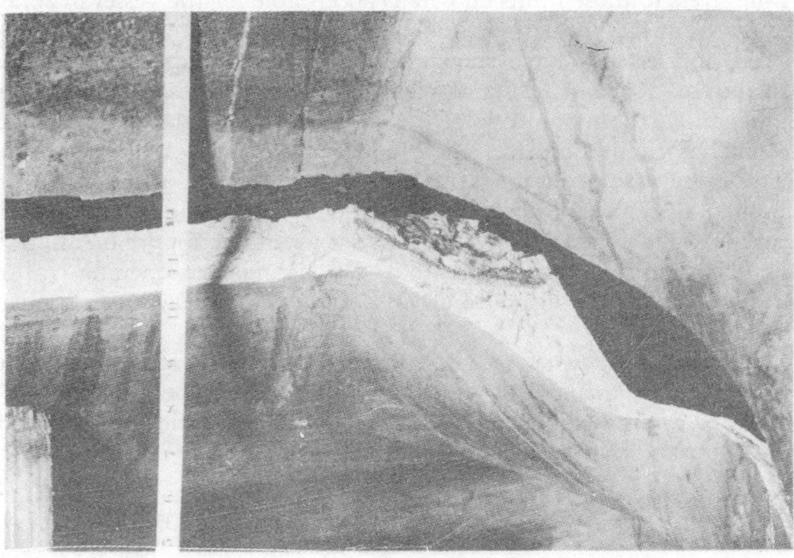

(b)

Fig. 6.23. Fatigue Failure of Test Pressure Vessel
(a) Nature of Propagation Throughout Nozzle and Vessel, and
(b) Origin of Fracture at Inside Corner of Nozzle[41]

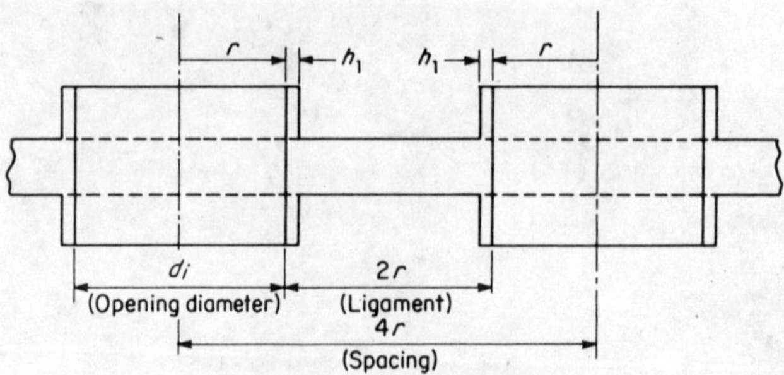

Fig. 6.24. Minimum Spacing of Adjacent Nozzles to Maintain Basic Average Membrane Stress in Vessel Wall

When reinforced nozzles are spaced closer than this, the ligament membrane stress is increased over that of the basic membrane stress in the vessel wall. This is a primary stress, therefore, it is necessary to increase the thickness of the vessel wall in this region in the proportion of 1.0/(actual ligament-opening ratio) to reduce the average membrane stress to those limits chosen as acceptable for the vessel in zones remote from this influence. Studies of single and closely spaced reinforced nozzles also have shown that while the shape of the stress distribution for a single and closely spaced nozzles do not differ significantly, the magnitude of the stresses for closely spaced nozzles are higher than those for single comparable nozzles. Further, the relative increase is in the same proportion as that associated with closely spaced unreinforced openings (holes) to a single hole for which analytical and experimental solutions are available. Figure 6.25 shows this effect and substantiates that for ligament-to-opening ratios greater than 1.0 the ligament effect is nil (less than 10 per cent) and is not significant.[44, 128, 129, 179]

The case of a row of radial reinforced nozzles located in a single plane[45,46] on a cylindrical vessel has also been experimentally investigated and has shown that these do not increase the stress beyond that associated with the same isolated nozzle when

$$\text{Ratio } \frac{\text{Circumferential ligament}}{\text{Opening diameter}} = \frac{L_c}{d_i} \geqq 0.5 \qquad (6.7.2)$$

3. *Non-radial Nozzles*

It is not always possible to use radial nozzles, but frequently nozzles must be installed non-radially for functional purposes. A non-radial

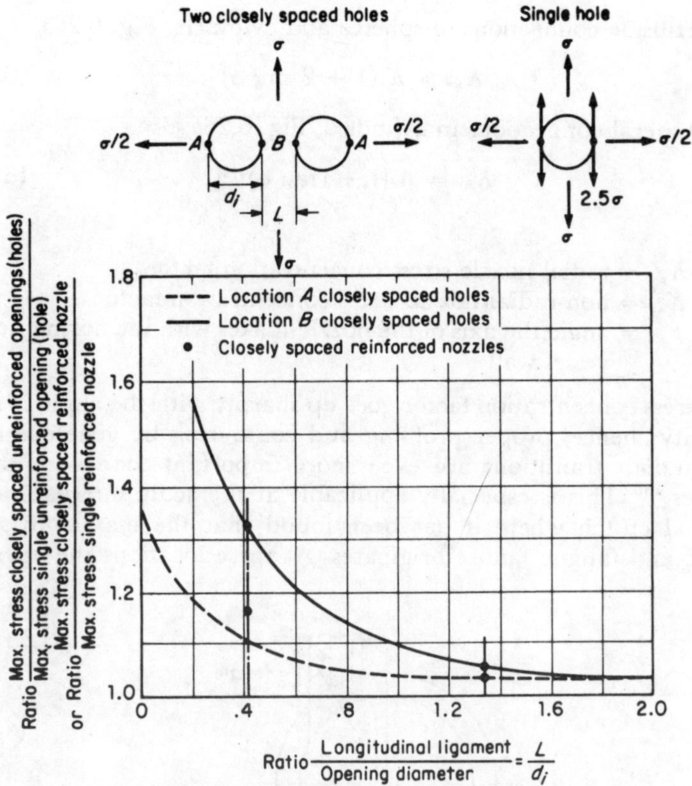

Fig. 6.25. Comparison of Stresses at Points *A* and *B* for two Closely Spaced Reinforced Openings in a Pressurized Cylinder with the Corresponding Stresses at Two Unreinforced Holes[44]

circular nozzle makes an elliptical opening in the vessel, and just as an elliptical hole in a plate gives rise to a higher stress concentration factor than does a circular hole, so does a non-radial nozzle have higher stress concentration factor than its comparable radial one,[47,48,49] paragraph 6.4. Analytical and experimental investigations of the stress distribution around non-radial holes in flat plates,[50,51] and nozzles in vessels have shown that maximum stresses occur in the vessel on the major axis of the elliptical opening close to the nozzle,[52,53] and are greater with nozzles of increased non-radiality. In spherical and cylindrical vessels their stress concentration factor can be approximately related to that for the same radial nozzle by the following relations:[54]

1. Hillside connections in spheres and cylinders, Fig. 6.26*a*,

$$K_{nr} = K_r(1 + 2 \sin^2 \phi) \tag{6.7.3}$$

2. Lateral connections in cylinders, Fig. 6.26*b*,

$$K_{nr} = K_r[1 + (\tan \phi)^{4/3}] \tag{6.7.4}$$

where

K_r = radial nozzle stress concentration factor
K_{nr} = non-radial nozzle stress concentration factor
ϕ = angle the axis of the nozzle makes with the normal to the vessel wall

The stress concentration factor goes up sharply with the angle of non-radiality; hence, proper profiling and contouring by generous radii and smooth transitions are even more important than with radial nozzles.[130] This is especially applicable at the acute internal lip and external crotch where it has been found that the maximum stress occurs and fatigue failure originates. At these locations the apex lip

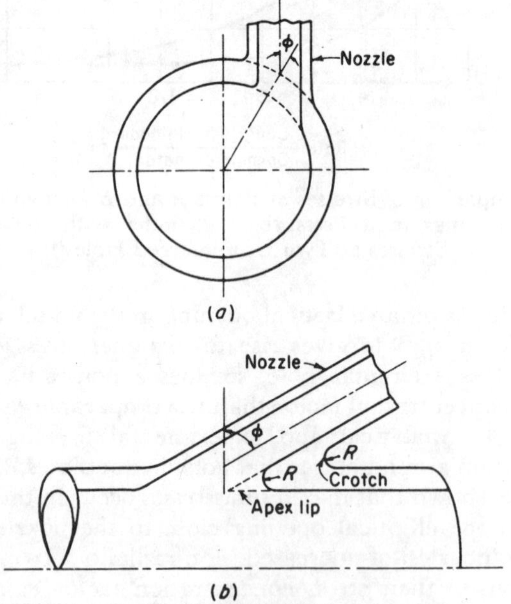

Fig. 6.26. Non-Radial Nozzle. (*a*) Hillside Nozzle in a Sphere or Cylinder, (*b*) Lateral Nozzle in a Cylinder.

material should be removed, and a generous concave surface used at the exterior crotch,[180, 181, 182] Fig. 6.26.

6.8 Fatigue and Stress Concentration

It has been mentioned in paragraph 5.11 that stress concentrations produced by section irregularities are expecially damaging in cases of fluctuating stresses and that proper allowance for their effect is the most important item in design to resist fatigue failure. Virtually all failures are a result of fatigue—fatigue in areas of high localized stress—so it behooves the designer to take into account all stresses, even though of a local nature.

Fatigue tests of specimens with sharp changes in cross section have shown that there is a reduction in life due to the stress concentration, but this is not always of the magnitude expected from the theoretical stress concentration factor. This effect is called the *fatigue-notch factor** or *fatigue-strength reduction factor* and is defined by the ratio:

$$K_f = \frac{\text{Endurance limit of unnotched specimen at } N \text{ cycles}}{\text{Endurance limit of notched specimen at } N \text{ cycles}} \tag{6.8.1}$$

As an example, the fatigue-notch factor fot the specimen[55] in Fig. 6.27 is $K_f = 70/38 = 1.8$, which is somewhat less than the theoretical stress concentration factor $K_t = 2$. Measured values have further shown that the actual amount of weakening depends not only on the peak stress but also on the stress gradient, the type of material, and the type of loading. These variations are ascribed to notch sensitivity of the material. The notch-sensitivity factor, q, is defined as

$$q = \frac{K_f - 1}{K_t - 1} \tag{6.8.2}$$

Therefore,

$$K_f = 1 + q(K_t - 1) \tag{6.8.3}$$

Notch sensitivity is a property of the material and is also a function of the radius of the notch, r. Figure 6.28 demonstrates this effect[56] and shows that, as the notch radius becomes smaller, K_t increases but q decreases, indicating that a compensating effect is taking place. The sensitivity factor is not a constant and depends both on material and size of specimen. In the case of fine-grained materials, such as

*The term "notch" is used in fatigue studies to imply any disturbance which produces a stress concentration.

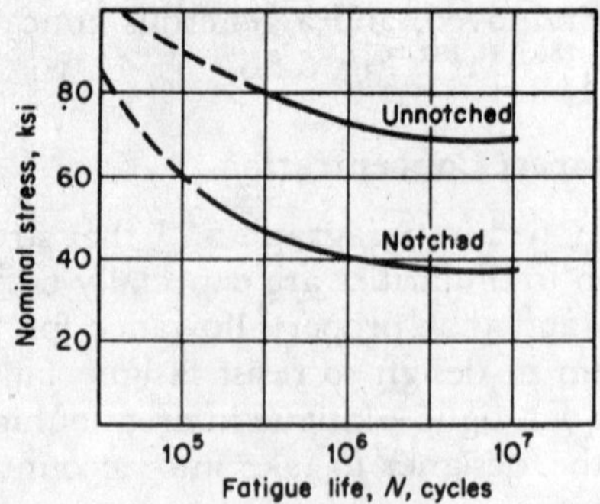

Fig. 6.27. Effect of Semicircular Groove ($K_t = 2.0$) on Rotating–Bending Fatigue Strength of a Round Bar, SA-4130 Steel[55]

quenched and tempered steels, and for larger specimens, q approaches unity much more rapidly than for the relatively coarse-grained, annealed, or normalized materials. This is highly significant in the design of pressure vessels using high-strength, heat-treated material, and emphasizes the paramount importance of minimizing geometries giving rise to stress raisers.

These quantitative studies have been based on the facts that the material is not perfectly homogeneous, but actually has a grain structure, and that in the region of a peak stress, different results are obtained when only a few grains are contained in that region as compared to the results obtained if many are contained in the same region. For instance, the stress distribution across a section containing a circular hole, Fig. 6.29a, has a high stress gradient at the edge of the hole. If the load is just sufficient to bring the peak stress up to the endurance limit, a fatigue failure would hardly be expected since the volume of material at this stress is zero. A finite volume of material must be at the

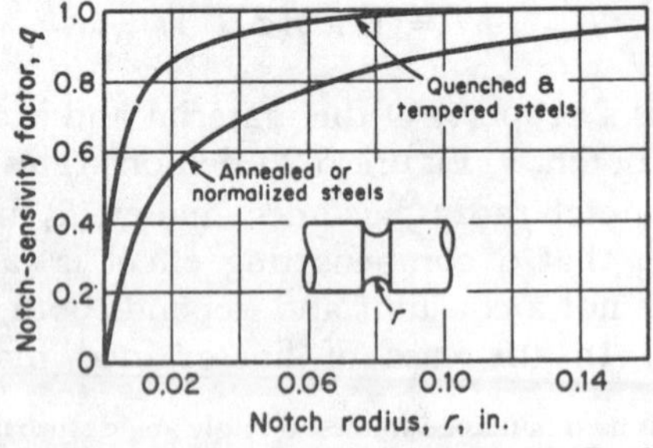

Fig. 6.28. Variation of Notch Sensitivity with Radius for Steel[56]

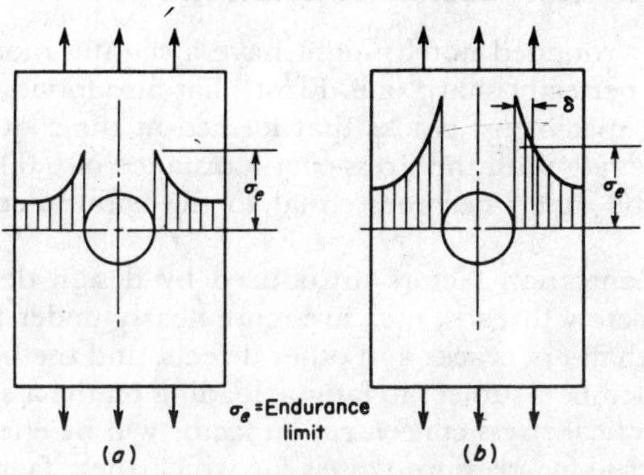

Fig. 6.29. Stress Distribution about a Circular Hole in a Bar

endurance limit before a crack will form, and to obtain this volume of material the endurance limit stress must exist at some finite depth, δ, below the surface; therefore, the steeper the stress gradient, the higher the load required to produce fatigue failure, Fig. 6.29*b*.

The dimension, δ, is a property of the material; and, in general, hard, fine-grained materials have small values of δ, whereas soft, coarse-grained materials have larger values. The relationship between δ and steel tensile strengths, based on correlating fatigue data and the shear theory of failure is shown in Fig. 6.30. Hence, to estimate the value of K_f it is necessary only to determine the stress gradient by analysis or photoelasticity and multiply K_t by the ratio of the stress at depth δ to the peak stress, in applying Eq. 6.8.3. Langer[57] has investigated the strength reduction factor of circular and elliptical holes and cavities for medium strength steel and has found the highest value of K_f to fall in the range $0 < r < 0.010$ in. for small r values. This means

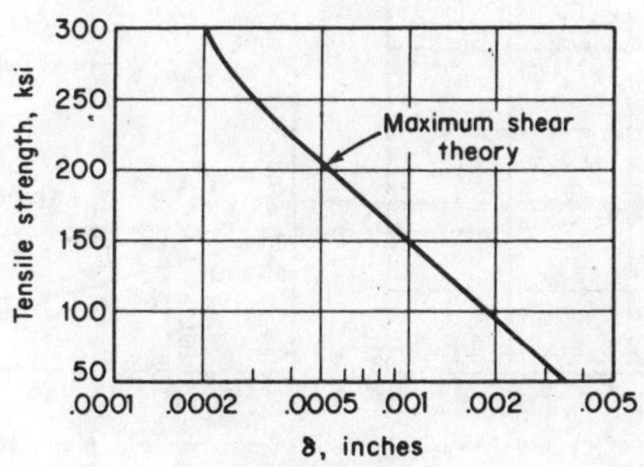

Fig. 6.30. Material Constant δ vs. Tensile Strength for Steel[57]

that a slightly rounded notch might have a slightly more deleterious effect than a perfectly sharp one. Frost[58] has also found that for mild steel notched specimens, cracks that formed at the root of the notch will not propagate until the stress over a distance of 0.002 in. in from the root of the notch becomes equal to the endurance limit of the material.

Stress concentration factors introduced by design details, such as holes, fillets, screw threads, etc., are more nearly under the control of the designer than are cracks and other defects, and the best procedure is to design members subject to fatigue loading on the assumption that the full theoretical stress concentration factor will be effective. This is particularly true for pressure vessels for which these factors are fairly well established.[59]

To estimate the fatigue life of a member subject to fluctuating stress, i.e., a mean stress plus an alternating component, a type of Goodman fatigue diagram is frequently used in which mean stress is plotted as the abscissa and the amplitude of the alternating component is plotted as the ordinate, Fig. 6.31. In using this diagram two considerations are important in the choice of variables. First, the fatigue strength reduction factor, K_f, should be applied to both the mean stress and the alternating component of the fluctuating stress. These values can be obtained from test results or the method described above. They are always smaller than the theoretical stress concentration factor, K_t, but approach K_t when the notch is large or when the material is hard. Second, the true value of the mean stress, considering the shift when these stresses enter the plastic range, should be used. This is best illustrated by Fig. 6.32 which shows a fluctuating stress that exceeds the yield point and the accompanying shift that occurs to the mean

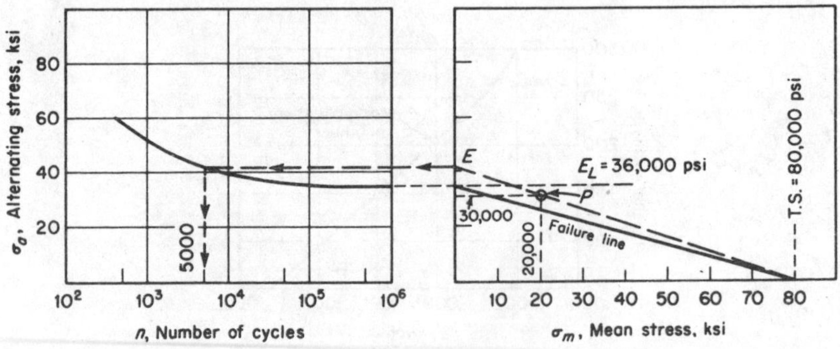

Fig. 6.31. Method of Determining Fatigue Life

component. In the first half cycle, the maximum stress OX is not realized, but the material strains elastically from 0 to A, and plastically from A to B. In the next half cycle, the material recovers elastically from B to C and is forced into compression from C to D because of the lower stress region which surrounds it. After this first cycle, it follows the line DB back and forth. This has the important effect of adjusting the mean stress component downward and giving a truer picture of the prevailing condition. This adjustment consists of a reduction in mean stress by the amount the maximum stress exceeds the yield stress. If the total stress range exceeds twice the yield stress, the mean stress always adjusts itself to zero.

As an example, consider a notched bar with $K_f = 3.0$ made of material of tensile strength = 80,000 psi, yield strength = 50,000 psi, endurance limit for complete stress reversal = 36,000 psi, and a $\sigma - n$ curve as given in Fig. 6.31, is cycled between nominal tensile stress values of 0 and 20,000 psi.

The modified Goodman diagram is constructed, Fig. 6.31, with the amplitude of the alternating stress component equal to $\frac{1}{2}$ (3 × 20,000) = 30,000 psi, and the mean stress equal to $\frac{1}{2}$ [50,000 − (3 × 20,000 − 50,000)] = 20,000 psi. The failure line is constructed by connecting the value of the maximum mean stress for no alternating stress which is the ultimate tensile strength of the material, 80,000 psi, and the value of the maximum alternating stress for zero mean stress which is the endurance limit. Combinations of mean and alternating stress falling below this line are considered safe. Combinations falling beyond this line are considered safe for a limited number of cycles which can be estimated by extending a line from the abscissa point of maximum

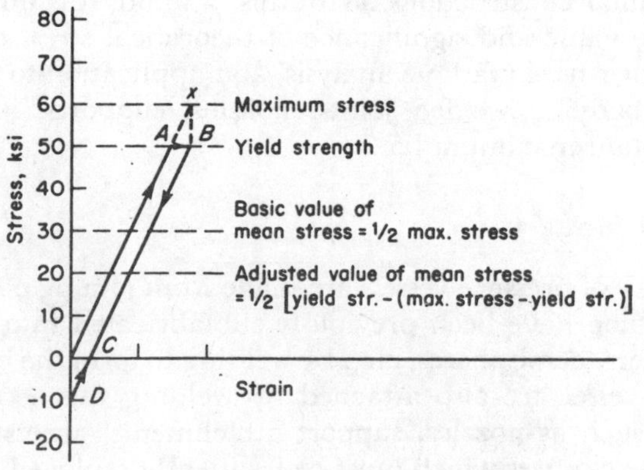

Fig. 6.32. Stress–Strain History

mean stress, 80,000 psi, through the evaluated stress point, P, to the alternating stress ordinate intercept E and reading the corresponding fatigue cycle life directly from the $\sigma - n$ curve, 5,000 cycles.

6.9 Pressure Vessel Design

It is not practical to design a vessel without stress concentrations, because vessels must have openings, supports, etc., as necessary operating features and these give rise to geometric discontinuities, hence, abnormal local stresses. They produce effects on vessels that are somewhat like those from pricking a balloon—for instance, a sharp point will immediately rupture it, whereas it will resist a blunt probe more, due to the less severe nature of the local induced stress. Operating conditions associated with nuclear vessels have placed new emphasis and importance on brittle fracture at low temperatures, fatigue failure, and material irradiation damage. It is rather paradoxical that the one most fruitful way of precluding such failures is by minimizing stress concentrations—a factor that has until recently been greatly ignored in vessel design. A reduction in the basic allowable stress level throughout the entire vessel may do little toward increasing the life of a vessel, yet add grossly to its weight and cost, if proper attention is not given to minimizing the stress concentrations at a nozzle, lug, or support.

Unquestionably, because of the many factors influencing the fatigue life of a complete structure, a test of a prototype would be desirable.[60] Lacking this, the usual design approach is to analyze the stresses throughout the structure and include the effect of stress concentrations by using the results of laboratory or full-size test data from specimens containing similar constructions. With this in mind, it is important to appreciate the value and significance of theoretical stress concentration factors, their use in fatigue analysis, and application to the design of vessels embodying welded joints, nozzles, supports, and bolted closures to obtain maximum life.

6.10 Welded Joints

The majority of pressure vessels are made from joining parts or subassemblies, which have been previously subfabricated into segments, such as cylinders, hemispheres, etc., by welding to form the base vessel. To this base vessel are also attached by welding the necessary appurtenances, such as nozzles, support attachments, access openings, etc. Only those closures which must be frequently removed for service or maintenance are attached by bolts, studs, or other mechanical closure devices. This gives a vessel of maximum leak-proofness, since the number of mechanical-gasket joints prone to such are kept to a minimum, and is a paramount design consideration for nuclear and other lethal service. Most of all, however, it gives the vessel constructor a fabricating tool—welding—by which he can comply with one of the designer's prime concerns, namely, to eliminate or control the effect of stress concentration factors. Stress concentration factors in welded joints, irrespective of the type of welding process used, i.e., whether by an electric, gas, magnetic, electron, or forging process, arise through three means:

1. The metallurgical structure of the weld metal with respect to adjacent parent metal.
2. The occurrence of welding defects, such as porosity, slag inclusions, and shrinkage cracks.
3. The weldment geometry—profile and contour of reinforcement, fillets, transitions, etc., including surface finish.

The first of these arises from an intrinsic change in the homogeneity of the weld-parent-plate metallurgical structure in passing from the weld, through the heat affected zone, to the parent plate. In many cases the whole welded joint can be heat treated so that the metallurgical structure of the weld metal and parent metal become nearly identical. In other cases this may not be feasible, and there remains a gradation in metallurgical condition from parent metal through heat affected zone to weld metal. The second gives rise to both metallurgical and geometric stress raisers through weld metal porosity, slag or oxide inclusions, poor penetrations, shrinkage cracks, weld surface roughness, etc. These first two sources are subject in great measure to quality control welding procedures, leaving the third as the main source of stress concentrations and one which is totally controllable by the design engineer.

1. *Strength of Weldments*

a. *Weldments Transverse to Load or Stress*

The basic design of welded joints is the butt joint shown in Fig. 6.33, together with the effect of weld profile on the fatigue strength as summarized by Grover[55] from the extensive literature on this subject. This brings out the tremendous beneficial effect of removing what might at first be considered "weld joint reinforcement" to obtain a flush joint

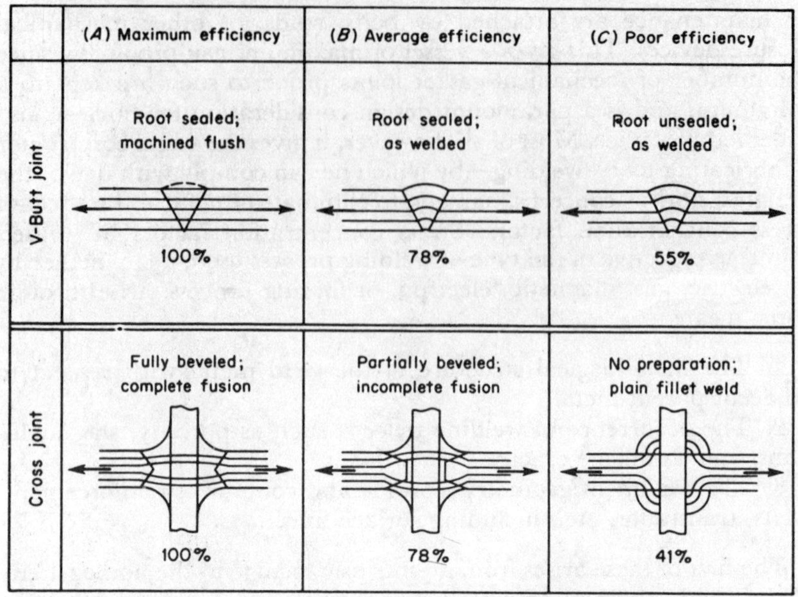

Fig. 6.33. Effect of Welded Joint Design on Fatigue Strength[55] (Load Transverse to Joint)

thereby increasing the fatigue efficiency of the joint from 78 percent to 100 percent; and the even further effect of eliminating undercut or re-entry corners which further reduced the joint efficiency to 55 percent. These fatigue joint efficiencies become even more pronounced when a member, such as a lug support or nozzle, is introduced to form a cross joint. Here the lack of full weld penetration and the more devious path the stress must take creates high stress concentrations with a consequent lowering of fatigue life.

b. *Weldments Parallel to Load or Stress*

Figure 6.33 gave the fatigue reduction factors for geometric discontinuities in welded joints located transverse to the applied load or stress. Changes in cross section, or surface finish, parallel to the direction of load or stress also result in a reduction in joint fatigue strength.[61] The cause is the same; namely, the local stress raising effect of sharp changes in cross section of the structure. For instance, in welded joints the weld bead ripple and electrode change points offer sources of fatigue crack initiation from which cracks spread across the joint perpendicular to the weld seam direction,[131,132] Fig. 6.34*a* and *b*. Auto-

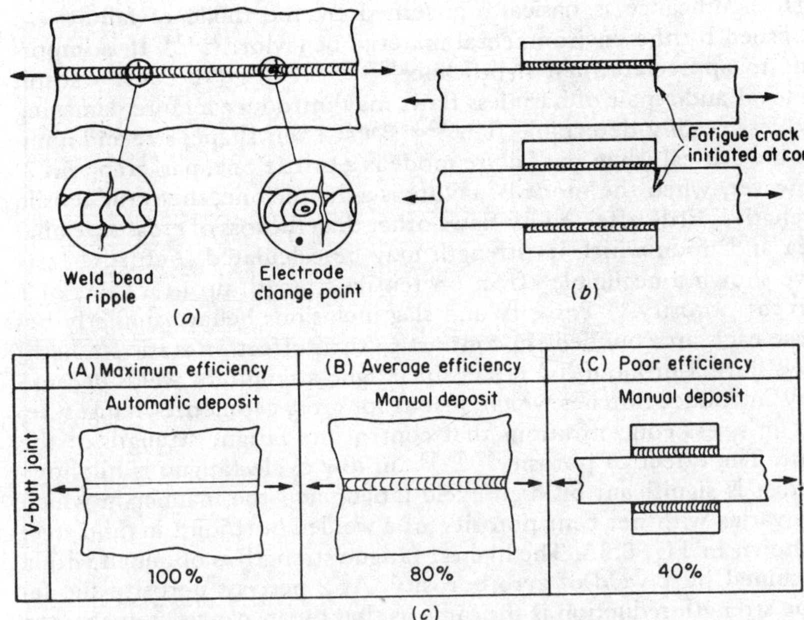

Fig. 6.34. Effect of Welded Joint Design on Fatigue Strength (Load Parallel to Joint)[61]

matic deposited welds have a much smoother surface and fewer starts-and-stops than those made manually, and tests show the former to have a greater fatigue life. The relative fatigue strength efficiency of such longitudinal butt welded joints, together with those longitudinal discontinuities as are encountered with vessel lifting and support lug attachments, are given in Fig. 6.34*c*. The fatigue strength efficiency for welded joints loaded transversely or parallel are essentially the same for like stress concentrating mechanisms; and the abruptness in change of geometric shape, whether on a minute or gross scale, in the controlling factor.

2. *Significance of Weld Joint Metal Flaws*

The most widely used process for joining parts together to form pressure vessels is that of welding. The quality of the deposited weld metal, as measured both by its porosity or slag inclusions and its shape or placement, influences the service performance of these joints. These flaws (porosity, slag inclusions, poor weld shape, misalignment and lack of fusion) are usually a result of faulty welding techniques and their significance is basically governed by the mode of failure established by the environmental material behavior.[133,134] It is important to appreciate their significance[135,136,137,138,139] because wanton removal and repair of harmless flaws may introduce a more damaging and less readily detectable flaw.[62,63,64,65] Their shape, size and number are critical when the failure mode is a brittle one, paragraph 5.22. However, when the mode of failure is a ductile one, the static tensile strength is little affected by flaws other than the loss of cross-sectional area and upon which its strength may be calculated. Actually, tests have shown a negligible effect on tensile strength up to a value of 7 percent porosity.[66] Porosity and slag inclusions behave similarly because each are rounded. In contrast to their effect on static strength, flaws have a pronounced effect on fatigue strength of welds because they introduce notches which, just as for gross geometric changes, result in stress concentrations that control the fatigue strength of the joint. The effect of porosity[67,68,69] on low cycle fatigue is nil; however, it is significant on high cycle fatigue and the manner in which this varies with per cent porosity of a welded butt joint in mild steel is shown in Fig. 6.35. The highest fatigue strength is obtained with a machined butt weld of zero porosity. At 2 percent porosity the fatigue strength reduction is the same as that for an average unmachined butt welded joint of zero per cent porosity metal. This means that

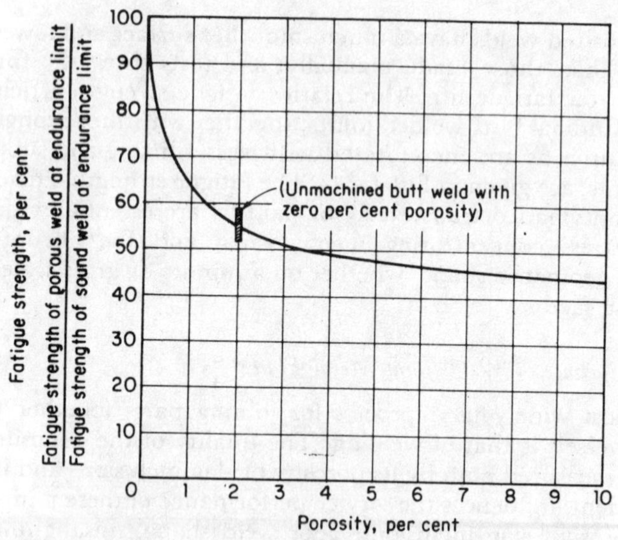

Fig. 6.35. Effect of Porosity on the Endurance Limit Fatigue Strength of Machined or Ground Butt Welds in Mild Steel[66]

both the shape of the weldment and quality of the weld metal must be considered in establishing fatigue life, and weld metal porosity requirements below 2 percent are unwarranted for joints not prepared by maching or grinding. Internal pores and slag inclusions become increasingly significant in both low and high cycle fatigue when they are close to the surface and especially when finish machining exposes them so they can act as nucleating sources for surface cracking, paragraphs 5.11, 5.12 and 5.26; hence, the value of periodic surface inspection of finish machined parts in such service. In the elastic temperature range the effect of porosity on tensile strength is nil, and on fatigue strength is modest. However, it is much more detrimental at elevated temperatures where pores at the surface particularly influence crack initiation and the associated strain concentration at these locations accentuates crack propagation. The result can be a radical reduction in creep rupture and fatigue strength.[140, 183, 207]

The fatigue strength of defect free butt welds is also dependent upon the amount of overfill or reinforcement of the joint.[70,71,141] The typical manner in which this varies with the angle θ formed between the plate surface and the reinforcement is shown in Fig. 6.36. The maximum fatigue strength is obtained with a flush butt weld, $\theta = 180°$, and decreases linearly for angles less than 180°.

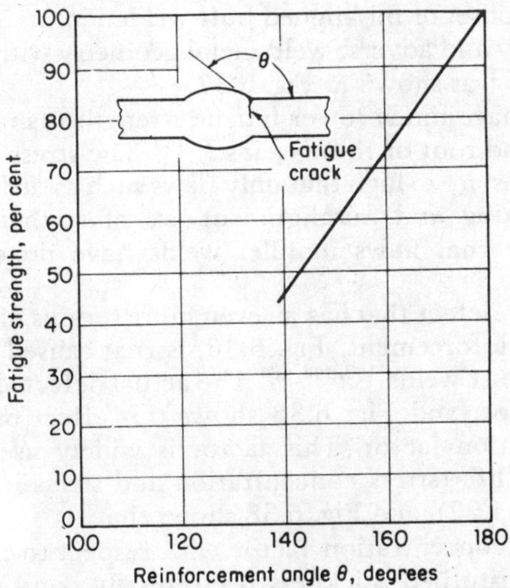

Fig. 6.36. Effect of Weld Overfill (reinforcement) on the Fatigue Strength of Butt Welds[70,71]

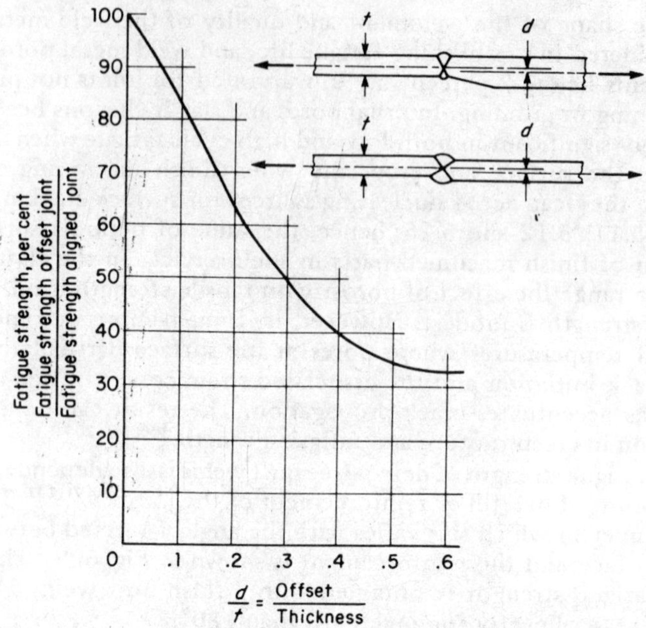

Fig. 6.37. Fatigue Strength of Offset or Misaligned Butt-welded Joints[72]

The effect of offset or misaligned butt-welded joints is to introduce a local eccentricity and adverse weld metal geometry with a reduction in fatigue strength[72] as shown in Fig. 6.37.

Fillet welds have much lower fatigue strength than butt welds and fail at the toe and root of these welds.[73,142] The stress concentrations at these locations are so high that only flaws such as surface cracks, and severe undercutting and misalignment are of further consequence. Accordingly, internal flaws in fillet welds have no effect on their fatigue strength.

One common defect that has an even more serious effect on fatigue life than over-reinforcement, Fig. 6.36, is that caused by incomplete penetration in butt welds.[143, 144, 184] The actual defect shape resembles a square-edge slot, and Fig. 6.38 shows this effect on the resulting stress concentration factor. This factor is widely used in design to evaluate fatigue life (stress concentration and stress intensity are related, Par. 5.12, 5.22), and Fig. 6.38 shows that:

1. The stress concentration factor with respect to the mean stress in the undisturbed thickness is essentially constant and can be taken as 4-5 for ratios of lack-of-penetration void size to full

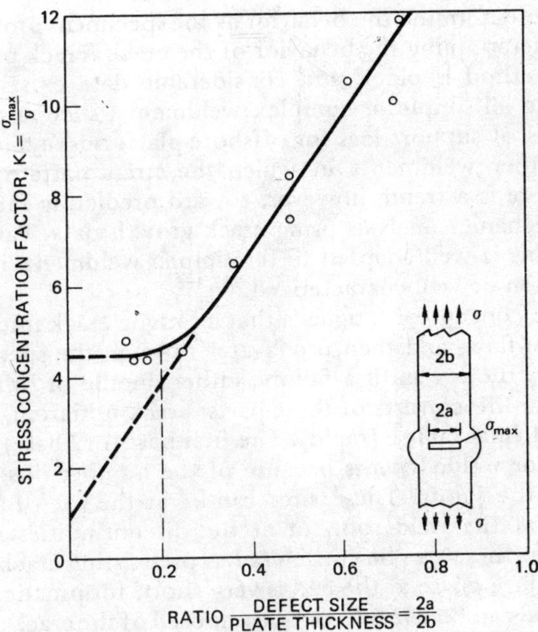

Fig. 6.38. Stress Concentration Factors for Lack-Of-Penetration Defects in Butt Welds[143]

thickness, a/b, up to approximately 0.2. Hence, the initiation period in fatigue will be the same for all defects up to a/b = 0.2.

2. For lack-of-penetration ratios larger than 0.2, the geometry at the ends of the narrow void defect has little effect on the stress concentration factor. The dominant parameter is the void size, and above this ratio the concentration factor increases linearly as the net ligament stress increases.
3. If welds in non-critically loaded areas are to be accepted on a "fitness for purpose" basis, this "acceptable limit" is in the region of a/b = 0.2.

3. *Assessing the Fatigue Behavior of Weldments*

There are two basic ways of assessing fatigue behavior. The first, and by far the most common, is to test a series of specimens to failure and then to analyze the results of these tests in the traditional manner using a diagram showing stress vs. number of cycles (σ–N). The second method is to determine the behavior as the specimens progress toward failure by determining the behavior of the crack (crack propagation). The first method is older, and considerable data exists for it. It is applicable to all simple or complex weldments, such as the multiple intersections of support legs for offshore platforms, cylinder bifurcations, or other weldments in which the stress pattern is not well known. There is a trend, however, toward predicting fatigue life by fracture mechanics analysis using crack growth data, Pars. 5.12 and 5.22, and this is well adapted to the simpler weldments in which the stress field can be well characterized.[185, 186]

The basic concept of fatigue is that a fatigue crack initiates, generally at some flaw, and then propagates through the specimen, component, or structure until a failure, either ductile or brittle, occurs. Thus, fatigue life consists of three parts: crack initiation, crack propagation, and final failure (rapid). The first part (or Phase) is generally very short for welded joints because of the number of sharp discontinuities in the joints. These sites can be at the toe of a weld reinforcement, at the weld root, or at the discontinuities in the weld itself. Hence, these can be considered as preexisting cracks, and since the rapid failure phase at the end is very short, for practical purposes, the entire life can be considered as composed of the crack propagation phase, Par. 5.12.

6.11 Bolted Joints and Gaskets

Many structures, such as buildings, bridges, airplanes, etc., rely wholely upon bolted joints, but in vessels these are confined to flanged closures or covers which must be removable for servicing. In nuclear applications, the number of flanged-gasket joints are generally kept to an absolute minimum because of the difficulty of guaranteeing their leak-proofness, especially under temperature cycling which alternately heats and cools the joint.[187] This design trend also exists for other lethal vessels in the chemical and process industries. Frequently, when bolted joints are employed in nuclear vessels, a welded seal membrane is used alone or in tandem with a gasket (which is intended to act as a flow check in the event of seal membrane failure), Figs. 6.39*a* and *b*. A double annular gasket may also be employed with a space between the two annuli used to monitor and collect any leakage from the inner gasket for safe disposal, Fig. 6.39*c*.

Bolts or studs must be designed to take the full pressure force plus that necessary to maintain gasket compression[147, 148, 210] when these are used. When pressures are high it becomes difficult to get sufficient bolting strength (number of bolts and their cross-sectional area) with

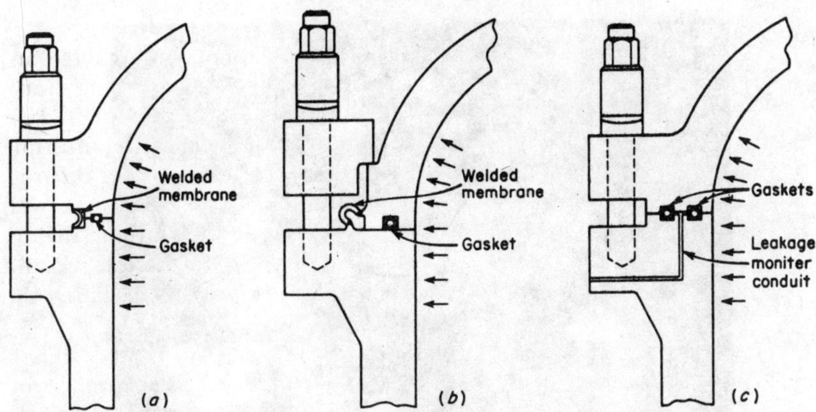

Fig. 6.39. Flanged Vessel Closure Joints of Welded Seal Membrane, and Gasket Designs

conventioanl nuts and bolts on the bolt pitch circle. It then becomes necessary to use sleeve type nuts in order to place them close together, and also to use very high tensile strength bolting material, Fig. 6.40.

Bolts and studs can fail in fatigue just as can any other structural member; in fact, they are more prone to such failure since they are by necessity basically highly stressed, embody high stress concentration factors by virtue of the threads, and are made of high-strength materials which are notch sensitive. Accordingly, the importance of surface finish, cold work, and, most of all, stress concentration, reaches its fullest here.[211]

Analysis of fatigue failures of bolts and their stress distribution by Hetényi[74,75] have shown that these failures occur at three major stress concentration locations; namely, (1) 65 percent at the first thread engagement, (2) 20 percent at the thread-to-shank run-out, and (3) 15 percent at the shank-to-bolt-head juncture, Fig. 6.41. Each of these failure categories is caused by high stress concentrations.

1. *Thread and Nut Design*

The first of these failure sources arises from two conditions: the high stress concentration factor due to the geometry of the thread form itself, and the high unit loading on the first threads of nut engagement as a major part of the load tends to transfer from bolt to nut through these first threads, Fig. 6.42. Obviously, the first can be optimized[76, 77] through profile of thread form by incorporating a generous root radius

Fig. 6.40. Pressure Vessel Flange Showing the Use of Sleeve Type Nuts and Spherical Washers for the large Studs, and Conventional Nuts and Flat Washers for the Small Studs. The Rods Protruding Through the Nuts Extend Through a Central Stud Hole for the Purpose of Measuring Stud Elongation and Stresses

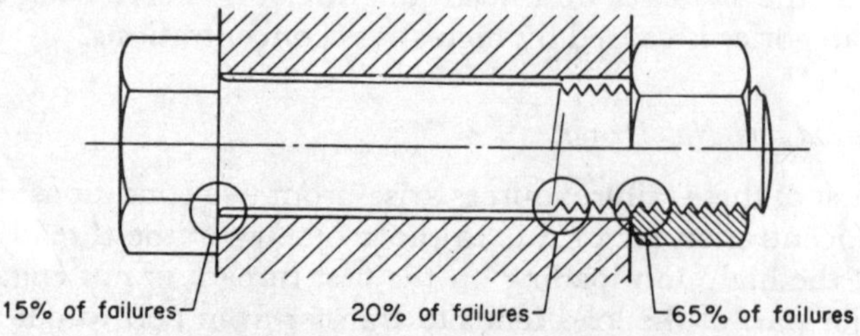

Fig. 6.41. Three Stress Concentration Locations Where Conventional Bolt and Nut Fasteners Most Often Fail[74]

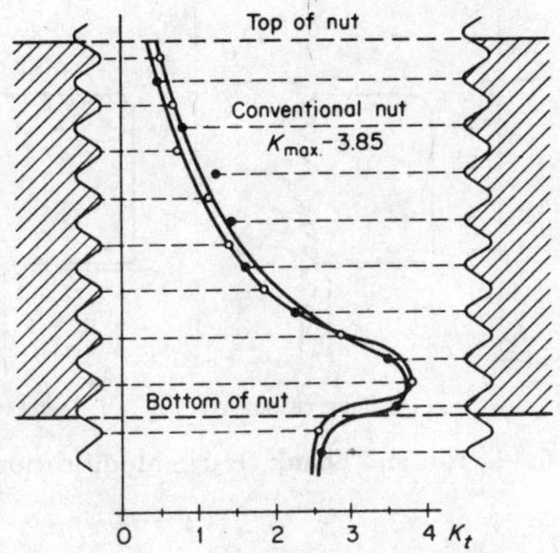

Fig. 6.42. Variation in Thread Root Stress Concentration Factors Throughout the Engaged Threads of a Conventional Nut and Bolt[74]

and thus is the direction that all thread profiles showing increased fatigue life have taken, Fig. 6.43. Whitworth threads have shown increased fatigue life over American Standard threads of 16 per cent[78] because of their rounded roots, where special larger root radii modifications to the American Standard threads have shown still further increases in fatigue life.[79] The second condition can be improved by using very high nuts so that, even though the tendency prevails to load highly the first few threads of engagement, the actual stress on these threads is reduced over that existing for a standard height nut because of the greater total number of threads involved. Fatigue life has been doubled by this means.[80] Another method of countering this

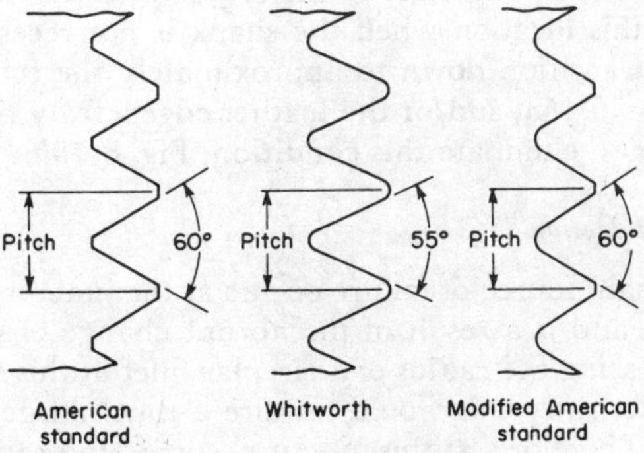

Fig. 6.43. Thread Profiles

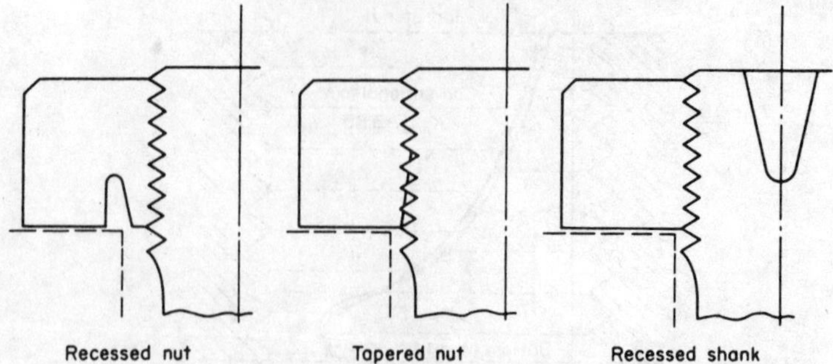

Fig. 6.44. Nut and Shank Design Modifications

effect, Fig. 6.44, is to increase the flexibility of either the nut by recessing it, or the bolt threads by cutting a taper on the first few threads of the nut so the thread bearing takes place farther out toward its peak. Either means introduces more cantilever bending action in these first few engaged threads with the result that more load is transferred to the upper ones. Reducing the net bolt area in the region of the upper threads, or a slight negative lead on the nut threads, will also tend to moderate this effect. Nut and bolt design modifications of these types have shown fatigue life increase of 30 percent or more.[81] Figure 6.56 (given subsequently) shows the application of a deep variable depth thread to the first few threads of a retaining ring for the cover of a 6,000 psi ammonia converter.

2. *Bolt Shank Design*

The second main source of failure occurs in the shank at the thread[82,83] run-out and is directly attributable to the high stress concentration at this location when the shank is not recessed to form a long gradual transition down to approximately the root diameter of the thread, Fig. 6.45*a*, and/or the feather edge left by the first thread is not cut back to eliminate this condition, Fig. 6.45*b*.

3. *Bolt Head Design*

The third main source of failure occurs at the juncture of the shank and bolt head and is a result of the abrupt change of section at this location. Increasing the radius of a circular fillet at this location is not usually possible, since this would require a much larger bolt hole to seat the head. Therefore, at this juncture, compound radii or elliptical profile fillets are used to advantage.[6,43,149] Recessing can also be ap-

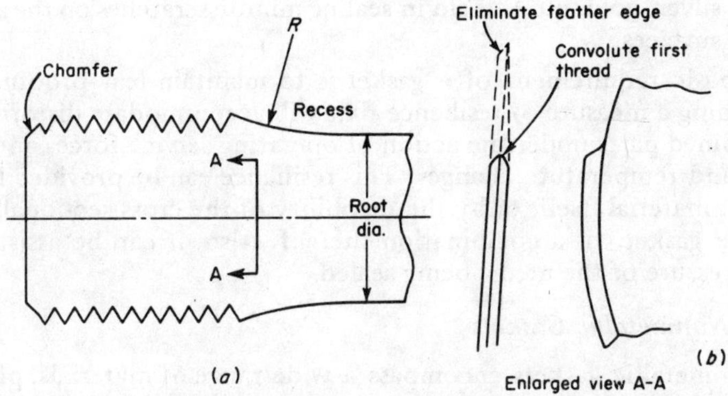

Fig. 6.45. Shank-to-Thread Transition to Minimize Stress Concentration

plied to the bolt head to accommodate this gradual transition, Fig. 6.46.

4. *Gaskets*

A gasket is a device used to maintain a barrier against fluid transfer across mating surfaces such as a pipe flange. Choice of the gasket material is established by the nature of the media being sealed, and of which the prime selection parameter is service temperature.[188, 212]

1. Low temperature service: Nonmetallic material, such as rubber, plastics, paper, and cork, individually or in combination with asbestos products, fibers, etc.
2. High temperature service: Metallic material of flat face, solid ring or resilient shape consistent with the service temperature. In addition, these are normally plated with a soft metal (copper,

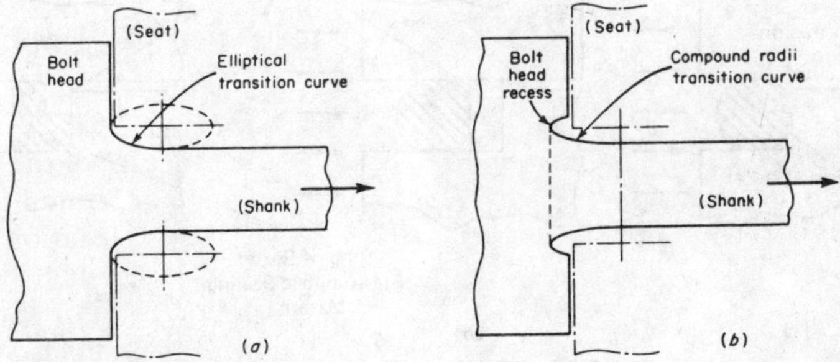

Fig. 6.46. Bolt Head-to-Shank Transitions to Minimize Stress Concentration

silver, gold, etc.) to aid in sealing minute scratches on the seating surfaces.

The basic requirement of a gasket is to maintain leak-proofness by providing a measure of resilience that will accommodate distortion of the joined parts under the action of operating service forces and pressure and temperature changes. This resilience can be provided by the gasket material itself, or by the flexibility of the cross-sectional shape of the gasket, or a combination thereof. Also, it can be assisted by the pressure of the media being sealed.

a. *Nonmetallic Gaskets*

Nonmetallic gaskets encompass a wide range of materials, plain or in metal jackets, with or without corrugations,[189] etc. Compatibility with the service pressure and temperature, and chemical stability in the media environment determine the choice.[190, 191] One of the most versatile of these is the elastomeric O-ring, which is often used in place of other gaskets in flanges, fittings, cylinder end caps, plugs, etc. A flat gasket contained between the flat faces of a pipe flange depends primarily upon the resilience, or spring-back, of the gasket material itself to maintain tightness. But if a round elastomeric gasket is confined in a recess in the flange face, the media pressure will try to extrude it out, thereby effecting a seal, Fig. 6.47*a*. The tightness of the seal is dependent upon the viscous strength of the gasket material and the exit gap created by machining tolerances and separation of the flange and shoulder. This extrusion opening can be further decreased by the use of a second, precisely dimensioned ring of a more viscous material, such as nylon or even metal, to restrict the extrusion of the softer O-ring material sealing the fitting, Fig. 6.47*b*

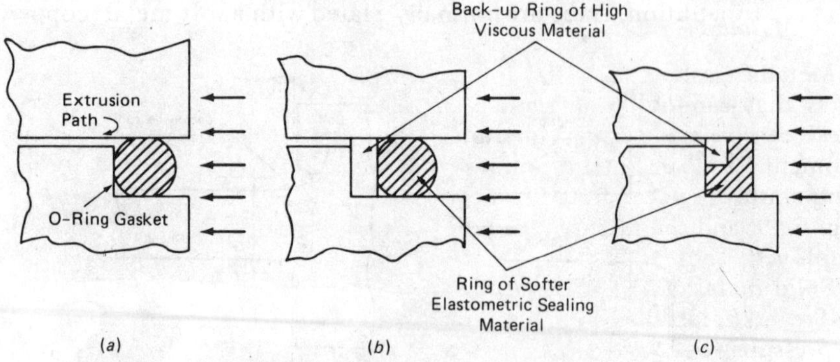

Fig. 6.47. Nonmetallic O-Ring Gaskets

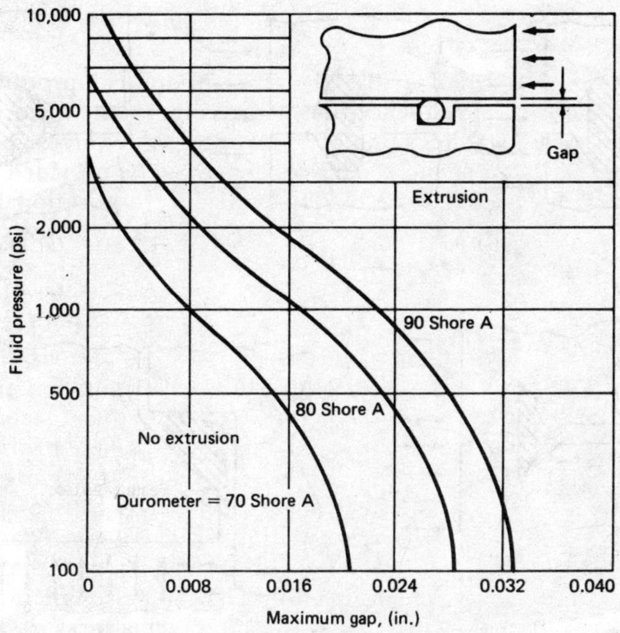

Fig. 6.48. Maximum Permissible Gap for Plastic O-Ring Construction[192]

and *c*. Pressures to 30,000 psi and above can be sealed in this manner; and because they are activated by the media pressure, flange contact pressures need not be as high as with conventional flat gaskets. Figure 6.48 shows the relationship of the gap and extrusion pressure with material viscosity.[192]

b. *Metallic Gaskets*

Metallic gaskets are used for high pressure and for high temperatures that cannot be adequately handled by nonmetallic ones, and these can be made of any material compatible with the service environment.[193, 194] All-metal joining rings or gaskets are required where temperatures exceed 930°F. They are available in a variety of shapes, Fig. 6.49, and their temperature limit depends upon the type of metal employed, Table 6.2.

Solid metal gaskets of oval, octogonal, and lens cross-section, Fig. 6.49*a* and *b*, installed in tapered or spherical seating surfaces, are used in pressure vessel and piping systems to 10,000 psi and above. Generally, these gaskets form a line contact seal, and the ring gasket should

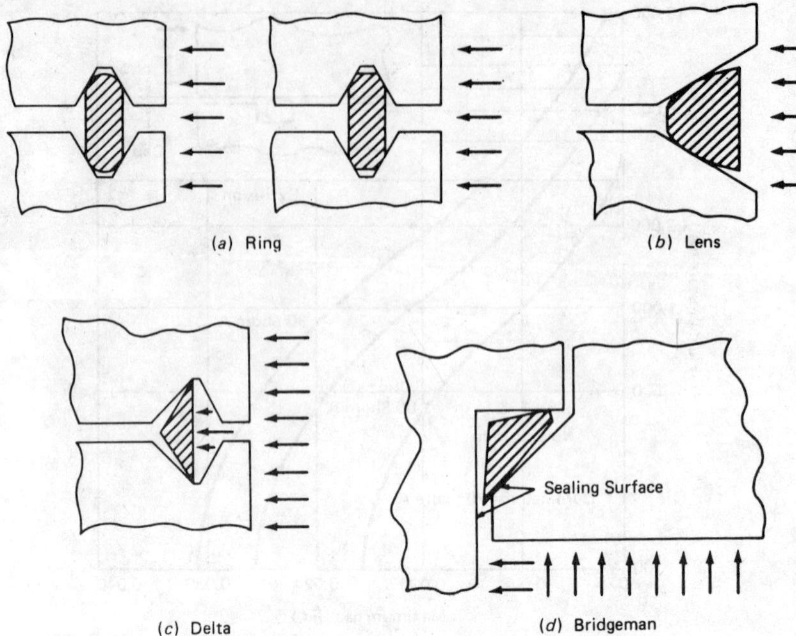

Fig. 6.49. Types of Solid Metal Gaskets. (*a*) Ring. (*b*) Lens. (*c*) Delta. (*d*) Bridgeman

be made of a softer material than the seating surface. In this range, these rings have heavy cross sections, require flange grooves of accurate location and tolerance, and do not respond rapidly to movements and distortion of the flange joint. However, when they are hollowed out to permit the media pressure to actuate and aid them in following movements, they are used for pressures of 20,000 psi and above.

Two frequently used solid ring gaskets that take advantage of the media pressure to aid seating are the Delta and Bridgeman type. The Delta gasket, Fig. 6.49*c*, has a triangular cross-section and requires precise control of gasket and groove dimensions. When pressure is applied, the gasket flexes about the two contact points and the sides wedge into the groove. They are mostly used in vessels and valve bonnets for pressures above 5,000 psi.

Bridgeman gaskets, Fig. 6.49*d*, are used in vessels, valves, etc., at pressures upwards of 1,500 psi, and these gaskets and their seats likewise demand fine machining tolerances.[195] They utilize the force resulting from the pressure applied to the closure to seat the gasket.

TABLE 6.2. METAL GASKET AND PLATING MAXIMUM TEMPERATURE LIMITS

Metals for Spiral-Wound and Jacketed Gaskets		*Metal O-Rings*		*Plating Materials for Metal O-Rings*	
Material	*Maximum temperature (°F)*	*Material*	*Maximum Temperature (°F)*	*Material*	*Maximum temperature (°F)*
Lead	210	Copper	750	Indium	270
Brass	480	Mild steel	1,020	Cadmium	390
Aluminum	750	Cupro-nickel	1,110	Silver	1,470
Stainless steels	930–1,560	Monel	1,110	Gold	1,560
Titanium	930	Nickel	1,290	Plastic	570
Nickel	1,380	Stainless steel	1,470		
Inconel	1,830	Inconel	1,560		
Hastelloy	1,830				

It is a non-bolted closure, Par. 6.13. In order to facilitate removal of these gaskets from the right-angle seat upon disassembly, their cross-section mating angle is rotated so as to permit the gasket to spring back to its original position upon release of the seating force; thereby, forestalling gasket sticking, Fig. 6.49*d* and Fig. 6.56*a*.

c. *Resilient Metal Gaskets*

Solid metal gaskets generally require thick flanges. Thinner flanges can be used with resilient metal gaskets. Resilient metal gaskets combine the efficiency of elastomeric O-rings with the temperature capability of metal gaskets. They get their resilience by virtue of their structural or geometric configuration. Accordingly, the basic element in a high strength structural element is a flexible cross section. These may be coated or plated with a soft metal or plastic to enhance the actual surface contact sealing. When coatings are not employed, seating surfaces require at least a 32 micro-inch finish; whereas, coatings bring this requirement to a 32–90 micro-inch finish range. Spiral-wound gaskets are an example of this type that have good resilience and sealability, Fig. 6.50. They consist of V-shaped strips of metal interwound in a spiral with an asbestos or plastic filler. These general purpose gaskets are widely used in metal-to-metal joints, in recesses, or with flat seats in combination with inner and/or outer gasket retainer-compression rings which act to space the gasket relative to the bolting and also limit the amount of applied compression. The temperature limits for the materials in spiral-wound or jacketed gaskets are shown in Table 6.2. Filler materials also affect maximum gasket temperature; for instance, asbestors loses its water of hydration at 790–930°F, and with plastic fillers the limit is in the range of 500–550°F.

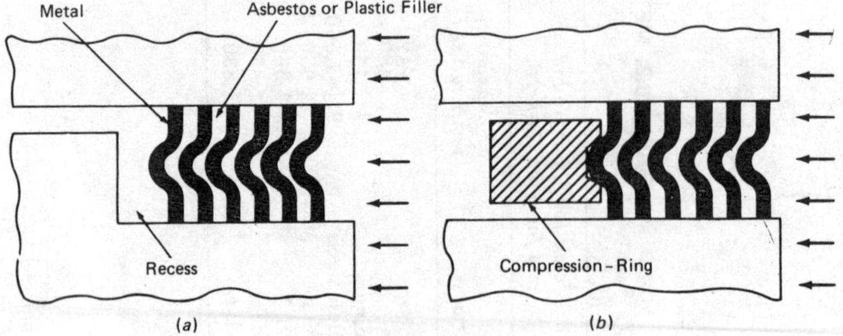

Fig. 6.50. Spiral-Wound Gasket

Hollow metal O-rings are also a resilient-type gasket that are widely used for sealing flanges, fittings, etc. They are constructed by welding together the ends of a piece of bent small-diameter tubing so as to form a hollow ring. They are compressed a predetermined amount to obtain maximum elastic resilience. Unplated gaskets are satisfactory in many liquid sealing applications, but platings are needed for some liquids and most gas applications. This type of gasket is usually used in full or semi-confined grooves in the flange face or seating surface. They are availabe in three basic types: plain, self-energized and pressure-filled. All three types can be used at cryogenic and elevated temperatures consistent with the chosen material and/or plating, Table 6.2.

1. Plain O-rings are used in both fully confined and semi-confined joints, Fig. 6.51*a*. In confined joints, these O-rings have been used from cryogenic temperatures of −420°F to elevated temperatures of +3,000°F; while in semi-confined joints, these temperatures are somewhat reduced.[189]
2. Self-energizing O-rings, so-called because they utilize vent holes or slots on the pressure side of the ring to permit equalization of pressure on the inside of the ring with the system service pressure, thereby forestalling buckling collapse of the O-ring, are a widely used type, Fig. 6.51*b*. Increasing the system pressure forces the ring further into the groove.

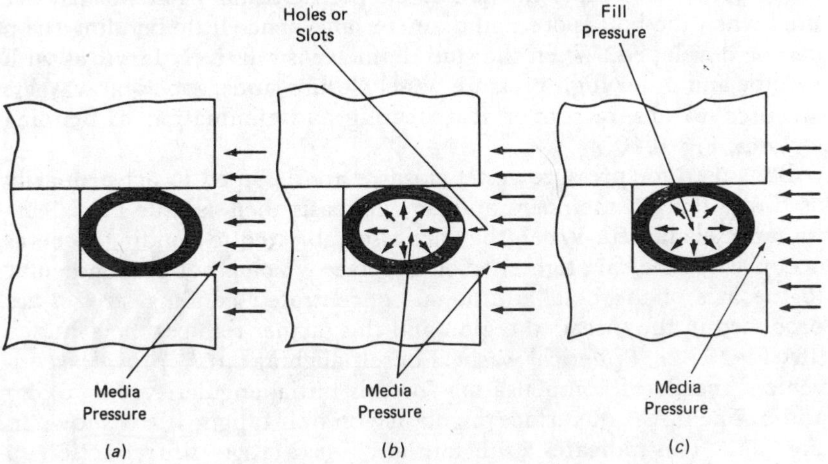

Fig. 6.51. Metallic O-Ring Gaskets. (*a*) Plain. (*b*) Self-Energized. (*c*) Pressure-Filled

3. Pressure-filled O-rings are filled with inert gas at about 600 psi, Fig. 6.51*c*. At high temperatures, this gas pressure increases, to some extent offsetting loss of tubing strength and increasing resilience of the gasket. Pressure-filled O-rings cannot support pressures as high as self-energized ones, but have special useful applications at temperatures from 800–1,500°F.

d. *Gasket Design*

As a rough gasket design guide, if the product of the operating pressure in psi and temperature in degrees F exceeds a value of 250,000, metal gaskets should be used. Metal gaskets are also used for temperatures below −65°F and above 850°F, and for pressure above 1,200 psi. Other operating conditions calling for metallic gaskets include vacuum below 10^{-4} torr and radiation above 10^{8} rad. These guides are useful design criteria, but the final choice of gasket material and type is largely established by trial-and-error proof testing, with the latter minimized by experience and experiment.[206, 208]

6.12 Bolt Seating and Tightening

Washers, made of a material harder than either the nut or seat, are used under the nuts of pressure vessel bolts* in order to forestall gouging or biting into the seating surface when the nut is turned and/or reducing the bending stress resulting from the seating surface not being perpendicular to the axis of the bolt. Ordinary flat washers are used when the bolt is of small diameter and hence little bending stress can be developed. When the stud diameter is relatively large, as with reactor and other high pressure vessel closure studs, spherical washers are used to ensure proper seat bearing and elimination of bending stresses, Fig. 6.40.

Bolts used for pressure vessel closures are designed to act primarily as direct tension members and on this basis their fatigue life is customarily evaluated. When they are also subjected to bending, such as occurs when they are tightened on a surface which is not perpendicular to the axis of the bolt, additional concentrated bending stresses are produced in the threaded region and this further reduces their fatigue life, Fig. 6.52*a*. Spherical washers or self-aligning nuts[84] provide a convenient means of compensating for this initial angularity, Fig. 6.52*b* and *c*. The effect of surface angularity on bolt fatigue life is shown in Fig. 6.53. This indicates a substantially linear fatigue life reduction up

*Either bolts or studs are implied.

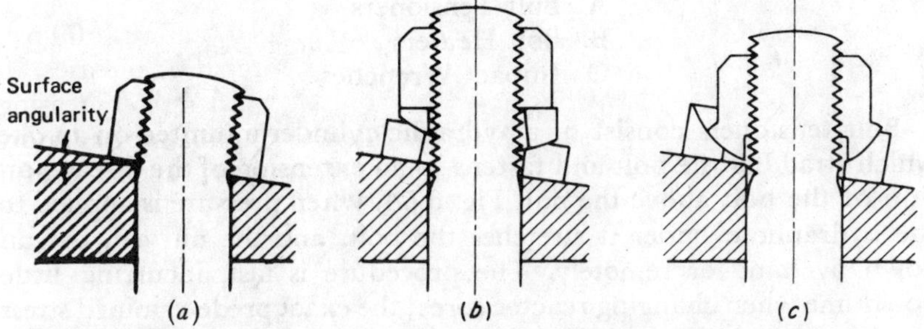

Fig. 6.52. Effect of Surface Angularity on Bolt Seating. (*a*) Conventional Nut Seat, (*b*) Spherical Mated Washer Seat, (*c*) Self-aligning Nut Seat

to a value of 50 percent for a misalignment of 2°. Above this the reduction increases rapidly.

It is difficult to develop the full strength of bolts larger than 2 in. diameter by hand wrenches. Beyond this size there are three ways, all of which have been employed to tighten the large size cover-head studs of nuclear reactor vessels, that can be used:

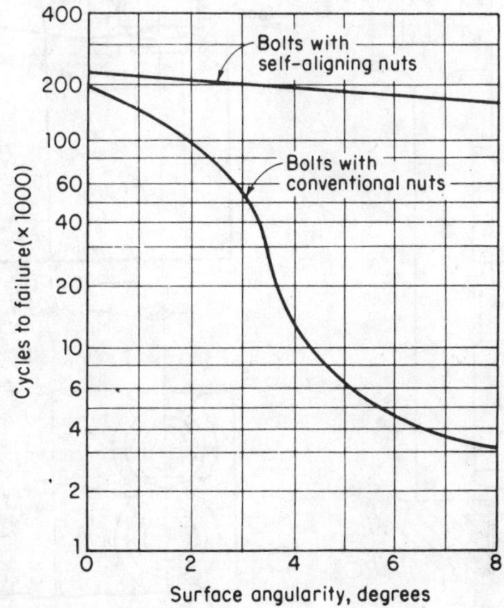

Fig. 6.53. Effect of Surface Angularity on the Fatigue Life of High-Strength Bolts.[84] (Test Cycle Stress Range: 80,000 psi maximum, 8,300 psi minimum.)

A. Bolt Tensioners
B. Bolt Heaters
C. Impact Wrenches

Bolt tensioners consist of a hydraulic cylinder mounted on a yoke which straddles the bolt and fastens to an extension of the thread portion of the bolt above the nut, Fig. 6.54. When pressure is applied to the hydraulic cylinder it stretches the bolt, and the nut can be run down by hand or remotely. The procedure is fast, incurring little downtime when changing reactor cores; the exact predetermined stress can be put into the studs by cross calibration with the bolt tensioner hydraulic pressure; no residual torsion stress is left in the bolts as with wrench tightening; and the operation can be done under water which is a radiation shielding requirement with some nuclear reactors.

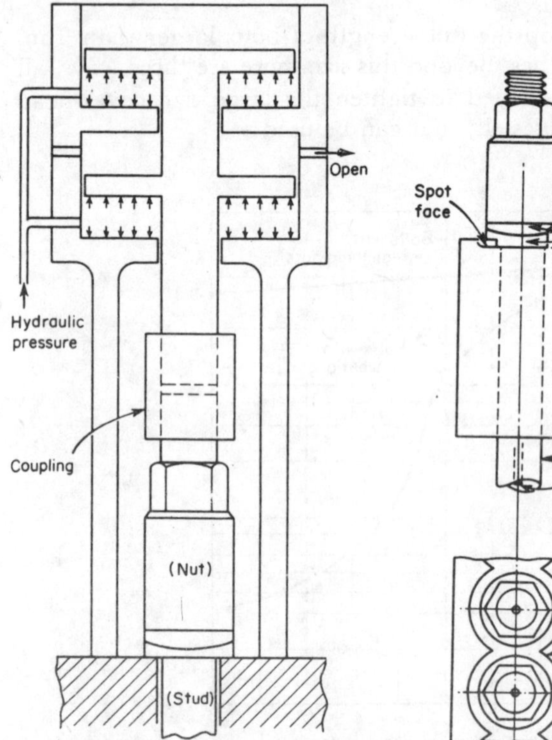

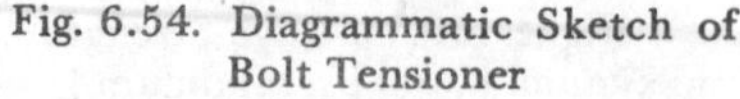

Fig. 6.54. Diagrammatic Sketch of Bolt Tensioner

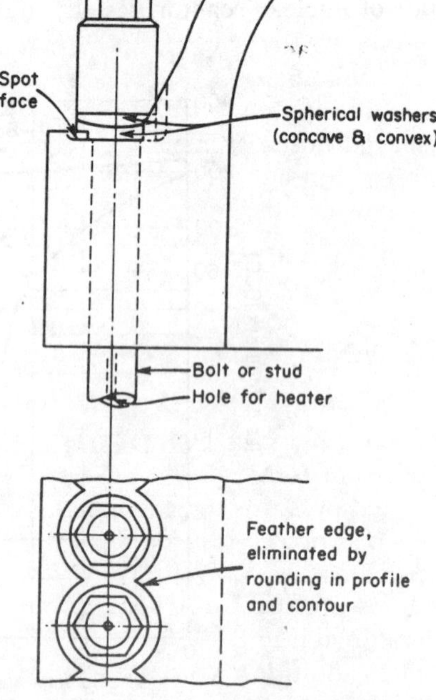

Fig. 6.55. Feather Edge Condition Occurring at Overlapping Spot Faces on Vessel Flanges

Bolt heaters consist of an electrical heating element which is placed in a small central hole running the length of the bolt, Fig. 6.55. When the bolt expands under heat, the nut is run down by hand. The only drawback to this method is the long downtime required; since the bolt assembly must be allowed to cool in order to check the bolt stress and with the associated heat loss to the surrounding seat, several adjustments are usually required. Also it cannot be used under water on nuclear vessels when this is a radiation safeguard requirement.

Impact wrenches induce bolt tension by turning the nut under load; hence, they introduce the problem of thread galling and leave a residual shear stress in the bolt. To prevent galling and reduce wrench turning friction, which may consume as much as 90 percent of the applied torque, lubricants may be used. These reduce friction by separating the sliding surfaces. Oil and grease are the most common lubricants but they are often inadequate, particularly at high temperatures, because they oxidize, thereby losing their lubricity. For applications where lubricants must remain stable at high temperatures, solid lubricants are an alternative. These include organic fluoropolymers[196] such as polytetrafluoroethylene (PTFE) and fluorinated ethylene propylene (FEP) for temperatures to 500°F, and inorganic materials such as molybdenum disulfide for temperatures to 900°F and colloidal graphite for higher temperatures.[85]

Occasionally it is necessary to spot face flanges to receive closely spaced bolts. This is in a region of high stress due to the abrupt change in section and discontinuity stress existing at this location. Hence, it is particularly important to avoid any further stress concentrations such as can arise from the feather edges of overlapping spotfaces, Fig. 6.55.

6.13 Non-Bolted Closures

When pressures are so high that it becomes impossible to get sufficient bolts in the pitch circle to take the pressure and gasket compression force,[150,151,152,153] other mechanical designs utilizing the internal pressure to seat the gasket and employing other type fasteners such as breech-blocks, shear pins, split-rings, etc., to take the pressure force are used.[86,87,88,89,154,155,156] This type of joint is often called a "pressure seal" or "Bridgeman"[90] joint, a typical example of which is shown in Fig. 6.56.

In addition to providing adequate mechanical strength in a closure, leak tightness must also be embodied. Since the inherent resilience of gaskets, particularly metallic ones which are frequently required for high temperatures or other environmental effects, is limited, designs

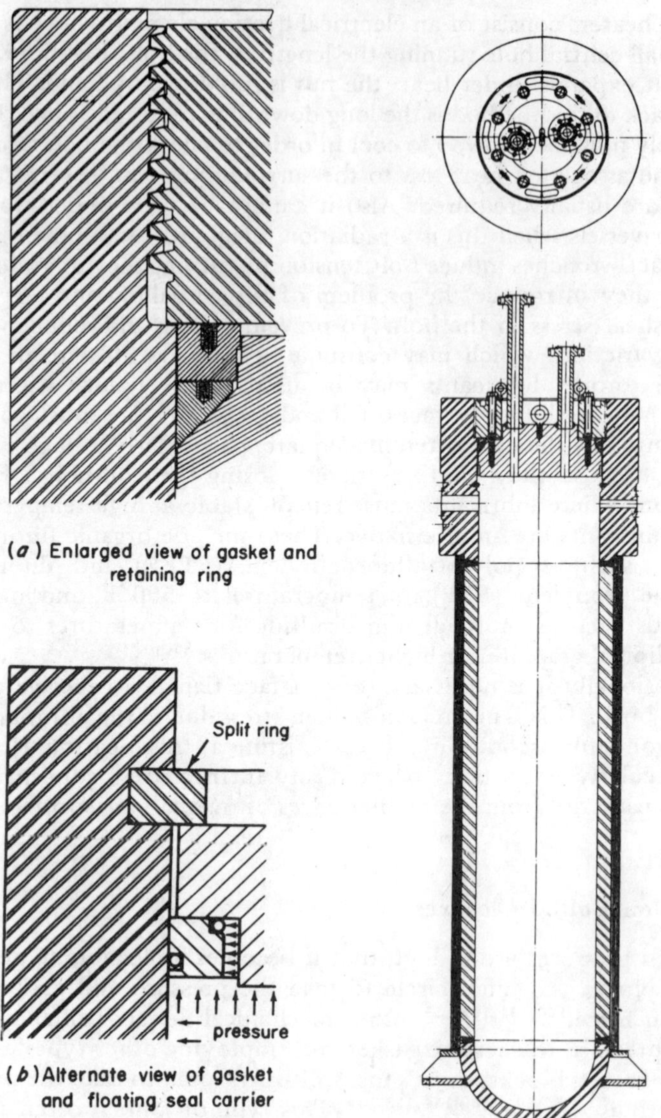

Fig. 6.56. Pressure Seal Closure for High-Pressure, Full, Open-End Vessel

that use the contained pressure to force the gasket to follow the dilation of the seating surfaces are often employed. One of the numerous gasket design and arrangements for accomplishing this is the triangular wedge gasket shown in Fig. 6.56*a* which is designed to rotate under the internal pressure load transmitted to it from the head closure, hence, it follows the dilation of the vessel wall which is subjected to the same internal pressure thereby maintaining the gasket seal. Another type of gasket arrangement for accomplishing this is the deflection compensating seal shown in Fig. 6.56*b*. This uses a pair of O-ring gaskets in a floating seal carrier. The internal pressure forces the seal carrier to expand and follow the dilation of the inside diameter of the vessel thereby precluding differential movement of the gasket seating surfaces hence maintaining tightness.

When gasket closure sealing arrangements are subject to wide temperature differences across the gasket seating surfaces or large thermal shocks, such that the inherent resilience of the gasket provided by its material properties and geometric cross section is insufficient to maintain gasket seat contact, welded membrane seals, Fig. 6.39, are used to insure a leakproof closure.

6.14 Vessel Supports and Attachments

1. *Design Approaches and Loadings*

Vessels may be supported by hangers, saddles, skirts, multiple brackets or nozzles, or combination of these,[91, 92, 93, 157] Fig. 6.57. Skirt construction permits radial growth of the pressure vessel due to pressure and temperature through bending of the skirt in the manner of a beam on an elastic foundation, and its length is chosen so as to permit this bending to take place safely. Bracket or heavy nozzle supports, on the other hand, provide for radial growth of the vessel by sliding on lubricated surfaces or radial pins, or via rockers or rollers. When these create external bending moments in the vessel, their effect may be evaluated by the method of Chapter 4 for continuous skirts, or by the method of Bijlaard for local discontinuous supports, paragraph 6.16.

Support brackets are widely used for land installations, but on shipboard where high horizontal and vertical shock loadings, as well as roll and pitch are involved, skirt supports must be used. When available space limitations allow insufficient length of skirt for flexing, a hybrid construction consisting of a partially longitudinally slotted skirt can be used, Fig. 6.58. The slotted portion acts as a multitude of cantilevers, while the unslotted portion behaves as a cyliner under the imposition of the moments and forces transmitted by the series of

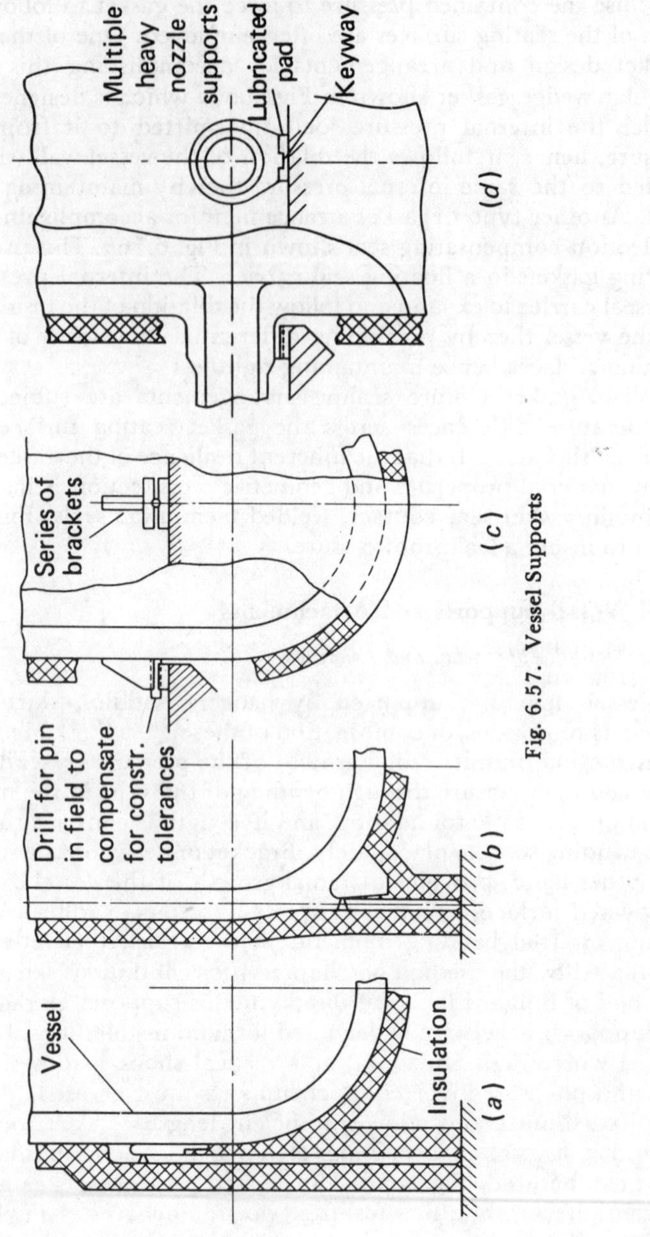

Fig. 6.57. Vessel Supports

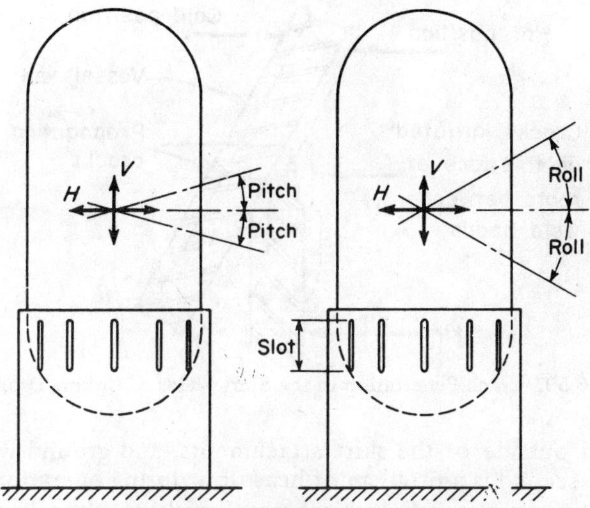

Fig. 6.58. Vessel Slotted Skirt Support

cantilever beams.[94] The narrow slots are placed uniformly around the circumference and terminate in drilled holes in order to minimize stress concentrations at the change of section from slotted to solid construction. They are also spaced so as to give relatively wide cantilevers in order to provide adequately for horizontal shear.

The main problem in the design of support skirts is a thermal stress problem introduced by the temperature gradient of the skirt at its contact with the vessel, and its difference in temperature from that of the cold support base. This condition imposes tension bending stresses on the outside of the skirt juncture during heatup of the vessel of a magnitude dependent upon the severity of the thermal gradient along the length of the skirt—the steeper this gradient the higher the stress. This type of problem is often encountered in the skirt attachment of coking drums which receive frequent severe heating and cooling cycles.[95] Figure 6.59 shows such a typical skirt attachment and the fatigue cracks that originate on the tensile face of the weld at weld bead roots and run-out, and propagate through the weld and into the vessel wall thickness. To avoid this type of fatigue failure in vessels it is necessary to avoid attachments embodying high stress concentrations that make it susceptible to high fatigue stresses, and to minimize the thermal stress by reducing the temperature gradient[96] along the length of the skirt. The first is essentially a geometric problem that can be optimized by using full penetration weldments, transitions of ample radius both

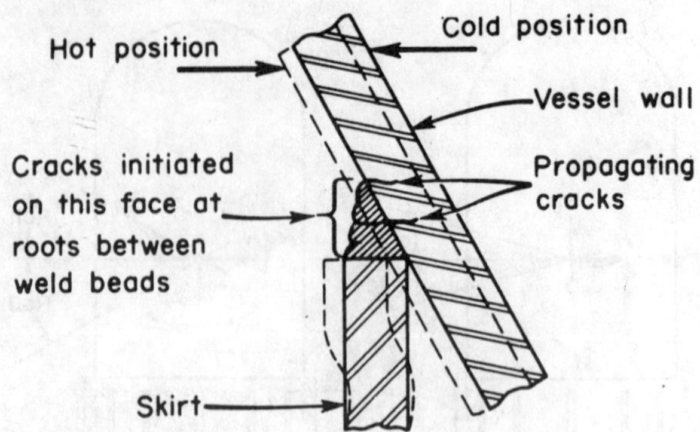

Fig. 6.59. Crack Penetration in the Skirt Welds of Coking Drums[95]

inside and outside of the skirt attachments, and ground welds, Fig. 6.57. The second is a problem of heat flow during operation, and the most effective ways to reduce this thermal gradient at the skirt juncture is by (1) full penetration weldments at the skirt-to-shell junction to permit maximum heat flow by conduction through the metal at this point, and (2) selective use or removal of insulation in the crotch region so as to permit heat flow from shell to skirt by both convection and radiation, Figs. 6.57*a* and *b*. Polished reflectors and high heat conductivity metals, such as copper and aluminum, have been used at this location to improve this condition.

2. *Axial Thermal Stresses in Support Skirts*

If a vessel is brought up to operating temperature uniformly it will undergo thermal expansion, and if there is no restraint to this expansion no thermal stresses[158] will result. If there is a change in temperature along its length, thermal stresses will be induced in the region of the change.[97,98] This is the problem that is encountered in the design of cylindrical skirts for the support of vessels which have a different temperature than that of the foundation, Fig. 6.57*a*, *b*. In this case the metal in the skirt and in the vessel at their juncture have the same temperature. However, from the juncture to the foundation an axial thermal gradient will exist and produce a varying radial thermal expansion, and it is this differential expansion between adjacent elements or parts that causes the thermal stress.

(a) *Linear Axial Thermal Gradient in a Cylinder*

If the temperature is constant through the thickness of a cylindrical vessel wall but varies along its length, the resulting thermal stress can

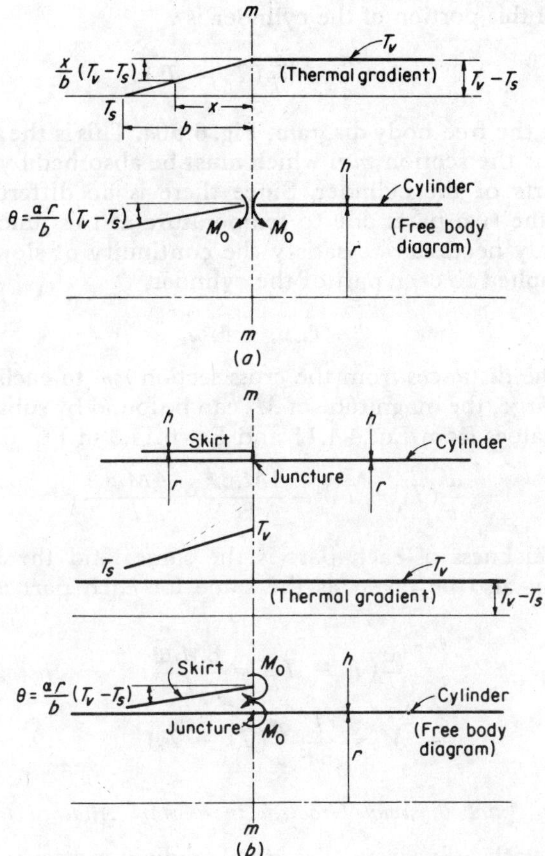

Fig. 6.60. Axial Thermal Gradients. (*a*) Linear Axial Gradient in a Cylinder, (*b*) Linear Axial Gradient in a Skirt Attached to a Cylinder

be determined by use of the elastic foundation equations of Chapter 4. Consider as an example the long cylinder shown in Fig. 6.60*a* in which the temperature to the right of section *mm* is constant, T_v, while that to the left decreases linearly to a temperature T_s at the end $x = b$ so that the difference in temperature between any location in this part and the constant temperature part is

$$t_x = \frac{x}{b}(T_v - T_s) \tag{6.14.1}$$

The accompanying differential displacement is

$$y = \frac{\alpha r x}{b}(T_v - T_s) \tag{6.14.2}$$

The slope of this portion of the cylinder is

$$\theta = \frac{dy}{dx} = \frac{\alpha r}{b}(T_v - T_s) \tag{6.14.3}$$

as shown in the free body diagram, Fig. 6.60*a*. This is the angle of discontinuity at the section *mm* which must be absorbed by rotation of the two parts of the cylinder. Since there is no differential radial growth of the two parts due to temperature at this juncture $P_0 = 0$ and it is only necessary to satisfy the continuity of slope by a moment M_0 applied to each part of the cylinder,

$$\theta = \theta_{s,M_0} + \theta_{v,M_0} \tag{6.14.4}$$

Assuming the distances from the cross section *mm* to each end of the cylinder is large, the magnitude of M_0 can be found by substituting the respective values from Eq. 4.4.11 and Eq. 6.14.3 in Eq. 6.14.4

$$\frac{\alpha r}{b}(T_v - T_s) = \frac{4M_0\beta_s^3}{k_s} + \frac{4M_0\beta_v^3}{k_v} \tag{6.14.5}$$

Since the thickness of each part is the same, and the modulus of elasticity can also be taken as the same for each part $\beta_s = \beta_v$ and $k_s = k_v$ so

$$\frac{\alpha r}{b}(T_v - T_s) = \frac{8M_0\beta^3}{k} \tag{6.14.6}$$

$$M_0 = \frac{k}{8\beta^3}\frac{\alpha r}{b}(T_v - T_s) \tag{6.14.7}$$

(b) *Linear Axial Thermal Gradient in a Skirt Attached to a Cylinder*

If a skirt with a linear axial thermal gradient is attached along the length of a long cylinder of equal thickness and uniform temperature, Fig. 6.60*b*, each have the same temperature at the juncture point. The juncture connection between skirt and cylinder is short and rigid so that the radius of skirt and cylinder may be assumed equal. In this case there is also no differential deflection at the juncture due to temperature so $P_0 = 0$, and the continuity rotation of the skirt as given by Eq. 4.4.11 and cylinder as given by Eq. 4.3.22

$$\theta = \theta_s + \theta_v \tag{6.14.8}$$

$$\frac{\alpha r}{b}(T_v - T_s) = \frac{4M_0\beta_s^3}{k_s} + \frac{M_0\beta_v^3}{k_v} \tag{6.14.9}$$

$$\frac{\alpha r}{b}(T_v - T_s) = M_0\left[\frac{4\beta_s^3}{k_s} + \frac{\beta_v^3}{k_v}\right] \tag{6.14.10}$$

For equal thickness of skirt and cylinder $\beta_s = \beta_v$ and $k_s = k_v$, so

$$M_0 = \frac{k}{5\beta^3}\frac{\alpha r}{b}(T_v - T_s) \tag{6.14.11}$$

When the thickness of the cylinder is appreciably thicker than the skirt, the rotation of the cylinder is nil and the skirt must absorb the entire rotation given by the thermal gradient Eq. 6.14.3. This means that the second term in the right-hand side of Eq. 6.14.10 is small compared to the first term and can be neglected. This gives

$$M_0 = \frac{k}{4\beta^3}\frac{\alpha r}{b}(T_v - T_s) \tag{6.14.12}$$

which is a 25 percent increase in the moment occurring on the skirt as compared to that when the skirt and cylinder are of equal thickness, Eq. 6.14.11. The maximum thermal stress in the skirt is

$$\sigma_{\text{max.}} = \frac{6M_0}{h^2} \tag{6.14.13}$$

(c) *Nonlinear Thermal Gradient in a Skirt Attached to a Cylinder*

The method used above for the calculation of thermal stresses in the case of a linear thermal gradient can also be used when the gradient is other than linear. Wolosewick[99] has given an approximate equation for the axial temperature distribution in vessel support skirts which are insulated on both the inside and outside as

$$T_x = (T_v - 50) - 6.037x - 0.289x^2 + 0.009x^3 \tag{6.14.14}$$

This states that the temperature is best described by a temperature difference at the juncture of skirt and cylinder of 50°F in addition to an axial gradient. Its differential radial deflection relative to the cylinder is

$$y = \alpha r(T_v - T_x) \tag{6.14.15}$$

Combining Eqs. 6.14.14 and 6.14.15 gives

$$y = \alpha r(50 + 6.037x + 0.289x^2 - 0.009x^3) \tag{6.14.16}$$

and differentiating gives the slope of the skirt

$$\theta = \frac{dy}{dx} = \alpha r(6.037 + 0.578x - 0.027x^2) \tag{6.14.17}$$

At the juncture of the skirt and cylinder $x = 0$,

$$y = 50\alpha r \tag{6.14.18}$$

and

$$\theta = 6.037\alpha r \tag{6.14.19}$$

If the cylinder is appreciably thicker than the skirt, as is the usual condition with nuclear and process equipment, the cylinder may be assumed rigid so that both the total thermal differential expansion and thermal rotation at the skirt juncture must be absorbed by the skirt itself. The deflection continuity condition at the juncture then from Eqs. 4.4.10 and 6.14.18 becomes

$$50\alpha r = \frac{2P_0\beta}{k} - \frac{2M_0\beta^2}{k} \tag{6.14.20}$$

and the slope continuity condition from Eqs. 4.4.11 and 6.14.19 is

$$6.037\alpha r = \frac{-2P_0\beta^2}{k} + \frac{4M_0\beta^3}{k} \tag{6.14.21}$$

The simultaneous solution of Eqs. 6.14.20 and 6.14.21 gives the value of the force P_0 and moment M_0 at the juncture induced by the thermal gradient as

$$M_0 = \frac{\alpha rk(50\beta + 6.037)}{2\beta^3} \tag{6.14.22}$$

$$P_0 = \frac{\alpha rk(100\beta + 6.037)}{2\beta^2} \tag{6.14.23}$$

The thermal stresses in the skirt can be found in the manner described in paragraph 4.8.2. Figure 6.61 shows a support skirt attached to a steam generator.

3. *Saddle Support of Horizontal Vessels*

Saddle supports for horizontal vessels are located so as not to:

1. Overstress the vessel in the longitudinal direction as it behaves as a beam in bending, Fig. 6.62*a*; and
2. Overstress the vessel in a circumferential direction in the region of the supports as the load from the vessel is transferred to the saddle support, Fig. 6.62*b*.

Most vessels, boiler drums, heat exchangers, etc. employ two supports because any settlement of the supporting structure does not change

Fig. 6.61. Support Skirt Attached to the Bottom Spherical Head of a Steam Generator (*Courtesy of The Babcock & Wilcox Co.*) The Vessel Will Be Up-Ended to Rest on the Circular Skirt Support

the distribution of the load to each saddle. Extremely long vessels, such as penstocks and pipelines, use multiple supports and the support reactions are computed as with continuous beams. These reactions are normally increased by 20-50 percent to compensate for potential differential settling of multiple supports.[197]

The total longitudinal stress in a horizontally supported cylindrical

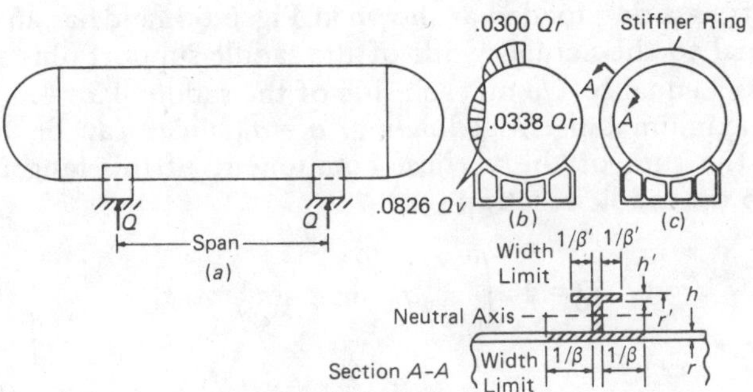

Fig. 6.62. Horizontal Vessel Saddle Support

vessel is composed of that due to the internal pressure stress, $pr/2h$, and that due to the (beam) bending moment, M/Z, where the section modulus for a thin shell of thickness, h, can be taken as $\pi r^2 h$:

$$\sigma = \frac{pr}{2h} \pm \frac{M}{\pi r^2 h} \tag{6.14.24}$$

A frequently occurring case of Eq. 6.14.24 is that in which the cylindrical shell thickness is established by the maximum hoop stress due to the pressure. Since the longitudinal pressure stress is one-half the hoop stress; the remaining one-half of the vessel thickness is available for the longitudinal bending stress due to the weight and other loadings on the vessel,

$$\sigma = \pm \frac{M}{\pi r^2 (h/2)} \tag{6.14.25}$$

or the permissible bending moment is

$$M = \pm \frac{\sigma \pi r^2 h}{2} \tag{6.14.26}$$

From which the maximum span between supports can be determined.

Saddle supports partially encompass the vessel circumferentially. The vessel behaves as a beam between the saddle supports, and the load (weight of the vessel, contents, and other imposed loads or forces) is transferred to the saddle through the wall of the vessel. In thin vessels this is by shear acting tangential to the circumference of the vessel and varying as the sine of the angle φ measured from the top, Fig. 6.63, where it is zero, to a maximum at the horizontal centerline. This is analogous to an I-beam in which the load is carried by shear through the thin web, and is a maximum at the centerline of the web. The result is that a section of the cylindrical vessel in the region of the saddle acts as a ring loaded as shown in Fig. 6.63, and has an effective width equal to the actual width of the saddle support plus a portion of the vessel equal to $1/\beta$ on each side of the saddle, Par. 4.6.

The maximum tangential loading, $q = q_0 \sin \varphi$, can be found by equating the sum of the vertical component of this tangential sine loading to the saddle reaction Q:

$$Q = 4 \int_0^{\pi/2} q_0 \sin \varphi \sin \varphi r \, d\varphi \tag{6.14.27}$$

$$q_0 = \frac{Q}{\pi r} \tag{6.14.28}$$

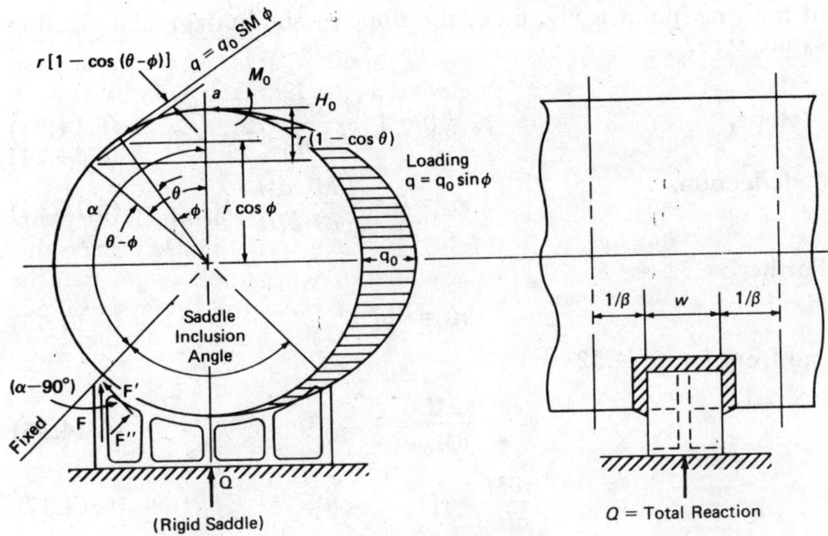

Fig. 6.63. Rigid Saddle

This loading causes tangential bending in a ring section of the cylindrical vessel, Fig. 6.63. The bending moment due to this loading is

$$dM_q = qr[1 - \cos(\theta - \varphi)]\, r d\varphi \tag{6.14.29}$$

Substituting $q = q_0 \sin\varphi = Q \sin\varphi/\pi r$ in Eq. 6.14.29 and noting that $\cos(\theta - \varphi) = \cos\theta\cos\varphi + \sin\theta\sin\varphi$,

$$dM_q = \frac{Q \sin\varphi}{\pi}[1 - \cos\theta\cos\varphi - \sin\theta\sin\varphi]\, r d\varphi \tag{6.14.30}$$

and integrating Eq. 6.14.30 between the limits 0 and θ gives

$$M_q = \frac{Qr}{\pi}\left[1 - \cos\theta - \frac{\theta \sin\theta}{2}\right] \tag{6.14.31}$$

The total moment at any point is then,

$$M = M_0 + \frac{Qr}{\pi}\left[1 - \cos\theta - \frac{\theta\sin\theta}{2}\right] + H_0 r(1 - \cos\theta) \tag{6.14.32}$$

The circumferential bending stress throughout can be analyzed by the Castigliano elastic strain energy theorem. Noting that at the zenith of the ring, point a Fig. 6.63, the slope φ_a and horizontal deflection δ_a are zero,

slope,
$$\varphi_a = 0 = \int \frac{M}{EI}\frac{\partial M}{\partial M_0}\,ds \tag{6.14.33}$$

deflection,
$$\delta_a = 0 = \int \frac{M}{EI}\frac{\partial M}{\partial H_0}\,ds \tag{6.14.34}$$

Further

$$ds = r d\theta \tag{6.14.35}$$

and from Eq. 6.14.32

$$\frac{\partial M}{\partial M_0} = 1 \tag{6.14.36}$$

$$\frac{\partial M}{\partial H_0} = r(1 - \cos\theta) \tag{6.14.37}$$

Substituting from Eqs. 6.14.32, 6.14.35, and 6.14.36 into Eq. 6.14.33 and integrating between the limits of 0 to α gives

$$0 = \int_0^\alpha \left[M_0 + \frac{Q}{\pi}\left(1 - \cos\theta - \frac{\theta\sin\theta}{2}\right) + H_0 r(1 - \cos\theta)\right] r d\theta \tag{6.14.38}$$

$$0 = \left[M_0\theta + \frac{Qr\theta}{\pi} - \frac{Qr\sin\theta}{\pi} - \frac{Qr}{2\pi}(\sin\theta - \theta\cos\theta) + H_0 r\theta - H_0 r\sin\theta\right]_0^\alpha \tag{6.14.39}$$

$$0 = \frac{M_0\alpha}{r} + \frac{Q}{2\pi}(2\alpha - 3\sin\alpha + \alpha\cos\alpha) + H_0(\alpha - \sin\alpha) \tag{6.14.40}$$

Similarly, substituting from Eqs. 6.14.32 and 6.14.35 and 6.14.37 into Eq. 6.14.34 and integrating between the limits 0 and α gives

$$0 = \int_0^\alpha \left[M_0 + \frac{Qr}{\pi}\left(1 - \cos\theta - \frac{\theta\sin\theta}{2}\right) + H_0 r(1 - \cos\theta)\right] [1 - \cos\theta]\, r^2 d\theta \tag{6.14.41}$$

$$0 = M_0(\alpha - \sin\alpha) + \frac{Qr}{8\pi}(11\alpha - 20\sin\alpha + 4\alpha\cos\alpha + 5\sin\alpha\cos\alpha$$

$$+ 2\alpha\sin^2\alpha) + \frac{H_0 r}{2}(3\alpha - 4\sin\alpha + \sin\alpha\cos\alpha) \qquad (6.14.42)$$

The value of M_0 and H_0 can be determined by solving Eqs. 6.14.40 and 6.14.42 simultaneously, and substituting these values in Eq. 6.14.32 to give the bending moment throughout the ring.

In addition to this bending moment, the reaction at the tip of the saddle due to the tangential loading q produces a direct tangential compressive stress and a radial shear stress. The vertical reaction, F, at this location, Fig. 6.63, is

$$F = q_0 r \int_0^\alpha \sin^2\theta \, d\theta \qquad (6.14.43)$$

and substituting the value of $q_0 = Q/r\pi$ from Eq. 6.14.28 gives

$$F = \frac{Q}{\pi}\left[\frac{\alpha}{2} - \frac{\sin 2\alpha}{4}\right] \qquad (6.14.44)$$

from which

$$F' = F\cos(\alpha - 90) \qquad (6.14.45)$$

and

$$F'' = F\sin(\alpha - 90) \qquad (6.14.46)$$

The direct compressive stress is

$$\sigma_c = -\frac{F'}{hb} \qquad (6.14.47)$$

The effective width, b, of the ring resisting the bending moment and the direct compressive stress is that included within the saddle width, w, plus a distance $1/\beta$, Par. 4.6, on each side of the saddle,[198] or

$$b = w + 2\left(\frac{\sqrt{rh}}{1.285}\right) \qquad (6.14.48)$$

The total stress (pressure plus bending plus direct compression) in the ring is

$$\sigma = \frac{pr}{h} \pm \frac{6M}{bh^2} - \frac{F'}{bh} \qquad (6.14.49)$$

and the shear stress is

$$\tau = \frac{F''}{bh} \tag{6.14.50}$$

Example: A thin cylindrical vessel, $h = 1$ in., $r = 30$ in., subject to an internal pressure of 300 psi, is supported on two rigid 90°-included-angle saddles, $w = 10$ in., welded to the vessel, Fig. 6.62*b*. The saddle is sufficiently removed from the stiffening effect of the vessel end closure so that the portion above the saddle acts as a ring in resisting the loading. What is the circumferential bending moment throughout this ring, and what is the maximum circumferential stress occurring at the saddle tip for a saddle reaction $Q = 20{,}000$ lb.

Substituting the following values in Eqs. 6.14.40 and 6.14.42:

$$\alpha = 135^\circ = 2.356, \ \sin 135^\circ = .707, \ \cos 135^\circ = -\ .707, \ \pi = 3.14$$

and solving simultaneously gives

$$M_0 = 0.0300Qr \tag{6.14.51}$$

$$H_0 = -\ 0.1322Q \tag{6.14.52}$$

Substituting Eqs. 6.14.51 and 6.14.52 into Eq. 6.14.32 gives the circumferential moment

$$M = 0.030Qr + \frac{Qr}{3.14}\left(1 - \cos\theta - \frac{\theta \sin\theta}{2}\right) - 0.1322\ \ Qr(1 - \cos\theta) \tag{6.14.53}$$

from which the moment diagram shown in Fig. 6.62*b* is obtained. The maximum circumferential bending moment occurs at the saddle tip or horn and is

$$\text{M} = .0826Qr = .0826 \times 20{,}000 \times 30 = 49{,}589 \text{ in. lb} \tag{6.14.54}$$

The reaction at the tip of the saddle due to the tangential loading q causes a direct tangential compressive stress and a radial shear stress at this location, Fig. 6.63. The vertical reaction, F, here is found from Eq. 6.14.44 as

$$F = q_0 r\left[\frac{\alpha}{2} - \frac{\sin 2\alpha}{4}\right]_0^{135^\circ} = 1.43\ q_0 r \tag{6.14.55}$$

and substituting the value of q_0 from Eq. 6.14.28 gives

$$F = .455\ Q \tag{6.14.56}$$

from which

$$F' = F\cos(\alpha - 90) = .322Q \tag{6.14.57}$$

$$F'' = F\sin(\alpha - 90) = .322Q \tag{6.14.58}$$

The total stress at the saddle tip is found from Eq. 6.14.49 and substituting from Eqs. 6.14.48, 6.14.54 and 6.14.57 to give

$$\sigma = \frac{300 \times 30}{1} \pm \frac{6 \times 49{,}589}{\left[10 + 2\left(\frac{\sqrt{1 \times 30}}{1.285}\right)\right] 1^2} - \frac{.322 \times 20{,}000}{\left[10 + 2\left(\frac{\sqrt{1 \times 30}}{1.285}\right)\right] 1}$$

$$\sigma = 9{,}000 \pm 16{,}003 - 336 = 24{,}667 \text{ psi}; - 7{,}339 \text{ psi} \quad (6.14.59)$$

The maximum compressive stress occurs when the internal pressure is zero. The maximum shear stress is from Eq. 6.14.48 and 6.14.50 is

$$\tau = \frac{F''}{bh} = \frac{.322 \times 20{,}000}{\left[10 + 2\left(\frac{\sqrt{1 \times 30}}{1.285}\right)\right] 1} = 336 \text{ psi} \quad (6.14.60)$$

The circumferential ring bending moment can be reduced throughout by increasing the saddle inclusion angle. Saddles that provide a minimum bearing of one-quarter to one-third of the shell circumference are normally required. Many vessel construction codes, such as the ASME Section VIII, Unfired Pressure Vessels, specifies a minimum saddle inclusion angle of 120° except for small diameter or heavy wall vessels such as may occur in the drums of boilers or heat exchangers.

The maximum or peak stress occurring in the region of a rigid saddle tip, Figs. 6.62*b* and 6.64*a*, can be reduced by the use of wear plates that extend beyond the edge of the saddle, thereby acting like

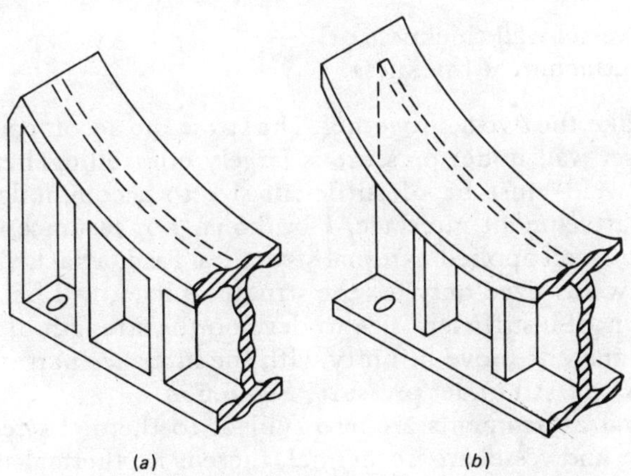

Fig. 6.64. (*a*) Rigid Saddle. (*b*) Flexible Saddle

a leaf-spring to mitigate the local bending moment.[199] Figure 6.64*b* shows a flexible saddle tip to accomplish a similar effect.

When the vessel wall itself is unable to sustain the circumferential bending stress and prevent instability in the saddle region, it may be reinforced by full circumferential stiffener rings. These consist of bars or structural shapes welded to the vessel so as to form a continuous ring. Such support rings can be constructed as a part of the saddle, Figure 6.62*c*, or complete rings supported by columns at the horizontal diameter. The latter method of support is frequently used for penstocks and large diameter pipelines. Stiffener rings combine with a portion of the vessel wall to form a composite shape of high resistance to bending, and the method of ascertaining its moment of inertia and section modulus is shown in Fig. 6.62*c*. Resistance to circumferential bending can also be achieved by utilizing the rigidity of the vessel head itself by placing the saddle close to it, i.e., within a distance $1/\beta$, Paragraph 4.6. This is the preferred support for thin-walled vessels of large diameter.

When there is a circumferential ligament efficiency present due to tube holes, such as occurs in the drums of boilers and heat exchangers, this factor must be introduced in Eq. 6.14.23 as applicable.

4. *Structural Attachments*

The fundametal requirement in the welding of attachments (lugs, saddles, stiffeners, etc.) to pressure vessels is to develop the strength of the attachment so that they act in union. That is, the connecting weld must be of sufficient size to develop either the strength of:

1. The vessel wall thickness or
2. The attachment thickness

so as to make the two act together. The stress and accompanying strain in the vessel wall under pressure is largely controlling; hence, the attachment weld must be of sufficient size to accommodate these at the vessel-attachment interface, Fig. 6.65*a*. For instance, a large size lug, with a small applied external structural load, attached to a vessel with fillet welds sized only for the structural load, will fail at the weld because it is of insufficient size to develop the strength of the attachment and make it move in unity with the interface part of the vessel wall as it is stressed under pressure, Fig. 6.65*b*.

Structural attachments are also subject to thermal stresses. If the attachment and vessel are isothermal, there is no thermal stress. However, if there is a thermal gradient across the attachment-vessel inter-

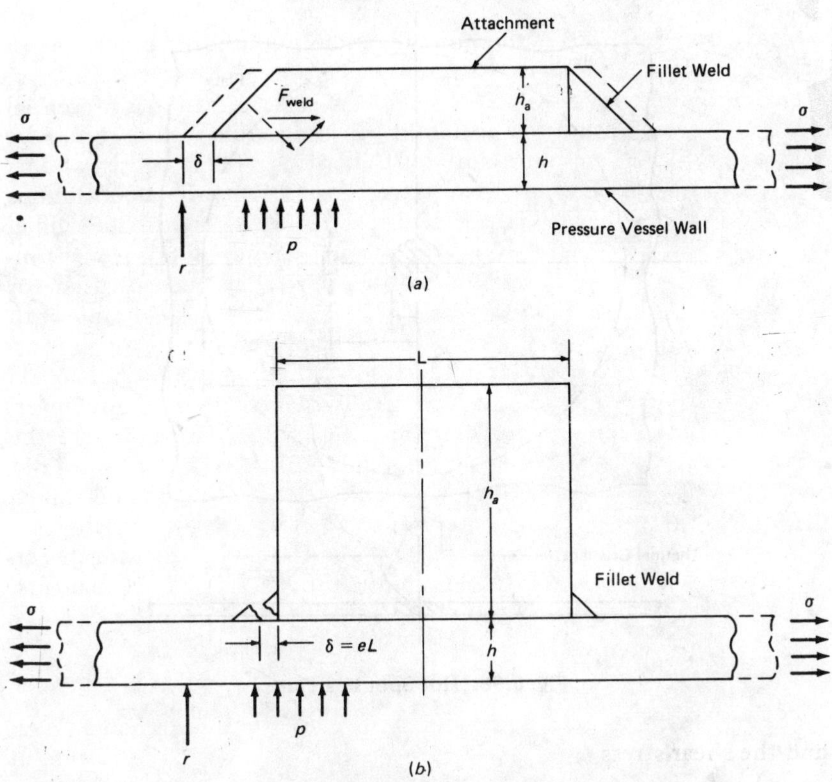

Fig. 6.65. Vessel-Attachment Interface. (*a*) Attachment Strength Developed. (*b*) Insufficient Attachment Weld

face, a thermal stress is created in the attachment and adjacent vessel wall. These stresses can be obtained by first noting that for a thin circular hollow disk, Fig. 6.66, where there is no surface forces or restriction to axial displacement, $\sigma_z = 0$ (Eq. 2.12.3). Since the degree of restraint has thereby been reduced by one, $\alpha E/(1 - \mu)$ is replaced by αE in Eqs. 2.12.1 and 2.12.2 and further, for the case of a thin solid disk, the inner radius a is set equal to zero to give

$$\sigma_r = \alpha E \left[\frac{1}{b^2} \int_0^b Trdr - \frac{1}{r^2} \int_0^r Trdr \right] \qquad (6.14.61)$$

$$\sigma_t = \alpha E \left[\frac{1}{b^2} \int_0^b Trdr + \frac{1}{r^2} \int_0^r Trdr - T \right] \qquad (6.14.62)$$

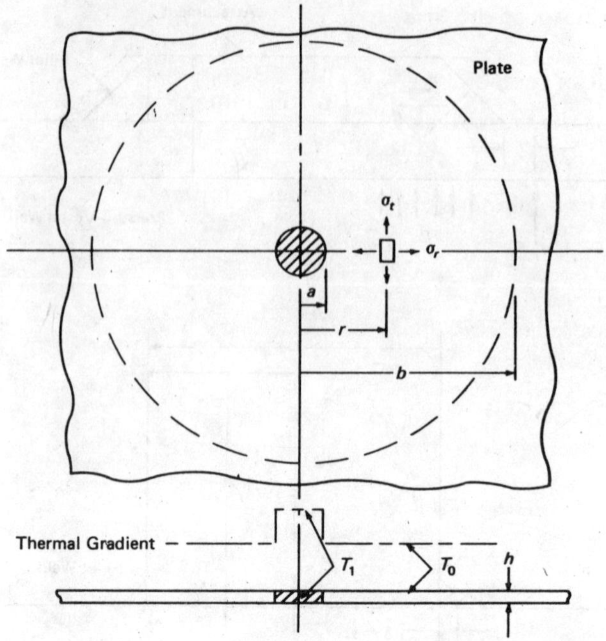

Fig. 6.66. Hot Spot in a Plate

and the shear stress

$$\tau = \frac{\sigma_t - \sigma_r}{2} = \alpha E \left[\frac{1}{r^2} \int_0^b Trdr - \frac{T}{2} \right] \tag{6.14.63}$$

T is the temperature at any point in excess of a certain uniform initial temperature of a body, measured from any convenient zero. It may be replaced by $T - T_0$, where T_0 is the temperature of the coldest part, Par. 2.11 and Par. 2.12, since that portion of the temperature gradient that is uniform throughout induces no thermal stress when there are no external restraints.

As in Par. 2.13, the first integral in Eqs. 6.14.61 and 6.14.62 can be written

$$\frac{1}{2\pi b^2} \int_0^b T2\pi rdr$$

and is equal to half the mean temperature of the entire disk. Also, the second integral is equal to half the mean temperature of the inner portion of radius r of the disk, so

$$\sigma_r = \frac{\alpha E}{2}\left[\begin{pmatrix}\text{Mean temperature}\\ \text{of the entire disk}\end{pmatrix} - \begin{pmatrix}\text{Mean temperature}\\ \text{of the inner portion}\\ \text{of radius } r\end{pmatrix}\right] \quad (6.14.64)$$

$$\sigma_t = \frac{\alpha E}{2}\left[\begin{pmatrix}\text{Mean temperature}\\ \text{of the entire disk}\end{pmatrix} + \begin{pmatrix}\text{Mean temperature}\\ \text{of the inner portion}\\ \text{of radius } r\end{pmatrix} - 2\begin{pmatrix}\text{Temperature of}\\ \text{desired stress}\\ \text{location}\end{pmatrix}\right] \quad (6.14.65)$$

A hot spot, or cold spot, can develop in the wall of a vessel due to the loss of internal refractory, introduction of fluid media through nozzles, and heat conduction through structural attachments, etc. If these are assumed small and circular of radius a and temperature T_1 in a very large thin circular disk or plate of radius b of uniform temperature T_0, the condition exists in which the temperature is a step-change one that is a function of the radius r only, Fig. 6.66, and the stresses are given by Eqs. 6.14.64 and 6.14.65. Noting that the temperature T_1 within the radius a has a nil effect on the mean temperature of the entire plate, and letting the uniform base temperature $T_0 = 0$, $T_1 - T_0 = \Delta T$; the maximum thermal stresses within the hot spot are:

$$\sigma_{r_a} = -\frac{\alpha E}{2}(T_1 - T_0) = -\frac{\alpha E}{2}\Delta T \quad (6.14.66)$$

$$\sigma_{t_a} = -\frac{\alpha E}{2}(T_1 - T_0) = -\frac{\alpha E}{2}\Delta T \quad (6.14.67)$$

$$\tau = \frac{\sigma_{t_a} - \sigma_{r_a}}{2} = 0 \quad (6.14.68)$$

and outside the hot spot

$$\sigma_{r_r} = \frac{\alpha E}{2}\left[T_0 - (T_1 - T_0)\frac{a^2}{r^2}\right] = -\frac{\alpha E a^2 \Delta T}{2r^2} \quad (6.14.69)$$

$$\sigma_{t_r} = \frac{\alpha E}{2}\left[T_0 + (T_1 - T_0)\frac{a^2}{r^2} - 2T_0\right] = \frac{\alpha E a^2 \Delta T}{2r^2} \quad (6.14.70)$$

$$\tau = \frac{\sigma_{t_r} - \sigma_{r_r}}{2} = \frac{\alpha E a^2}{2r^2}\Delta T \quad (6.14.71)$$

The maximum shear stress is $\frac{1}{2}\alpha E \Delta T$ just outside the hot spot. Any mitigation of this sharp step-change thermal gradient by heating around the spot, as occurs in actual practice, tends to relieve this stress.

If the hot spot is rectangular, the direct thermal stress components, σ_x and σ_y, do not exceed $\frac{1}{2}\alpha E \Delta T$; however, the sharp corners cause very high local shear stresses.

Structural attachments of the type shown in Figs. 6.67 and 6.68 are frequently used on pressure components. Their thermal stress effects can be evaluated by assuming the attachment provides a pseudo-hot spot or cold spot, in a large plate; and further, that the mean temperature of the plate in this adjacent region of radius r is, Fig. 6.67,

$$T_1 = \frac{T_0 h + T_a h_a}{h + h_a} \tag{6.14.72}$$

where T_a is the mean temperature of the attachment. Substituting Eq. 6.14.72 in Eqs. 6.14.64 and 6.14.65 give the stresses in the plate adjacent to the attachment as

$$\sigma_r = \sigma_t = \frac{\alpha E}{2}\left[T_0 - \frac{T_0 h + T_a h_a}{h + h_a}\right] = -\frac{\alpha E}{2}\left[(T_a - T_0)\frac{h_a}{h + h_a}\right] \tag{6.14.73}$$

If the attachment is very thin compared to the plate, it creates little thermal stress in the plate; whereas, if the attachment is appreciably thicker than the plate, a maximum value of

$$\sigma_{\text{max.}} = -\frac{\alpha E}{2}(T_a - T_0) \tag{6.14.74}$$

is approached. When a cold spot occurs, the sign of the thermal gradient is changed in the foregoing equations.

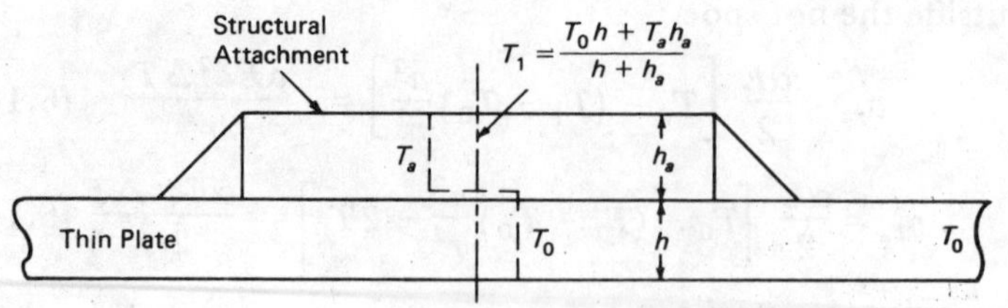

Fig. 6.67. Structural Attachment Thermal Gradient

The thermal force imposed on the fillet weld is

$$F_{t,\text{weld}} = \sigma_r h \tag{6.14.75}$$

$$F_{t,\text{weld}} = -\frac{\alpha E h}{2}\left[(T_a - T_0)\frac{h_a}{h + h_a}\right] \tag{6.14.76}$$

These thermal stresses can be reduced by the use of full penetration welds to create an integral attachment to the vessel wall, thereby eliminating the discontinuous interface with its large step-change gradient, and also provide a full cross-sectional attachment weldment. Figure 6.68 shows such full penetration weldment lugs on the head of a vessel subject to high pressure and thermal transients.

Fig. 6.68. Integral Attachment Lugs on the Head of a Pressure Component Subject to Pressure and Thermal Transients

6.15 Nozzle Thermal Sleeves

Vessels are often used that contain heat sources or heat sinks. Nuclear reactor vessels contain cores which generate heat, and their associated steam generators contain a heat exchanger tube bank for absorbing this heat. When fluids or gases are introduced into or discharged from a vessel nozzle causing a temperature higher or lower than that of the vessel material, thermal stresses are set up at the nozzle opening directly proportional to this temperature difference equal to $E\alpha\Delta T/(1 - \mu)$.

A small temperature difference presents no problem to the vessel material, even though it be cyclic, and the difference can be determined for the particular material involved. As an example, the allowable ΔT and alternating stress for a carbon steel vessel material of these properties can be computed as follows:

T.S. = 70,000 psi
Y.P. = 38,000 psi
σ_E = 0.5 × T.S. = 35,000 psi (Endurance Limit)
α = 0.000007 in./in./°F
$E_{t=500}$ = 28,500,000 psi
μ = 0.3 for steel
σ = T.S./3 = 23,333 psi (allowable design hoop stress)
K_t = 2.0
$K_t\sigma$ = 46,666 psi

The stress-strain diagram is constructed in Fig. 6.69*a*, and the mean stress found to be 14,667 psi. Plotting this on the modified Goodman diagram, Fig. 6.69*b*, for safe cycles gives a corresponding allowable alternating stress σ_a = 27,000 psi, or an equivalent normal allowable alternating stress σ_a/K_t = 13,500 psi. Equating this allowable stress to the maximum developed by the temperature difference ΔT gives

$$\frac{\sigma_a}{K_t} = \frac{E\alpha\Delta T}{(1-\mu)} \tag{6.15.1}$$

$$\Delta T = \frac{\sigma_a(1-\mu)}{K_t E\alpha} \tag{6.15.2}$$

or for a typical pressurized or boiling water reactor vessel temperature of 500°F, a metal thermal shock of

$$\Delta T = \frac{27{,}000 \times 0.7}{2 \times 28{,}500{,}000 \times 0.000007} = 48°\text{F} \tag{6.15.3}$$

is permissible.

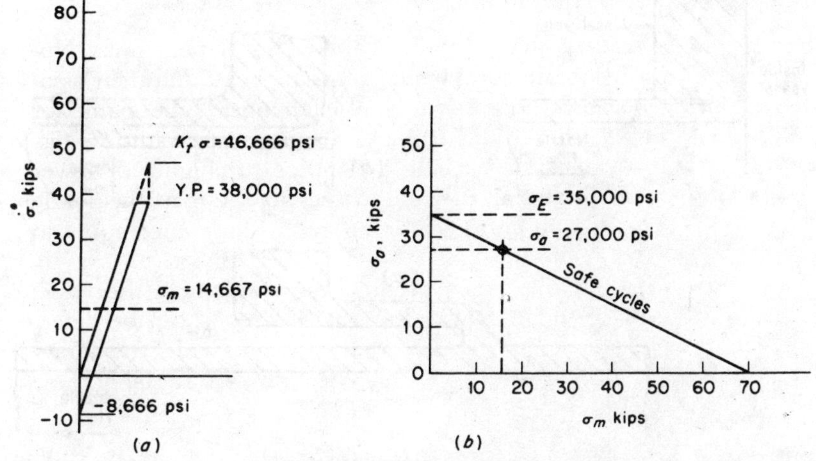

Fig. 6.69. Fatigue Life Evaluation

When thermal shocks exceed those that can be safely taken by the vessel material, it must be protected. Protecton of the vessel wall proper is the fundamental design requirement, since failure here is the most critical. Accordingly, excessive thermal drops should be taken in the thinner and more flexible wall of the nozzle than in the usually heavier and more rigid wall of the vessel. This is accomplished by the use of so-called "thermal sleeves," Fig. 6.70. For relatively small temperature difference these may consist of a series of grooves (*a*) machined in the nozzle neck which reduces the metal contact or heat flow area between nozzle and vessel wall; thereby mitigating any thermal disturbance to the vessel wall. When these temperature differences are greater, thermal sleeves of the type shown in (*b*) and (*c*) are used; and for extremely large thermal drops, combinations of these must be used in series to obtain a safe design, (*d*) and (*e*). The design principle is the same for each—namely, the introduction of a heat conductance resistance between the nozzle media and vessel wall.[213, 214]

6.16 External Loading on Shells

Pressure vessels are not only acted upon by the forces of internal and external pressures, but since most vessels are elements of a system they are also subject to loadings at the supports, nozzles, and attachments, Fig. 6.71. These loadings produce deflections, edge rotations, shears, bending moments and memberane forces in both the circumferential

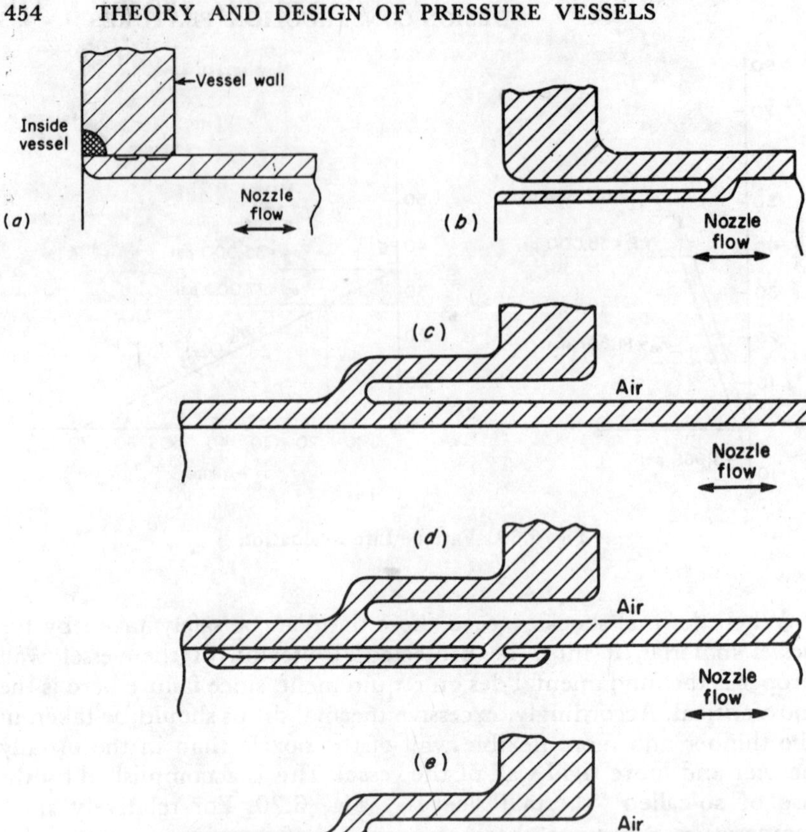

Fig. 6.70. Thermal Sleeves

and longitudinal planes.[100–110] The effect is a rapidly damped one with the maximum stress occurring in the immediate vicinity of the load.[159, 160, 216]

In practical applications the number of variables is considerable, and great selectivity must be used to choose the important ones and eliminate those of minor importance. The ensuing discussion of these uses the following nomenclature:

τ = shear stress, psi

ρ = $0.875r/R$ (Attachment parameter)

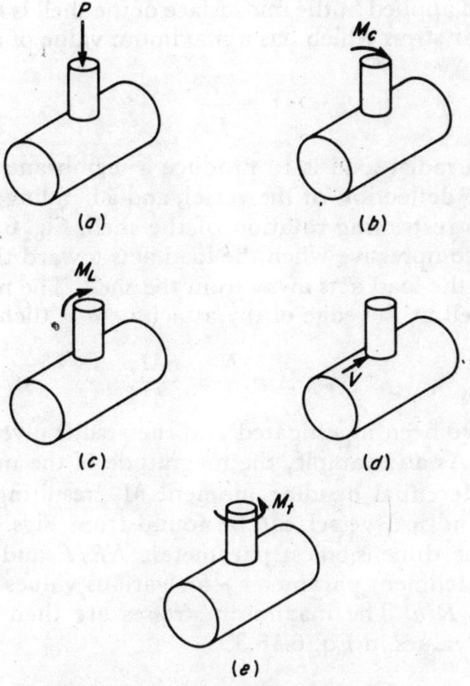

Fig. 6.71. Types of External Loading

σ_y = normal stress in the circumferential direction, psi
h = thickness of shell, in.
r = radius of attachment, in.
P = radial load, lb.
R = mean radius of shell, in.
V = tangential load, lb.
M = external moment on a spherical shell, in. lb.
M_t = twisting moment, lb in.
M_y = bending moment per unit length in the circumferential direction, in. lb/in., for a cylindrical shell
N_y = membrane force per unit length in the circumferential direction, lb/in.

The stress at the juncture of a circular attachment produced by a twisting moment is

$$\tau = \frac{M_t}{2\pi r^2 h} \tag{6.16.1}$$

Tangential load applied at the midsurface of the shell is transmitted by membrane shear stress which has a maximum value of approximately

$$\tau = \frac{V}{\pi r h} \tag{6.16.2}$$

The effect of a radial load is to produce a membrane stress equal to N/h due to the deflection of the vessel, and a bending stress equal to $6M_y/h^2$ due to restricting rotation of the shell, Fig. 6.72. The membrane stress is compressive when the load acts toward the shell, and it is tensile when the load acts away from the shell. The maximum total stress in the shell at the edge of the attachment is then

$$\sigma_y = (\pm)\frac{N}{h} \pm \frac{6M_y}{h^2} \tag{6.16.3}$$

The effects have been investigated and the results given in a series of design curves. As an example, the magnitude of the membrane force N and circumferential bending moment M_y resulting from a radial load on a cylindrical vessel can be found from Figs. 6.73 and 6.74 which show the dimensionless parameters NR/P and M_y/P plotted against the attachment parameter ρ for various values of the shell-to-thickness ratio R/h. The maximum stresses are then found by substituting these values in Eq. 6.16.3.

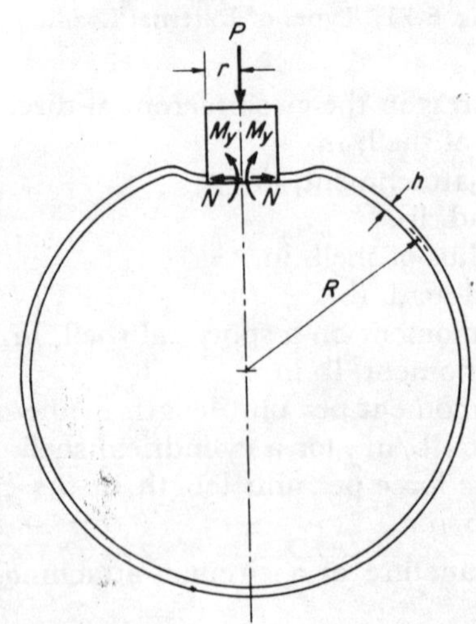

Fig. 6.72. Distortion of a Cylindrical Shell Under an External Radial Load

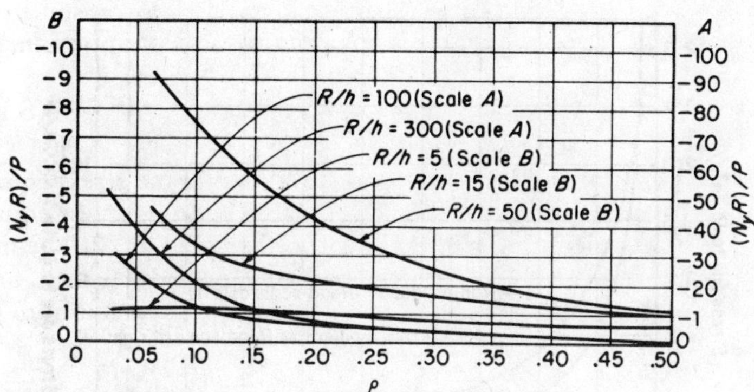

Fig. 6.73. Circumferential Membrane Force Due to an External Radial Force on a Cylinder[105]

Illustrative problem: Determine the maximum stress in a 100 in. diameter, 1 in. thick cylindrical shell at a 10 in. diameter attachment which imposes a 10,000 lb radially inward load.

Solution:

$$\frac{R}{h} = \frac{50}{1} = 50$$

$$\rho = \frac{0.875r}{R} = \frac{0.875 \times 5}{50} = 0.0875$$

From Fig. 6.73

$$\frac{N_y R}{P} = -8, \; N_y = \frac{-8P}{R}$$

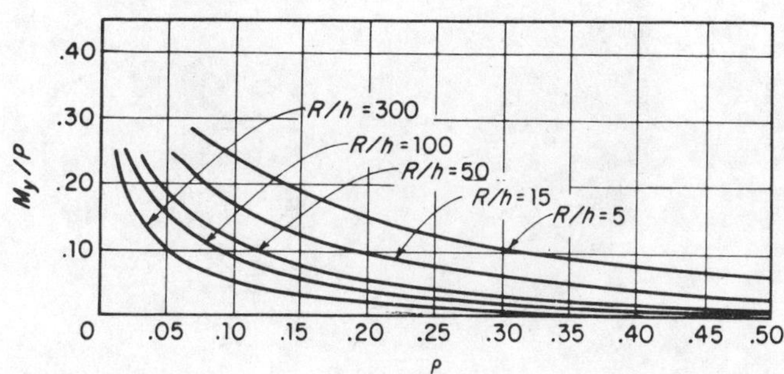

Fig. 6.74. Circumferential Bending Moment Due to a Radial Force on a Cylinder[105]

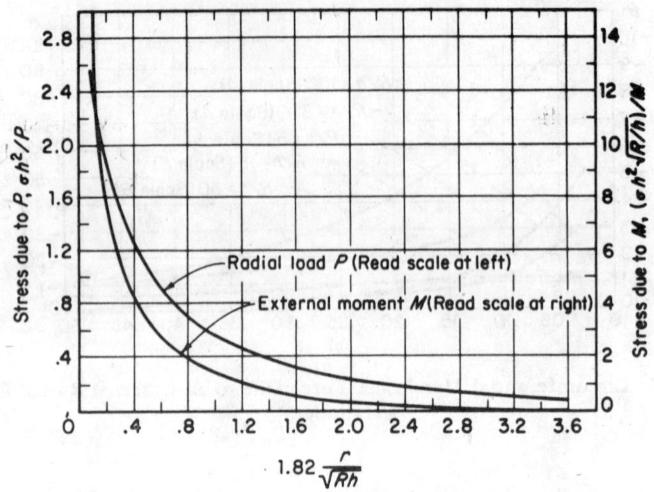

Fig. 6.75. Maximum Stress Due to External Loading on a Spherical Shell[105]

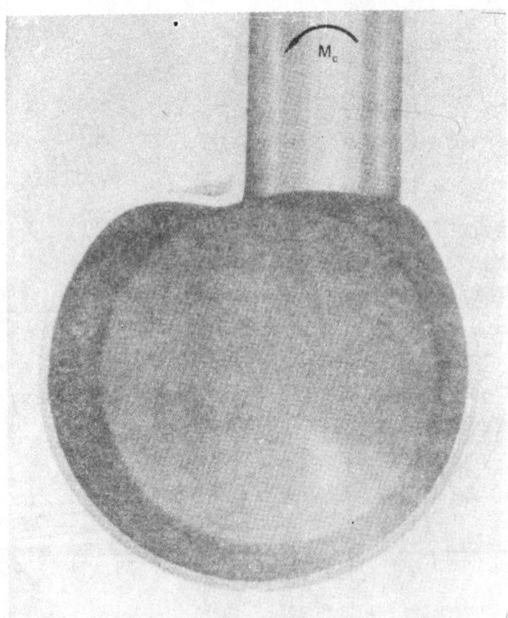

Fig. 6.76. Section Through a Tee Plastically Deformed by an External Out-of-plane Moment

and substituting the value of $P = 10{,}000$ and $R = 50$ gives

$$N = \frac{-8 \times 10{,}000}{50} = -1{,}600 \text{ lb}$$

From Fig. 6.74

$$\frac{M_y}{P} = 0.14, \ M_y = 0.14P$$

and substituting the value of $P = 10{,}000$ gives

$$M_y = 0.14 \times 10{,}000 = 1{,}400 \text{ in lb/in.}$$

The total maximum stress is found from Eq. 6.16.3

$$\sigma_y = -\frac{1600}{1} \pm \frac{6 \times 1400}{1^2} = -10{,}000 \text{ psi}$$

The effect of external loads and moments on spherical shells has also been investigated.[161, 162] Here the presence of axial symmetry allows the use of fewer parameters and Fig. 6.75 permits determining the values of the stresses directly, in dimensionless form. The value of $\sigma h^2/P$ need only be multiplied by the actual P/h^2 to give the maximum stress, σ, produced by the radial load P. The value of $\sigma h^2\sqrt{Rh}/M$, when multiplied by the actual $M/h^2\sqrt{Rh}$, gives the maximum stress produced by a moment M.

Figure 6.76 shows the nature of the deformation occurring at a nozzle-vessel attachment (or branch-pipe tee connection) when is has been subjected to an out-of-plane external moment sufficient to cause plastic distortion.

6.17 Brittle Fracture Crack Arrest Constructions

The three requirements for the occurrence of brittle fracture given in paragraph 5.22 also present the basic means for the design of construction features to arrest this type of fracture. Crack arresters work on the principle of placing a roadblock across the path which a crack must propagate to cause failure. This roadblock can consist of a zone of low stress material and/or a zone of high toughness material, Fig. 6.77, or crack interceptors such as the holes[111] of riveted joints or the noncontinuum of bolted flanges. The primary purpose of crack arresters is to prevent catastrophic failure and they need not necessarily prevent leakage.[166, 167, 169, 200, 201]

In a cylindrical vessel subject to internal pressure only the maximum stress is in the hoop direction; hence, the path of the fracture will be longitudinal. Figure 6.77*a* shows a cylindrical vessel in which alternate

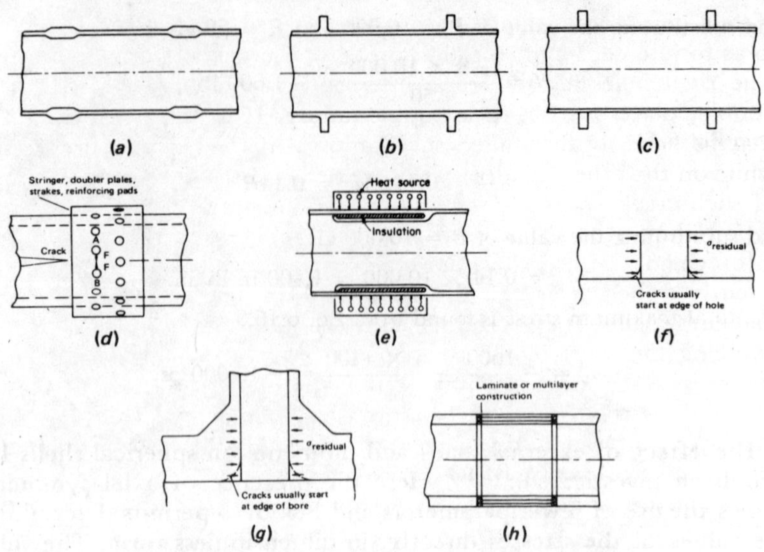

Fig. 6.77. Crack Arrest Constructions

thick cylindrical courses provide a low stress level and/or high material toughness field, while in Fig. 6.77*b*, *c* and *d* annular rings are provided to lower the hoop stress at these locations.[112] The toughness of the vessel material can also be increased by raising its temperature by an external heat source, Fig. 6.77*e*. This is the reverse thermal sleeve approach and care must be taken to minimize the induced thermal stresses. In gas transmission pipelines, brittle fractures have propagated for many miles before being arrested by a flow-control valve which presented a low stress field of high toughness material.

An environment change that reduces material toughness can also present a condition prone to brittle failure. This can result from a temperature drop, chemical embrittlement, or irradiation damage. The first of these conditions is the most prevalent. It was widely encountered in the welded ships of World War II which operated satisfactorily in tropical and temperate waters but frequently broke in two when operating in frigid waters. (Many pressure vessels have experienced the same type casualties.[113]) This susceptibility of ship decks and hulls to brittle fracture was moderated on new constructions by the use of greater toughness materials and better reinforcement of hatch openings to minimize stress concentrations. On existing ships the condition was alleviated by the use of crack arresters. These consisted of long heavy doubler plates or strakes riveted to their decks and hulls so as to intercept and stop the spread of a crack should one develop. The rivet holes act as crack interceptors, and the hybrid riveted doubler plates restrict catastrophic failure. It is this arrest capability that has greatly forestalled such failures in riveted pressure vessels; while on the other hand, those of welded construction which are void of such crack arrest constructions have been more vulnerable to this type failure. Holes, per se, without accompanying stress reduction reinforcement, offer a delaying crack arrest device and they are frequently drilled at crack tips as a provisory means pending more permanent repairs. They accomplish this by blunting the crack tip thereby reducing the stress concentration. However, the beneficial effect of this is in some measure mitigated by the increased crack size due to the hole as it becomes part of the crack. Holes can offer more permanent arrest capability when they are cold worked by rolling, coining, peening or mandrelizing to institute a residual compressive stress. This method of introducing a residual compressive stress field to oppose the applied tensile stress field has been extensively used to delay fatigue crack initiation and crack propagation, Par. 5.17.3. It may be applied to unreinforced or reinforced openings to optimize their crack resistance, Fig. 6.77*f* and *g*. When stringers, strakes, or reinforcing pads are riveted or bolted to the structure these take the load from the cracked member and thus reduce the stress intensity factor, Par. 5.22. In the absence of the reinforcement, points *A* and *B* can move freely apart as the crack approaches line AB, Fig. 6.77*d*; however, at this location the reinforcement constrains the displacement[169] transmitting the force F through the rivets or bolts to the cracked member thereby reducing the acting stress intensity factor, Par. 5.22. Just as rivet or bolt holes can act as crack arresters in two-dimensional construction, spherical voids[170] can accomplish a similar effect in three-dimensional construction; i.e., propagating cracks are arrested by the spherical void. This principle has been used in composite fly-wheels where cracks initiating or propagating from sharp crack-like defects are arrested by blunt spherical voids intentionally placed in the material. Here, flaws, rather than being detrimental, are helpful. This also accounts for the limited effect of weld porosity, with its spherical or rounded shape, on the fatigue strength of welded joints, Par. 6.10.2.

Many constructions designed to forestall brittle fracture also enhance fatigue life. For instance, the severity of the stress concentration at the edge of a hole in a plate may be reduced by applying a thermal shock to a narrow annulus a short distance removed from the edge so as to upset the metal (stress beyond the yield strength) in this narrow region, Fig. 6.78. The effect is to create a favorable residual stress pattern consisting of a compressive stress at the edge of the hole which opposes the applied tensile stress in this region. Thus the fatigue life is improved. This method is well adapted to thin perforated sheets such as used in aircraft.[203]

Composite, sandwich, laminate or layer constructions can also be used effectively as crack arresters, Fig. 6.77*h*. These can be of the same material composition as the adjoining thicker main structure, and they obtain their increased fracture toughness, K_c, Par. 5.22.3, Fig. 5.101, by virtue of the thinner material enforcing a plane stress condition on the crack tip, rather than one of plain strain; thereby presenting a higher resistance to fracture. When the fracture toughness of the thin section material is further enhanced by alloying and heat treatment, its crack arrest capabilities also increase.

Crack arrest constructions are becoming increasingly important as the criticality of weight or the economics of design continually seek out high strength materials and high working stress—each of which increases the susceptibility of the structure to brittle fracture.

A special case of thermal stress[171] crack arrest[172] occurs in the case of a nuclear reactor vessel in which the toughness of the material varies with its proximity to the nuclear core, Par. 5.10. Figure 6.79 shows

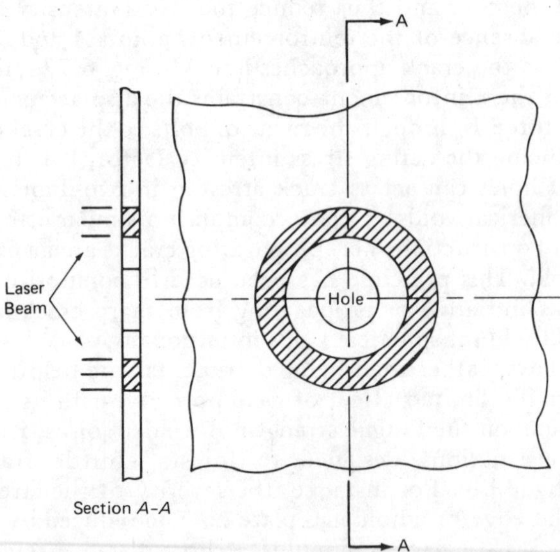

Fig. 6.78. Selective Residual Stress Induced Around a Hole in a Thin Plate[203]

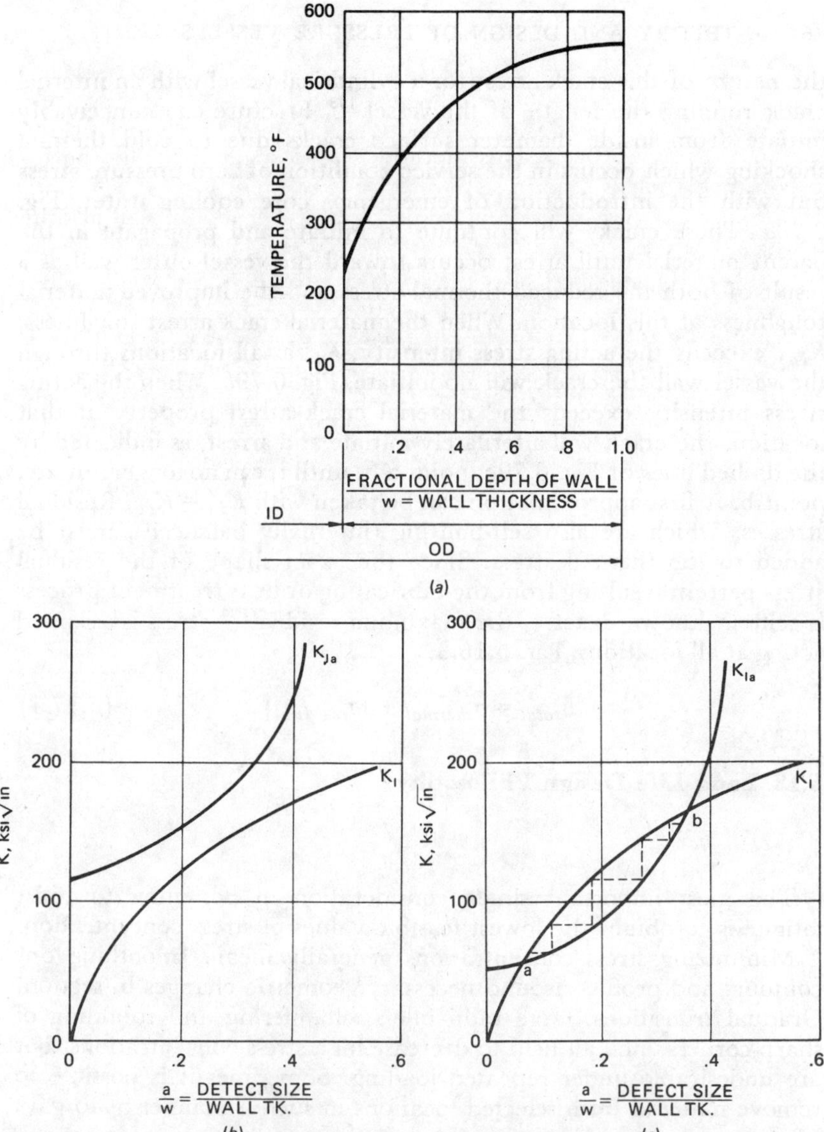

Fig. 6.79. (a) Temperature Distribution in the Wall of a Typical Nuclear Reactor Vessel at Time $t = 5$ minutes During a Cooling Thermal Shock to the Inside Surface with 70°F Water; (b) Variation in K_I and K_{Ia} with a/w Ratio for Typical Reactor Conditions of Thermal Stress an Material Irradiation Toughness Variation Exhibiting no Crack Initiation; (c) Variation in K_I and K_{Ia} with a/w Ratio for Typical Reactor Conditions of Thermal Shock Stress and Material Irradiation Toughness Variation Exhibiting Crack Initiation and Arrest.

the nature of this crack arrest for a cylindrical vessel with an internal crack running the length of the vessel.[173] Fracture can conceivably initiate from inside diameter surface cracks due to cold thermal shocking which occurs in the service condition of zero pressure stress but with the introduction of emergency core cooling water, Fig. 6.79*a*. These cracks will continue to initiate and propagate in the parent material until arrest occurs toward the vessel outer wall as a result of both the reduced thermal stress and the improved material toughness at this location. When the material crack arrest toughness, K_{Ia}, exceeds the acting stress intensity, K_I, at all locations through the vessel wall the crack will no initiate, Fig. 6.79*b*. When the acting stress intensity exceeds the material crack arrest property at that location, the crack will alternately initiate and arrest, as indicated by the dashed lines of Fig. 6.79*c*, point a-b, until it can no longer initiate, point b. A first approximation can be taken with $K_{Ia} = K_{Ic}$. Residual stresses, which are also self-limiting (internally balanced), must be added to the thermal stress. Since the exact shape of the residual stress pattern resulting from the fabricating or heat treatment process is seldom known exactly, the maximum residual[174] stress is assumed acting at all locations, Par. 5.16.3,

$$\sigma_{total} = \sigma_{thermal} + |\sigma_{residual}| \tag{6.17.1}$$

6.18 Long-Life Design Philosophy

1. *Vessels*

The most important single consideration in designing to resist fatigue is to obtain the lowest feasible values of stress concentration.

Minimizing stress concentrations generally means smoothing out contours and profiles around necessary geometric changes in section. Gradual transitions, large radii fillets, chamfering and rounding of sharp corners, etc., all help to decrease high stress concentrations that are undesirable under repeated loading. Sometimes it is possible to remove material from selected locations in such a manner as to gain more by a decrease in K_t than is lost by increase in nominal stress due to decreased net section. This is well illustrated by the attachment of nozzles to vessels whereby removal of the sharp interior corner metal and replacement with a radius equal to 50 percent of the vessel wall thickness results in a reduction in stress concentration for this location in the order of 30 percent, Fig. 6.80*a*. Likewise it may be desirable to add additional stress raisers to mitigate the effect of a particularly severe one. Figure 6.80*b* illustrates this effect by the placement of more shallow grooves on each side of a deep groove to reduce the stress gradient in its immediate vicinity, causing the stress flow pattern to approach the major restriction more gradually than would occur without the additional grooves. Minimizing stress concentrations around a circular hole in an uniaxially loaded plate by introducing auxiliary smaller holes on either side of the large one is another example, Fig. 6.80*b*. Here the fundamental principle is the same—the use of any device or means to reduce the stress gradient. There are also examples of removing material to reduce primary stress such as a square bar in simple bending in a plane through its diagonal, or a round bar in bending. Here by removing the small amount of material shown in Fig. 6.80*c* the stress is reduced due to the fact that the moment of inertia of the total cross section is diminished in a smaller

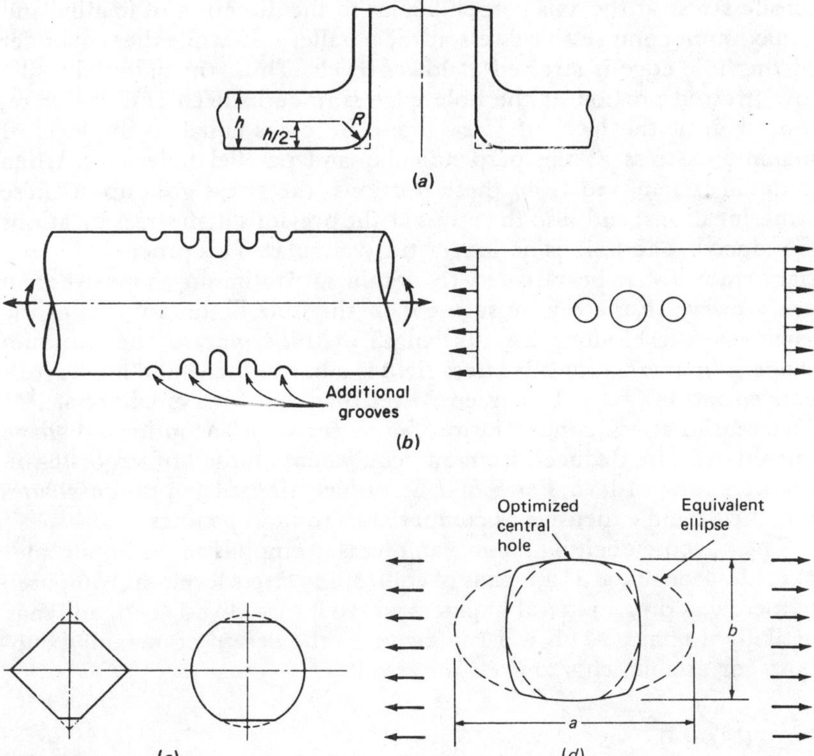

Fig. 6.80. Examples of Removing Material to Improve the Stress Condition

proportion than the depth; hence, the section modulus increases and stress decreases.[114,209]

Keeping weight down is not only healthy for human beings, but the concept is also applicable to the fatigue life of man-made structures. For instance, a reduction in the maximum tensile stress by a design modification also leads to an increase in fatigue life. The conventional design philosophy, and usually the most direct and least costly, is to add more material to high stress regions, thereby reducing their levels; for example, by adding reinforcement around holes, Par. 6.6. Another way of accomplishing this is by the removal of unnecessary material from low stress regions with an accompanying beneficial reduction of the high stress levels at critical locations. The basic principle is a reduction of peak stresses by a shape change. For instance, the stress distribution at the edge of a circular hole in a plate under uniaxial tension, Eq. 5.5.2 and Fig. 6.5, exhibits a maximum tensile stress at the axis perpendicular to the direction of loading and a maximum compressive stress at the parallel axis, while the remainder of the hole edge is stressed at lower levels. Thus, the material at the low stressed portion at the hole edge is underutilized and can be removed until the level of stress there becomes equal to the level of maximum stress at the perpendicular and parallel hole axes. When material is removed from these portions, the stress goes up at these same locations; and also the stress at the previous high stress locations is reduced. The hole is no longer truly circular. This process of material removal can be pursued to obtain an optimum shape wherein every point on the tensile segment of the hole boundary attains the same stress level along the full length of this segment. The optimum shape in a uniaxial tensile stress field is a barrel shape and has a stress concentration factor 16 percent lower than that for a circular hole.[215] The tensile stress concentration factor for such an optimized shape opening can be deduced from an "equivalent ellipse" of major-to-minor axes ratio of 1.75, Fig. 6.80*d*. Manufacturing such optimum shapes is difficult and expensive as compared to round openings.

The purpose of citing these examples is to emphasize the importance of contour or shape as a means of controlling stress levels, as compared to merely adding material in places where it may do no good, and may actually do harm. This is a key factor in the design of openings, supports, or attachments to pressure vessels.

2. *Vessels for Nuclear Service*

The significant difference in the design of vessels for nuclear service,[115,116,117,175] as compared to other pressure-temperature conditions, is the material irradiation damage environment. This is generally limited to the reactor vessel proper, and even more so to these high neutron flux regions immediately adjacent to the nuclear core. In some reactors this may at first seem unimportant because the operating temperature is above the material transition temperature (ductile behavior), but low temperature excursions, startups, and hydrostatic tests must be considered. Consequently, it behoves the designer to employ all mitigating and control procedures at his disposal to cope with this conditions.[204] Some of these are:

A. Provide sufficient distance between vessel wall and core to reduce fast neutron absorption, nvt, consistent with the specific material irradiation properties that will give the required life. It is also important that the material irradiation test data be comparable to the actual environment to preclude the unknown behavior of contributing variables and inaccuracies of measuring neutron absorption which can vary as much as plus or minus 30 percent.

B. Eliminate all structural discontinuities in the high neutron flux region, such as internal or external supports, lugs, etc., that might give rise to high stress concentrations.

C. Provide a number of surveillance samples, made from the same material as the vessel, and located immediately adjacent to the wall in the maximum neutron flux zone, that can be periodically removed and tested as a monitor for the actual vessel material damage.[176]

D. Provide in the design suitable means for irradiation damage annealing should the surveillance samples indicate this necessity, and use only vessel material amenable to this treatment.

6.19 Vessel Construction Codes

Federal, state, and insurance company codes play a part in all vessel design; and rightfully so, for there is hardly an application where the public would not be involved from either a health, property damage, or personal liability viewpoint in the event of a failure. The fast moving developments in this field, and the associated potential hazard, have focused attention on the need of applicable vessel construction codes to provide a sound engineering and legal means of procedure.

Codes or sets of rules can set only minimum standards and offer general guidance. This allows continual development of the tools of progress in engineering, material, and fabrication toward the end of safety and economy.[205] The construction of vessels are governed by jurisdictional authorities of the site location and/or accepted applicable codes. In the United States the codes of the American Society of Mechanical Engineers for pressure vessel construction are widely used. They cover vessels made of metal, concrete and composites for power, chemical, petroleum, and nuclear applications. Each of these codes has recognized that the safest vessel is one with the lowest total stress throughout all parts of the vessel and does not attempt to employ a large factor of safety on the basic pressure stress only, and ignore the effect of severe local stress risers, etc., that can well result in an inferior design prone to low fatigue life. Virtually all failures are a result of fatigue—fatigue in the area of high localized stress— and in the elimination or minimizing of these lies the solution to safe pressure vessel construction.

REFERENCES

1. M. Hetényi, *Handbook of Experimental Stress Analysis*, John Wiley & Sons, Inc., New York, 1950.

2. S. Timoshenko, *Strength of Materials*, Part II, D. Van Nostrand, Princeton, N.J., 1956.

3. ASME Boiler and Pressure Vessels Code, Section I, III and VIII.

4. R. B. Heywood, *Designing by Photoelasticity*, Chapman and Hall, London, 1952.

5. R. E. Peterson, *Stress Concentration Design Factors*, John Wiley & Sons, Inc., New York, 1953.

6. M. M. Frocht, "Factors of Stress Concentration Photoelastically Determined," *ASME Journal of Applied Mechanics*, Vol. 2, 1935.

7. J. H. Heifetz and I. Berman, "Measurement of Stress Concentrations in the External Fillets of a Cylindrical Pressure Vessel," *Experimental Mechanics*, December, 1967.

8. R. C. Gwaltney and J. W. Bryson, "Effect of Fillets on Cylindrical Shells with Step Changes in Outside Diameters," Oak Ridge National Laboratory, Report ORNL-TM-3766, May, 1972.

9. H. Neuber, "Theory of Notch Stresses," English Translation, Edwards Brothers, Ann Arbor, Michigan, 1946.

10. J. C. Gerdeen and R. E. Smith, "Experimental Determination of Stress-concentration Factors in Thick-walled Cylinders with Crossholes and Sideholes, *Experimental Mechanics*, Vol. 12, No. 11, November, 1972.

11. C. R. Smith, "Effective Stress Concentration for Fillets in Landed Structures," *Experimental Mechanics*, Vol. 11, No. 4, April, 1971.

12. S. Timoshenko and J. N. Goodier, *Theory of Elasticity*, McGraw-Hill Book Co., Inc., New York, 1951.

13. R. G. Belie and F. J. Appl, "Stress Concentration in Tensile Strips with Large Circular Holes," *Experimental Mechanics*, Vol. 12, No. 4, April, 1972.

14. A. J. Durelli and I. M. Daniel, "A Nondestructive Three-dimensional Strain-analysis Method," *ASME Transactions*, Paper 60-WA-16, 1960.

15. H. Tramposch and G. Gerard, "An Exploratory Study of Three-dimensional Photothermoelasticity," *ASME Transactions*, Paper 60-WA-69, 1969.

16. M. V. V. Murthy, "Stresses Around an Elliptic Hole in a Cylindrical Shell," *ASME Publication Paper No.* 69-APM-G, 1969.

17. D. N. Pierce and S. I. Chou, "Stresses Around Elliptic Holes in Circular Cylindrical Shells," *Experimental Mechanics*, Vol. 13, No. 11, November, 1973.

18. C. H. Tsao, A. Ching, and S. Okubo, "Stress-concentration Factors for Semi-elliptical Notches in Beams Under Pure Bending," *Experimental Mechanics*, March, 1965.

19. R. Hicks, "The Design of Reinforced Elliptical Holes in Pressure Vessels," *British Welding Journal*, March, 1958.

20. G. J. Schoessow and E. A. Brooks, "Analysis of Experimental Data Regarding Certain Design Features of Pressure Vessels," *ASME Transactions*, Vol. 72, No. 5, July, 1950.

21. V. Vicentini, "Stress-concentration Factors for Superposed Notches," *Experimental Mechanics*, Vol. 7, No. 3, March, 1967.

22. R. Shea, "Dynamic Stress-concentration Factors," *Experimental Mechanics*, Vol. 4, No. 1, January, 1964.

23. S. J. Fang, J. H. Hemann, and J. D. Achembach, "Experimental and Analytical Investigation of Dynamic Stress Concentrations at a Circular Hole," *ASME Publication Paper* 73-WA/APM-12, 1973.

24. H. Becker, "An Exploratory Study of Stress Concentrations in Thermal Shock Fields," *ASME Transactions*, August, 1962.

25. A. F. Emery, J. A. Williams and J. Avery, "Thermal-Stress Concentration Caused by Structural Discontinuities," *Experimental Mechanics*, December, 1969.

26. C. E. Taylor, N. C. Lind and J. W. Schweiker, "A Three-dimensional Study of Stresses Around Reinforced Outlets in Pressure Vessels," *ASME Transactions*, Paper 58-A-148, 1958.

27. E. C. Rodabaugh and R. L. Cloud, "Proposed Reinforcement Design Procedure for Radial Nozzles in Spherical and Cylindrical Shells with Internal Pressure," *Welding Research Council Bulletin* 133, September, 1968.

28. F. Ellyin and N. Turkkan, "Limit Analysis of Nozzles in Cylindrical Shells," *ASME Publication Paper* 72-WA/PVP-1, 1972.

29. T. J. Atterbury and C. W. Bert, "A New Approach for Design of Reinforcement for Thin Shells," *ASME Paper No.* I-4, Second International Conference on Pressure Vessel Technology, San Antonio, Texas, 1973.

30. W. L. Greenstreet, R. C. Gwaltney, J. P. Callahan, and J. F. Bryson, "ORNL Nozzles Analysis Program," Third Annual Progress Report on Studies in Applied Solid Mechanics, ORNL-4821, December, 1972.

31. K. Lingaiah, W. P. T. North, and J. B. Mantle, "Photoelastic Analysis of an Asymmetrically Reinforced Circular Cut-out in a Flat Plate Subjected to Uniform Unidirectional Stress," *Experimental Mechanics*, December, 1966.

32. P. G. Hodge, "Full-strength Reinforcement of a Cutout in a Cylindrical Shell" *ASME Publication Paper* 64-APM-41, 1964.

33. I. Berman and D. H. Pai, "An Experimental Investigation of Stresses in an HY-80 Marine Boiler Drum," *Welding Research Council Supplement*, July, 1962.

34. W. F. Riley and H. Kraus, "Experimental Determination of Stress Distribution in Thin-Walled Cylindrical and Spherical Pressure Vessels with Circular Nozzles and a Review and Evaluation of Computer Programs for the Stress Analysis of Pressure Vessels," *Welding Research Council Bulletin* 108, September, 1965.

35. R. T. Rose, "Stress Analysis of Nozzles in Thin Walled Cylindrical Pressure Vessels," *British Welding Journal*, Vol. 12, No. 2, February, 1965.

36. E. C. Rodabaugh and E. O. Walters, "Review of Service Experience and Test

Data on Openings in Pressure Vessels with Non-Integral Reinforcing," *Welding Research Council Bulletin* 166, October, 1971.

37. L. M. Cassidy and C. H. Coogan, Jr., "Stress Concentrations at Reinforced Openings in Ellipsoidal Pressure Vessel Heads," *ASME Publication Paper* 64-PET-3, 1964.

38. C. E. Taylor, N. C. Lind, M. M. Leven, and J. L. Mershon, "Photoelastic Study of the Stresses Near Openings in Pressure Vessels," *Welding Research Council Bulletin* 133, April, 1966.

39. F. J. Witt, R. C. Gwaltney, R. L. Maxwell, and R. W. Holland, "A Comparison of Theoretical and Experimental Results from Spherical Shells with a Single Radially Attached Nozzle," *ASME Publication Paper* 66-WA/PVP-4, 1966.

40. J. Decock, "Determination of Stress Concentration Factors and Fatigue Assessment of Flush and Extruded Nozzles in Welded Pressure Vessels," *ASME Paper No.* II-59, Second International Conference on Pressure Vessel Technology, San Antonio, Texas, 1973.

41. M. M. Lemcoe, "Cyclic Pressure Tests of Large Size Pressure Vessels," Report No. 17, March 15, 1960, Southwest Research Institute.

42. M. M. Frocht and D. Landsberg, "Factors of Stress Concentration Due to Elliptical Fillets," *ASME Transaction*, September, 1959.

43. J. B. Mahoney and V. L. Salerno, "Analysis of a Circular Cylindrical Shell Perforated by a Large Number of Radial Holes," *ASME Publication Paper* 66-WA/PVP-1, 1966.

44. M. M. Leven, "Stress Distribution at Two Closely-spaced Reinforced Openings in a Pressurized Cylinder" Westinghouse Research Laboratories Report 71-9E7-PHOTO-R1, April 9, 1971, Third Annual Progress Report on Studies in Applied Solid Mechanics, Oak Ridge National Laboratories, December, 1972.

45. R. M. Stone and Hochschild, "The Effect of Nozzle Spacing on the Pressure Stresses at the Intersection of Cylindrical Nozzles and Shells," *ASME Publication Paper* 66-WA/PVP-7, 1966.

46. D. Decock, "Fatigue Tests on Headers," ASME Paper No. II-64, Second International Conference on Pressure Vessel Technology, San Antonio, Texas, 1973.

47. J. L. Mershon, F. Ellyin, M. M. Leven, and R. Fidler, "Interpretive Report on Oblique Nozzle Connections in Pressure Vessel Heads and Shells Under Internal Pressure Loadings," *Welding Research Council Bulletin* 153, August, 1970.

48. R. C. Gwaltney and W. L. Greenstreet, "Comparisons of Theoretical and Experimental Stresses for Spherical Shells Having Single Non-radial Nozzles," ASME Paper No. I-7, Second International Conference on Pressure Vessel Technology, San Antonio, Texas, 1973.

49. D. E. Johnson, "Stresses in a Spherical Shell with a Non-radial Nozzle," *ASME Publication Paper* 67-APM-H, 1967.

50. I. M. Daniel, "Photoelastic Analysis of Stresses Around Oblique Holes," *Experimental Mechanics*, Vol. 10, No. 11, November, 1970.

51. C. A. Rav, Jr., "Elastic-Plastic Strain Concentrations Produced by Various Skew Holes in a Flat Plate Under Uniaxial Tension," *Experimental Mechanics*, March, 1971.

52. J. C. M. Yu and W. A. Shaw, "Stress Distribution of a Cylindrical Shell Non-radially Penetrated into a Spherical Pressure Vessel," ASME Paper No. I-9, Second International Conference on Pressure Vessel Technology, San Antonio, Texas, 1973.

53. P. Stanley and B. V. Day, "Stress Concentrations at Offset-Oblique Holes in Thick-walled Cylindrical Pressure Vessels," ASME Paper No. I-12, Second International Conference on Pressure Vessel Technology, San Antonio, Texas, 1973.

54. ASME Boiler and Pressure Vessel Code, Section III.

55. H. J. Grover, S. A. Gordon, and L. R. Jackson, "Fatigue of Metals and Structures," U. S. Government Printing Office, Washington, D. C., 1960.

56. R. E. Peterson, "Relation Between Life Testing and Conventional Tests of Materials," *ASTM Bulletin No.* 133, March, 1945.

57. B. F. Langer, "Application of Stress Concentration Factors," Paper TID-4500, Department of Commerce, April, 1960.

58. N. E. Frost, "Correlation of the Alternating Stresses Required to Initiate and Propagate a Fatigue Crack in Mild Steel," *Nature*, pp. 223–234, July, 1960.

59. M. Terao and Y. Yamashita, "Low Cycle Fatigue Strength of Thick-Walled Cylinders," ASME Paper II-55, Second International Conference on Pressure Vessel Technology, San Antonio, Texas, 1973.

60. L. F. Kooistra, E. A. Lange, and A. G. Pickett, "Full-size Pressure-vessel Testing and Its Application to Design," *ASME Publication Paper* 63-WA-293, 1963.

61. R. F. Newman, "Fatigue—The Problem in Relation to Welded Structures, Part IV," *The Welder*, pp. 8–15, January-March, 1961.

62. A. W. Pense and R. D. Stout, "Influence of Weld Defects on the Mechanical Properties of Aluminum Alloy Weldments," *Welding Research Council Bulletin* 152, July, 1970.

63. S. T. Carpenter and R. F. Linsenmeyer, "Weld Flaw Evaluation," *Welding Research Council Bulletin* 42, September, 1958.

64. H. Thielsch, "When are Weld Defects Rejectable?" Second Conference on the Significance of Defects in Welds, London, 1968, The Welding Institute, Abington Hall, Abington, Cambridge, England.

65. H. Greenberg, "An Engineering Basis for Establishing Radiographic Acceptance Standards for Porosity in Steel Weldments," *ASME Transactions*, December, 1965.

66. J. D. Harrison, "The Basis for a Proposed Acceptance Standard for Weld Defects, Part 1: Porosity, March, 1971, Part 2: Slag Inclusions, July, 1972. The Welding Institute, Abington Hall, Abington, Cambridge.

67. D. V. Lindh and G. M. Peshak, "The Influence of Weld Defects on Performance," *Welding Council Supplement*, February, 1969.

68. H. Kihara, M. Watanabe, Y. Tada, and Y. Ishii, "Effect of the Flaws in Welds on Their Strength," Welding Research Council, Welding Research Abroad, April, 1962.

69. A. G. Pickett, S. C. Grigory and C. G. Robinson, "Effect of Weld Defects on The Fatigue Strength of A302-B Steel," Southwest Research Institute, San Antonio, Texas, January, 1966.

70. R. Weck, "A Rational Approach to Standards for Welded Construction," *Welding Research Council, Welding Research Abroad,* Vol. XIII, No. 2, February, 1967.

71. S. Lundin, "Consequences of Defects in Welds," *Welding Research Abroad, Welding Research Council*, Vol. XVII, No. 2, February, 1971.

72. J. H. Rogerson, "Defects in Aluminum Welds and Their Influence on Quality," Second Conference on the Significance of Defects on Welds, London, 1968. The Welding Institute, Abington Hall, Abington, Cambridge, England.

73. S. J. Maddox, "An Analysis of Fatigue Cracks in Fillet Welded Joints," The Welding Institute, January, 1973, Abington Hall, Abington, Cambridge, England.

74. M. Hetényi, "A Photoelastic Study of Bolt and Nut Fasteners," *ASME Journal of Applied Mechanics*, June, 1943.

75. J. C. Wilhoit, Jr. and J. E. Merwin, "An Experimental Detemination of Load Distribution in Threads," *ASME Publication Paper* 64-PET-21, 1964.

76. R. L. Marino and W. F. Riley, "Optimizing Thread-root Contours Using Photoelastic Methods," *Experimental Mechancs*, Vol. 4, No. 1, January, 1964.

77. J. D. Chalupnik, "Stress Concentrations in Bolt-thread Roots," *Experimental Mechanics*, September, 1968.

78. A. Schwartz, Jr., "New Thread Form Reduces Bolt Breakage," *Steel*, Vol. 127, pp. 86–94, September, 1950.

79. Fastener Symposium, Standard Pressed Steel Co., Jenkintown, Pa., December, 1958.

80. S. M. Arnold, "Effect of Screw Threads on Fatigue," *Mechanical Engineering*, July, 1943.

81. "New Thread Design Doubles Life of Nut-bolt Combinations," *The Iron Age*, January 14, 1960.

82. R. E. Weigle and R. R. Lasselle, "Experimental Techniques in Predicting Fatigue Failure of Cannon-breech Mechanisms," *Experimental Mechanics*, February, 1965.

83. W. O. Clinedinst, "Strength of Threaded Joints for Steel Pipe," *ASME Publication Paper* 64-PET-1, 1964.

84. R. Ollis, Jr., "Self-aligning Nuts," *Machine Design*, June 21, 1962.

85. H. E. Sliney, "High Temperature Solid Lubricants-When and Where to Use Them," *ASME Publication Paper* 73-DE-9, 1973.

86. I. Berman and N. M. Kramarow, "Large Steel Vessels for Deep-Submergence Simulation," *ASME Publication Paper* 66-WA/UNT-13, 1966.

87. S. M. Jorgensen, "Closures and Shell Joints for Large High Pressure Cylinders," *ASME Publication Paper* 68-PVP-9, 1968.

88. C. Rviz and M. I. R. El Magrissy, "Flangeless Closure Joints: Experimental Validation of an Unconventional Design," ASME Paper No. I-44, Second International Conference on Pressure Vessel Technology, San Antonio, Texas, 1973.

89. I. McFarland, "A New Fabrication Method for Large High-Pressure Vessels," ASME Paper II-84, Second International Conference on Pressure Vessel Technology, San Antonio, Texas, 1973.

90. P. W. Bridgeman, *Studies on Large Plastic Flow and Fracture*, McGraw-Hill Book Co., Inc., New York, 1952.

91. L. P. Zick, "Stresses in Large Horizontal Cylindrical Pressure Vessels on Two Saddle Supports," *Journal American Society of Welding*, Vol. 30, 1951.

92. R. S. Wozniak, "Seismic Design of Thin Shell Supports," *ASME Publication Paper* 72-PET-23, 1972.

93. J. D. Wilson and A. S. Tooth, "The Support of Unstiffened Cylindrical Vessels," ASME Paper I-6, Second International Conference on Pressure Vessel Technology, San Antonio, Texas, 1973.

94. D. H. Cheng and N. A. Weil, "The Junction Problem of Solid-Slotted Cylindrical Shells," *ASME Transactions, Paper* 60-APM-2, 1960.

95. N. A. Weil and F. S. Rapasky, "Experience with Vessels of Delayed-coking Units," American Petroleum Institute, May 13, 1958.

96. P. J. Heggs, "Design of the Thermal Distribution in a Reactor Vessel Wall Resulting in Acceptable Thermal Stresses," *ASME Publication Paper* 71-WA/HT-23, 1971.

97. D. H. Cheng and N. A. Weil, "Axial Thermal Gradient Stresses in Thin Cylinders of Finite Length," *ASME Publication Paper* 65-WA/MET-17, 1965.

98. L. C. Shao and P. S. Hunt, "Effect of Thermal Stresses on Design of Steam Generators for 330-MW(e) HTGR Plant of Public Service Company of Colorado," *ASME Publication Paper* 71-WA/HT-27, 1971.

99. F. E. Wolosewick, "Supports for Vertical Pressure Vessels," *Petroleum Refiner*, No. 8, 1951.

100. P. P. Bijlaard, "Stresses from Radial Loads in Cylindrical Pressure Vessels," *The Welding Journal*, 33 (12) *Research Suppl.* 615-s to 623-s (1954).

101. P. P. Bijlaard, "Stresses from Radial Loads and External Moments in Cylindrical Pressure Vessels," *The Welding Journal* 32 (12) *Research Suppl.* 608-s–617-s (1955).

102. P. P. Bijlaard, "(1) Stresses in a Spherical Vessel from Local Loads Transferred by a Pipe; (2) Additional Data on Stresses in Cylindrical Shells Under Local Loading," *Welding Research Council Bulletin* 50, May, 1959.

103. P. P. Bijlaard and E. T. Cranch, "An Experimental Investigation of Stresses in the Neighborhood of Attachments to a Cylindrical Shell," *Welding Research Council Bulletin* 60, May, 1960.

104. F. A. Leckie and R. K. Penny, "(1) A Critical Study of the Solutions for the Asymmetric Bending of Spherical Shells; (2) Solutions for the Stresses at Nozzles in Pressure Vessels; (3) Stress Concentration Factors for the Stresses at Nozzle Interrections in Pressure Vessels," *Welding Research Council Bulletin* 90, September, 1963.

105. B. F. Langer, "PVRC Interpretive Report of Pressure Vessel Research, Section 1—Design Considerations," *Welding Research Council Bulletin* 95, April, 1964, New York.

106. P. P. Bijlaard, "Computation of the Stresses from Local Loads in Spherical Pressure Vessels or Pressure Vessel Heads," *Welding Research Council Bulletin* 34, March, 1957.

107. P. P. Bijlaard, "(1) Stresses in a Spherical Vessel from Radial Loads Acting on a Pipe; (2) Stresses in a Spherical Vessel from External Moments Acting on a Pipe; and (3) Influence of a Reinforcing Pad on the Stresses in a Spherical Vessel Under Local Loading," *Welding Research Council Bulletin* 49, April, 1959.

108. K. R. Wichman, A. G. Hooper, and J. L. Mershon, "Local Stresses in Spherical and Cylindrical Shells due to External Loadings," *Welding Research Council Bulletin* 107, August, 1965.

109. D. H. White, "Experimental Determination of Plastic Collapse Loads for Cylindrical Shells Loaded Radially Through Rigid Supports," ASME Paper I-25, Second International Conference on Pressure Vessel Technology, San Antonio, Texas, 1973.

110. J. W. Hansberry and N. Jones, "Elastic Stresses Due to Axial Loads on a Nozzle which Intersects a Cylindrical Shell," ASME Paper I-10, Second International Conference on Pressure Vessel Technology, San Antonio, Texas, 1973.

111. A. S. Kobayashi, B. G. Wade and D. E. Maiden, "Photoelastic Investigation on the Crack-arrest Capability of a Hole," *Experimental Mechanics*, January, 1972.

112. J. I. Bluhm and M. M. Mardirosian, "Fracture-Arrest Capabilities of Annulary Reinforced Cylindrical Pressure Vessels" *Experimental Mechanics*, March, 1963.

113. W. S. Pellini and P. P. Puzak, "Fracture Analysis Diagram Procedures for the

Fracture-Safe Engineering Design of Steel Structures," Welding Research Council Bulletin 88, May 1963.

114. S. Timoshenko, *Strength of Materials*, Part I, D.Van Nostrand Co., Inc., Princeton, N. J., 1955.

115. B. F. Langer and W. L. Harding, "Material Requirements for Long-Life Pressure Vessels," *ASME Publication Paper* 63-WA-194, 1963.

116. A. Cowan and R. W. Nichols, "Assessment of Steels for Nuclear Reactor Pressure Vessels," *ASME Transactions*, October, 1964.

117. O. Kellermann, E. Kraegeloh, K. Kussmaul, and D. Sturm, "Considerations About the Reliability of Nuclear Pressure Vessels Status and Research Planning," ASME Paper No. I-2, Second International Conference on Pressure Vessel Technology, San Antonio, Texas, 1973.

118. P. E. Erickson and W. F. Riley, "Minimizing Stress Concentrations Around Circular Holes in Uniaxially Loaded Plates," *Experimental Mechanics*, March 1978.

119. S. I. Chou and Om P. Chaudhary, "Stresses Around a Rib–Reinforced Elliptic Hole in a Circular Shell Under Tension," *ASME Publication Paper* 76-PVP-13, 1976.

120. D. R. Hayhurst, C. J. Morrison and F. A. Leckie, "The Effect of Stress Concentrations on the Creep Rupture of Tension Panels," *ASME Publication Paper* 75-APM-R, 1975.

122. ASME Boiler and Pressure Vessel Code, Section III, Nuclear Power Plant Components.

121. J. DeCock, "Reinforcement Method of Openings in Cylindrical Pressure Vessels Subjected to Internal Pressure," Welding Research Abroad, Welding Research Council, Nov. 1975, New York, N.Y.

122. R. A. Heller and T. Chiba, "Alleviation of the Stress Concentration with Analogue Reinforcement," *Experimental Mechanics*, December, 1973.

123. J. W. Bryson, W. G. Johnson and B. R. Bass, "Stresses in Reinforced Nozzle-Cylinder Attachments Under Internal Pressure Loading Analyzed by the Finite-Element Method–A Parameter Study," Oak Ridge National Laboratory Report ORNL/NUREG-4, October, 1977.

124. M. B. Bickell and S. H. Dance, "Stress Concentration Factors for Nozzles in Cylindrical Vessels Subjected to Internal Pressure," Third International Conference on Pressure Vessel Technology, Tokyo, Japan, 1977.

125. A. S. Kobayashi, N. Polvanich, A. F. Emery and W. J. Love, "Corner Crack at a Nozzle," Third International Conference on Pressure Vessel Technology, Tokyo, Japan, 1977.

126. M. J. G. Broekhoven, "Fatigue and Fracture Behavior of Cracks at Nozzle Corners; Comparison of Theoretical Predictions with Experimental Data," Third International Conference on Pressure Vessel Technology, Tokyo, Japan, 1977.

127. S. Miyazono, S. Ueda, T. Kodaira, K. Shibata, T. Isozaki and N. Nakajima, "Fatigue Behavior of Nozzles of Light Water Reactor Pressure Vessel Model," Third International Conference on Pressure Vessel Technology, Tokyo, Japan, 1977.

128. G. D. Whitman and R. H. Bryan, "Heavy-Section Steel Technology Program Quarterly Progress Report," ORNL/NUREG/TM-194, May, 1978.

129. F. K. W. Tso, J. W. Bryson, R. A. Weed, and S. E. Moore, "Stress Analysis of Cylindrical Pressure Vessels with Closely Space Nozzles by the Finite-

Element Method," Oak Ridge National Laboratory Report ORNL/NUREG-18/VI, November, 1977.

130. F. Arav and J. G. Bolten, "The Strength of Y-Branch Pieces Under Internal Pressure," Third International Conference on Pressure Vessel Technology, Tokyo, Japan, 1977.

131. J. D. Harrison, "The Analysis of Fatigue Test Results for Butt Welds and Penetration Defects Using a Fracture Mechanics Approach," Fracture, 1969, *Proc. 2nd Int. Conf. on Fracture,* Brighton, Chapman and Hall, 1969.

132. T. R. Gurney, "Finite Element Analysis of Some Joints with the Welds Transverse to the Direction of Stress," The Welding Research Institute, March, 1975.

133. G. C. Robinson, J. E. Smith, R. W. Derby and G. D. Whitman, "Destructive Testing of Flawed 6-In.-Thick Pressure Vessel," *ASME Publication Paper No.* 74-Mat-5, 1974.

134. S. T. Rolfe and J. M. Barsom, *Fracture and Fatigue Control in Structures, Applications of Fracture Mechanics,* Prentice-Hall, New Jersey, 1977.

135. T. Kanazana and A. S. Kobayashi, *Significance of Defects in Welded Structures,* University of Tokyo Press, Japan, 1974.

136. G. Barthdlome and G. Vasoukis, "Analysis of Defects in Welds," *ASME Publication Paper No.* 75-PVP-14, 1975.

137. C. D. Lundin, "The Significance of Weld Discontinuities-A Review of Current Literature," *Welding Research Council Bulletin 222,* December, 1976.

138. C. F. Boulton, "Acceptance Levels of Weld Defects for Fatigue Service," American Welding Society, January, 1977.

139. B. R. Ellingwood, "An Improved Basis for Formulating Acceptance Criteria for Structural Welds," *ASME Publication Paper* 78-PVP-2, 1978.

140. R. A. Buchanan and D. M. Young, "Effect of Porosity on Elevated Temperature Fatigue Properties of 2-1/4-1 Mo Steel Weldments," *Welding Journal,* September, 1975.

141. K. Iida, T. Kawai, K. Hirakawa, T. Okazawa, K. Fujisawa, "Low Cycle Fatigue Test of Hemispherical Pressure Vessel with Misaligned Joint," Third International Conference on Pressure Vessel Technology, Tokyo, Japan, 1977.

142. J. W. Knight, "Improving the Strength of Fillet Welded Joints by Grinding and Peening," The Welding Institute, 8/1976/E, March, 1976, Cambridge, England.

143. C. P. Burger, L. W. Zachary and W. F. Riley, "Application of Photoelasticity to a Weld-Penetration Problem," *Experimental Mechanics,* February, 1975.

144. J. D. Burk, and F. V. Lawrence, Jr., "Effects of Lack-of-Penetration and Lack-of-Fusion on the Fatigue Properties of 5083 Aluminum Alloy Welds," *Welding Research Council Bulletin 234,* January, 1978.

145. T. J. Kiernun and K. Nishida, "The Buckling Strength of Fabricated HY-80 Steel Spherical Shells," David Taylor Model Basin Report 1721, July 1966, Washington, D.C.

146. S. R. Heller, Jr., "Structural Design of Spherical Shells Subjected to Uniform External Pressure," *Naval Engineers Journal,* December, 1974, Washington, D.C.

147. J. W. Fisher and J. H. A. Struik, *Guide to Design Criteria for Bolted and Riveted Joints,* John Wiley & Sons, New York, 1974.

148. K. P. Singh, "Study of Bolted Joint Integrity and Inter-Tube-Pass Leak-

age in U-Tube Heat Exchangers, Part I and Part II," *ASME Publication Papers Nos.* 77-WA/NE-6 and 7, 1977.

149. R. C. Landt, "Criteria for Evaluating Bolt Head Design," *ASME Publication Paper No.* 76-DE-30, 1976.

150. P. D. Stevens-Guille and W. A. Crago, "Application of Spiral Wound Gaskets for Leak-Tight Joints," *ASME Publication Paper No.* 74-WA/PVP-5, 1974.

151. C. Ruiz, "Static and Dynamic Seals: Part I-Static Seals," *ASME Publication Paper No.* 76-DET-83, 1976.

152. M. O. M. Osman, W. M. Mansour and R. V. Dukkipati, "On the Design of Bolted Connections with Gaskets Subject to Fatigue Loading," *ASME Publication Paper No.* 76-DET-57, 1976.

153. R. J. Pick and H. D. Harris, "Morrison and Parry Seals for Water Pressures up to 345 MPa," *ASME Publication Paper No.* 78-PVP-13, 1978.

154. S. M. Jorgensen, "The Shear Stud Closure," *ASME Publication Paper No.* 74-PVP-17, 1974.

155. R. Pechacek, "High-Pressure, Quick Acting Closure for Large Diameter, Full Opening, Nuclear and Petro-Chem Pressure Vessels," *ASME Publication Paper No.* 78-PVP-74, 1978.

156. P. D. Stevens-Guille, J. E. Newmarch, C. R. Thorpe and R. J. Eccleston, "Sealing Forces for Leak-Tight Operation of a Self-Energized Pressure Vessel Closure," *ASME Publication Paper No.* 78-PVP-16, 1978.

157. G. Duthie and A. S. Tooth, "The Analysis of Horizontal Cylindrical Vessels Supported by Saddles Welded to the Vessel—A Comparison of Theory and Experiment," Third International Conference on Pressure Vessel Technology, Tokyo, Japan, 1977.

158. W. L. Greenstreet, "Structural Analysis Technology for High Temperature Design," *Nuclear Engineering and Design*, May, 1977.

159. A. Blake, "Formulas for Canister and Pipe Design in Underground and Nuclear Emplacement," *ASME Publication Paper* 74-PVP-2, 1974.

160. E. C. Rodabaugh, S. K. Iskander and S. E. Moore, "End Effects on Elbows Subjected to Moment Loading," Oak Ridge National Laboratory Report ORNL/SUB-2913/7, March 1978.

161. E. C. Rodabaugh and R. C. Gwaltney, "Elastic Stresses at Reinforced Nozzles in Spherical Shells with Pressure and Moment Loading," Oak Ridge National Laboratory Report ORNL/Sub/2913-4, October, 1976.

162. R. C. Gwaltney, M. Richardson and R. L. Battiste, "Room-Temperature Inelastic Test of a Nozzle-to-Spherical-Shell Model Subjected to Pressure and Moment Loading," Oak Ridge National Laboratory Report ORNL/TM-6170, April 1978.

163. F. Ellyin, "Experimental Investigation of Limit Loads of Nozzles in Cylindrical Vessels," *Welding Research Council Bulletin 219*, September, 1976.

164. F. Ellyin, "An Experimental Study of Elasto-Plastic Response of Branch-Pipe Tee Connections Subjected to Internal Pressure, External Couples and Combined Loadings."

R. L. Maxwell and R. W. Holland, "Collapse Test of a Thin-Walled Cylindrical Pressure Vessel with Radially Attached Nozzle," *Welding Research Council Bulletin, 230*, September 1977.

165. J. Schroeder and P. Tuğcu, "Plastic Stability of Pipes and Tees Exposed to External Couples," *Welding Research Council Bulletin 238*, June 1978.

166. D. Brock, *Elementary Engineering Fracture Mechanics*, Noordhoff International Publishing, Leyden, The Netherlands, 1974.

167. J. R. Maison and W. L. Server, "Comparative Evaluation of Fracture Analytical Methods; A Case Study," *ASME Publication Paper No.* 73-WA/Oct-22, 1973.

168. M. Brumovsky, R. Filip and S. Stepanak, "Initiation and Arrest—Two Approaches to Pressure Vessel Safety," Third International Conference on Pressure Vessel Technology, Tokyo, Japan, 1977.

•169. B. G. Wade and A. S. Kobayashi, "Photoelastic Investigation on the Crack-Arrest Capability of a Pretensioned Stiffened Plate," *Experimental Mechanics*, February, 1975.

170. B. J. Hogen, "Dispersed Rubber Particles Toughen Flywheels," *Design News*, January 17, 1977.

171. C. P. Burger, "Photoelastic Modeling of Stresses Caused by Thermal Shock," *Experimental Mechanics*, March, 1976.

172. J. M. Bloom and W. A. Van Der Sluys, "Determination of Stress Intensity Factors for Gradient Stress Fields," *ASME Publication Paper No.* 76-WA/PVP-10, 1976.

173. R. D. Cheverton "Thermal Shock Investigations," Quarterly Progress Report on Reactor Safety Programs sponsored by the NRC Division of Reactor Safety Research for October-December 1974, ORNL-TM-4805, Vol. II, March, 1975.

174. Y. Ueda and K. Fukuda, "Application of Finite Element Method for Analysis on Process of Stress Relief Annealing," *Transactions of the Japan Welding Society*, Vol. 8, No. 1, April 1977.

175. G. R. Egan and J. N. Robinson, "The Application of Elastic-Plastic Fracture Mechanics Parameters in Fracture Safe Design," *Nuclear Engineering and Design*, January, 1978.

176. D. P. Johnson, "Cost Risk Optimization of Nondestructive Inspection Level," *Nuclear Engineering and Design*, January, 1978.

177. J. B. Hanus and C. P. Burger, "Stress-Concentration Factors for Elliptical Holes Near an Edge", *Experimental Mechanics*, September 1981.

178. W. L. McBride and W. S. Jacobs, "Design of Radial Nozzles in Cylindrical Shells for Internal Pressure," ASME Publication Paper No. 79-PVP-14, 1979.

179. F. K. W. Tso and R. A. Weed, "Stress Analysis of Cylindrical Pressure Vessels with Closely Spaced Nozzles by the Finite-Element Method," Oak Ridge National Laboratory, NUREG/CR-0507, May 1979.

180. E. C. Rodabaugh "Review of Data Relevant to the Design of Tubular Joints for Use in Fixed Offshore Platforms," Welding Research Bulletin 256, January 1980.

181. D. J. Walsh and G. E. Woods, "Determination of Stress Intensification Factors for Integrally Reinforced 45° Lateral Branch Correction," ASME Publication Paper No. 79-PVP-98, 1979.

182. J. Schroeder, "Limit and Burst Pressures of Fabricated Branch Connections," ASME Publication 82-PVP-16, 1982.

183. W. W. Sanders and R. H. Day, "Fatigue Behavior of Aluminum Alloy Weldments," Welding Research Council Bulletin 286, August 1983.

184. R. C. Cipolla and J. L. Grover, "Significance of Centerline Lack-of-Penetration Defects in Double Seam Welded SA-312 Pipe", ASME Publication Paper No. 81-PVP-33, 1981.

185. T. R. Gurney, "Revised Fatigue Design Rules," WRC Welding Research Abroad, October 1983.

186. K. Masubuchi, *Analysis of Welded Structures*, Pergamon Press, New York, 1980.

187. R. B. Davis, "Experimental Study of Fatigue Crack Initiation Due to Rapid Thermal Cycling in Pressure Vessel Studs," ASME Publication Paper No. 79-PVP-109, 1979.

188. H. Kraus, "Interpretive Report on Gasket Leakage Testing," Welding Research Council Bulletin 262, October 1980.

189. "Seals," *Machine Design*, Vol. 51, No. 15, June 28, 1979.

190. T. J. McCuistion, "O-Ring and Gland Design for High Temperature Seals," *Product Engineering*, January 1956.

191. H. R. Stephens, "High Pressure Teflon Seals and Closures," *Plastics Technology*, February 1957.

192. J. Wilcox, "Nonstandard Groove Dimensions for O-Rings," *Machine Design*, March 25, 1982.

193. A. R. Freeman, "Gaskets for High Pressure Vessels", *Mechanical Engineering*, December 1952.

194. W. Coopey, "Holding High-Pressure Joints," *Petroleum Refiner*, May 1956.

195. W. B. Daniels and A. A. Hruschka, "Seals for Pressures to 10,000 Atmospheres," *The Review of Scientific Instruments*, Volume 28, Number 12, December 1957.

196. D. P. Willis, "Fighting Friction with Bonded Coatings," *Machine Design*, April 10, 1980.

197. K. P. Singh and V. K. Lik, "An Approximate Analysis of Foundation Stresses in Horizontal Pressure Vessels," ASME Publication Paper No. 79-NE-1, 1979.

198. "Penstock Analysis and Stiffener Design, Boulder Canyon Project, Part V—Technical Investigations," Bulletin 5, United States Department of the Interior, Bureau of Recalmation, 1940.

199. G. Duthie and A. S. Tooth, "The Analysis of Horizontal Cylindrical Vessels Supported by Saddles Welded to the Vessel—A Comparsion of Theroy and Experiment," Third International Conference on Pressure Vessel Technology, Tokyo, 1977.

200. A. K. Shoemaker, "Relationships Between Mechanical Properties and the Extension and Arrest of Unstable Cracks in Line Pipe Steels," ASME Publication Paper No. 79-PVP-76, 1979.

201. G. T. Hahn and M. F. Kanninen, "Crack Arrest Methodology and Applications", American Society for Testing and Materials, STP-711, 1980.

202. W. G. Clark, Jr., "Effect of Spherical Discontinuities on Fatigue Crack Growth Rate Performance," *Fracture Mechanics*, ASTM STP 677, American Society for Testing and Materials, 1979.

203. T. R. Hsu, W. J. McAllister, and D. A. Scarth, "Reinforcement of Perforated Metal Sheets by Thermal Shock with Laser Beams," *Experimental Mechanics*, August 1982.

204. N. J. I. Adams, "A Damage Tolerant Design and Inspection Philosophy for Nuclear and Other Pressure Vessels," ASME Publication Paper No. 79-PVP-129, 1979.

205. M. Inagaki, "Approaches to Quality Assurance Scheme of Pressure Vessels

and Some Cases for Evaluation of Their Welding Quality in Japan," *Transactions of National Research Institute for Metals*, Vol. 22, No. 4, 1980; *WRC Welding Research Abroad*, October 1981.

206. M. H. Jawad and J. R. Farr, *Structural Analysis and Design of Process Equipment*, Wiley-Interscience, New York, 1983.

207. J. H. Devietian and W. E. Wood, "Factors Affecting Porosity in Aluminum Welds—A Review," Welding Research Council Bulletin 290, December, 1983.

208. A. Bazergui and L. Marchand, "PVRC Milestone Gasket Tests—First Results," Welding Research Council Bulletin 292, February, 1984.

209. W. F. Marks, "Visualizing Stress Discontinuities by Flow Analogy", Machine Design, March 9, 1984.

210. J. H. Bickford, *An Introduction to the Design and Behavior of Bolted Joints*, Marcel Dekker, Inc. New York, 1981.

211. C. Crispell, "New Data on Fastener Fatigue," *Machine Design*, April 22, 1982.

212. A. Bazergui, "Short Term Creep and Relaxation Behavior of Gaskets," Welding Research Council Bulletin 294, May 1984.

213. F. Ellyin and D. Kujawski," Plastic Strain Energy in Fatigue Failure," ASME publication Paper 84-PVP-28, 1984.

214. D. Green, R. Parker and D. Marsh, "Comparison of Theoretical Estimates and Experimental Measurements of Fatigue Crack Growth Under Severe Thermal Shock Conditions," ASME Publication Paper 84-PVP-35, 1984.

215. K. Rajaiah and N. K. Naik, "Hole-Shape Optimization in a Finite Plate in the Presence of Auxiliary Holes," *Experimental Mechanics*, June, 1984.

216. J. L. Mershon, K. Mokhtarian, G. V. Ranjan, and E. C. Rodabaugh, "Local Stresses in Cylindrical Shells Due to External Loadings on Nozzles—Supplement to WRC Bulletin No. 107," Welding Research Bulletin 297, August 1984.

7

Fabrication, Innovation and Economics

7.1 Pressure Vessel Economics

Engineering advancements and developments are axiomatically economic in nature, since these are the only commercial incentives that bear fruit. The definition of an engineer as one who can accomplish with cents what others can with dollars, keynotes his introduction to engineering but frequently loses its impact prior to starting his professional career.

Pressure vessels made before 1930 were fabricated by riveting together plate courses and butt straps which previously had been punched, drilled, and reamed to form various shapes, Fig. 7.1. The labor and time consumed were expensive. Such procedure was ex-

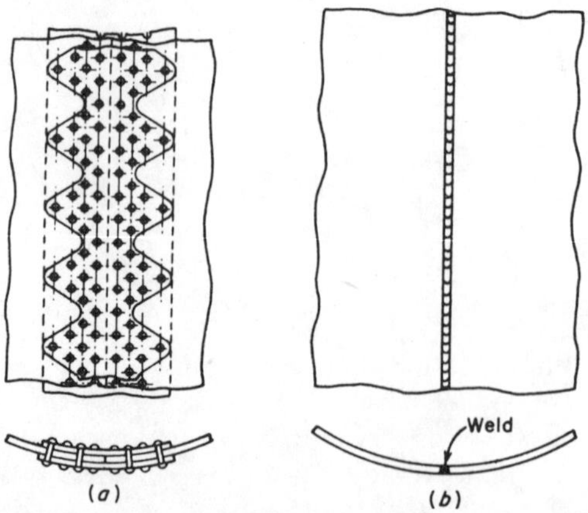

Fig. 7.1. (*a*) Typical Riveted Joint of the Butt and Double-Strap Type Used in Vessels Prior to the Advent of Welded Construction, (*b*) Typical Welded Joint Design of Comparable Vessel

pensive also from a material viewpoint; since the efficiency of a riveted joint seldom exceeds 85 percent, the entire basic shell thickness beyond the riveted joint was appreciably thicker than required. This posed an economic incentive to the vessel manufacturer which was augmented by the requirements of the power industry for higher pressures to reduce the cost per kilowatt of energy, all of which led to the development of welded construction. This trend continues throughout the process industries, and now extends to the nuclear power field with the same prime motivator–reduction in cost.[130]

The keen business competition in today's markets has encouraged a tendency to emphasize initial price as the major or sole criterion, and to overlook the test of shrewd purchasing or product design which lies in achieving the lowest total cost: initial investment plus operating and maintenance cost over useful life. The second part of this equation is easily overlooked, yet may be far greater than the initial investment. This is an increasingly important factor from three viewpoints:

1. The major expense item in operating and maintenance costs is manpower, and wages will continue to increase.

2. Downtime, or total operating loss of a facility, is becoming a major loss source. As production rates increase and manufacturing operations become more complex, even the loss of small periods of production time becomes intolerable. This is particularly applicable to the continuous process industries, such as those in the chemical and petroleum field, and to power plants for utilities supplying large populations, or for ships at sea.

3. The nuclear power vessel, with its extremely limited maintenance accessibility because of radiation hazards, has made maximum quality of prime consideration for these vessels.

Engineering is the profession of accomplishing the most for the least, but with full cognizance of usage requirements, both technical and economic.

7.2 Economic Considerations and Cost Reduction

The economic trend in all industrial plants is toward large-size equipment in order to reduce capital investment and operating costs by eliminating the duplicity of servicing appurtenances, controls, etc. associated with equivalent capacity multiple separate small units. Pressure vessels follow this same economic trend.[1,2] Their construction must be sufficiently strong and reliable, yet embody the maximum saving in materials. This, together with present-day commercial competition, presents a formidable economic factor in pressure vessel design. Reduction in weight involves an increase in the allowable working stresses of present materials, or the use of new high-strength materials now under development. This can be permitted with safety only on the basis of a thorough stress analysis of the structure, and after careful experimental investigation of the physical properties of the material in its environment.

Three general avenues of approach to cost reduction are:

A. Engineering Design
B. Materials of Construction
C. Methods of Fabrication

Cost reduction is not a single avenue approach, but must be chosen consistent with the result desired; for instance,

A. Maintaining present costs is a normal cost accounting province consisting of a continual review to ensure adherence to an established procedure.
B. Reducing costs moderately to offset predetermined direct labor increases, etc., is a procedure of "Upgrading" all existing practices. This is the perfection of a method and consists of a critical review of each item under the "Cost-of-function" microscope. This is often called "value engineering."
C. Slashing costs radically to revive a dwindling product market is a new solution to the cost compatibility equation, SIMPLICITY = STRENGTH = ECONOMY, employing creative engineering, newest materials and latest fabrication methods as input data. It must be remembered that all the input data to this equation is time-dependent; hence, so is the answer.

The important thing in an economically competitive market is the speed with which a new engineering design, new material or new fabrication method is put into effect because a favorable position lasts for only the very short time it takes for competition to catch up. Frequently, the slowness with which a new concept is adopted is explained as a case of "walking before you run"—perhaps, but it can be a case of walking backwards with the disastrous end result of high costs.

7.3 Engineering Design and Innovation

Pressure component design is both an art and a science, and the responsibility of the design engineer is to conceive the whole as an economical combination of simple components.[131] The applicable engineering economics is a full circle reiterative process which is illustrated in Fig. 7.2—and has three basic spokes:

1. Definition: define the problem
2. Creation: provide solutions by analysis
3. Selection: establish the most economical solution

Since there are many comparable engineering design approaches, procedures and formulas for quickly establishing basic sizes, thicknesses, profiles, materials, etc., are essential. These are necessary to permit selection of more promising designs and to allow a thorough stress analysis of these to be made, since the evaluation of indeterminate stresses requires finished geometrics to work from. One way that the designer, sitting at his desk or drawing board, can improve the economy and time schedule of design is to maintain an up-to-date, at-hand collection of the material used daily, supplemented by an extensive general reference library. He should acquaint himself with the technical periodicals giving the latest developments in the mechanics of materials, so that advantage can be taken of the work already done, thus giving time for further efforts. The cheapest research is that which has already been done—and much remains to be done.[131, 132]

Another main avenue of approach to cost reduction is through the very cost of engineering design itself; i.e., the cost and time of obtaining and applying knowledge to the optimum solution of a problem. When a structure can be analyzed by the simple equations of static

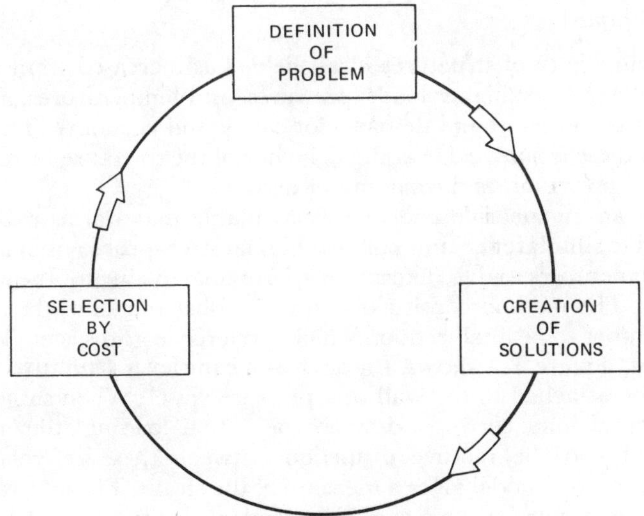

Fig. 7.2 Engineering Design Process

equilibrium, namely,

$$\sum F_x = 0$$
$$\sum F_y = 0$$
$$\sum M = 0$$

standardized procedures and formulas are available to establish quickly basic sizes, thicknesses, etc. Many structures, however, are not adaptable to such direct analytical solutions and are said to be indeterminate. They must first be sized geometrically in order for a stress analysis to be made, with the optimum size of members determined by successive approximations. The procedure is usually long, tedious and time consuming.[134]

One method approach, whether to the solution of a new problem or the review of an old one, that has much to recommend it from the viewpoint of engineering design economy is the utilization of rubber models to observe strain distribution. It is admittedly approximate, but applicable to establishing first sizing of indeterminate structures. It has a limited order of accuracy, but better than none at all when tackling a new problem. It may be thought to be more qualitative than quantitative, but:

1. It is cheap.
2. It is quick.
3. It is visual.
4. It is available to the engineer right at his desk or drawing board.

The complexity of structures in all fields has increased tremendously with demands for higher loads, pressures, and temperatures, and with it exists the ever-present demand for safety and economy. The key to each of these is knowledge of the behavior of the structure to guarantee safety of operation and economy of design.

Such an inexpensive and readily available material as rubber can be used to simulate the uniaxial and biaxial stress problems in irregular shaped members and to determine approximate values of the localized stresses. The method consists of using a rubber model of the shape of the member to be analyzed, on which a reference grid system has been laid out. Figure 7.3 shows a model of a cantilever structural lug that might be attached to the wall of a pressure vessel. When subjected to an external load corresponding to the actual loading, the model is distorted and the relative distortion between these reference lines throughout the model gives a measure of the strain. The general strain pattern is apparent, and the high strains occurring at the abrupt

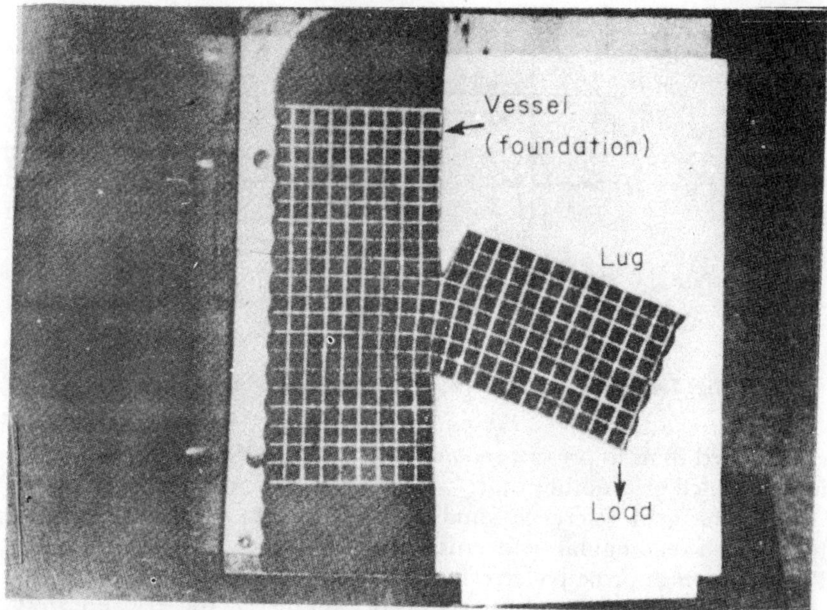

Fig. 7.3 Strain Distribution in a Flat Rubber Model of a Short Cantilever Structural Attachment with an Undercut Subjected to External Loading. This Also Visually Depicts the Behavior of a Crack Tip in a Ductile Material, Par. 5.22.3B

change in section are noticeable as is also their attenuation into the vessel wall (foundation).*

Tensile or compressive strains are measured by the change of distance between lines, and the shearing strain by the angular distortion of the grid squares for reference lines, Fig. 7.4. The stress values throughout the model can be determined in the usual manner from these strain measurements. A plot of the principal stress trajectories throughout the entire member to give a graphic location and appraisal of critical regions is also often helpful. This can be done by starting with equally spaced principal stress trajectories at the boundaries or locations where uniform strain, hence stress, exists, i.e., where the grid reference lines are not distorted angularly but are only stretched or

*A technique of this type is also used to make direct visual observations of the interaction of soil with various structural members such as foundations, tunnels, and retaining walls. See E. T. Selig, "A Technique for Observing Structure–Soil Interaction," *Materials Research and Standards*, September, 1961; also C. Brémond and A. J. Durelli, "Experimental Analysis of Displacements and Shears at Surfaces of Contact," *Experimental Mechanics*, March 1981.

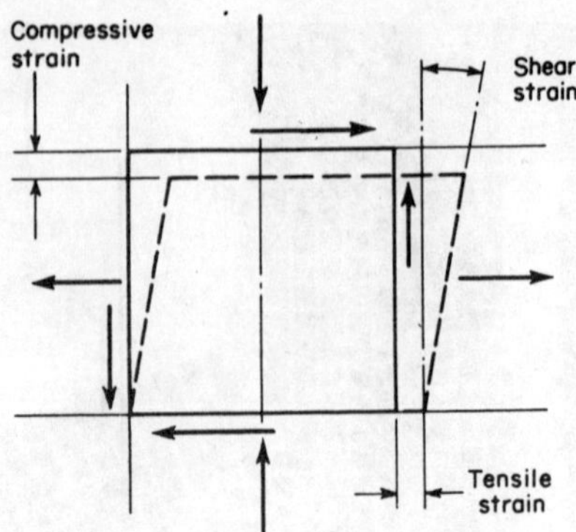

Fig. 7.4. Combined Tensile, Compressive, and Shear Strains

compressed so as to present rectangles. Then proceed to the surrounding reference grid noting that:

1. If the grid reference squares merely stretch or shrink so as to maintain a rectangular grid unit, the grid lines correspond locally to the direction of principal stress.

2. If the grid reference squares distort angularly, the grid lines are not coincident with the principal stress but are at an angle thereto. This is due to shear action; and the direction it causes the principal stress trajectories to rotate may be determined by noting that, if the shearing stresses were the only ones acting, the principal stresses would be on one of the 45 degree planes, Fig. 7.5*a*, and if the normal stress were the only one acting, it would be the maximum or principal stress, and hence its plane with the normal stress is equal to zero, Fig. 7.5*b*. If both shearing and normal stresses act, the plane of the principal stress will be between the plane on which the normal stress acts and the 45 degree plane—the greater the value of the shear stress relative to the normal stress, the closer to the 45 degree plane the principal stress trajectory moves.

The strain measurements can be made directly on the model, or if more convenient, on a photograph of the model. The latter has the advantage of also serving as a record for review or extended application to future projects. The rubber model should be as large as possible so that the grid spacing can be relatively small compared to the model geometry, and the distortion readily measurable between grid lines. Although this method is subject to inaccuracies caused by errors arising from the nonlinear stress-strain curve of rubber and by the fact that the gross distortion necessary to render the strains measurable may in itself alter the strain distribution throughout the model, it

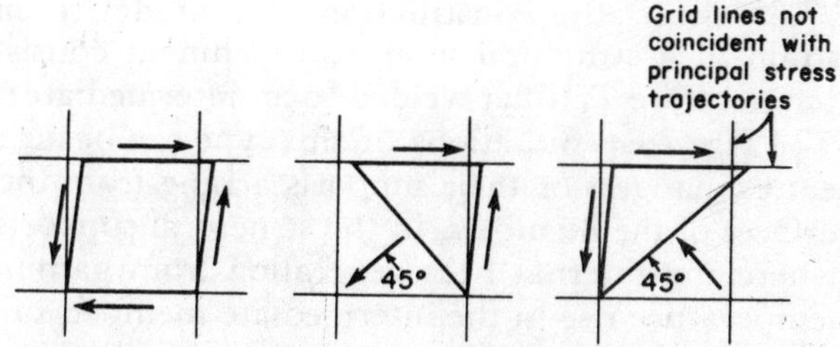

(a) Principal stress direction due to simple shear only.

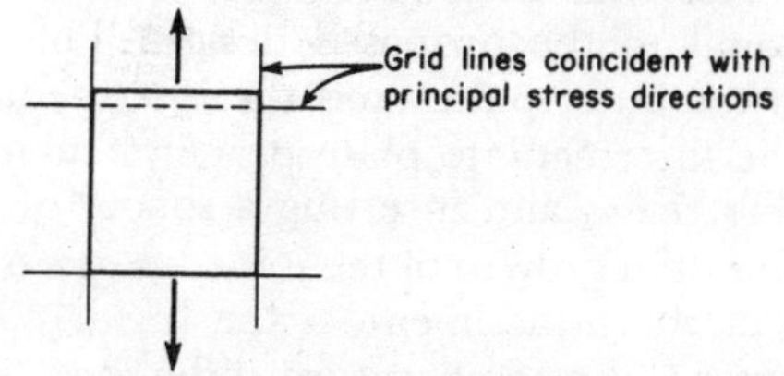

(b) Normal stress direction coincident with principal stress.

Fig. 7.5. Relation of Grid System to Principal Stress Trajectories

gives a vivid graphic view of the general stress pattern over the entire surface of the model, pinpoints the location of maximum stress, and offers approximate values of these stresses. The closer the stress trajectories, the higher the principal stress in this region; and the sharper their curvature, the higher the shear stress. This method is adaptable to complicated structures, such as vessel saddles or attachments that can be readily modeled with sheet rubber and cement.

Models of this type are also readily adaptable to the solution of problems of thermal stress, which are becoming of increasing importance in the design of many structures, including pressure vessels for the petroleum, chemical, and nuclear industries. Such stresses are produced by restricting the natural growth of a material induced by temperature. The restriction may be an external constraint that prevents free expansion of the entire body, or, as is more generally the case, a varying temperature throughout the structure whereby the free thermal growth of each fiber is influenced by that of those surrounding it. The result is that fibers at high temperatures are compressed and those at lesser temperatures are stretched. Herein lies the fundamental difference between thermal and mechanical stresses, namely, that the differential thermal expansion requires only that a prescribed strain pattern be satisfied, and the accompanying stress pattern need satisfy only the requirements for equilibrium of the internal forces.

Figure 7.6*b* shows the construction of a model to simulate the thermal strain in a structural support attachment consisting of two plates at temperature T_1 fillet welded to an intermediate plate at temperature T_2, Fig. 7.6*a*. Situations of this type can occur in extended surface heat exchangers or their supports arising from the touch-gap contact surfaces of the members, in the structural supports of nuclear reactors where the internal heat generation from gamma radiation causes a temperature rise in the intermediate member, or in the connection of bimetallic materials.

The model of Fig. 7.6*b* was constructed of rubber $\frac{1}{2}$ in. thick, geometrically proportional to the proposed design. The differential thermal strain between the outer plates and the intermediate plate is simulated by cutting the intermediate plate perpendicular to its axis (along the line *ab* of Fig. 7.6*a*) and inserting a spacer of a thickness proportional to the differential growth of the joined parts, $\Delta e = \alpha B \Delta T$, where B is the length of the attachment, α the linear coefficient of thermal expansion, and ΔT the temperature difference of the outer and innerplates $(T_1 - T_2)$, Fig. 7.6*c*. The model shows visually the thermal strain pattern throughout the member including the fillet welds. It also allows an estimation of the relative amount of strain absorbed in the welds as compared to the straight portions of the members, i.e., an estimation of the flexibility of the welds. For instance, the difference between the thickness of the spacer insert and the sum of the total elongation in the outer plates, e_1, and shortening the intermediate plate, e_2, is attributable to the weld strain, e_w, i.e.,

$$\Delta e_{\text{total}} = e_1 + e_2 + e_w$$

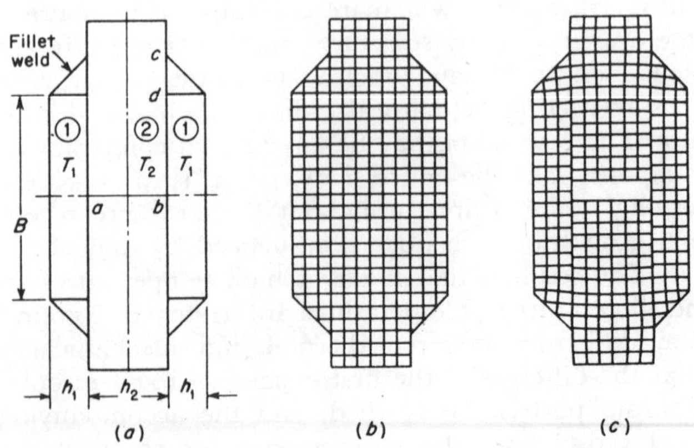

Fig. 7.6. Rubber Model of Fillet Welded Attachment. (*a*) Plan of Model, (*b*) Model Grid System, (*c*) Simulated Thermal Strain Distribution

When the attachment is short, weld strain or flexibility is an important item in establishing the thermal gradient that the joint can endure. As the joint is made longer the effect of weld flexibility becomes proportionally less and can be neglected for very long joints.

The manner in which the thermal stress load is transferred through the fillet weld can also be determined from the model. The relative variation in load transfer across the weldment, line *cd*, Fig. 7.6*a*, can be approximated from the slope of the grid lines (measuring shear) at this location, Fig. 7.6*c*, or from the proximity of the principal stress trajectories at this location, Fig. 7.7. The closer together the stress trajectories are, the higher the stress at this location. It is seen that the load is higher near the root of the weld and diminishes somewhat toward the toe much in a trapezoidal manner, Fig. 7.8*a*. This means that the usual assumption that all bodies are rigid so that the load would be transferred uniformly along the length of the weld is not strictly correct, but that elastic deformation alters this basic concept of mechanics.[3] It is similar to the manner in which the major portion of the load in a bolt is transferred to the nut by the first few engaged threads, though to a lesser degree, since clearances, fits, etc., play no part in a welded connection. To the design engineer it points out that, although for nonrepetitive loading it is satisfactory to design the weldment on the basis of uniform load per unit length of weld, for fatigue loading it may be preferable to reduce this peak stress. This can be

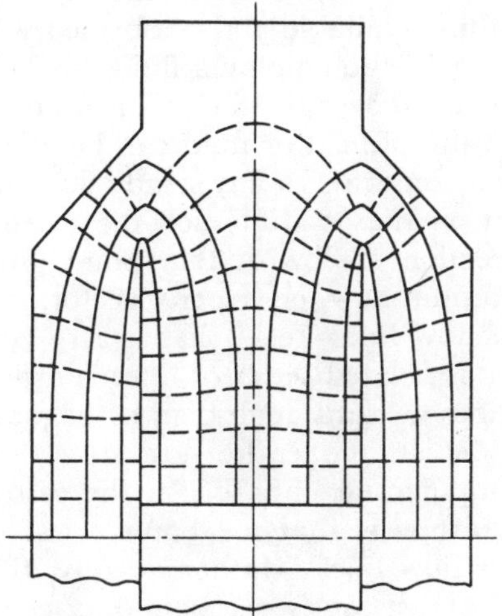

Fig. 7.7. Principal Thermal Stress Trajectories Established from the Strain Distribution of Fig. 7.6*c*

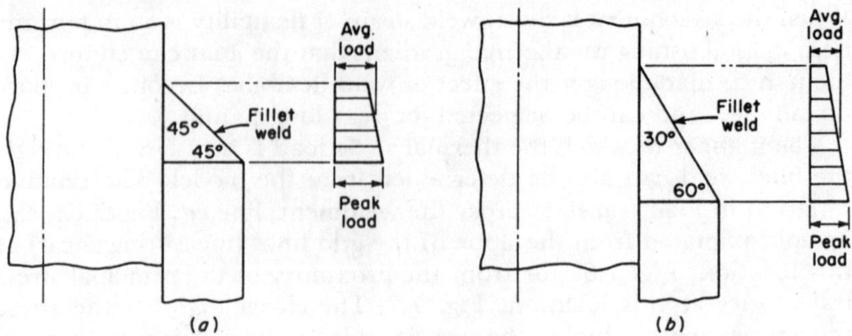

Fig. 7.8. Approximate Variation in Load Along the Transfer Surface of Fillet Welds

done in a manner similar to that which has been used successfully in bolts, namely, increasing the length of engagement by changing from the conventional 45° fillet weld to a 30° to 60° angle, Fig. 7.8*b*. This does not eliminate the load variation along the weldment, but it does reduce the peak value at the root of the weldment whence fatigue failures are prone to originate.[92]

Solutions to problems in two-dimensional heat flow are required in order to determine temperature gradients and are a prerequisite to thermal stress analysis. Exact analytical solutions are difficult and time consuming, even with the aid of computer equipment, but they are susceptible to approximate inexpensive and quick model evaluations using materials readily available and performed with office or home facilities. One approach is to employ a fluid flow analogy.

Basic equations which cover the flow of heat and fluid are similar.[4] In each case a quantity of heat or fluid can be equated to a variable term multiplied by a constant. In any steady flow problem, a series of lines can be drawn on the conduit such that their direction at any point gives the direction of flow at that point. Since flow lines are tangent to the maximum flow, or velocity vector, it follows that there is no flow across a flow line; i.e., it has a zero component at right angles. If, then, an associated series of lines is drawn such that they intersect the flow lines at right angles, an orthogonal network results which graphically depicts the entire flow state. The latter series of lines are called equipotential lines, or, in the case of heat flow, isotherms. It is a requirement that the boundaries coincide with particular isotherms, or flow lines. Mathematically, this means that the general solution also must satisfy the boundary conditions. Since the flow pattern is orthogonal, this gives rise to the fact that it applies equally as well when the source and sink are interchanged, or when the flow paths of one problem are identical to the equipotential lines of the other. Thus, the solution of any one problem always implies the solution of three others.

Using the following nomenclature:

A = flow area, sq. ft.
A_m = mean flow area, sq. ft.
h_s = thickness of shell, ft.
h_t = thickness of nozzle, ft.
k = thermal conductivity at temperature t, Btu per hr. ft.°F
L = depth, ft. (perpendicular to xy plane)
q = steady heat flow, Btu per hr.
t = temperature, °F
t_1 = temperature, hot face, °F
t_2 = temperature, cold face, °F
V = velocity, ft. per sec.
x = length, ft.
y = length, ft.
Δt = temperature difference, °F.

Figure 7.9 shows a conduit of variable thickness, bounded on the surface EF by an isothermal plate of temperature t_1, and on the other surface by a colder isothermal plate CD of temperature t_2, so that heat flow is from EF to CD. There is no heat flow in direction L, which is perpendicular to the plane of the figure. The heat may be visualized as flowing through a series of narrow lanes bounded by flow lines and flowing, with an equal quantity q, in each lane, from the isotherm EF to some point on the isotherm CD.

The quantity of heat flow at each lane is:[5]

$$q = -\frac{\int_{t_1}^{t_2} kdt}{\int_1^2 \frac{dx}{A}} = -\frac{A_m}{x}\int_{t_1}^{t_2} kdt$$

The integral $\int kdt$ may be divided into any number of equal parts, $k\Delta t_x$. Considering a small segment $OPQR$, the heat flows at right angles to the mean area $A_m = yL$, where y is the mean width of the quadrilateral $OPQR$ and L is the length (perpendicular to the plane of Fig. 7.9) of the conduit. The flow of heat throughout the lane is steady, $q = kA_m \Delta t_x/x = kyL \Delta t_x/x$, and since $k\Delta t_x$ is the same from one iso-

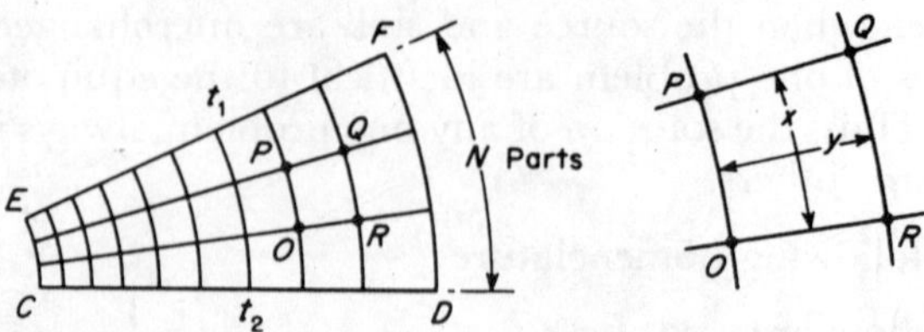

Fig. 7.9. Diagram Illustrating Steady Two-dimensional Flow Through Conduit of Variable Length

therm to the next, and L is constant, the ratio y/x must be constant throughout the lane.

Even with such knowledge of the individual nature of the quadrilaterals of the orthogonal flow system, the plotting of these for an irregular conduit is a lengthy process. This can be greatly aided, however, by resorting to a comparable fluid analogy and the use of a simple model to observe the flow lines as brought out by threads, confetti, or dye introduced into the stream. Referring to a conduit of the same shape in Fig. 7.9, fluid flow is also considered from EF to CD. Again the fluid flow is considered as made up of a number of lanes bounded by flow lines and of quantity $q = AV$, where $A = yL$ is the flow area and V is the velocity. If now the velocity is represented by the distance between equipotential lines x, the product xy must be constant throughout the lane. Consequently, it follows that the flow patterns for both heat flow and fluid flow will be identical only if the individual quadrilaterals are made curvilinear squares, since the only condition under which $y/x = xy$ is when $x = 1$, $y = 1$. Thus, in plotting the fluid flow and equipotential lines, they should be made to intersect at right angles to give quadrilaterals such that y is substantially equal to x if a direct graphical comparison to the heat flow pattern is desired.

Figure 7.10 shows a simple fluid flow model consisting of a broad-crested right-angle weir of a shape change such as would occur in the region of a nozzle in a vessel.[6] The weir crest widths are made proportional to the nozzle and vessel wall thicknesses at their juncture. The flow lines are observed by attaching short silk threads to the weir face, Fig. 7.11*a*. When equipotential lines are drawn perpendicular to these flow lines, starting from regions removed from the crotch where the lines are equally spaced, their intersection with the downstream edge establishes the temperature of this boundary surface, Fig. 7.11*b*.

It was pointed out above that heat-flow and fluid-flow solutions are analogous, since the governing equations and boundary conditions

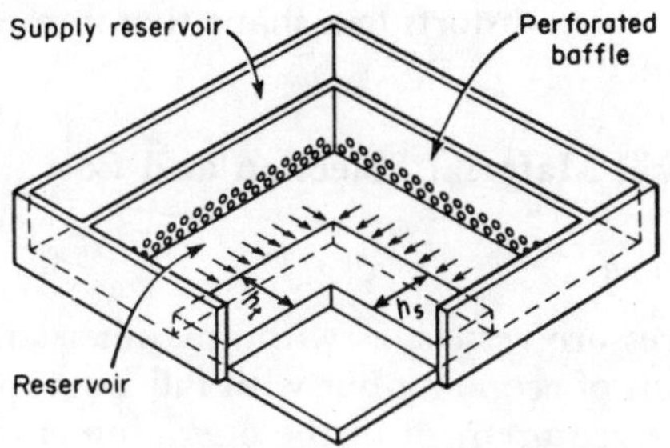

Fig. 7.10. Broad-Crested Weir Model for Observing Fluid Flow Pattern in the Region of a Shape Change

are equivalent. Although these are the required conditions of true analog, the practice of drawing superficial analogies, parallels, or aids from nature in establishing structural designs, appraising their worth, or fathoming the great amount of technical literature being generated daily is a helpful one. This is particularly so for the young engineer as he coalesces his knowledge with experience to wisdom. The human body, and the products of nature, subscribe to the same principle of adjusting to do the least amount of work, fewer aches, as does a

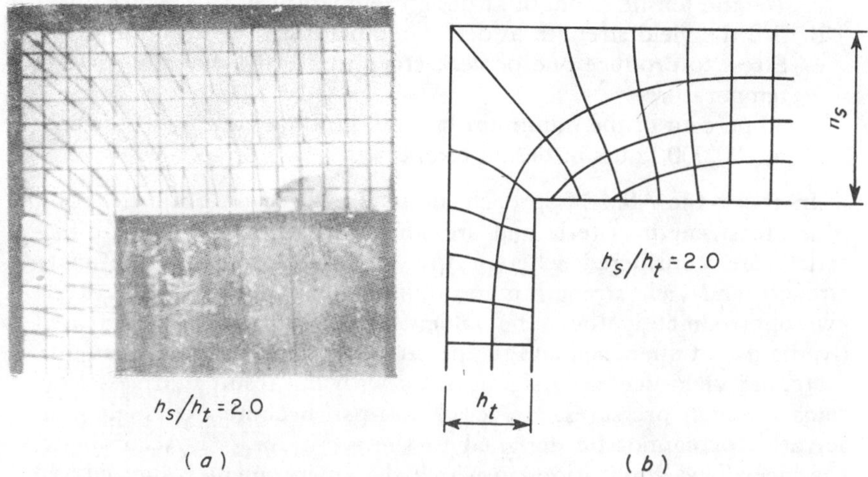

Fig. 7.11. (*a*) Photograph of Flow Path Traces Around a Nozzle Opening as Indicated by Silk Threads Attached to the Weir Face of an Analogous Fluid Flow Model, Ratio of Intersecting Thickness $h_s/h_t = 2.0$, (*b*) Orthogonal Flow Pattern as Plotted from the Traces of (*a*)

material structure which distorts to a shape that minimizes the stresses imposed by its load.

7.4 Pressure Vessel Material Selection and Cost

1. *Material Selection*

Materials for pressure vessels, as with any other structure, must be selected on the basis of economy but with full assurance of safely performing the service requirement in the operating environment. Since these demands are multiple, allowable design stresses are not based on a single material property, but on a combination of several that are considered paramount to safety. Each specification for a type of material embodies many physical property requirements such as tensile strength, yield strength, elongation, reduction in area, etc., and from these material properties construction codes establish allowable design stress values. For instance, the American Society of Mechanical Engineers Boiler and Pressure Vessel Code Section, VIII, Division I, Rules for Construction of Pressure Vessels, uses as the basis for establishing allowable design stresses for steels the lesser value of:

a. 1/4 the specified minimum tensile strength at room temperature.
b. 5/8 the specified minimum yield strength at room temperature.
c. 1/4 the tensile strength at design temperature.
d. 5/8 the yield strength at design temperature.
e. Stress to produce one percent creep in 100,000 hours at design temperature.
f. 80 percent of the minimum stress to produce rupture at the end of 100,000 hours at design temperature.

In this basic ASME approach, limits are based on both yield and ultimate strength criteria and are consistent when low-strength materials are considered wherein the relationship between ultimate strength and yield strength is approximately 2:1 and the two criteria give approximately the same value. As strength levels are increased by the use of quenched and tempered steels, the spread between ultimate and yield decreases significantly with the result that vessel designs for high pressures, above 10,000 psi, become extremely conservative or cannot be designed to these criteria.[93,94] Accordingly, for specialized applications in which the environment is limited and controlled, such as a hydraulic pressure accumulator vessel where no temperature effects are present, a criteria based solely on yield strength is in order because this is the one single material property that controls operation function (distortion) and is the precursor to failure.

This yield criteria is frequently employed in European countries and a factor of safety of 1.5 is used.

2. *Material Costs*

Materials costs, which account for 50 to 60 percent of the total cost of a pressure vessel, represent a possible major cost reduction source. This avenue of approach can be appraised by establishing a comparative material price index (allowable working stress)/(cost per pound of material), for materials that are suitable for the intended service environment. The higher this index value, the lower the base material cost. Such material cost comparisons generally point out two things. First, the lowest grade material that will satisfy the design requirements should be chosen because the higher the alloy constituents of the material, the higher the vessel cost. Second, the high strength quenched and tempered materials can offer a potential saving as compared to their annealed counterpart in acceptable environments..

The material comparative cost index does not tell the entire story since each material has its own fabricating peculiarities or required quality control measures which must also be considered in appraising its economic choice. Generally, cost goes up with increase in alloy content of the material and a higher total price vessel may result, even though a higher comparative index material is used giving a lower weight vessel. This trend is shown in the tabulation of Table 7.1 giving the approximate relative fabrication cost factors of several commonly used pressure vessel steels. These factors are largely independent of the type of fabrication machinery used, but will vary somewhat with the production process and operating efficiency of the individual fabricator.

The application of the material comparative cost index and fabrication cost factor to determine the break-even metal design temperature of two commonly used pressure vessel materials, carbon steel SA-212B and manganese-molybdenum steel SA-302B, is illustrated in Fig. 7.12. The strength of steel decreases with increase in temperature so the material cost index curve for these materials has been plotted using the allowable design stresses at temperature as given by the ASME Code Section VIII, Division 1, (In actual practice the allowable stress values of the applicable design specification or jurisdictional rules would be used). Curve *A* for the carbon steel and curve *B* for the manganese-molybdenum steel are based on the material comparative cost index only (stress/dollars per pound). These intersect at 750F indicating that based on material cost only it is more econom-

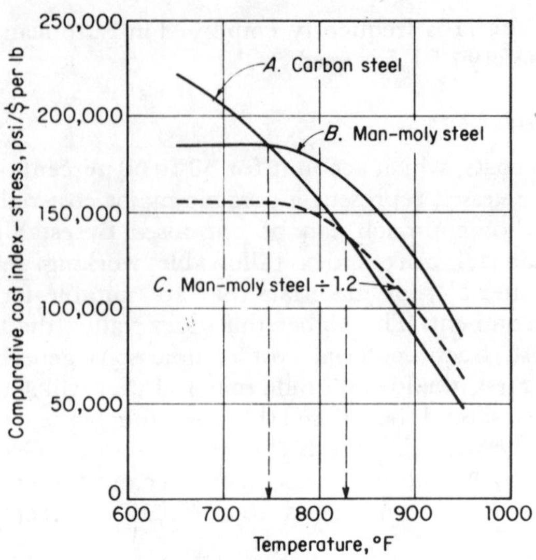

Fig. 7.12. Comparative Material Cost Index, Carbon Steel vs Manganese-Molybdenum Steel

ical to use manganese-molybdenum material above this temperature. However, when the relative fabrication cost factor of 1.2 from Table 7.1 is applied to the manganese-molybdenum material to reflect the actual fabricated vessel cost, curve *C*, the break-even temperature becomes 825F showing that only above this temperature do the economics favor the higher alloy manganese-molybdenum material vessel.

It is not always possible to select a vessel material solely on the basis of material cost and fabrication economy for the operating pressure and temperature, since other nonstress related requirements may govern. For instance, if the contained media is very corrosive to carbon steel but not to a higher alloy steel, the material choice is

TABLE 7.1. RELATIVE FABRICATION COST FACTORS FOR SEVERAL COMMONLY USED PRESSURE VESSEL MATERIALS

Type Material	*Relative Fabrication Cost Factor*
Carbon Steel	1.00
Manganese-Moly	1.20
1 Cr–1 Ni-Moly	1.25
$2\frac{1}{4}$ Cr-Moly	1.30
18 Cr–8 Ni Alloy	1.35

based on ensuring integrity of the metal. The reverse condition also exists, as in the case of vessels for nuclear service, when it is necessary to protect the media from contamination with the material corrosion crud in order not to endanger coolant flow by plugging any parts of the small nuclear fuel element passages. These requirements can be met by making the entire vessel of the necessary environmental resistant material. When this material is unduly expensive, the inside surface of the vessel in contact with the media can be clad with a thin layer of this material, and the base metal can be a less expensive but equally strong one for pressure. Cladding can be applied by forging, bonding or welding methods. Generally, the cost of cladding processes is more favorable for large diameter thick-walled vessels where material tonnage is high, and there is ready access for repairs and clad overlays at the joints of the shell courses. The thickness break-even point of vessels made with solid alloy versus alloy protected or clad material varies somewhat with the type and cladding process, but is about one inch. Other things being equal, economics favor the use of clad materials for vessel wall thickness greater than one inch. Once again, special environmental requirements, such as hydrogen embrittlement may not permit the use of clad materials even though the economics may point in this direction. Cladding thicknesses are kept to a small percentage of the vessel wall thickness—less than 10 per cent—and do not influence the behavior of the vessel. Tests have shown that the cladding process itself does not present a source of fatigue failures.[7] However, claddings are subject to thermal fatigue due to the differential thermal expansion coefficients of the thin cladding and the much thicker base material, Fig. 7.13*a*. With a uniform increase in temperature, the cladding thermal growth is entirely restricted, Eq. 2.11.3, by the base material so that

$$\sigma_{CL} = -\frac{ET(\alpha_{CL} - \alpha_B)}{1 - \mu} \tag{7.4.1}$$

$$T = -\frac{\sigma_{CL}(1 - \mu)}{E(\alpha_{CL} - \alpha_B)} \tag{7.4.2}$$

From Eq. 7.4.2, the temperature at which the cladding reaches its yield strength, or the shake-down temperature at which the cladding reaches twice yield strength and will then cycle elastically between an initial tensile prestress yield stress and a compressive yield stress, can be found.

Example: A very thick ferritic steel component has a thin protective cladding layer of stainless steel. Assuming the temperature is raised uniformly: (A) at

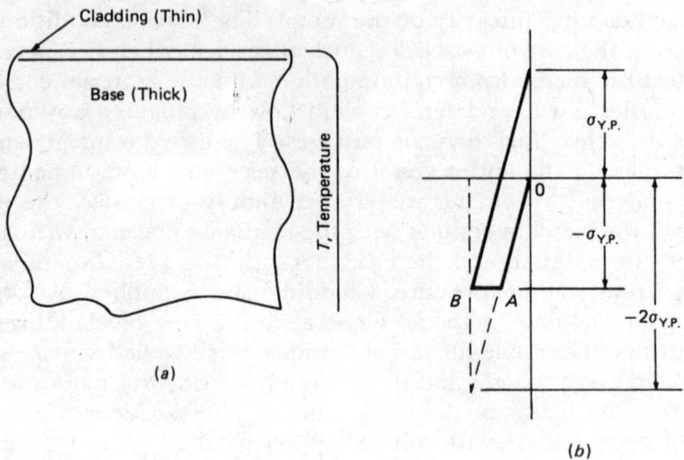

Fig. 7.13. (*a*) Cladding on Thick Component, (*b*) Cladding Stress-Strain Diagram

what temperature will the cladding material yield strength of $\sigma_{CL} = 25{,}000$ psi be reached; and (B) at what temperature does a condition of shake-down occur (twice yield strength) and through which the cladding will cycle elastically? Take $E = 25{,}000{,}000$ psi, $\mu = 0.3$, $\alpha_S = 0.00001$ in./in.-°F and $\alpha_f = 0.000008$ in./in.-°F.

Substituting these values in Eq. 7.4.2 gives:

$$T_A = \frac{\sigma_{CL}(1-\mu)}{E(\alpha_s - \alpha_f)} = \frac{25{,}000(1-0.3)}{25{,}000{,}000 \times 0.000002} = 350°\text{F}$$

This is the temperature at which the cladding yield strength is reached. When the component is returned to its initial temperature, it will have cycled through a stress range of 0 to −25,000 psi.

$$T_B = 2 \times T_A = 700°\text{F}$$

At this temperature, the cladding has become plastic and permanently strained so that upon return to its initial temperature, it has a residual prestress state of 25,000 psi, Fig. 7.13*b*. During subsequent increases in temperature to 700°F and return to the initial temperature, it will cycle elastically through a stress range of 25,000 psi to −25,000 psi.

2.1 Optimum Factors of Safety

Factors of safety are a trade-off means of establishing equal performance credibility and safety by consigning to a single parameter varying degrees of quality assurance (design analysis, material testing, fabrication control, and inservice inspection). This is the basis upon which Codes and Standards are based. The ASME has several codes based on different factors of safety, Table 7.2. For instance, it re-

TABLE 7.2. FACTORS OF SAFETY

ASME Code Section	*Factor of Safety, FS*	*Basis*	
		Theory of Failure	*Material Property*
Section VIII-3, Pressure Vessels	1.75	Maximum Shear	Material yield shear strength first reached throughout wall thickness
Section III, Nuclear Components Section VIII-2, Pressure Vessels	3	Maximum Shear	Average shear stress in wall thickness reaches material ultimate shear stress
Section I, Power Boilers Section VIII-1, Pressure Vessels	4	Maximum Stress	Average tensile stress in wall thickness reaches material ultimate tensile strength
Section IV, Heating Boilers	5	Maximum Stress	Average tensile stress in wall thickness reaches material ultimate tensile strength

quires an $SF = 5$ when the degree of design analysis, material testing, and fabrication control is very limited; while it permits a lowering of the FS to 4, to 3, and to 1.75 as the degree of quality assurance is successively increased. The justification for reducing the factor of safety is, to some extent, a matter of engineering judgement and may vary in detail from code to code, but is based on filling the gaps in our reliability knowledge, Fig. 1.17. Estimates of the cost of this knowledge vary somewhat with each plant facility and the resources at its disposal. Figure 7.14 shows such a trend based on the criteria listed. Starting from a base of $FS = 4$ at which point the cost breakdown of a typical vessel is approximately: Material = 55%, Fabrication = 35%, and Engineering = 10%; this shows the optimum FS to be in the range of $2\frac{1}{2}$ to $3\frac{1}{2}$. For factors of safety below this, the engineering cost of establishing credibility increases rapidly and assumes the major part of the total cost, while higher ones have an overriding material cost. Cost estimates of this type are based on commercial vessel constructions with availability of technical support in the material, fabrication and engineering disciplines; however, it must be pointed out that particular circumstances may well dictate otherwise. For instance, in fabricating simple low pressure heating boilers it is usually more economical of manufacturing plant operation to compensate for engineering skills by additional material costs. On the other hand, this approach could not be used for a deep-diving submersible where entire loss of function would result. The criticality of weight demands extremely low factors of safety with their attendant sophisticated engineering nondestructive examination and evaluation. Here, knowledge is expensive but essential.

Likewise, in the design of many chemical process vessels the optimum economical factor of safety of $2\frac{1}{2}$–$3\frac{1}{2}$ must be foregone in order to obtain their high pressure requirements, Fig. 7.14. Here equal safety with a lower FS is justified on the basis of increased knowledge, Fig. 1.17. The ASME Pressure Vessel Code Section VIII-3 is an example of this approach wherein FS = 1.75 is justified on the basis that[135] (1) in addition to the primary hoop stress requirements a fracture mechanics and fatigue analysis is made, (2) extended material qualification and examinations are performed, (3) fabrication quality control is mandatory, and (4) periodic in-service inspection is required to detect crack initation or propagation. For instance, by requiring a fracture mechanics analysis, the use of an extremely high strength but low toughness material that is inconsistent with the vessel's service requirements is avoided. This is illustrated in the following example.

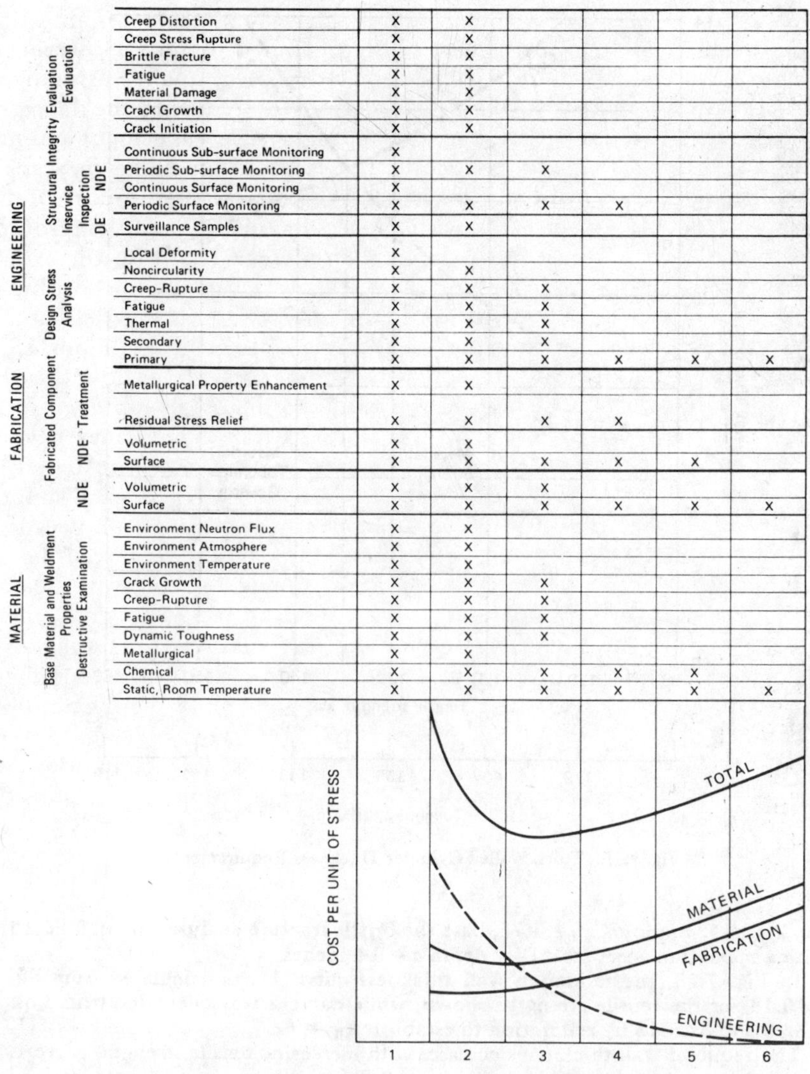

Category	Item	1	2	3	4	5	6
ENGINEERING — Structural Integrity Evaluation	Creep Distortion	X	X				
	Creep Stress Rupture	X	X				
	Brittle Fracture	X	X				
	Fatigue	X	X				
	Material Damage	X	X				
	Crack Growth	X	X				
	Crack Initiation	X	X				
ENGINEERING — Inservice Inspection Evaluation, NDE	Continuous Sub-surface Monitoring	X					
	Periodic Sub-surface Monitoring	X	X	X			
	Continuous Surface Monitoring	X					
	Periodic Surface Monitoring	X	X	X	X		
ENGINEERING — Inservice Inspection, DE	Surveillance Samples	X	X				
ENGINEERING — Design Stress Analysis	Local Deformity	X					
	Noncircularity	X	X				
	Creep-Rupture	X	X	X			
	Fatigue	X	X	X			
	Thermal	X	X	X			
	Secondary	X	X	X			
	Primary	X	X	X	X	X	X
FABRICATION — Fabricated Component, Treatment	Metallurgical Property Enhancement	X	X				
	Residual Stress Relief	X	X	X	X		
FABRICATION — Fabricated Component, NDE	Volumetric	X	X				
	Surface	X	X	X	X	X	
MATERIAL — Base Material and Weldment Properties, NDE	Volumetric	X	X	X			
	Surface	X	X	X	X	X	X
MATERIAL — Base Material and Weldment Properties, Destructive Examination	Environment Neutron Flux	X	X				
	Environment Atmosphere	X	X				
	Environment Temperature	X	X				
	Crack Growth	X	X	X			
	Creep-Rupture	X	X	X			
	Fatigue	X	X	X			
	Dynamic Toughness	X	X	X			
	Metallurgical	X	X				
	Chemical	X	X	X	X	X	
	Static, Room Temperature	X	X	X	X	X	X

Fig. 7.14. Effect of Factor of Safety on Cost of Pressure Components

Example: Determine the required thicknesses for a 24 inch inside diameter cylindrical vessel, subjected to an internal pressure of 43,500 psi, and fabricated of a $3\frac{1}{2}$% Ni-Cr-Mo-V steel obtainable in the tensile strength range and corresponding toughness values shown in Fig. 7.15. Base the wall burst pressure on

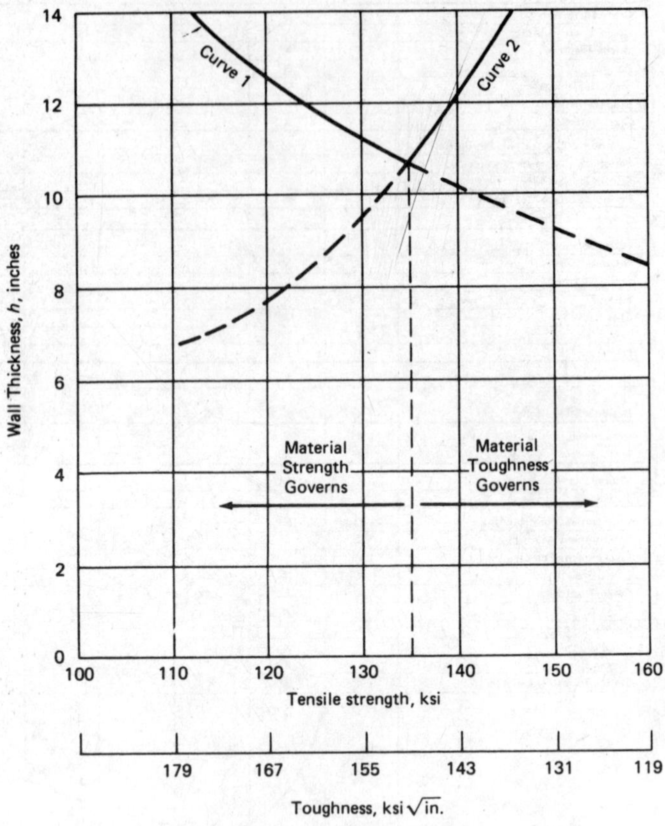

Fig. 7.15. Thick Walled Cylinder Thickness Requirements

Eq. 2.10.14, employ a FS = 1.75, base the brittle fracture analysis on Eq. 5.25.13 with a maximum acceptable flaw depth a = 0.4 inches.

In Fig. 7.15, the minimum wall thickness curve 1 was calculated from Eq. 2.10.14 for the tensile strengths shown, while curve 2 was calculated from Eqs. 2.8.20 and 5.25.13 by reiteration to establish $K_{IC} = K_I$.

The required wall thickness decreases with increasing tensile strength, curve 1, up to the point of intersection with curve 2. Beyond here toughness governs and the required thickness increases so as to reduce the applied stress in compensation for lower toughness. Hence, the optimum (minimum) thickness is found at the intersection point.

The exact equations for curves 1 and 2, Fig. 7.15, will vary according to the applicable design specification, code or regulation. For instance, the ASME code for High Pressure Vessels, Section VIII-3, employs for curve 1 a minimum thickness based on full plastic yielding of the entire wall thickness, Eq. 2.10.9, the maximum shear theory of failure, Eq. 5.15.28 and a FS = 1.75 to give

$$p = \frac{\sigma_{y.p.}}{1.75} \log_e K \tag{7.4.3}$$

where K = outside radius/inside radius; and for curve 2 a fracture analysis based on Eq. 5.25.14 is prescribed. The method of analysis is then the same as given above using toughness values for the specific material employed.

2.2 Rupture of Thick Walled Cylinders Subjected to Internal Pressure

Ruptures of thick walled monobloc cylinders are of three types characterized by the failure mode of the material at the time of fracture, Fig. 5.112.

1. *Brittle fracture failure mode.* Here the stress remains elastic up to the time of failure and fracture is brittle. The material toughness is low and this property, together with the size of the existing crack, governs failure; namely, linear elastic fracture mechanics, Par. 5.22.3A. Figures 5.92 and 5.93 illustrate this type of fracture with its characteristic flat surface exhibiting little ductility.
2. *Quasibrittle-ductile fracture failure mode.* Here the stress exceeds the yield strength in a local region adjacent to the crack edge at the time of failure. This elastic-plastic material behavior is of a higher toughness due to the increased plastic deformation at the crack tip, and failure is governed by yield in the presence of a crack; namely, crack opening displacement, COD, or general yield fracture mechanics, Par. 5.22.3B. Figure 7.16*a* shows the elastic-plastic fracture of a very thick cylinder, ratio OD/ID = 5.0. The fracture contains both brittle and plastic portions. The failure started at the inside surface where the tensile stress is the highest, and an existing radial crack prevailed, propagating in a radial direction and manner characterized by an elastic-plastic fracture mechanics evaluation, Par. 5.22.3B. The primarily flat radial fracture changes to a typical plastic 45° shear one exhibiting the typical jagged fracture face of an ordinary ductile tensile failure, Fig. 5.25. This latter feature is not necessarily a characteristic of elastic-plastic behavior, but the

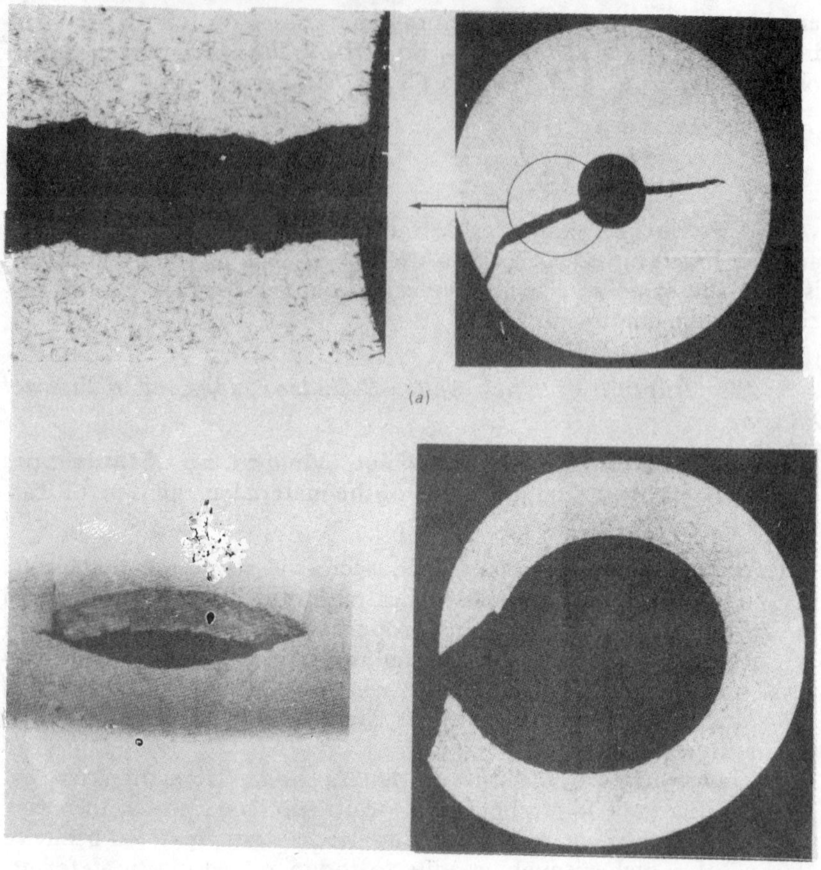

(a)

(b)

Fig. 7.16. Rupture of Thick Walled Monobloc Steel Cylinders Subjected to Internal Pressure. (*a*) Elastic-Plastic Behavior of a OD/ID = 5.0 Ratio Cylinder Showing a Mixed Brittle–Ductile Failure in the Presence of a Crack and Subscribing to General Yield Fracture Mechanics Evaluation. (*b*) Totally Plastic Behavior of a OD/ID = 3.0 Ratio Cylinder Showing a Ductile Failure. (Courtesy Battelle Columbus Laboratories)

fracture may be void of this final partial 45° degree shear behavior.

3. *Ductile fracture failure mode.* Here the stress is totally plastic throughout at the time of failure. Toughness of the material is high and failure is attributable to plastic instability; i.e., it is

independent of toughness. The plastic instability failure is governed by the flow stress, $\bar{\sigma}$, of the material. Flow stress is the stress level at which the material exhibits considerable plasticity. Two relationships for the calculation of the flow stress, as a function of the uniaxial yield and ultimate strength, that are frequently used are

$$\bar{\sigma} = \sigma_u , \tag{7.4.4}$$

or a midpoint between the yield and ultimate strength

$$\bar{\sigma} = (\sigma_{y.p.} + \sigma_u)/2, \tag{7.4.5}$$

Figure 7.16*b* shows the ductile fracture of a thick cylinder, ratio OD/ID = 3.0. The fracture is a ductile shear one starting at the inside where the stress is the highest, Eq. 2.8.21, and progressing throughout the wall thickness following a 45° direction to the maximum principle stress. It shows the typical jagged surface of a ductile failure, Par. 5.8, Figs. 5.24 and 5.25. The material of Fig. 7.16*b* is the same chemical composition as that of Fig. 7.16*a* but it has been annealed to reduce its strength properties and correspondingly enhance its toughness property. Hence, the same material can exhibit a brittle or ductile failure mode dependent upon its metallurgical condition.

3. *Composite Materials*

Steels and other metals used in the construction of pressure vessels are essentially isotropic, exhibiting similar mechanical properties in all directions, Par. 5.3. In such materials it is impossible to develop the full molecular bond strength, which is in excess of a million pounds per square inch, because it is impractical to manufacture these materials with perfect uniformity of structure or entirely free of defects. It is around these defects that stresses concentrate, and cracks opened up by externally induced stress propagate throughout the material so that fracture results at a fraction of the theoretical strength. In the late 1920's A. A. Griffith in England observed that the smaller the piece of material being fabricated, the closer these defects can be controlled and the higher the strength properties obtainable. For instance, glass fibers 0.04 in. diameter having a tensile strength of 25,000 psi, increased to 500,000 psi at a diameter of 0.0001 in., and to over a million psi for even finer diameter fibers.[8,9,10] It is believed that the extremely small diameters inhibit the size of surface defects and limit the directions a crack may follow. The theoretical strength of materials has been approximated in whiskers which are elongated single crystals grown under controlled conditions to minimize the occurrence of defects. These high strength materials, together with the selective orientation of their fibers, form the basis of composites. Composites are materials in which the properties of one component enhance those of another. They are largely fulfilling the continuing need for materials with exceptionally high strength-to-weight ratios and other unique material combination properties such as used in the high-strength facings and low density cores of honeycomb[11,12] panels and the comparable flanges and webs of beams,[13] filament-wound vessels, etc.

Composite materials combine a substance of high tensile strength (strength to break) and high modulus of elasticity (resistance to stretch prior to break) with a substance of comparatively low modulus of elasticity. If high-strength, high-modulus fibers are embedded in and bonded to a low modulus matrix so that both strain equally when loaded, the fibers will carry most of the load as indicated by the relationship[14,15,95] shown in Fig. 7.17 and given by Eq. 7.4.1. At the same time the isolation of the imperfections present in individual small fibers of the high-strength substance prevents the propagation of cracks from these imperfections, Fig. 7.18. Individual fibers may rupture,[16,96] but the weaker matrix serves as a medium for transferring the load to the stronger adjacent fibers by the mecha-

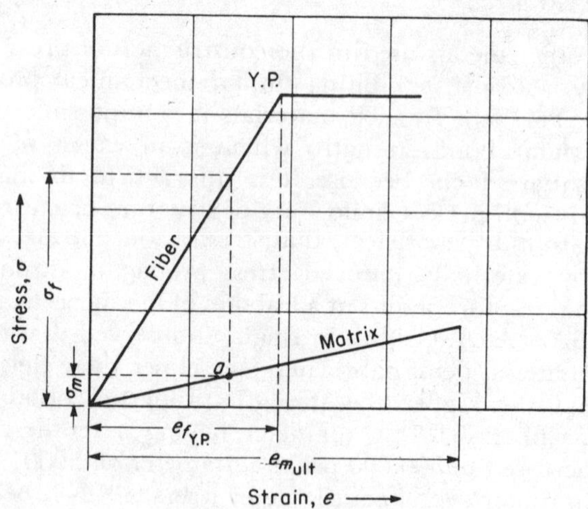

Fig. 7.17. Relation of the Stress and Strain Properties of the Fiber and Matrix of a Composite Material

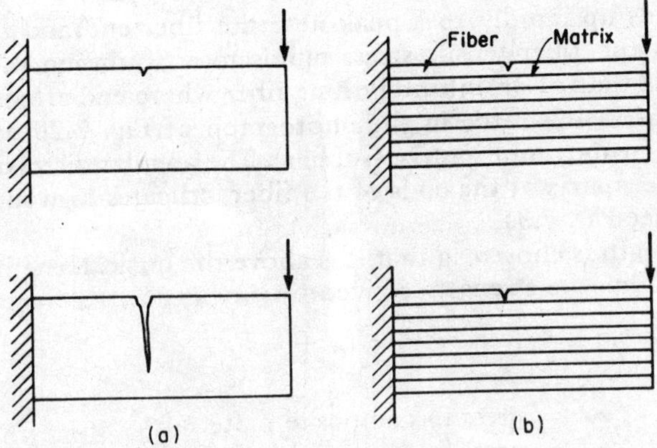

Fig. 7.18. Propagation of a Crack in a (*a*) Homogeneous Material as Compared to a (*b*) Composite Material

nism of shear at the fiber-matrix interface. This is the basic objective in the development of a fiber reinforced composite. The extent to which this is achieved determines the strength of the composite.[17,18] Figure 7.19 schematically represents the elastic distribution of shear and tensile stresses in the vicinity of a discontinuous fiber embedded in a weaker matrix. When the matrix deforms elastically, the shear

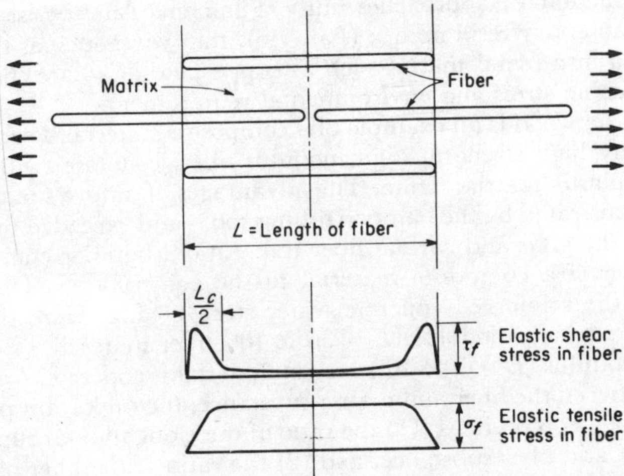

Fig. 7.19. Representation of the Interaction of Fiber and Matrix in a Composite Material and Resulting Stress Pattern

stress builds up rapidly to a peak near the fiber end and then drops sharply as the fiber tensile stress builds to a maximum value which approaches that of an infinitely long fiber where end effects are negligible. This is illustrated in the photograph of Fig. 7.20 of a model of a fiber (wood) in a matrix (rubber). The angular distorsion of the grid on the matrix at the ends of the fiber indicates high shear at this location (see Par. 7.3).

Fiber length is chosen so that it is above the critical length in which case the composite strength is given by

$$\sigma_c = \sigma_f V_f + \sigma_m V_m \tag{7.4.6}$$

where

σ_c = stress in composite material
σ_f = stress in fiber at the given strain
σ_m = stress in matrix at the given strain
V_f = volume fraction of fiber material
V_m = volume fraction of matrix material
L_c = critical length, length of fiber required to develop in shear the full tensile strength of the fiber

From Eq. 7.4.6 it is seen that the higher the fiber volume fraction, the stronger the composite material. The continuous filament wound vessel is a commonly used example of a composite vessel in which the volume fraction V_f approaches unity. Filaments can consist of glass fibers in an epoxy resin matrix (Par. 7.6), thin wires such as tungsten embedded in a metal matrix such as copper, etc.; and are chosen for the applicable stress and environmental requirements.

In nature, wood is an example of a composite material. It combines a relatively high strength, high modulus fiber, cellulose, with a low modulus plastic matrix, lignin. The advantages of nature's material is well demonstrated by the bamboo fishing rod—and exceeded by man's material, the glass and carbon fiber rod. An old familiar engineering application of a composite material to the construction of pressure vessels is the reinforced concrete water storage tank. Here the high strength and high modulus ($E = 30 \times 10^6$) fiber material is steel and the low modulus ($E = 5 \times 10^6$) matrix material is concrete. Assuming no slip between the fiber and matrix substance, the important physical property of a composite is: (1) the ratio of the elongation at rupture of the matrix and fiber substance, and (2) the ratio of the fiber modulus to the matrix modulus. As an example, Fig. 7.17 shows the stress-strain curve of the fiber and matrix for such a material. It is seen that

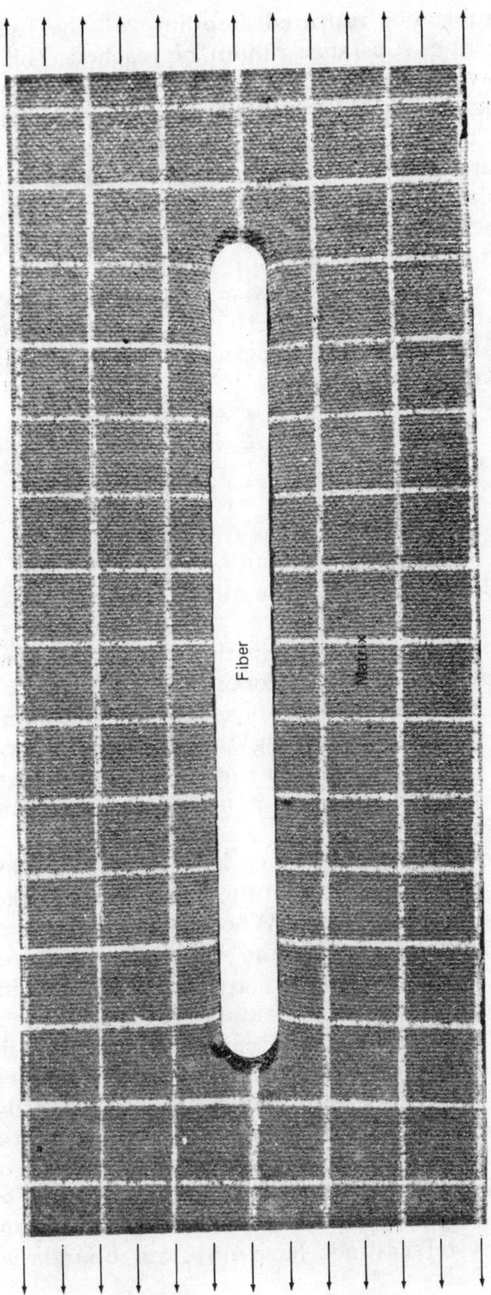

Fig. 7.20. Model Showing the Elastic Strain Distribution in the Vicinity of a Wood Fiber in a Rubber Matrix

if the matrix substance ruptured at point "*a*" the full yield point strength of the fiber substance cannot be reached. This can only be accomplished when the elongation of the matrix at rupture exceeds that of the fiber at its yield point, Eq. 7.4.7. Practically, it is a requirement that the low modulus substance must have a greater elongation at rupture than the high modulus substance. It is also noted that if the modulus of both the fiber and matrix were identical the latter could not perform its function of distributing stress to all of the fiber. It has been found that for satisfactory results the modulus of the fiber should be at least twice that of the matrix, Eq. 7.4.8. This explains why steel, $E = 30 \times 10^6$, can be used to reinforce concrete, $E = 5 \times 10^6$, while glass fibers, $E = 5 \times 10^6$, cannot be used. These two requirements; namely,

$$\frac{e_{m\,\text{ult.}}}{e_{f\,y.p.}} \geqq 1 \tag{7.4.7}$$

$$\frac{E_f}{E_m} \geqq 2 \tag{7.4.8}$$

are basic ones for the fiber and matrix substances of a composite material.

The arrangement of the fibers in the matrix must be such as to develop maximum strength and economy for particular applications. An example of this is the drawing of glass into fine filaments of extremely high strengths, and laying these filaments parallel to the direction of the applied stress. In the fabrication of a pressure vessel[97] the filaments can be arranged in patterns to withstand maximum hoop and longitudinal stresses, Par. 7.6.

Two advanced fibers[19,20] that have been under extensive development and are finding their place in the commercial market are those of boron and carbon (graphite). When they[21] are embedded in an epoxy or polyimide resin matrix the service temperature limitation is 400–600° F, in an aluminum[98] matrix to 800° F and in a titanium matrix to 1200°F. As service temperature requirements increase, both the fiber and matrix are selected for environment compatibility. High temperature alloys of great tensile strength can be produced by composite materials having their reinforcing fibers directionally solidified. This has been accomplished by sintering, and also by casting whereby unidirectional solidification of eutectic alloy systems is obtained by practices involving directional cooling heat flow and/or magnetic fields.[22] The optimum in directional solidification is a single-crystal casting. Single-crystal castings have no grain boundaries and have high temperature properties superior to those with multiple grains. The reason for this is that with grain boundaries eliminated there is no need for the addition of alloying elements to partially strengthen these boundaries. The prime application of such castings is in the high temperature gas turbine field where high creep rupture resistance is the key to thermal efficiency and service life. Fibers of tungsten, carbon, silicon, titanium, boron and alumina, etc. in metal and ceramic matrixes[23] have exhibited high temperature capability and chemical compatibility to 2400F, and the ultimate promise of composities lies in the utilization of the oxides, carbides and nitrides of these materials as fibers, Table 7.3. Hybrid types of composites with whiskers of several types of material in a single matrix have shown strength increases over that of either of the single type composites. The joining of dissimilar metals in welded structures is extremely difficult and gives an interface region of high stress, Par. 4.8, and metallurgical concern. However, structures made of composites may readily have parts fabricated from different composites to withstand various stress levels or offer different properties. The method of joining[24] the two different composites may be made by interleaving the plies of one composite with those of another and utilizing the mutually compatible matrix to effect the transition joint. The science of materials is at the root of all design objectives, and advances in materials and their applications are often the key to success, Fig. 1.17. Yesterday's threshold materials will be common place tomorrow—and considering the pace of technological advancement, tomorrow is now! The fiberglass polyester petroleum and chemical storage tank has found an economical and wide place in the pressure vessel field, as has also the glass filament wound vessel for high pressure vessels, etc. Hence, the experienced sophisticated materials and designs developed for a specific high technology application, such as aerospace, find their place in the commercial market in an economic basis of supply-and-demand—with the designer who is best able to optimize the use of "advanced composites" having a significant edge of the ever-present competition.[25,26] Fibers have invoked high interest not only because they are most promising for exploiting the ultra-high strengths inherent in structural materials, but the fiber form lends itself to a wide range of packing modes[99] and geometrical patterns so that engineers can design structures to meet specific service requirements.[136]

In order to produce thick-walled pressure vessels, constructions composed of a multitude of concentric layers (Fig. 2.5), coil-layer[27] or wire wrappings,[28] Fig. 7.25*e*, are frequently employed. In these

TABLE 7.3. WHISKER AND MATRIX PROPERTIES[10]

WHISKER PROPERTIES

Material	Melting Point (deg F)	Density (lb/in.^{3})	Modulus of Elasticity (psi × 10^6)	Modulus to Density Ratio (in. × 10^6)
Aluminum Oxide	3720	0.143	60	419
Boron Carbide	4442	0.090	65	722
Carbon (Diamond)	6692	0.076	140	1842
Silicon Carbide	5162	0.116	65	560
Silicon Nitride	3452	0.116	55	474
Titanium Carbide	5684	0.178	64	359

MATRIX PROPERTIES

Material	Melting Point (deg F)	Density (lb/in.^{3})	Modulus of Elasticity (psi × 10^6)	Modulus to Density Ratio (in. × 10^6)
Aluminum	1220	0.098	9	91.2
Iron	2799	0.284	29	102.1
Magnesium	1202	0.063	6.4	101.5
Nickel	2647	0.322	30	93.1
Titanium	3038	0.163	16.8	103

STRENGTH OF ALUMINA WHISKERS

Temperature (deg F)	Tensile Strength (psi)
Room Temperature	4,000,000
1200	1,500,000
2000	1,300,000
3400	200,000

constructions the same type material, though of varying tensile strengths, is used throughout the vessel. However, such vessels do exhibit the composite principle of crack-arrest since a crack in one layer, coil, or wire does not propagate to others. Cylindrical and spherical vessels[29,30,31] are frequently built in this manner which may have particular advantages when the environment is conducive to stress-corrosion cracking or the material is prone to brittle fracture.

7.5 Fabrication Costs

The economic gain from a change in fabrication method is generally a radical one, as compared to the gradual though significant gain from the continual perfection of a method. The change from riveted to welded vessel construction accomplished a marked reduction in fabrication costs, to say nothing of the engineering advancements it made possible. Second only to material costs is fabrication cost. This amounts to approximately 35 per cent of the total vessel cost, and is principally accounted for by the joining of the parts to effect a final assembled vessel. Hence, methods to (1) eliminate joints, or (2) to make them more economically establishes the direction fabrication must go.

Engineering constructions are a compromise between man's imagination and what it is possible to do with economically available materials. This compromise can be narrowed in amount to the extent that fabrication methods can be developed, and in time scheduled to the extent that the inertia of old methods is dissipated. This is well illustrated by two familiar examples. Considerations of cost and availability have generally confined structural designs to the limited dimensions and single materials of rolled structural shapes. These shapes are produced by a massive rolling operation, in both procedure and capital investment, to successively reduce a solid piece to an *I*-beam, Fig. 7.21*a*. This process is now giving way to the continuous welding machine that makes custom shapes out of plate steel faster than standard shapes can be rolled fron an ingot, Fig. 7.21*b*. Not only can an infinite variety of shapes be readily made, but they can be made from a mixture of steels which cannot be obtained by rolling. Such "hybrid" shapes get a maximum amount of work from each pound of their weight through an intricate match of various strength level steels to the required stress levels in the tension flange, compression flange and the web of the beam. A second example is that of tube or pipe fabrication. Here a solid bar is first pierced to form a cylinder after which it is successively rolled, beat and drawn to the required final size, Fig. 7.22*a*. This is an operation much like that of rolling structural shapes both in extent and investment—and is reflected in the final price. Here again, this method is being supplanted by the full utilization of welding as a fabrication method. It permits a tubular pressure vessel to be made from a long ribbon of flat plate by forming it in a circular configuration as it is welded together in a single longitudinal weldment, Fig. 7.22*b*. The product has not only economic advantages over rolled seamless tubes, but may have other engineering attributes; such as, unlimited tube length,

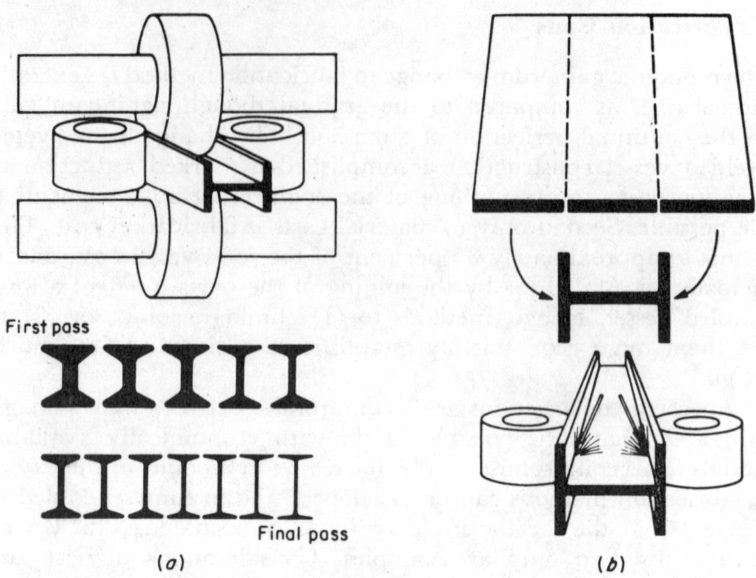

Fig. 7.21. Fabricating a Structural Shape by (*a*) Series Rolling From a Solid Piece of One Type Material, (*b*) Continuous Welded Components of a Preferentially Selected Material Combinations

uniform wall thickness obtained from the greater accuracy control of rolled plate, and ready inside and outisde material surface inspection prior to welding for quality control.

1. *Castings*

In the trend toward large size vessels, steel castings will play an increasingly important role from both economic and design viewpoints. The physical size of large flanged vessel heads seriously limit their procurement sources as forgings, which coupled with the high cost of the forging process itself, creates a formidable economic hurdle. The mechanical properties generally associated with plate and other relatively thin forgings cannot be obtained in heavy forged sections; hence, their mechanical properties closely approach those of a casting. Therefore, one solution to these material and procurement problems is that of castings. There is no objection to the use of steel castings manufactured to established quality, mechanical and chemical specifications. They offer a ready and economic solution of the ideal reinforcement of openings and internal and external attachments. The proper reinforcement, suitably contoured and profiled, can be ob-

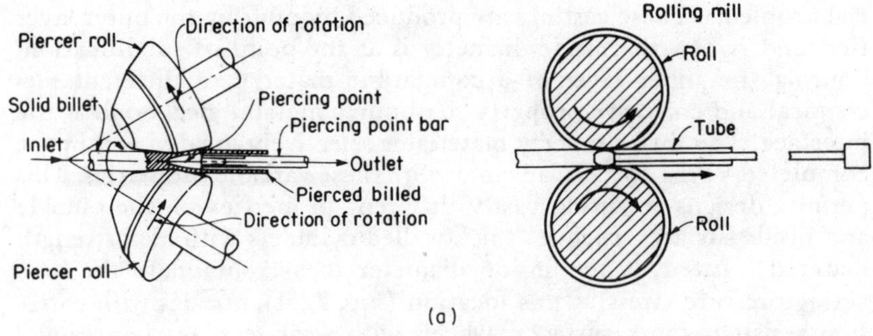

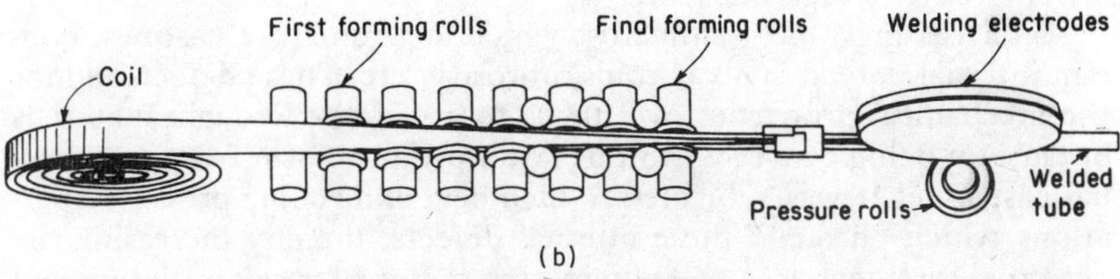

Fig. 7.22. (*a*) Piercing a Solid Piece of Metal to Form a Cylinder and Rolling to Final Tube Size, (*b*) Fabrication of Electric-Resistance Welded Tubing From a Coil of Flat Strip

tained integral with the vessel without the more devious, expensive and time-consuming means of a multitude of small weldments[32,33] machined and ground to reduce stress concentrations. Figure 7.23 shows a comparison of a cast or cast-and-weldment versus a typical weldment flanged vessel closure head to illustrate the design and economic potential of quality controlled steel castings.

Centrifugally cast cylindrical structures of dual-property material (material in which the mechanical properties vary throughout the thickness) have the economic and engineering potential to solve spe-

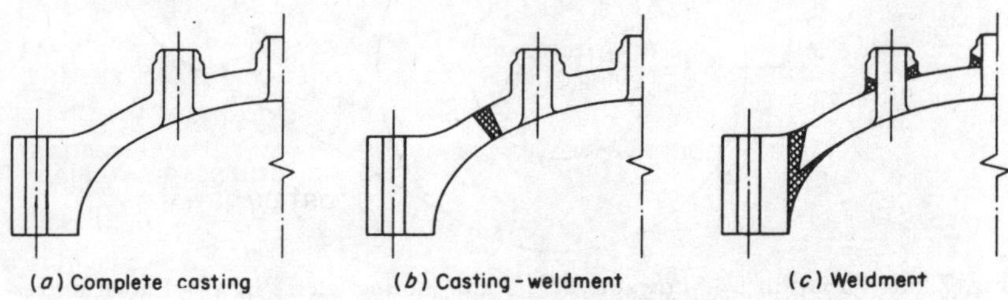

Fig. 7.23. Closure Head Constructions

cial problems. These castings are produced by pouring the outer layer first and, when the inside diameter is at the point of solidification, pouring the inner layer of a compatible material of different mechanical and corrosive property to obtain a metallurgical bond at the interface. The dual-property material is selectively located within the completed vessel to take advantage of these variable properties. This permits designs requiring vastly different properties on the outside and inside surfaces, such as: thick-walled cylinders with high strength material located at the inside diameter to accommodate the high acting pressure stress at this location (Fig. 7.24), nozzles with corrosion resistant inner surfaces, wheels with wear resistant flanges and lower strength tough hubs, etc.

Metal castings and weldments, which are miniature castings, contain internal defects such as voids, porosity, etc. These defects reduce the mechanical properties over those found in like wrought materials because in being converted to this latter product form from the original cast billet it was subjected to high heat and rolling pressure operations which "healed" these internal defects, thereby increasing the net cross-sectional area and eliminating points of crack initiation and growth. A similar healing effect can be obtained in individual castings or weldments by placing them in a vessel containing a compatible inert gas, such as argon, and an internal furnace so that they may be subjected simultaneously to high pressure and high temperature. This

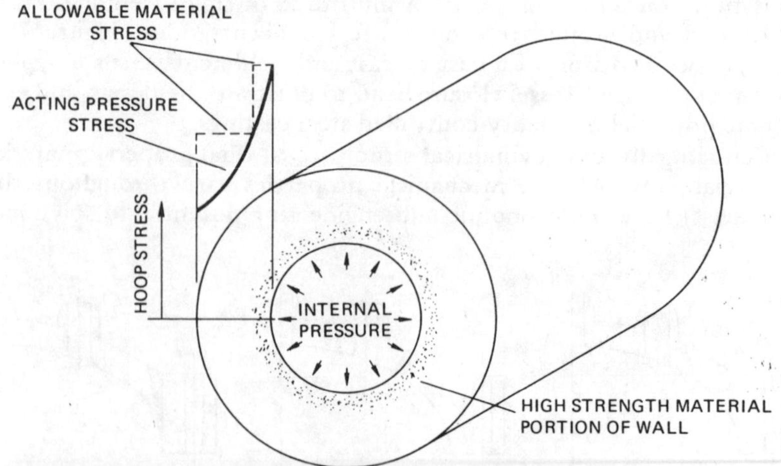

Fig. 7.24. Variation in Allowable Stress and Acting Stress in a Cylindrical Pressure Vessel Constructed of Dual-Property Material

is called "hot isostatic pressing," and it collapses the voids and metallurgically bonds across what were formerly free surfaces. Pressures in the order of 10,000-25,000 psi and temperatures of 1,000-2,300°F are generally used. Surface cracks are not healed by this process, but only internal ones which can be collapsed under pressure. There is a tradeoff in the pressure-temperature relationship, with the minimum pressure being that of the yield strength of the material at the operating temperature. The ultimate strength, fatigue strength, and creep rupture strength are increased to values closely approaching those of the wrought material. This can be done to the most complex configurations without the use of expensive tooling.

2. *New Methods of Fabrication and Joining Processes*

Fabrication methods used today will be replaced by those now under development, and these in turn by those unknown. The basic incentive that has accomplished this in the past, and will ensure its continuance in the future, is economic competition. Improvements in the present welding techniques of joining vessel parts by multiple layers of metal to the required thickness, Fig. 7.25*a*, will give way to other simpler, quicker and cheaper methods now in development.[34, 35,36] Some of these are:

a. Full thickness weld metal pour techniques,[37,38] Fig. 7.25*b*.
b. Joint welding by ultrasonics,[39,40] electron beam,[41,42,43] plasma,[44] inertia or friction,[45,46] lasers,[47,48,49] explosives,[50] solid state diffusion bonding,[51,52] high-frequency,[53,54] etc., with or without the aid of heat and pressure, Fig. 7.25*c*.
c. Bonded joints made with a second but metallurgically and chemically compatible material, Fig. 7.25*d* . Adhesives are an example of such an increasingly promising method, Par. 7.10.

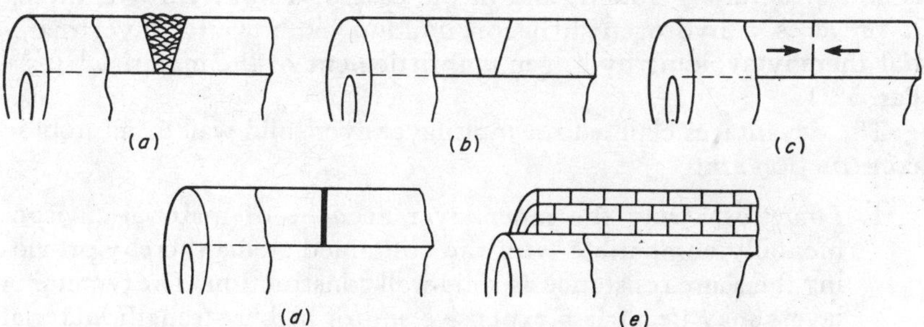

Fig. 7.25. Methods of Fabrication

Other construction and fabrication methods also may well replace or supplement these now in use due to either their economic potential or engineering necessity. Some approaches that are being advanced are prestressed[55] mesh or strand wrapped[28] vessels of high-strength metal[100] or composite fibers,[56] Fig. 7.25*e*, coil wrapped vessels,[27, 101] etc. While the basic construction principle of wrapping is not new but has been used in making wooden cannons, penstocks and storage tanks, the extremely high strength materials presently available in small thicknesses offers a new economic incentive. The wrappings can be of plain round or rectangular cross-section, or they can be of an indented interlocking ribbon wrapping[57] (Wickelofen) which when spirally wound can substain both hoop and longitudinal forces. In the forming of metal parts for vessels the use of high-rate forming processes[58,59] using chemical explosives, high voltage electrical discharge,[60] electromagnetic hammers, water and gas rams, etc. are also showing increasing technical and economic promise particularly in forming high alloy exotic materials that are difficult to form by present conventional methods. These forming methods are extremely rapid and are adaptable to the size of the part to be formed. For instance, the shock waves obtained from the underwater detonation of explosives are used to for extremely large vessel heads, etc.

3. *Multilayer Construction*

One of the most widely used wrapped type vessels, Par. 2.2, is that of multilayer construction,[102] Fig. 7.26*a*. It consists of an inner shell on which formed concentric layers, approximately $\frac{1}{4}$" thick, are wrapped and welded to obtain the required thickness. The inner shell is the media containing membrane; whereas, vent holes are drilled through the remaining layers to the inner shell. This permits constant leakage monitoring and detection while the vessel itself is still structurally sound, and in the case of a hot hydrogen media it prevents a hydrogen diffusion buildup in the outer layer material thereby avoiding hydrogen embrittlement of the material,[137, 138] Par. 5.21.

The advantages claimed for multilayer over solid wall or monobloc construction are:

1. *Low Cost:* Only the inner layer need be of material environmentally compatible with the contained media thereby providing the same resistance as solid wall construction. The remaining layers may be of less expensive and/or higher-strength material in order to comply with the stress gradient through the wall, Par. 2.8, and also facilitate a weight and shipping advantage.

2. *Brittle Fracture Resistant:* Thin plates exhibit superior mechanical properties to thick ones made to the same specification. This applies to both strength and notch toughness properties, Par. 5.22.3.
3. *Crack Propagation Arrestant:* The layer interfaces provide a crack arrest construction by confining crack propagation to the individual distress layer, Par. 7.4.3.

Some limitation of multilayer construction arise from the layers not being attached to each other; hence act independently under some loading conditions.[139,140,141] These are:

1. *High Local Discontinuity Stresses:* These occur at points of locally applied forces and moments, such as at nozzles, supports, etc., and are due to the reduced rigidity of a multitude of small thicknesses as compared to that of a single solid wall of the same total thickness, Eq. 3.2.9. This can be improved by:

a. Installing additional layers at these locations, Fig. 7.26*b*.

b. Employing solid wall construction in these local areas. For instance, flare insert type nozzles remove the multilayers from high local bending stress location, Par. 6.7, Fig. 7.26*c*. Likewise, solid wall shell construction in the local shell-to-head junction region can be employed, or means provided for the transfer of longitudinal shear from layer-to-layer thereby making the layers behave as a solid wall in bending. The latter can be accomplished by introducing additional through thickness weld seams in this local high shear area, Par. 4.4, Fig. 7.26*d*.

2. *Restricted Thermal Gradients:* Rapid internal cool-down rates must be avoided to prevent the innter layer from shrinking away from the adjacent supporting ones thereby becoming overstressed.[103] As a corollary, high heat-up rates must be limited to prevent the inner layer from buckling under thermal expansion stresses. Here, since thermal expansion of the inner layer is restricted from expanding radially outward by the surrounding outer cooler ones, a compressive tangential stress is produced. The inner layer is, in effect, a cylindrical shell subjected to a uniform external pressure and encased in a rigid cavity so that only radial inward displacements can occur, Fig. 8.2. The buckling strength of such a constrained cylindrical layer is approximately twice that of the same layer acting as an ordinary unrestrained shell,[149] Eq. 8.2.38. Thin corrosion resistant liners inserted in monobloc vessels also behave in the same manner.

3. *Limited External Pressure or Vacuum Service:* Since the layer vent holes subject only the inner layer to the total external collapse pressure, a limited pressure based on the thickness of the inner layer can be accommodated.

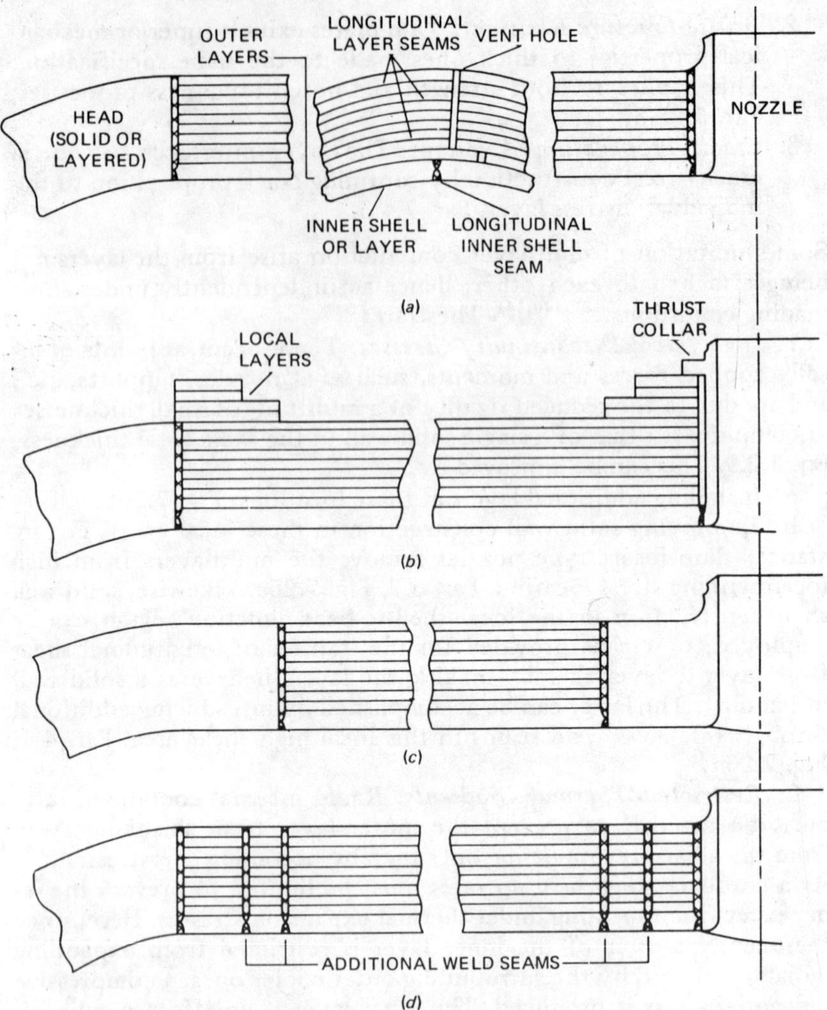

Fig. 7.26. Multilayer Constructions

4. *Banded and Wire-Wrapped Construction:* A variation of the multilayer vessel is that of banded or wire-wrapped construction, Fig. 2.5 (a) and Fig. 7.27. Here, only the inner shell takes longitudinal stress as well as hoop stress and is subject to a bi-axial stress pattern. Whereas, the outer wrappings are subject to only a circumferential uniaxial hoop stress, Fig. 7.27 and 7.28. Using the following nomenclature,

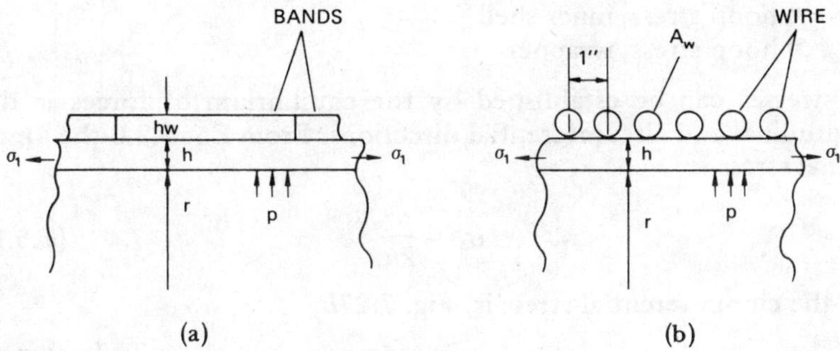

Fig. 7.27. Banded and Wire-Wrapped Vessel Construction, (a) Banded, (b) Wire-Wrapped

A_w = cross-sectional area of wire-wrapping per inch in the longitudinal direction
E = modulus of elasticity of inner shell material
E_w = modulus of elasticity of wrapping material
e = strain, inner shell
e_w = strain, wrapping
h = thickness inner shell
h_w = thickness, wrapper band
σ_1 = longitudinal stress, inner shell

Fig. 7.28. Banded Vessel, Showing the Spherical Head and Banded Cylinder, being Installed at a Site

σ_2 = hoop stress, inner shell
σ_w = hoop stress, wrapper

the stresses can be established by the equilibrium of forces in the longitudinal and circumferential directions. From Eq. 2.2.6 the longitudinal stress is

$$\sigma_1 = \frac{pr}{2h} \tag{7.5.1}$$

and the circumferential stress is, Fig. 7.27*b*,

$$\sigma_2 h + \sigma_w A_w = pr \tag{7.5.2}$$

From Eq. 2.3.2

$$e = \frac{\sigma_2}{E} - \frac{\mu\sigma_1}{E} \tag{7.5.3}$$

and

$$e_w = \frac{\sigma_w}{E_w} \tag{7.5.4}$$

But since $e = e_w$, and assuming $E = E_w$, the right hand side of Eqs. 7.5.3 and 7.5.4 can be equated to give

$$\sigma_w = \sigma_2 - \mu\sigma_1 \tag{7.5.5}$$

Substituting this relation in Eq. 7.5.2, together with Eq. 7.5.1, gives the hoop stress in the inner shell as

$$\sigma_2 = \frac{pr}{h}\left[\frac{h + \frac{\mu}{2}A_w}{h + A_w}\right] \tag{7.5.6}$$

The stress in the wrapping can be found by substituting the value of σ_2 from Eq. 7.5.6 in Eq. 7.5.2 to give

$$\sigma_w = \frac{pr}{A_w}\left[\frac{\left(1 - \frac{\mu}{2}\right)A_w}{h + A_w}\right] \tag{7.5.7}$$

When wide bands are used for the wrappings, Fig. 7.27*a*, $h_w = A_w$ is placed in the above equations. These vessels may be autofrettaged to develop favorable prestress patterns[104] in the same manner as is done with solid wall vessels, Par. 2.10.

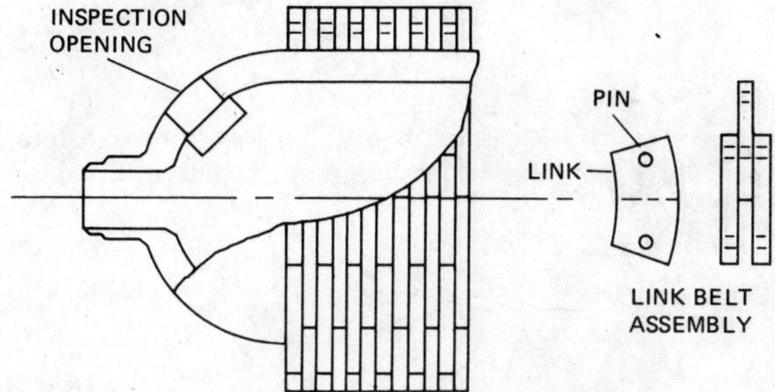

Fig. 7.29. Link Belt Banded Vessel Construction

5. *Link Belt Construction:* A derivative of the banded construction principle for thick-walled vessels utilizes high strength nonweldable steel clevis or link belts, Fig. 7.29. Here, a series of pin connected belts is substituted for the long welded bands of Fig. 7.27*a*. The links are stamped from thin inexpensive rolled plate (as compared to thick plate or forgoings) which affords high strength mechanical properties by heat treatment and alloying. The notch toughness of the thin material of the links is that accompanying the plane stress condition which is several times higher than that for thick section material which behaves in a plane strain manner, Par. 5.22. This, together with its high order of structural redundancy, offers a crack arrest construction that can be dismantled for transport or in-service inspection.

7.6 Filament-Wound Pressure Vessels

Filament-wound vessels are fabricated by continuously winding a filament over a mandrel of the shape and size of the desired vessel.[61] The mandrel can be a removable one, or one that remains a permanent part of the vessel. The matrix material is usually applied simultaneously with the winding operation. The winding of vessels can be accomplished by using two independent wrapping systems oriented in the direction of two principal stresses,[62,63] or a single helical winding with the filaments laid down upon the mandrel at an angle to the axis depending upon design consideration, such as access openings, Fig. 7.30. Filament-wound pressure vessel designs must be such that the filaments are placed only where they are needed and in such a manner that they are all uniformly stressed.[64,65] For instance, in

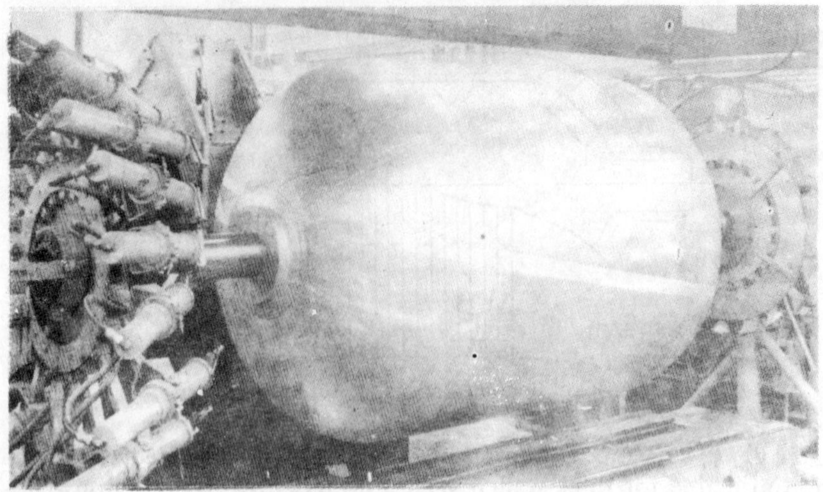

Fig. 7.30. Filament Winding of a Cylindrical Vessel[61]

a cylindrical pressure vessel where the hoop load is twice the longitudinal load, it is necessary to have twice as much circumferential winding thickness as longitudinal thickness in order for the stresses in the filaments to be uniform. Fig. 7.31 shows the basic geometry of a typical cylindrical pressure vessel with closure heads incorporating openings. The winding angle α is that which will cover the heads and provide longitudinal strength. The angle β is the winding one that will provide the necessary hoop strength in the cylindrical portion of the vessel.

The angle α can be determined from the geometry of the vessel as

$$\alpha = \tan^{-1}\left(\frac{r_a + r_b}{L}\right) \tag{7.6.1}$$

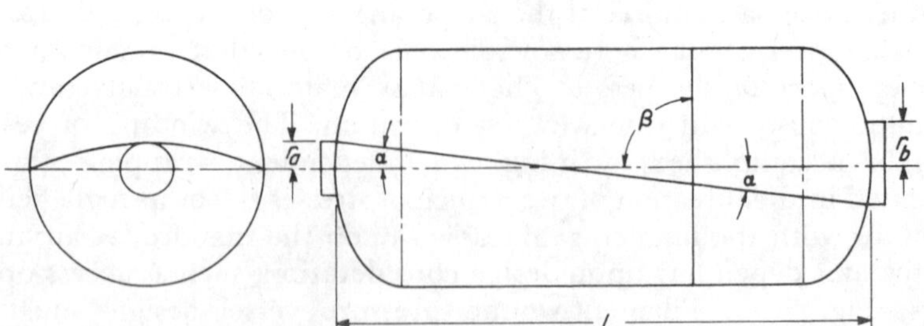

Fig. 7.31. Geometry of Filament-wound Cylindrical Vessel

Once the angle α is known, the thickness of the windings, h_α, at this angle can be calculated by equating the longitudinal vessel load to that sustained by the filaments at this angle of inclination, Fig. 7.32*a*.

$$\frac{F_1}{\cos \alpha} = \sigma_f \, h_\alpha \cos \alpha \tag{7.6.2}$$

$$\frac{F_1}{\sigma_f} = h_\alpha \cos^2 \alpha \tag{7.6.3}$$

where

F_1 = Longitudinal pressure load, lb per in. of circumference.
F_2 = Hoop pressure load, lb per in. of longitudinal length.
h_α = Total thickness of filaments in α direction.
h_β = Total thickness of filaments in β direction.
σ_f = Allowable unidirectional stress for filament material.

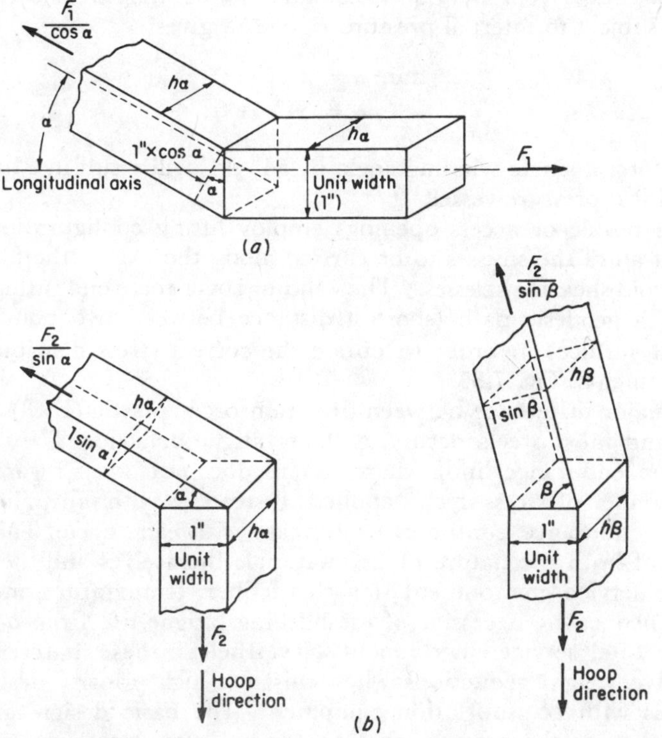

Fig. 7.32. Forces on Filaments of Filament-wound Cylindrical Vessel

The thickness of the winding in the hoop direction, h_β, of the cylindrical portion can be determined by equating the hoop load, F_2, to the sum of the hoop direction component of the α and β filament windings, Fig. 7.32*b*,

$$\frac{F_2}{\sigma_f} = h_\beta \sin^2 \beta + h_\alpha \sin^2 \alpha \tag{7.6.4}$$

A single filament winding system with a single angle α can also be used for cylindrical vessels. In this case it must withstand all of the hoop and longitudinal loads; i.e., it must satisfy these equations:

$$\frac{F_1}{\sigma_f} = h_\alpha \cos^2 \alpha \tag{7.6.5}$$

$$\frac{F_2}{\sigma_f} = h_\alpha \sin^2 \alpha \tag{7.6.6}$$

Dividing Eq. 7.6.6 by Eq. 7.6.5, and noting that in a cylindrical vessel subject to internal pressure $F_2 = 2F_1$, gives

$$\tan^2 \alpha = 2 \tag{7.6.7}$$

$$\alpha = 54^\circ\ 44' \tag{7.6.8}$$

Therefore, a single winding angle of 54° 44′ will result in a balanced cylindrical pressure vessel.[142]

End nozzle or access openings employ fitting configurations that permit all of the stresses to be carried along the axis of the filaments and avoid shearing stresses.[66] The windings over these end fittings must follow a geodesic path (shortest distance between two points on a curved surface) in order to obtain the correct stress distribution in the filaments, Fig. 7.31.

A major difference between fiber-reinforced plastic (FRP) composites and most steels occurs in their fatigue behavior. They do not have an endurance limit where failure does not occur regardless of the number of stress cycles applied. Instead, as with most aluminum alloys, endurance continues to decrease with each cycle. This is because of both the nature of the materials themselves and the effects of the service environment (length of fiber, temperature, moisture, etc.); hence, the necessity of establishing fatigue life from σ–n tests in the total service environment. Nevertheless, these materials offer effective weight reductions when this is a high priority design goal such as with transportation equipment. The basic design approach for fatigue is the same as that for direct stress; namely, orient the fibers in the direction of the principal stress,[143, 144] Fig. 7.33. As in

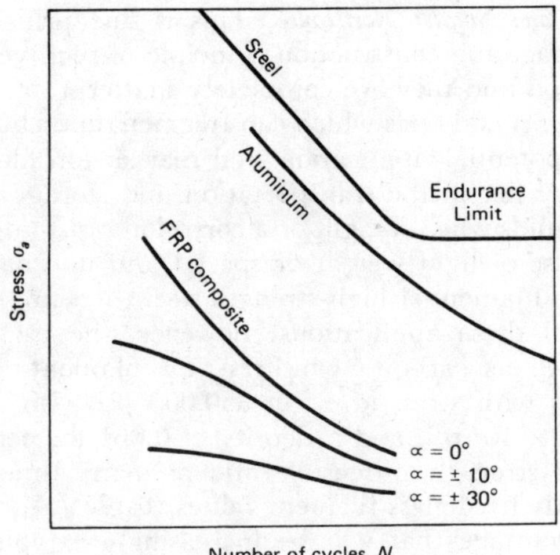

Fig. 7.33. Composite Fatigue Strength. (α = Angle of Fiber Orientation to Principal Stress)

metals, Para. 5.18.3, fatigue cycle frequency has a minor effect on fatigue performance.

Filament-wound pressure vessels are used in place of homogeneous metal ones for special applications or environments.[67,68] Some of these are:

Weight Saving. This is primarily due to the extremely high stress levels and low density[69] that can be obtained with this type of construction. It is the paramount reason for its wide usage in rocket and space flight devices.

Corrosion Resistance. This is due principally to the generally high corrosion resistance[70] of the plastic material used in the matrix. In many cases, the resins used in these materials are frequently applied to metallic structures to improve their corrosion resistance.

Low Thermal Conductivity. This is due to the general lower value of this property for plastics as compared to homogeneous metallic vessels. This has been found advantageous in cases involving solid-propellant rockets, and certain piping applications.

Low Notch-Sensitivity. Such things as scratches, poor welds, mismatched welds and other types of flaws in homogeneous metal vessels can result in notch sensitivity failures. This is avoided with filament-wound vessels due to the type construction, wherein failure of one filament does not spread to adjacent ones.[105]

High Impact and Shatter Resistance. This is due principally to the non-crack propagating construction principle of such vessels. Impact resistance is good and they are completely shatterproof, as compared to homogeneous metal ones which can fragment upon bursting.[106,107]

The prime potential application and market for filament-wound pressure vessels[71] lies in the transportation and storage of petroleum and chemical fluids where weight or a corrosion resistant environment demands the use of light weight or special and unusually expensive materials. Glass filament or high-strength steel wires presently account for the bulk of these applications; however, the use of advanced composites, such as carbon "whiskers" for filaments can result in pressure vessels with stress levels of 850,000 psi with a modulus of elasticity of 85×10^6 psi, and a density of 0.061 lb. per cu. in. This would mean a strength-to-density ratio of many times that of the highest presently used glass filament values, Table 7.4.

Another advantage that will be increasingly exploited as vessels increase beyond transportable size, is that of the low capital investment equipment costs and adaptability of this form of construction to large field fabricated vessels. Reinforced concrete in combination with prestressing is an example of a composite material that is being used for the construction of large on-site nuclear pressure vessels, paragraph 7.7, as well as the more conventional storage tanks, but need not remain restricted to these materials.

TABLE 7.4. PHYSICAL PROPERTIES[61]

	Filament Winding		Titanium	Steel
	"Type E" Glass	High Modulus Glass		
Density (Lb/In^3)	0.075	0.081	0.163	0.285
Tensile Yield Strength (psi)	130,000 Unidirectional Strength 270,000	115,000	150,000	220,000
Tensile Strength/Density (in)	174×10^4	142×10^4	92×10^4	78×10^4
Modulus of Elasticity (psi)	5×10^6	9.0×10^6	16.5×10^6	30×10^6
Elongation, %	2.36	1.28	0.91	0.73

7.7 Prestressed Concrete and Cast Iron Pressure Vessels

1. *Prestressed Concrete Pressure Vessels*

The prestressed concrete vessel, PCV, is particularly adaptable when the size that can be shipped and/or thickness that can be fabricated exceeds that of a metallic vessel.[72,73,74,75] It consists of a leakproof thin steel membrane liner surrounded by a thickness of concrete.[76,77] High strength steel wires, or tendons composed of a multitude of wire strands in a sheath, are embedded within the concrete, Fig. 7.34, and then prestressed by jacking or wedging so that under all loading conditions the concrete, which has little tensile strength, is kept in compression.[78] They take the full hoop stress resulting from the internal pressure, and may extend completely around the vessel or only a portion of the circumference.[79,80] In either case they overlap and are anchored to fittings on the outside surface of the concrete, Fig. 7.34. The tendons or wires may also be wrapped around the outside of the vessel and prestressed to obtain the same effect.[81,108,109,110,145,146]

When prestressed concrete vessels are used to contain a hot media, the concrete must be protected from the deleterious effects of high temperature,[111,112,113] and is usually kept to a temperature under 150°F. This is accomplished by placing insulation on the hot media side of the liner and also by placing auxiliary cooling coils in the concrete adjacent to the outside of the steel liner, Fig. 7.34. Section A-A. There is no saving in steel material tonnage in a PCV as com-

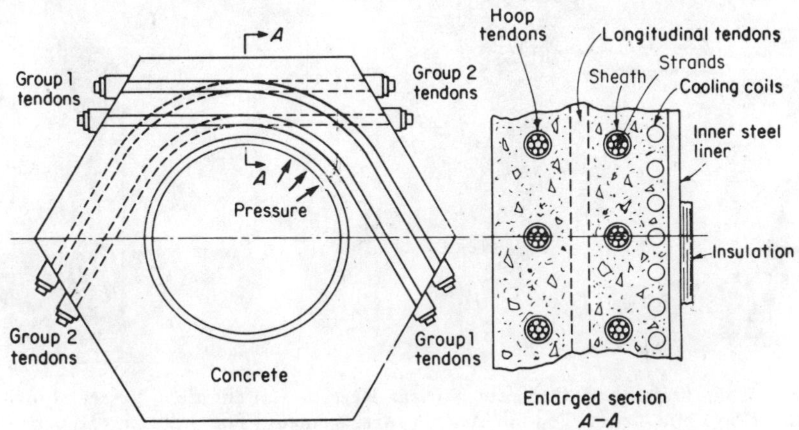

Fig. 7.34. Diagrammatic Arrangement of Prestressed Concrete Vessel

pared to a steel monobloc vessel for the same allowable stress level.[82] In fact, under such conditions there is an increase in the amount of steel required because in a monobloc vessel the steel is used in a biaxial stress state whereas in a PCV the same steel is used only in an uniaxial stress state. However, it is customary to employ high tensile steel for the small diameter wires and tendons thereby mitigating this condition. The main advantages claimed for the PCV are:

1. A construction method having no inherent size limitations and permitting on-site fabrication with ordinary skills and equipment.
2. A vessel with a high order of fail-safe redundancy by virtue of the many tendons making crack propagation from one tendon to another as a result of a local defect exceedingly unlikely.

The prestressed concrete vessel is not a vessel made of a two-phase material, paragraph 7.4.3. The steel tendons are placed in sheaths to permit prestressing, and there is no bond to the concrete by which load can be transferred from one strand to another by shear, Fig. 7.19. The concrete acts in direct compression to transfer the internal

Fig. 7.35. Prestressed Concrete Nuclear Reactor Containment Vessel Showing the Hoop and Dome Tendon Anchor Arrangement. The Adjacent Hyperbolic Cooling Towers May be Built of Concrete, Structural Shapes or Lobed Constructions Using Steel-Wire Nets

pressure load to the steel tendons which take the entire load resulting from the internal pressure, Fig. 7.35.

2. *Prestressed Concrete Storage Tanks*

The use of prestressed concrete tanks for the storage of water, petroleum, liquid gas, etc., is increasing as they offer an economical on-site construction alternate to steel ones.[114] These tanks may be internally coated with organic sealants or lined with a thin metal lining when it is necesssary to promote the integrity of the stored product, prevent deterioration of the concrete by the stored product, or prevent loss of the stored product by permeation through the concrete. Storage tanks normally operate at low internal pressures but since they are frequently placed underground or underwater they are subject not only to internal pressure but also uplift and hydrostatic external pressure when empty. Just as with steel vessels subject to external pressure, the buckling strength of concrete tanks against external pressure is also reduced by imperfections. Studies conducted on cylindrical concrete vessels have concluded that they are weakened in buckling by 20 percent due to noncircularity, 10 percent due to creep, and 6 percent due to flat spots at reinforcing bar locations for a total of 36 percent.[115, 147]

3. *Prestressed Cast Iron Pressure Vessel*

The functional purpose of concrete in a PCV is to accomodate the system of prestressing tendons that carry the pressure stress. In this system the concrete, which is kept under compression because it is a brittle material and can take little tension, can also be replaced by a comparable material such as cast iron.[83] Unlike concrete, which has the disadvantage that it is temperature limited to 150°F and for service above this temperature elaborate and costly cooling devices must be employed, Fig. 7.34, cast iron can resist temperatures to 750°F without appreciable loss of strength.

The prestressed cast iron vessel, PCIV, has the same high order of redundancy to brittle fracture and cracks arrest as does the PCV since the cast iron segments or building-blocks are made small, thereby acting as crack arrestors, and the multiple tendon system remains the same as used with the PCV.

The advantages claimed for cast iron, as compared to concrete, pressure vessels are:

a. A vessel suitable for high temperature service without the need of material cooling devices as required for concrete, and its greater compressive strength permits thinner walls with reduced associated occupancy space.

b. A construction that can be disassembled for inspection and repair.

c. A construction that allows complete shop fabrication of the cast iron parts, partial shipment, and speedy field assembly that can offer potential cost savings.

7.8 Lobed Pressure Vessels

The design of very large pressure vessels poses the practical limits of increasing the wall thicknesses proportionally to their increasing basic diametrical dimension.[84] The equation for the membrane stress in cylindrical and spherical vessels is

$$pr = K\sigma h$$

K being a constant depending upon geometry, from which it is seen that the product of pressure and radius reaches a limit fixed by the product of an allowable stress and wall thickness, h. One way of accomplishing this is by constructing the entire vessel of very high strength materials. Another is through the use of "lobed" pressure vessels of which the intersecting spheres analized in paragraph 2.5 is one. This is the construction that is used in gas flight balloons and airships, and the low air pressure supported plastic or metallic foil "bubble" buildings; wherein, a network of spacial cables or frames suspend or envelop a multitude of pressure tight membrane lobes.

In this construction the thickness, h, of the pressure tight lobe can be made thin by keeping its radius, r, small, relative to that of the basic vessel radius, R. This is achieved by arranging the membranes as shown in Fig. 7.36; hence, the term "lobed." The lobe has two radii of curvature but that in a direction parallel to the suspension supports is large compared to that perpendicular to these; hence, these lobes act substantially as segments of long cylinders.

The reaction forces from these lobes are transferred to the spacial cable or frame support system. Additional guy cables or ties between diametrically opposite support cables or frames can be used to strengthen lobed vessels in the same manner that stay-bolts or stay-tubes are used in some steam boilers.

Lobed vessels are well adapted to exploiting the economic advantages of several materials in the same structure. The "bubble" building employs a pressure tight plastic membrane and high strength steel cables. Correspondingly, metal vessels can employ pressure tight weldable membrane lobe material, in combination with high strength

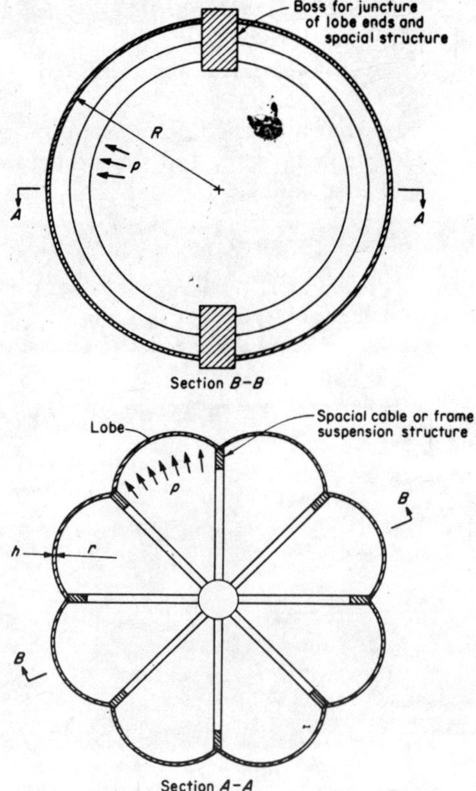

Fig. 7.36. Lobe Vessel Construction

non-weldable spacial suspension cable or support structures. Concrete water storage tanks are also frequently constructed using this principle to effect economy.

One of the most rapidly increasing uses of lobed vessel construction is that for so-called "fabric structures" because the membrane is made of fabric. They are especially adaptable to large spans requiring unobstructed areas, such as stadiums, greenhouses, amphitheaters and cooling towers. The membrane, or fabric, must be held in tension to be structurally stable. This is accomplished in two basic ways. The first way, called air supported, is that in which the primary net loading is an internal positive pressure one wherein the gravity and other external loads are supported by an internal air pressure. The lobes are concave inward as are also the lobe suspension support, Fig.

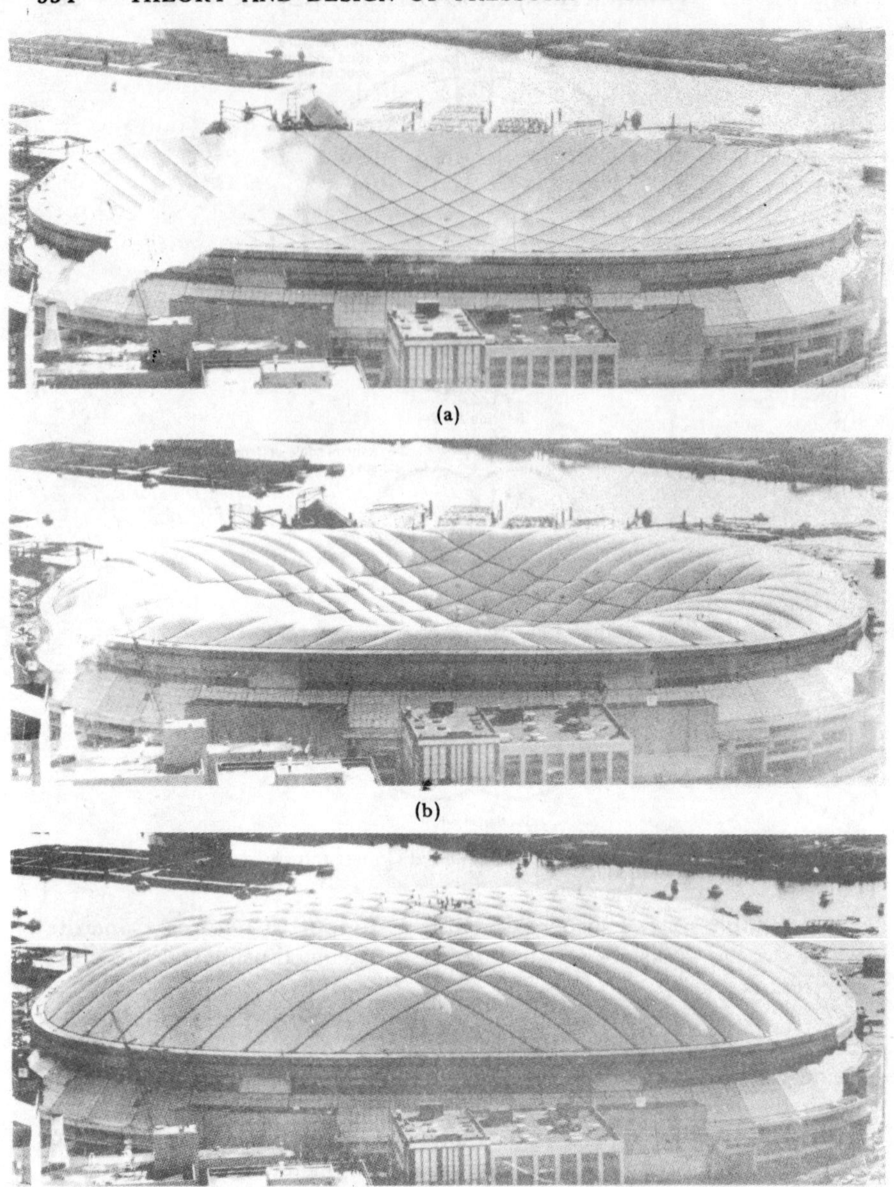

Fig. 7.37. Air Supported Amphitheater. (*a*) Prior to Inflation, (*b*) Partial Inflation, (*c*) Fully Inflated, (Courtesy Owens-Corning Fiberglas Corporation)

7.37. The air supported structure has a double curvature, and since these have like signs, the surface is called synclastic. Many stadiums and amphitheaters are air supported, Fig. 7.37. The second way this is accomplished is called tension supported. Tension structures also employ a double curvature to obtain their structural stability, but it is anticlastic, i.e., its principal curvatures have opposite signs, as in a saddle, Fig. 7.38. Here, the fabric is mechanically pretensioned into stable anticlastic shapes, and an applied load is resisted by a modified stress field. If the fabric is pretensioned to prevent live loads from relaxing the fabric stress to zero, and the structure is reinforced with steel tension cables, the fabric structure can withstand both dead and live loads (wind, snow, seismic). Tension structures are supported by masts, arches, cables, and edge beams which react to tensile stresses in the fabric. Unlike air supported structures, tension structures can be open to the outside because internal air pressure is not required for stability. Many pavilions, mall enclosures and airport terminals are tension supported structures. In this type of structure, as in all membrane structures, the fabric must remain in tension under all loading conditions because membranes are unstable in compression.

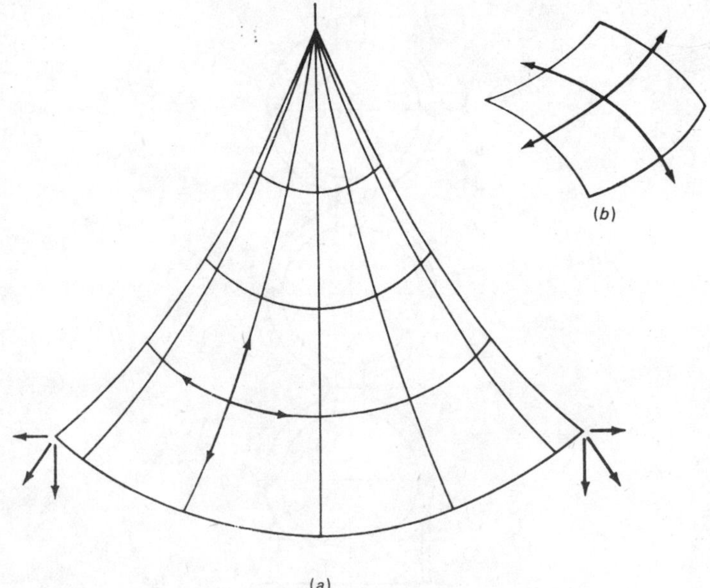

Fig. 7.38. (*a*) Tension Structure. (*b*) Anticlastic Element

The advancement in fabric structures is attributable to the high tensile strengths available in glass and other fiber fabrics, and their environmental protective silicone, teflon, fluopolymer, etc. coatings. Fabric structures offer an economic advantage over their more traditional steel and concrete counterpart for spans greater than approximately 150 feet.

7.9 Modular Construction

An exception to the economic trend from multiple small vessels to single large ones is that of modular construction. This consists of utilizing a single standardized mass-produced module, or building block, and changing total size of vessel required in constant increments by the addition or subtraction of a module. An example of this, which also subscribes to the ideal design of a compound vessel, is that of the multiple spheroid vessel shown in Fig. 7.39. The sphere is the

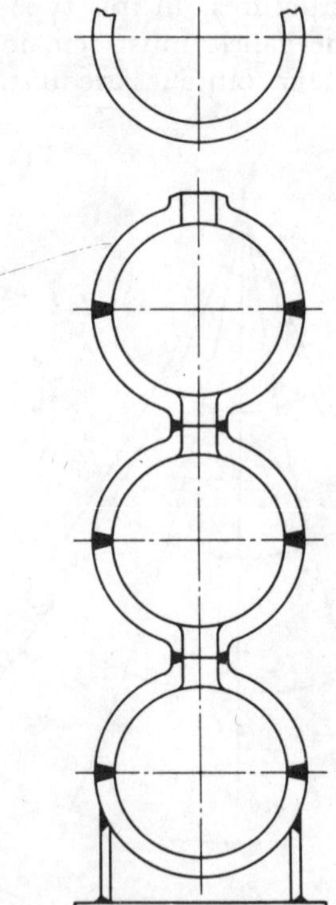

Fig. 7.39. Modular Spherical Pressure Vessel Construction

most favorable shape stress-wise for a pressure vessel. It has the maximum volume and minimum surface area, hence lowest material requirement, and the hemispherical modules are readily mass produced. The final assembly merely consists of joining hemispherical heads and connecting necks, which are formed in the same operation with the hemisphere itself, to make a pressure container of a required volume. Vessels of this type for the high-pressure storage of gases and liquids, and for hydraulic and pneumatic accumulators, are widely used.[116]

Undersea exploration for petroleum, natural gas, and minerals is a rapidly developing commercial activity calling for the use of subsea manned pressure vessels such as diving bells, undersea habitats, and deep-depth submarines.[117,118,119] These are subject to high external pressure, yet must be light in weight; hence, are frequently composed of spheres, intersecting spheres, or combinations thereof with cylinders, Fig. 7.40, and are well adapted to modular construction.

7.10 Adhesives

The attraction that makes particles of a body stick together is a molecular interaction phenomenon called adhesion.[85,86] These molecular attraction forces have an extremely short range, hence it is necessary to have the two particles or pieces within this molecular attraction range or provide a third body capable of supplying this contact relationship. For instance, initially the pages of a new book are frequently difficult to separate because they have been pressed

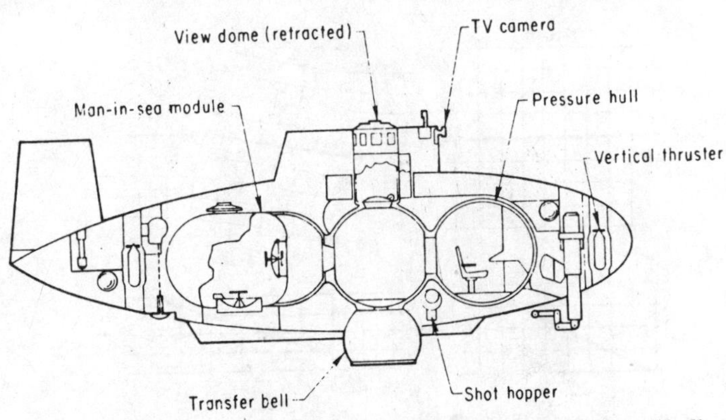

Fig. 7.40. Deep Diving Submersible Showing Pressure Retaining Shell of Intersecting Spheres and Cylinders

close together, but this disappears with use and wear. If however, the pages of this used book becomes wet with water they stick together more than ever. Here water has acted as an adhesive by filling the irregular space between the pages with molecules which allow these large interaction forces to take place between the pages. Because it is practically impossible to obtain molecular distances between large structure parts, adhesives are employed to accomplish this. Glue is an old familiar one, but of limited application.

There are two major requirements for an adhesive. The first is that it have good wetting qualities; i.e., stickiness or attraction for the materials being joined. The second is that the adhesive must not set up excessive residual stress upon solidifying. In addition to the stress[120,121,122] caused by solidification, these joints are also subject to those produced by externally applied loads. Differences in the elastic and plastic properties of the adhesive and the pieces it joins usually cause some concentration of stress. This is shown in Fig. 7.41 for three typical types of joints.[87, 88, 89] The butt joint of Fig. 7.41*a* can exhibit a tendency to contract laterally at the joint.[123,124,125] The lap joint of Fig. 7.41*b* shows the adhesive is strained mainly at the ends

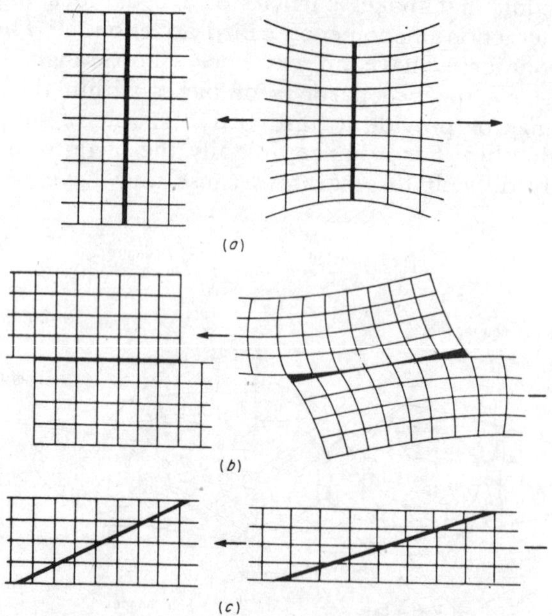

Fig. 7.41. Behavior of Typical Types of Joints Made with Adhesives. (*a*) Butt Joint, (*b*) Lap Joint, (*c*) Scarf Joint

under the action of a shearing force and a tearing force tending to open the joint. The scarf joint of Fig. 7.41*c* in which the two sides are tapered so that they stretch uniformly throughout the length of the joint is the strongest. One of the most widely used for this service is epoxy resin. It flows easily between the surfaces to be joined, and has excellent wetting qualities thereby penetrating surface pores and pits. The solidification process does not depend upon evaporation or oxidation hence it can take place throughout a large joint of non-porous metal. There is also negligible associated shrinkage upon solidification and therefore a minimum of residual stress.

Adhesives are used in aircraft construction to create new structural elements such as "honeycombs," bond reinforcement to riveted or welded joints, or place crack arrestors at strategic locations.[126,127] In pressure vessel design the filament-wound vessel, paragraph 7.6, utilizes adhesives for filament bonding. Joining materials by adhesives for structual and pressure vessel applications is rapidly growing in application and will continue to do so as its engineering and cost advantages are further exploited.[128,129]

In addition to similar materials, many dissimilar ones which cannot be joined by welding can be joined by one or more of the structural adhesives available, as shown on the adhesive selector chart, Fig. 7.42. The main reason for the increased popularity of adhesives is the production cost-cutting potential they offer over traditional fastening techniques.

7.11 Cryogenic Stretch Forming of Vessels

In paragraph 5.7 it was mentioned that work hardening or cold stretching of steels increases their yield strength. It may also increase the ultimate strength of some steels. This is particularly so of the austenitic stainless steels and is the primary way their physical properties are controlled. If cold working is done at extremely low or cryogenic temperatures, the room temperature strength is even further increased. This is the basis of a fabrication procedure for high strength stainless steel pressure vessels.[90] It consists of first soaking a vessel, prefabricated by conventional welding practice, in a liquid nitrogen bath. After the vessel has reached the minus 320°F of the nitrogen bath, it is internally pressurized with nitrogen gas to force it against the walls of a forming die so sized as to control stretching to 10–15 per cent. Unstretched prefabricated vessels with a tensile strength of 102,000 psi have been strengthened to 260,000 psi with a stretch of 13 per cent.[91,148]

Substrates	Urethane	ABS	Acrylic sheet	Poly-carbonate	Nylon	Modified poly-phenylene oxide	Aromatic polyester elastomer	PVC	Poly sulfone	Poly styrene
Urethane	U	A/U	A	A	U	A	U	U	A/U	U
ABS	A/U	A	A	A	U/A	A	U	A	A	A
Acrylic sheet	A/U	A	A	A	U	A	U	A	A/E	A
Poly-carbonate	A/U	A	A	A	U	A	U	A	A	A
Nylon	U	U/A	U	U	U	A/U	U	U	U/A	U
Modified poly-phenylene oxide	A/U	A	A	A	A/U	A	U	A	A	A/U
Aromatic polyester elastomer	U	U	U	U	U	U	U	U	U	U
PVC	U	A	A	A	U	A	U	A	A	U
Polysulfone	A/U	A	A	A	U/A	A	U	A	A	U
Polystyrene	U	A	A	A	U	A/U	U	U	U	U
Epoxy	A/E	A/E	A/E	A/E	U/E	A	U/E	A	A/E	A/E/U
Chlorinated polyethylene	U	U	U	U	U	U	U	U	U	U
Closed-cell urethane foam	A/U	A	A	A	E/U	A	U	A	A	A/U
Open-cell urethane foam	U	U/E	U/E	U/E	U/E	U	U/E	U/E	U/E	U/E
Expanded polystyrene	U/E	U/E	U/E	U/E	U/E	U	U	U/E	U/E	U/E
Cold-rolled steel	A	A	A	A	E	A	U	A	A	A/U
Stainless steel	A	A	A	A	E	A	U	A	A	A/U
Aluminum	A	A	A	A	E	A	U	A	A	A/U
Copper	E	E/A	E/A	E/A	E/U	A/E	U/E	E/A	E/A	E/U
Rubber	E	E	E	E	E	E/U	U/E	E/U	E/U	E/U
Polyester	E/U	A/U	A/U	A/U	U/E	A/U	U/E	A/U	A/U	A/U

A—modified acrylic structural adhesive: U—urethane

Fig. 7.42. Adhesive Selector Chart

There is no limit to the size of vessels that can be fabricated by this method. Neither is it limited to the type of material. While austenitic materials have been most widely used, satisfactory results with aluminum and titanium have also been obtained by proper control or selection of the amount of stretch best suited for each material. It is frequently impossible to realize the full strength of cold worked steels in

Epoxy	Chlorinated poly-ethylene	Closed-cell urethane foam	Open-cell urethane foam	Expanded polystyrene	Cold-rolled steel	Stainless steel	Aluminum	Copper	Rubber	Polyester
A/E	U	A/U	U	U/E	A	A	A	E	E	E/U
A/E	U	A	U/E	U/E	A	A	A	E/A	E	A/U
A/E	U	A	U/E	U/E	A	A	A	E/A	E	A/U
A/E	U	A	U/E	U/E	A	A	A	E/A	E	A/U
U/E	U	E/U	U/E	U/E	E	E	E	E/U	E	U/E
A	U	A	U	U	A	A	A	A/E	E	A/U
U/E	U	U	U/E	U	U	U	U	U/E	U/E	U/E
A	U	A	U/E	U/E	A	A	A	E/A	E/U	A/U
A/E	U	A	U/E	U/E	A	A	A	E/A	E/U	A/U
A/E/U	U	A/U	U/E	U/E	A/U	A/U	A/U	E/U	E/U	A/U
A/E	U	A/E	U/E	U/E	A/E	A/E	A/E	E/A	E	E
U	U	U	U	U	U	U	U	U	U/E	U
A/E	U	A	U/E	U	A	A	A	E/A	E/U	A
U/E	U	U	U/E	U/E	U/E	U/E	U/E	E/U	E/U	U/E
U/E	U	U	U/E	U/E	U/E	U/E	U/E	E/U	E/U	U/E
A/E	U	A	U/E	U/E	A	A	A	E/A	E	A/E
A/E	U	A	U/E	U/E	A	A	A	E/A	E	A
A/E	U	A	U/E	U/E	A	A	A	E/A	E	A
E/A	U	E/A	E/U	E/U	E/A	E/A	E/A	E	E	E/U
E	U/E	E/U	E/U	E/U	E	E	E	E	E	E
E	U	A	U/E	U/E	A/E	A	A	E/U	E	E/U

structural adhesive: E—epoxy structural adhesive

welded cylinders and spheres because the heat of welding at the seams considerably reduces the yield strength of these materials in this area. It is through remedial measures to obviate this to which the economy of this fabrication method is attributed. That is, complete welding of the vessel from annealed plates is undertaken without taking expensive precautions to control out-of-roundness, straightness, or other weld shrinkage effects, and reliance is placed on stretching to remove these distortions. Cost reductions of 15 per cent have been obtained by this method.

These steel vessels may be wound with filaments to produce a lightweight composite structure. Typical applications include solid propellant rocket-motor cases, and high-pressure storage containers for liquids and gases.

REFERENCES

1. V. Weiler, "Some Considerations of Cost in the Design of Pressure Vessels," *Chemical Engineering Progress*, December, 1954.

2. F. D. Clark and S. P. Terni, "The Complete Costs of Thick-Wall Pressure Vessels," *Chemical Engineering*, April 3, 1972.

3. K. Horikawa and T. Okumura, "Elastic and Plastic Stress Distribution on Base Plates in the Neighborhood of Fillet Welds in Lap Joints," *Transactions of Japan Welding Society*, Vol. 1, No. 1, April, 1970.

4. F. M. Flanigan, "Use of a Hydraulic Analogue," *American Society of Heating, Refrigerating and Air Conditioning Engineers Journal*, August, 1961.

5. W. H. McAdams, *Heat Transmission*, McGraw-Hill Book Co., Inc., New York, 1942.

6. J. F. Harvey, "Cooling Effect of Tube Holes in Thick Shells of Marine Boilers," *Transactions of the Society of Naval Architects and Marine Engineers*, November, 1950.

7. C. E. Bowman and T. J. Dolan, "Biaxial Fatigue Studies of High Strength Steels Clad with Stainless Steel," Report T&M 1964, University of Illinois, May, 1960.

8. R. H. Krock, "Whisker-Strengthened Materials," *International Science and Technology*, No. 59, November, 1966.

9. G. Slayter, "Two-Phase Materials," *Scientific American*, January, 1962, pp. 124–134.

10. A. P. Levitt, "Whisker-Strengthened Metals," *Mechanical Engineering*, pp. 36–42, January, 1967.

11. M. M. Schwartz, "Brazed Honeycomb Structures," *Welding Research Council Bulletin* 182, April, 1973.

12. N. O. Brink, "Sandwich Cylinder Construction for Underwater Pressure Vessels," *Naval Engineers Journal*, Vol. 76, No. 6, December, 1964.

13. D. W. Levinson, "Fiber Reinforced Structural Materials," *Machine Design*, 1964.

14. J. H. Faupel, *Engineering Design*, John Wiley and Sons, New York, 1964.

15. T. F. Maclaughlin, "Effect of Fiber Geometry on Stress in Fiber-Reinforced Composite Materials," *Experimental Mechanics*, Vol. 6, No. 10, pp. 481–491, October, 1960.

16. A. E. Armenakas and C. A. Sciammarella, "Experimental Investigation of the Failure Mechanism of Fiber-Reinforced Composites Subjected to Uniaxial Tension," *Experimental Mechanics*, Vol. 13, No. 2, February, 1973.

17. G. C. Sih, E. P. Chen, and S. L. Huang, "Fractures Mechanics of Plastic-Fiber Composites," *ASME Publication Paper No.* 73-DE-20, 1973.

18. R. J. Fedor and L. J. Ebert, "A Study of the Effects of Prestrain on the Tensile Properties of Filamentary Composites," *ASME Publication Paper No.* 72-Mat-K, 1972.

19. K. R. Berg and F. J. Fillippi, "Advanced Fiber-Resin Composites," *Machine Design*, April 1, 1971.

20. W. Wigotsky, "Advanced Materials," *Design News*, Vol. 28, No. 7, April 9, 1973.

21. J. W. Weeton, "Fiber-Metal Matrix Composites," *Machine Design*, Vol. 41, No. 4, February 20, 1969.

22. H. R. Clauser, "Advanced Composite Materials," *Scientific American*, Vol. 229, No. 1, July, 1973.

23. W. H. Sutton and J. Chrone, "Development of High-Strength, Heat Resistant Alloys by Whisker Reinforcement," *Metals Engineering Quarterly*, Vol. 3, No. 1, February, 1963.

24. W. D. Harris, "Joining of Plastics," *Welding Research Council Bulletin* 114, May, 1966.

25. K. R. Berg, "Advanced Composites: Cost Versus Design Improvements," *Mechanical Engineering*, December, 1972.

26. H. A. Sherwood, "Applying Creativity to Low Cost Design," *ASME Publication Paper No.* 73-DE-36, 1973.

27. T. Uno and K. Sudo, "Study of Endurance Strength of Coillayer Vessel with Nozzles," ASME Paper No. II-62, Second International Conference on Pressure Vessel Technology, San Antonio, Texas, 1973.

28. J. E. Bower, "Pressure Tests of Cylindrical Pressure Vessels Reinforced with Steel Wire Wrapping," *ASME Publishing Paper No.* 68-PVP-24, 1968.

29. H. E. Rubo, "New Vistas Opened Up by Welded Multi-Wall Pressure Vessels Having One Wall Isolated from the Next," *Welding Research Abroad*, Vol. IX, No. 10, December, 1963.

30. Pimshteim, Khimatulin and Lushpei, "The Strength of Welded Multi-Layer High Pressure Vessels," *Welding Research Abroad*, Vol. XIII, No. 6, June–July, 1967.

31. J. S. McCabe and E. W. Rothrock, "Multilayer Vessels for High Pressure," *Mechanical Engineering*, March, 1971.

32. H. S. Avery, "Cast Heat-Resistant Alloys for High-Temperature Weldments," *Welding Research Council Bulletin* 143, August, 1969.

33. E. J. Wellaver, "Steel Castings in Weldments Cut Costs," *Mechanical Engineering*, July, 1963.

34. R. R. Irving, "Welding, New Methods for New Materials," *The Iron Age*, March 29, 1962.

35. D. C. Martin, "Modern Welding," *International Science and Technology*, June, 1967.

36. H. E. Pattee, "Joining Ceramics to Metals and Other Materials," Welding Research Council Bulletin 178, November, 1972.

37. H. C. Campbell, "Electroslag, Electrogas and Related Welding Processes," *Welding Research Council Bulletin* 154, September, 1970.

38. L. Munener, "Fracture Toughness of Electroslag Welded Joints," ASME Paper II-82, Second International Conference on Pressure Vessel Technology, San Antonio, Texas, 1973.

39. S. W. Neville, "Ultrasonic Welding," British Welding Research Association Report C41/3/60, *Welding Research*, Vol. VII, No. 7, August–September, 1961.

40. "Ultrasonic Welding," *Machine Design*, April 9, 1964.

41. R. F. Bunshah, "High-Power Electron Beams," *International Science and Technology*, April, 1962.

42. K. J. Miller and T. Takenaka, "Electron Beam Welding," *Welding Research Council Bulletin* 100, October, 1964.

43. R. Bosel and G. Sepold, "Recent Developments in Electron Beam Welding of Unalloyed Steel Joints," ASME Paper II-81, Second International Conference on Pressure Vessels Technology, San Antonio, Texas, 1973.

44. R. L. O'Brien, "Arc Plasmas for Joining, Cutting and Surfacing," *Welding Research Council Bulletin* 131, July, 1968.

45. V. I. Vill, "Friction Welding of Metals," American Welding Society, New York, 1962.

46. M. B. Hollander, "Welding Metals by Friction," *Materials Engineering and Design*, February, 1962.

47. R. B. Aronson, "Lasers: What They Do," *Mechanical Engineering*, May, 1962.

48. M. M. Schwartz, "Laser Welding and Cutting," Welding Research Council Bulletin 167, November, 1971.

49. Y. Arata and I. Miyamoto, "Some Fundamental Properties of High Power Laser Beam as a Heat Source (3 reports)" *Transactions of the Japan Welding Society*, Vol. 3, No. 1, April, 1972.

50. A. H. Hultzman and G. R. Cowan, "Bonding of Metals with Explosives," *Welding Research Council Bulletin* 104, April, 1965.

51. J. M. Gerken and W. A. Owczarski, "A Review of Diffusion Welding," *Welding Research Council Bulletin* 109, October, 1965.

52. C. B. Boyer, J. E. Hatfield, and F. D. Orcutt, "Unique Applications of Pressure Equipment for Hot Isostatic Processing," *ASME Publication Paper* 70-PVP-2, 1970.

53. Applications Continue to Mount for High-frequency Welding, *The Iron Age*, April 9, 1964.

54. D. C. Martin, "High-frequency Resistance Welding," *Welding Research Council Bulletin* 160, April, 1971.

55. F. Wolff and A. Harvey, "Prestressed Fiber-metal Cylindrical Pressure Vessels," *ASME Publication Paper* 63-AHGT-70, 1970.

56. J. L. Mattavi and A. G. Seibert, "Feasibility Evaluation of Boron Filament-Wound Pressure Vessels," *ASME Publication Paper*, No. 68-PVP-21, 1968.

57. L. E. Brownell and E. H. Young, *Process Equipment Design-Vessel Design*, John Wiley & Sons, New York, 1959.

58. W. Johnson, "High-Rate Forming of Metals," *Metal Treatment and Deep Forging*, September, 1963.

59. F. Park, "High Energy Rate Metalworking," *International Science and Technology*, June, 1962.

60. W. F. Courtis, "Electrical Discharge Metal Forming," *Mechanical Engineering*, October, 1962.

61. R. Gorcey, "Filament-Wound Pressure Vessels," *ASME Publications Paper No.* 61-MD-17, 1961.

62. J. D. Marketos, "Optimum Toroidal Pressure Vessel Filament Wound Along Geodesic Lines," American Institute of Aeronautics and Astronautics, 1963.

63. ASME Boiler and Pressure Vessel Code, Section X, "Fiberglass Reinforced Plastic Pressure Vessels."

64. R. F. Lark, "Filament-Wound Composite Vessel Material Technology,"

ASME Paper I-42, Second International Conference on Pressure Vessel Technology, San Antonio, Texas, 1973.

65. J. M. Whitney and J. C. Halpin, "Analysis and Design of Fibrous Composite Pressure Vessels" ASME Paper I-39, Second International Conference on Pressure Vessel Technology, San Antonio, Texas, 1973.

66. J. T. Hofeditz, "Structural Design Considerations for Fiberous Glass Pressure Vessels," Modern Plastics, McGraw-Hill Book Company, New York, April, 1964.

67. K. Zander, "Some Safety Aspects of Wire-Wound Pressure Vessels and Press Frames for Isostatic Pressing in Particular and Industrial Applications in General," ASME Paper I-40, Second International Conference on Pressure Vessel Technology, San Antonio, Texas, 1973.

68. J. R. Faddoul, "Structural Considerations in Design of Lightweight Glass-Fiber Composite Pressure Vessels," ASME Paper I-41, Second International Conference on Pressure Vessel Technology, San Antonio, Texas, 1973.

69. K. Holm, J. E. Buhl, and A. R. Willner, "Glass-Reinforced Plastics for Submersible Pressure Hulls," *Naval Engineers Journal*, October, 1963.

70. F. Wolff and T. Siuta, "Factors Affecting the Performance and Aging of Filament Wound Fiberglass Structures," American Rocket Society Journal, June, 1962.

71. E. C. Young, "Fiber-Glass Tanks," *Mechanical Engineering*, October, 1964.

72. R. O. Marsh and W. Rockenhauser, "Prestressed Concrete Reactor Vessels," *Mechanical Engineering*, July, 1966.

73. T. C. Waters and N. T. Barrett, "Prestressed Concrete Pressure Vessels for Nuclear Reactors," *Journal British Nuclear Energy Society*, July, 1963.

74. M. Bender, "A Report on Prestressed Concrete Reactor Pressure Vessel Technology," *Nuclear Structural Engineering*, Vol. No. 1, July, 1965.

75. M. G. Suarez and N. M. Kramorow, "Large Prestressed Concrete Vessels for Deep Submergence Testing," *ASME Publication* 65-WA/UNT-8, 1965.

76. W. Gorholt and A. J. Neylan, "Design of a Composite Steel and Concrete Closure for a Prestressed Concrete Reactor Vessel," *ASME Publication Paper No.* 72-PVP-17, 1972.

77. R. J. Koerner, "A Simplified Overpressure Analysis of the Walls of a Cylindrical Prestressed Concrete Pressure Vessel," *Nuclear Engineering and Design*, Vol. 14(1970) No. 2, December 1970, North-Holland Publishing Co., Amsterdam, The Netherlands.

78. K. Hiraga, T. Takemori, K. Terada, T. Mogami, K. Watanabe, and M. Satake, "Summary of Experimental Investigations on the Wedge Winding Method for Prestressing Cylindrical Concrete Pressure Vessel Structures," *Nuclear Engineering and Design*, Vol. 22, (1972) No. 1, North-Holland Publishing Co., Amsterdam.

79. D. J. Lewis, J. Irving, and G. D. T. Carmichael, "Advances in the Analysis of Prestressed Concrete Pressure Vessels," *Nuclear Engineering and Design*, Vol. 20 (1972) No. 2, North-Holland Publishing Co., Amsterdam.

80. T. Akagi, T. Ohno, and M. Irobe, "Viscoelastic Analysis of Prestressed Concrete Shell of Revolution, ASME Paper I-34, Second International Conference on Pressure Vessel Technology, San Antonio, Texas, 1973.

81. I. Davidson, "Theorectical and Experimental Modes of Behavior of Cylindrical Model Prestressed Concrete Pressure Vessels When Pressurized to Failure Hydraulically and Pneumatically," *Nuclear Engineering and Design*, Vol. 20 (1972) No. 2, North-Holland Publishing Co., Amsterdam.

82. W. Fuerste, G. Hohnerlein, and H. G. Schafstall, "Prestressed Concrete Reactor Vessels for Nuclear Power Plants Compared to Thick-Walled and Multi-Layer

Steel Vessels," ASME Paper No. I-35, Second International Conference on Pressure Vessel Technology, San Antonio, Texas, 1973.

83. F. E. Schilling and G. Beine, "The Prestressed Cast Iron Reactor Pressure Vessel (PCIPV)," *Nuclear Engineering and Design*, Vol. 25 (1973), North-Holland Publishing Co., Amsterdam.

84. J. P. Duncan and N. W. Murray, "Lobed Pressure Vessels," *Welding Research*, Vol. VII, No. 7, August, 1961.

85. N. A. DeBruyne, "The Action of Adhesives," *Scientific American*, March, 1962.

86. L. H. Sharpe, H. Schonhorn, and C. J. Lynch, "Adhesives," *International Science and Technology*, April, 1964.

87. P. Swannell, "Some Theoretical Notes on the Influence of Joint Stiffness and Geometry on the Strength of Bonded Joints," *Australian Welding Journal*, March/April, 1973.

88. P. D. Hilton and G. D. Gupta, "Stress and Fracture Analysis of Adhesive Joints," *ASME Publication Paper No.* 73-DE-21, 1973.

89. E. F. Hess, "Adhesives for Fabricating High Strength Structures," *Metal Progress*, June, 1963.

90. J. Jonson, "Coldstretched Austenitic Stainless Steel Pressure Vessels," ASME Paper II-85, Second International Conference on Pressure Vessel Technology, San Antonio, Texas, 1973.

91. "Stretch Forming at −320°F Increases Strength of Welded Stainless Vessels," *Welding Design and Fabrication*, April, 1962.

92. O. W. Blodgett, "Paper and Scissors Show How Parts React to Stress," *Machine Design*, March 6, 1975.

93. H. Ford, B. Crossland and E. H. Watson, "Thoughts on a Code of Practice for Forged High Pressure Vessels of Monobloc Design," *ASME Publication Paper No.* 78-PVP-62, 1978.

94. G. J. Mraz and E. G. Nisbett, "Design, Manufacture and Safety Aspects of Forged Vessels for High Pressure Services," *ASME Publication Paper No.* 78-PVP-72, 1978.

95. J. R. Vinson and Tsu-Wei Chou, *Composite Materials and Their Use in Structures*, John Wiley & Sons, New York, 1975.

96. W. W. Stinchcomb, K. L. Reifsnider and R. S. Williams, "Critical Factors for Frequency-Dependent Fatigue Processes in Composite Materials," *Experimental Mechanics*, September, 1976.

97. F. P. Gerstle, "High Performance Advanced Composite Spherical Pressure Vessels," *ASME Publication Paper No.* 74-PVP-42, 1974.

98. F. A. Simonen, N. C. Henderson, R. D. Winegardner and K. Specht, "Analysis of Strength and Residual Stresses in Filament Reinforced Aluminum Cylinders," *ASME Publication Paper No.* 75-PVP-24, 1975.

99. I. M. Daniel, R. E. Rowlands and J. B. Whiteside, "Effects of Material and Stacking Sequence on Behavior of Composite Plates with Holes," *Experimental Mechanics*, January, 1974.

100. C. B. Boyer and J. C. Wahl, "Equipment for Cold Isostatic Pressing," *ASME Publication Paper No.* 74-PVP-21, 1974.

101. J. S. Porowski, W. J. O'Donnell and A. S. Roberts, "Theory of Free Coiled Pressure Vessels," *ASME Publication Paper No.* 78-PVP-73, 1978.

102. R. Pechacek, "Advanced Technology for Large, Thick-Wall High Pressure Vessels," *ASME Publication Paper No.* 76-Pet-77, 1977.

103. T. Maruyama, H. Togawa and S. Kimura, "Analysis of Transient Thermal

Stress and Thermal Fatigue Strength on Multiwall Vessel and Nonintegral Reinforcing Structure," Third International Conference on Pressure Vessel Technology, Tokyo, Japan, 1977.

104. R. L. Brockenbrough and J. E. Steiner, "Autofrettaged Wire-Wrapped Pressure Vessels," *ASME Publication Paper No*. 76-PVP-47, 1976.

105. W. D. Humphrey and R. F. Foral, "Fatigue and Residual Strength Tests of Fiberglass Composite Pressure Vessels," *ASME Publication Paper No*. 74-PVP-47, 1974.

106. M. W. Wilcox and O. B. Abhat, "Dynamic Response of Laminated Composite Shell under Radial or Hydrostatic Pressures," *ASME Publication Paper No*. 74-PVP-40, 1974.

107. H. J. Sutherland and H. H. Calvit, "A Dynamic Investigation of Fiber-Reinforced Viscoelastic Materials," *Experimental Mechanics*, August, 1974.

108. D. S. Mehta, H. W. Osgood, A. J. Bingaman and K. P. Buchert, "Trends in the Design of Pressurized Water Reactor Containment Structures and Systems," *Nuclear Safety*, Vol. 18, No. 2, March 1977, Oak Ridge National Laboratory.

109. W. G. Dodge, Z. P. Bazant and R. H. Gallagher, "A Review of Analysis Methods for Prestressed Concrete Reactor Vessels," Oak Ridge National Laboratory, ORNL-5173, February, 1977.

110. G. L. England, "Steady-State Stresses in Concrete Structures Subjected to Sustained and Cyclically Varying Temperatures," *Nuclear Engineering and Design*, October, 1977, North-Holland Publishing Co., Amsterdam.

111. W. L. Greenstreet, C. B. Oland, J. P. Callahan, and D. A. Canonico, "Feasibility Study of Prestressed Concrete Pressure Vessels for Coal Gasifiers," Oak Ridge National Laboratory Report ORNL/5312, August, 1977.

112. W. G. Dodge, G. C. Robinson, and J. P. Callahan, "Development of the PCRV Steam Generator Cavity Closure for the Gas Cooled Fast Reactor," *Nuclear Engineering and Design*, February, 1978.

113. J. P. Callahan, G. C. Robinson, and R. C. Burrow, "Uniaxial Compressive Strengths of Concrete for Temperatures Reaching 1033K," *Nuclear Engineering and Design*, February, 1978.

114. J. J. Closner, "Design and Application of Prestressed Concrete for Oil Storage," *ASME Publication Paper No*. 75-Pet-18, 1975.

115. O. Buyukozturk and P. V. Marcal, "Strength of Reinforced Concrete Chambers Under External Pressure," *ASME Publication Paper No*. 75-PVP-7, 1975.

116. S. Andersson, "Element-Built Pressure Vessel for Greater Safety," *ASME Publication Paper No*. 74-PVP-11, 1974.

117. F. E. Bryson, "Opening the Ocean Frontiers," *Machine Design*, December 27, 1973.

118. W. E. Schneider and J. A. Sasse, "12,000-Ft. Sea-Water Variable Ballast System for the Submersible 'Alvin'," *ASME Publication Paper No*. 73-WA/Oct-12, 1973.

119. J. F. Saunders and F. Hirschfeld, "The Growing Market for Commercial Submarines," *Mechanical Engineering*, October, 1976.

120. W. D. Bascom, R. L. Cottington and C. O. Timmons, "Fracture Design Criteria for Structural Adhesive Bonding, Promise and Problems," *Naval Engineering Journal*, Vol. 88, No. 4, August, 1976.

121. D. R. Mulville, D. L. Hunston, and P. W. Mast, "Developing a Failure Criteria for Adhesive Joints Under Complex Loading," *ASME Publication Paper No*. 77-WA/Met-10, 1977.

122. M. M. Ratwani, "A Parametric Study of Fatigue Crack Growth in Adhesively Bonded Metallic Structures," *ASME Publication Paper No.* 77-WA/Met-5, 1977.

123. K. L. DeVries, M. L. Williams and M. D. Chang, "Adhesive Fracture of a Lap Shear Joint," *Experimental Mechanics*, March, 1974.

124. T. Swift, "Fracture Analysis of Adhesively Bonded Cracked Panels," *ASME Publication Paper No.* 77-WA/Met-2, 1977.

125. L. J. Hart-Smith, "Adhesive Bonded Stresses and Strains at Discontinuities and Cracks in Bonded Structures," *ASME Publication Paper No.* 77-WA/Met-6, 1977.

126. M. R. Gecit, and F. Erdogan, "The Effect of Adhesive Layers on the Fracture of Laminated Structures," *ASME Publication Paper No.* 77-WA/Met-4, 1977.

127. K. R. Wentz, and H. F. Wolfe, "Development of Random Fatigue Data for Adhesively Bonded and Weld-Bonded Structures Subjected to Dynamic Excitation," *ASME Publication Paper No.* 77-WA/Met-1, 1977.

128. M. J. Heckenkamp and V. Pavelic, "New Directions for Adhesive Usage," *ASME Publication Paper No.* 76-DE-15, 1976.

129. J. P. McNally and C. R. Ronan, "Metal-to-Metal Adhesive Bonding," *Welding Research Council Bulletin 220*, October 1976.

130. S. B. Palmeter, "Power Plant Capital Costs—What's Behind the Upward Climb?" ASME Publication Paper No. 79-PVP-87, 1979.

131. E. Raudsepp, "Stimulating Creative Thinking," *Machine Design*, June 9, 1983.

132. T. Bakey, "Economic Benefits and Economic Impact of Interactive Computer-Aided Design," ASME Publication Paper No. 79-PVP-80, 1979.

133. N. Petrick, "Standardization as a Means of Reducing Power Plant Costs," ASME Publication Paper No. 79-PVP-108, 1979.

134. A. D. McCoy, "The Use of Design Models in the Power Industry," ASME Publication Paper No. 79-PV-P-7, 1979.

135. E. Karl and H. Lenz, "The Influence of the Safety Factor on the Wall Thickness and the Critical Crack Size," ASME Pressure Vessel and Piping Conference, San Antonio, 1984.

136. G. Lubin, *Handbook of Composites*, Von Nostrand Reinhold, New York, 1982.

137. H. C. Rauschenplat, "Design and Fabrication of Multiwall Pressure Vessels," ASME Publication PVP-Vol. 57, 1982.

138. C. P. Shen, A. R. Beim, O. M. Luz, and I. Metzger, "Bursting Experiment of a High Pressure Multiwall Test Vessel," ASME Publication Paper No. 79-PVP-96, 1979.

139. L. A. Van Gulick, "Slideline Verification for Multilayer Pressure Vessel and Piping Analysis," ASME Publication Paper 83-PVP-28, 1983.

140. R. A. Mancuso and H. C. Rauschenplat, "Thermal Transient Analysis in Layered Pressure Vessels," ASME Publication Paper No. 79-PVP-13, 1979.

141. M. Sabbaghian and E. J. Henriquez, "Optimum Design and Creep Relaxation of Dissimilar Material Multilayer Shrink Fitted Vessels," ASME Publication Paper No. 81-PVP-30, 1981.

142. G. E. O. Widera and D. L. Logan, "A Shell Theory for Composite Cylindrical Pressure Vessels and Tanks," ASME Publication PVP-Vol. 57, 1982

143. S. G. Ladkany, "Prestressed Circumferentially Reinforced Large Alumi-

num Fiberglass Composite Pressure Vessels," ASME Publication PVP-Vol. 57, 1982

144. D. A. Riegner and J. C. Hsu, "How Fiber-reinforced Plastics Behave in Fatigue," *Machine Design*, May 20, 1983.

145. D. C. Leigh, T. R. Tauchert, U. Chandra, and M. Podhorsky, "Selection of Structures for Large Diameter Thick-Wall Coal Conversion Pressure Vessels," ASME Publication 81-WA/PVP-1, 1981.

146. H. L. Gotschall, J. E. Roberts, and J. P. Callahan, "Plan for Development of an Optimized PCRV for an HTGR Power Plant," Oak Ridge National Laboratory, ORNL/TM-6606, June, 1979.

147. H. H. Haynes, W. F. Chen, T. Y. Chang, and H. Suzuki, "External Hydrostatic Pressure Loading of Concrete Cylinder Shells," ASME Publication Paper 79-PVP-125, 1979.

148. R. F. Foral, "Composite Spherical Pressure Vessels with Hardening Metal Liners," ASME Publication Paper No. 79-PVP-5, 1979.

149. Y. Shimizu and Y. Moriwaki, "Buckling Strength of Lining Shells of Pressure Vessels," *Fifth International Conference on Pressure Vessel Technology, Volume I, Design and Analysis, San Francisco, California, September 9–14, 1984*, ASME Publication, 1984.

8

Buckling of Vessels Under External Pressure

8.1 Buckling Phenomenon

The foregoing chapters have dealt mainly with vessels subject to internal pressure and the resulting primary stresses were tensile, with exceptions at local knuckle or discontinuity regions, Pars. 2.6.4 and 4.7. With few exceptions, such as brittle behavior, Par. 5.22, fracture is preceded by gross general distortion at stresses above the yield strength. When these same vessels are subject to external pressure and the primary stresses become compressive, buckling instability can occur at stresses well below the elastic limit or yield strength of the material. This type of failure gives little warning of the impending collapse. The basic elastic analysis premise was that stresses and deflections are proportional to the applied loads, and failure is a consequence of reaching the yield or ultimate strength of the material. However, experiments show that when the compressive force in a vessel approaches a critical value, radial deflections begin and increase rapidly with increase in the compressive force. A load equal to this critical value is usually sufficient to produce complete failure of the structure, Fig. 8.1*a*,*b*, and *c*. The significant characteristic of elastic buckling is that the elastic deflections are not proportional to the loads; hence, the problem is not one of determining a relationship between load and deflection but rather that of obtaining an expression for the critical collapse or buckling load.

The column buckling phenomenon is also exhibited in the failure of thin-walled cylinders subject to uniform external pressure. Thin-walled cylinders are those in which the thickness of the wall is small in comparison with the diameter of the vessel so that elastic buckling occurs prior to the wall material reaching its yield strength. Submersibles, fire-tube boilers, nuclear fuel elements, oil well casings, vacuum tanks, etc. are examples of vessels under external pressure. These will

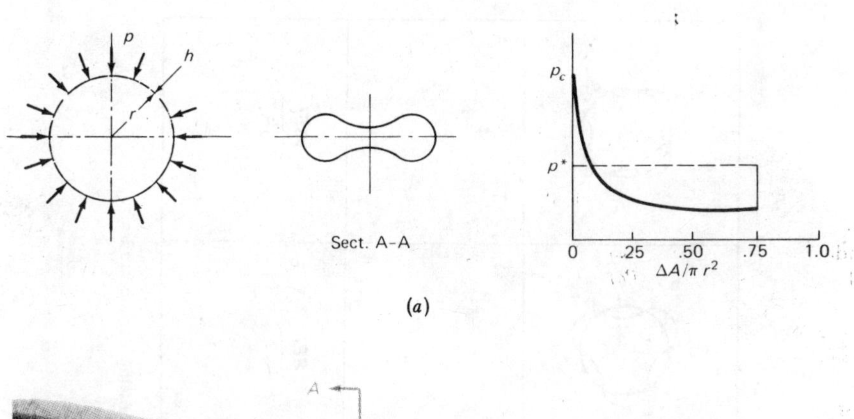

(*a*)

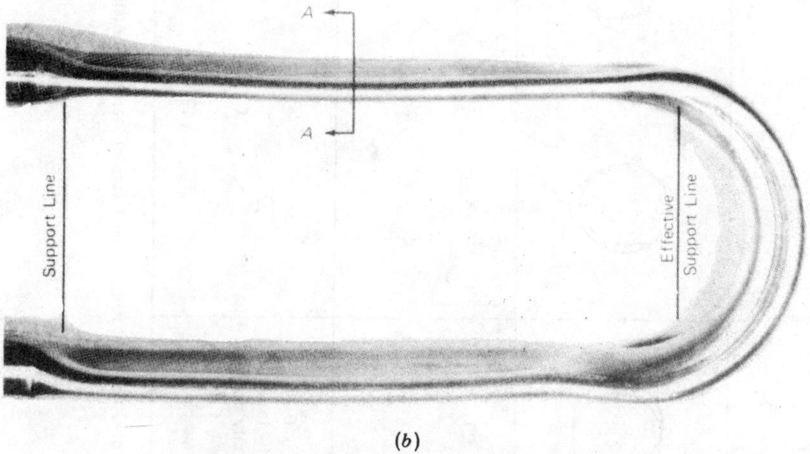

(*b*)

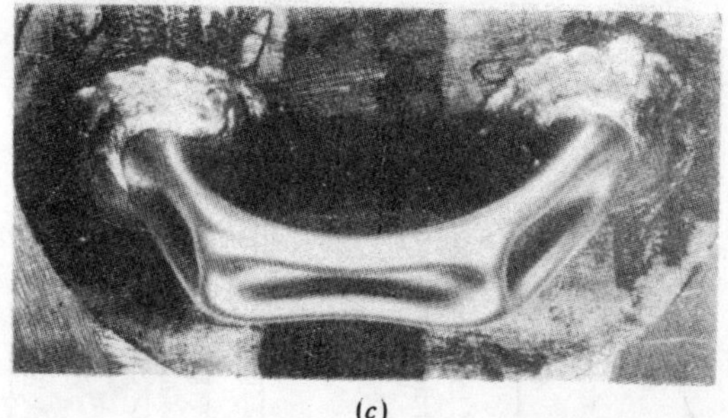

(*c*)

Fig. 8.1. (*a*) Straight Cylindrical Tube and Bend, $r = 0.311$ in., $h = 0.058$ in., $r_0 = 1.31$ in. Under Uniform External Pressure. (*b*) 2-Lobe Collapse Pattern of Long Straight Cylindrical Portion at $p_c = 6{,}800$ psi, and Stiffening Effect of U-Bend. (*c*) Two Stabilizing Longitudinal Nodes Occurring at Approximately 60° Intervals Around the Bend When the Ends of This Bend Have Been Welded to a Diaphragm to Force Collapse of the Bend at $p_c = 12{,}200$ psi.

Characteristics / Buckling Mode	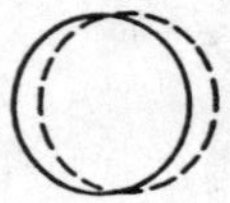	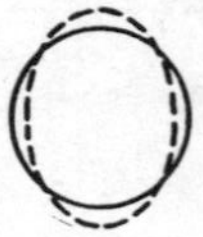	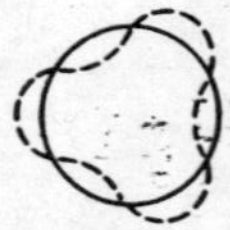	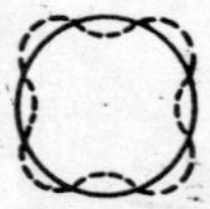
k = Number of lobes (k is the number of full sine waves around the periphery)	1	2	3	4
p_c = Critical pressure the elastic buckling or collapse	0 (No elastic buckling occurs, only sidewise displacement of undistorted circle)	$\frac{3EI}{r^3}$	$\frac{8EI}{r^3}$	$\frac{15EI}{r^3}$

Fig. 8.2. Buckling or Collapse of Cylindrical Ring Under External Pressure Showing Various Modes

fail with increasing external pressure when this reaches a critical value called the critical load or collapse pressure.

A thin-walled cylinder buckles in a definite pattern depending on its relative dimensions and the restraint conditions at the ends. Figure 8.2 shows the more common patterns taken by cylinders as they collapse. They are described by the number of lobes (2, 3, 4, etc.) with the smallest number giving the lowest collapsing pressure. Figure 8.1*b* shows the two-lobe pattern of a thin-walled cylinder at the minimum collapsing pressure. The number of lobes for a cylinder of given dimensions is largely dependent upon the type of restraint at the circumference or edge of the ends of the cylinder and the distance between these restraints. Restraints can take the form of structural rings or bends in the cylinder itself, as shown by Fig. 8.1*b*, wherein the two-lobe pattern collapse of the long straight section was prevented at the more rigid U-bend portion of the cylinder. As is the case with plates or beams, a cylinder which is maintained round (deflection prevented) by means of a stiffening ring is termed "simply supported"; however, if, in addition the section is prevented from rotating (changing slope), the cylinder is termed "clamped or fixed" at the section.

8.2 Elastic Buckling of Circular Rings and Cylinders under External Pressure

A thin-walled cylinder whose length is large in comparison to its diameter will buckle elastically in the form of two lobes when subject to uniform external pressure. In finding the elastic collapse pressure for long cylinders, that for an isolated ring of unit length is first obtained. This ring, Fig. 8.3*a* is assumed to be perfectly circular before the application of the external pressure. Each cross-section of the thickness of the arc *ad* is subjected to a direct compressive hoop force, per Eq. 2.2.1. This is similar to a straight slender column under a central axial load for which the buckling load is given by Euler's formula* $P = \pi^2 EI/\ell^2$. Neglecting the curvature of the ring; and further taking the length of the column equal to one-quarter the circumference of the ring (equivalent to four inflection points), arc *ad*, this formula gives:[1]

$$pr = \frac{\pi^2 EI}{(\pi r/2)^2} \quad \text{or} \quad p = \frac{4EI}{r^3} \tag{8.2.1}$$

$$p = \frac{E}{3}\left(\frac{h}{r}\right)^3 \quad \text{or} \quad \frac{8}{3}E\left(\frac{h}{d}\right)^3 \tag{8.2.2}$$

*See any textbook on strength of materials.

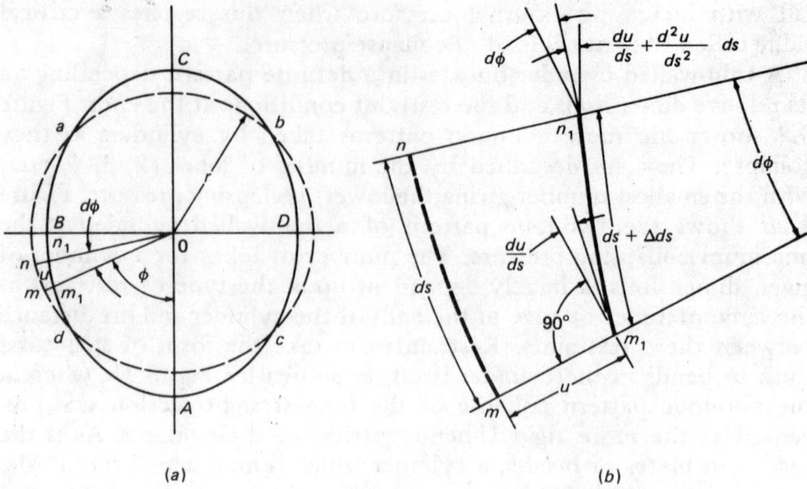

Fig. 8.3. Deformation of a Thin Circular Ring. (*a*) General Deformation into a Two-Lobe Shape. (*b*) Deformation of Small Element

where h = thickness of cylinder and I = moment of inertia of unit width of thickness.

Equations 8.2.1 and 8.2.2 show that the critical load does not depend upon the strength of the material but only on the stiffness of the material as measured by its modulus of elasticity, and on the geometric dimensions as measured by the slenderness ratio of perimeter length to cross-sectional moment of inertia. Hence, identical size and thickness vessels made of low and high strength steel will collapse at the same external pressure because the modulus of elasticity of all steels is approximately the same. An increase in collapse load or pressure can only be obtained by increasing the cross-sectional moment of inertia. Due to the assumptions made, this value of the collapse pressure is approximate, (the numerical constant in Eq. 8.2.1 is not the same as that for two lobes given in Fig. 8.2.), and a more exact equation for this curved member is developed in the following paragraph.

1. *Deflection Curve for a Circular Ring*

The differential equation of the deflection curve of a thin circular ring is analogous to that for a curved bar where the change in curvature due to an applied bending moment is expressed by:[2]

$$\frac{1}{r_1} - \frac{1}{r} = -\frac{M}{EI} \tag{8.2.3}$$

For a straight beam, r is infinite, and this reduces to

$$\frac{1}{r_1} = -\frac{M}{EI} \tag{8.2.4}$$

The minus sign on the right-hand side of Eqs. 8.2.3 and 8.2.4 is consistent with the sign of the bending moment, which is taken to be positive when it produces a decrease in the initial curvature of the ring. In Fig. 8.3*a*, *ABCD* represents a circular ring after deformation, in which u denotes the small radial displacement referred to the perfect initial circle. The variation in curvature during bending is found by comparing one element mn of the true circular ring and the corresponding element m_1n_1 of the deformed ring between the same radii, Fig. 8.3*b*. The initial length of the element mn is

$$ds = rd\phi \tag{8.2.5}$$

and its initial curvature is

$$\frac{d\phi}{ds} = \frac{1}{r} \tag{8.2.6}$$

For small deflections, the curvature of the same element after deformation is taken equal to the curvature of the element m_1n_1 or

$$\frac{1}{r_1} = \frac{d\phi + \Delta d\phi}{ds + \Delta ds} \tag{8.2.7}$$

in which $d\phi + \Delta d\phi$ is the angle between the normal cross sections at points m_1 and n_1 of the deformed ring; and $ds + \Delta ds$ is the length of the deformed element m_1n_1. The displacement u is assumed to be very small in comparison to the radius of the ring, and is taken as positive when it is toward the center of the ring. At m_1 the angle between the tangent to the deformed ring and a perpendicular to the cross section mm_1 (radius m_10) is du/ds. The corresponding angle at point n_1 on cross section nn_1 is

$$\frac{du}{ds} + \frac{d^2u}{ds^2}ds \tag{8.2.8}$$

Then

$$\Delta d\phi = \frac{du}{ds} + \frac{d^2u}{ds^2}ds - \frac{du}{ds} \tag{8.2.9}$$

$$\Delta d\phi = \frac{d^2u}{ds^2}\, ds \tag{8.2.10}$$

In comparing the length of the element m_1n_1 with that of the element mn, the small angle du/ds is neglected and the length of m_1n_1 is taken as $(r - u)\, d\phi$. Then

$$\Delta ds = -u d\phi = -\frac{uds}{r} \tag{8.2.11}$$

Substituting Eqs. 8.2.10 and 8.2.11 in Eq. 8.2.7 gives

$$\frac{1}{r_1} = \frac{d\phi + \dfrac{d^2u}{ds^2}\, ds}{ds\left(1 - \dfrac{u}{r}\right)} \tag{8.2.12}$$

$$\frac{1}{r_1} = \frac{d\phi}{ds\left(1 - \dfrac{u}{r}\right)} + \frac{d^2u}{ds^2\left(1 - \dfrac{u}{r}\right)} \tag{8.2.13}$$

Multiplying the right-hand parts of Eq. 8.2.13 by

$$\frac{1 + \dfrac{u}{r}}{1 + \dfrac{u}{r}}$$

gives

$$\frac{1}{r_1} = \frac{d\phi}{ds}\, \frac{\left(1 + \dfrac{u}{r}\right)}{1 - \left(\dfrac{u}{r}\right)^2} + \frac{d^2u}{ds^2}\, \frac{\left(1 + \dfrac{u}{r}\right)}{1 - \left(\dfrac{u}{r}\right)^2} \tag{8.2.14}$$

and neglecting the small quantities of higher order

$$\frac{1}{r_1} = \frac{d\phi}{ds}\left(1 + \frac{u}{r}\right) + \frac{d^2u}{ds^2} = \frac{1}{r}\left(1 + \frac{u}{r}\right) + \frac{d^2u}{ds^2} \tag{8.2.15}$$

or

$$\frac{1}{r_1} - \frac{1}{r} = \frac{u}{r^2} + \frac{d^2u}{ds^2} \tag{8.2.16}$$

Substituting from Eq. 8.2.3 in Eq. 8.2.16 gives

$$\frac{d^2u}{ds^2} + \frac{u}{r^2} = -\frac{M}{EI} \tag{8.2.17}$$

which is the differential equation for the deflection curve for a thin circular ring. For an infinitely large r this equation reduces to that for a straight beam.

2. *Buckling of a Circular Ring*

The method of obtaining the collapsing pressure is to assume that the true circular section is slightly deflected elastically, and then find the external pressure that will just hold the ring in equilibrium in this shape. Its value can be obtained by use of the equation for the deflection curve of a ring, Eq. 8.2.17. There are four inflection points (a, b, c, and d, Fig. 8.4a) at which the bending moment is zero and also the radial deflection is zero. The bending moment at any cross section B of the ring, Fig. 8.4a, is:

$$M = M_0 + p\overline{AO} \cdot \overline{AD} - \left(\frac{p}{2}\right)\overline{AB}^2 \tag{8.2.18}$$

In the triangle AOB, the law of cosines gives:

$$\overline{OB}^2 = \overline{AO}^2 + \overline{AB}^2 - 2\overline{AO} \cdot \overline{AB} \cos\theta \tag{8.2.19}$$

but $\cos\theta = \overline{AD}/\overline{AB}$, so

$$\overline{OB}^2 = \overline{AO}^2 + \overline{AB}^2 - 2\overline{AO} \cdot \overline{AD} \tag{8.2.20}$$

or

$$\frac{\overline{AB}^2}{2} - \overline{AO} \cdot \overline{AD} = \frac{1}{2}(\overline{OB}^2 - \overline{AO}^2) \tag{8.2.21}$$

and substituting the values $\overline{OB} = (r - u)$ and $AO = (r - u_0)$ gives:

$$\frac{\overline{AB}^2}{2} - \overline{AO} \cdot \overline{AD} = \frac{1}{2}[(r-u)^2 - (r-u_0)^2] \tag{8.2.22}$$

Noting that u is small in comparison to r, terms in u^2 and u_0^2 can be neglected to give

$$\frac{\overline{AB}^2}{2} - \overline{AO} \cdot \overline{AD} = r(u_0 - u) \tag{8.2.23}$$

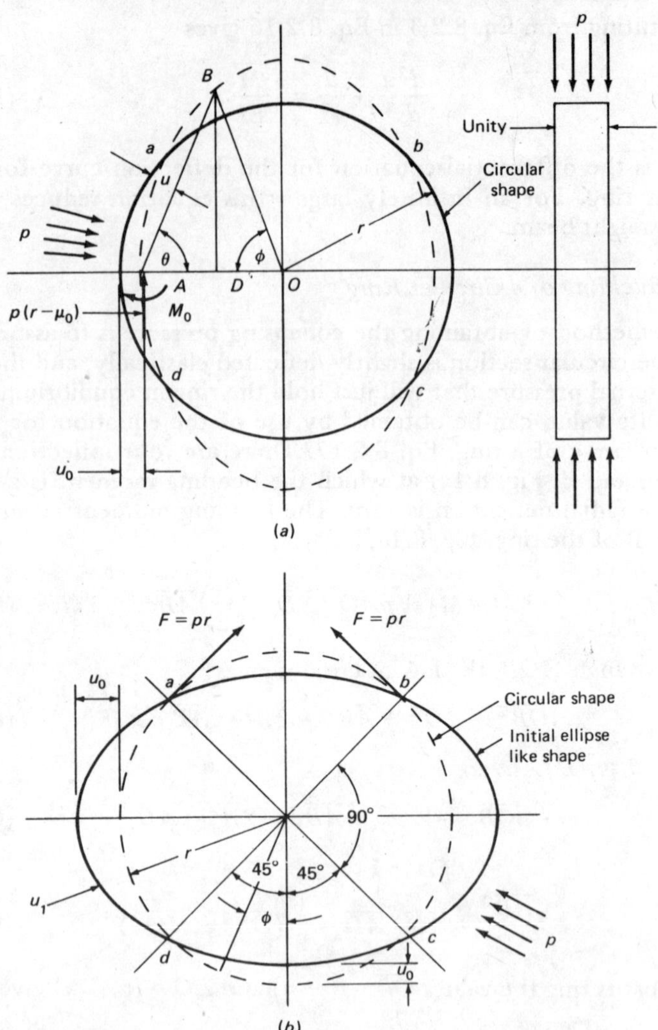

/Fig. 8.4. (*a*) Cylindrical Ring under External Pressure During Buckling to an Ellipselike Shape. (*b*) Cylinder with Initial Ellipticity under External Pressure

Substituting Eq. 8.2.23 in Eq. 8.2.18 gives

$$M = M_0 - pr(u_0 - u) \tag{8.2.24}$$

Further, substituting Eq. 8.2.24 in Eq. 8.2.17 noting $d^2s = r^2\,d\phi^2$ gives

$$\frac{d^2u}{d\phi^2} + u = \frac{-r^2[M_0 - pr(u_0 - u)]}{EI} \tag{8.2.25}$$

or

$$\frac{d^2u}{d\phi^2} + u\left(1 + \frac{pr^3}{EI}\right) = \frac{-r^2M_0 + pr^3u_0}{EI} \tag{8.2.26}$$

This differential equation is linear and has a constant right-hand member. Letting

$$k^2 = 1 + \frac{pr^2}{EI} \tag{8.2.27}$$

the solution of Eq. 8.2.26 is

$$u = C_1 \sin k\phi + C_2 \cos k\phi + \frac{-r^2M_0 + pr^3u_0}{EI + pr^3} \tag{8.2.28}$$

The constants C_1 and C_2 can be determined from the conditions at cross section A and F of the buckled ring where from symmetry,

$$\left(\frac{du}{d\phi}\right)_{\phi=0} = 0 \tag{8.2.29}$$

and

$$\left(\frac{du}{d\phi}\right)_{\phi=\pi/2} = 0 \tag{8.2.30}$$

from Eq. 8.2.29 $C_1 = 0$, and from Eq. 8.2.30

$$0 = C_2 \sin k\left(\frac{\pi}{2}\right) \tag{8.2.31}$$

But C_2 cannot be zero because this value would make u, Eq. 8.2.28, or its derivative, $du/d\phi$, constant for all values ϕ; hence,

$$0 = \sin k\left(\frac{\pi}{2}\right) \tag{8.2.32}$$

The smallest root of Eq. 8.2.32, other than zero, is

$$\frac{k\pi}{2} = \pi \tag{8.2.33}$$

hence

$$k = 2 \tag{8.2.34}$$

Substituting Eq. 8.2.34 in Eq. 8.2.27 and noting $I = h^3/12$ gives

$$p_c = \frac{3EI}{r^3} = \frac{E}{4}\left(\frac{h}{r}\right)^3 = 2E\left(\frac{h}{d}\right)^3 \tag{8.2.35}$$

The value of p_c, Eq. 8.2.35, is the critical uniform buckling or collapsing pressure (force per unit area) for the ring shown in Fig. 8.4, where h is the thickness, r is the outside radius, and d is the outside diameter. The basic premise in thin vessel analysis is that the difference between the inside radius and the outside radius is nil. However; when the pressure is external, the pressure boundary is established by the outside surface dimensions and these are used in satisfying static equilibrium. Correspondingly, when vessels are subject to internal pressure, Chapter 2, the pressure boundary is established by the inside surface dimensions and these are used to satisfy static equilibrium. The factor 2 in Eq. 8.2.35 is compared to the approximate value of 2.67, Eq. 8.2.2, when the arc *ad* is assumed to act as a straight column.

In Eq. 8.2.32 k is the number of full sine waves around the periphery, and other roots, such as, $k\pi/2 = 2\pi$, 3π, etc., correspond to a larger number of waves in the buckled ring and a greater value of the critical pressure, Eq. 8.2.27. Thus,

$$p_c = \frac{EI}{r^3}(k^2 - 1) = \frac{EI}{r^3}(0, 3, 8, 15, \cdots) \qquad \text{for} \quad k = 1, 2, 3, 4, \cdots \tag{8.2.36}$$

Mathematically, $k = 1$ means a small displacement of the undistorted ring which is in equilibrum without pressure.[3] At a value of $k = 2$ the ring goes into a two-lobe or ellipselike shape for the minimum critical pressure. Higher modes[4] of buckling gives higher critical pressures, Fig. 8.2, and the design method for obtaining this increased stability is through fastened ends, structured stiffeners, diaphragms, closure heads, or bends, Fig. 8.1*b*.

3. *Buckling of Long Cylinders or Tubes*

The theory of buckling for a circular ring can also be used for long cylinders or tubes subjected to external pressure because the ring may be considered cut from a long cylinder by two cross sections a unit distance apart,[5] Fig. 8.4*a*. Since the cross section of the long cylinder will not be distorted during ring bending, it is only necessary to substitute for the ring rigidity, EI, the comparable plate flexural rigidity, D, from Eq. 3.2.9 in Eq. 3.2.35 to give

$$p_c = \frac{3EI}{(1-\mu^2)r^3} = \frac{E}{4(1-\mu^2)}\left(\frac{h}{r}\right)^3 = \frac{2E}{(1-\mu^2)}\left(\frac{h}{d}\right)^3 \tag{8.2.37}$$

This equation gives the critical value of the pressure provided the corresponding compressive stress does not exceed the proportional limit of the material. The critical stress is obtained by substituting the value of p_c from Eq. 8.2.37 in Eq. 2.2.4 to give

$$\sigma_c = \frac{E}{4(1-\mu^2)}\left(\frac{h}{r}\right)^2 = \frac{E}{(1-\mu^2)}\left(\frac{h}{d}\right)^2 \tag{8.2.38}$$

The modulus of elasticity, E, for materials with a definite yield point, such as the steel of Fig. 5.1, remains substantially constant until the yield point is closely approached. The simplified elastic-plastic stress-strain diagram, Fig. 1.5*c* and Fig. 8.5*b*, is based on this assumption. The failure curve for a cylinder made of this type material is shown in Fig. 8.5*a* and consists of two parts:

1. Part *AB* for thin, high d/h ratio cylinders where elastic buckling is critical, Eq. 8.2.38, and
2. Part *BC* for thick, low d/h ratio cylinders where plastic yield strength is critical, Eq. 2.2.4.

The curve shows that for thick or low d/h ratio cylinders it may be economical to use high yield strength steels, whereas for thin ones

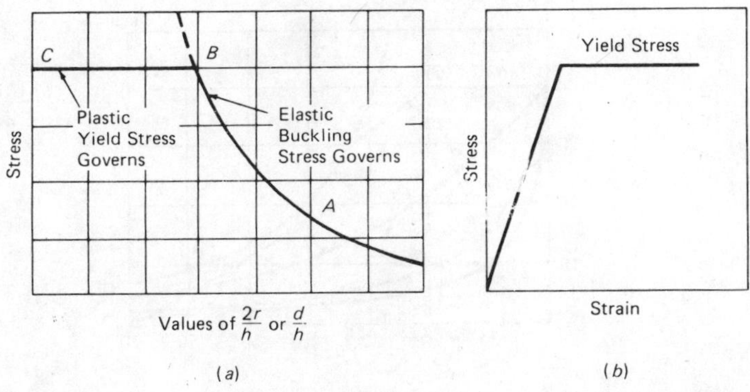

Fig. 8.5. (*a*) Buckling Failure of Cylinder with (*b*) Simplified Elastic-Plastic Relationship

with high d/h ratios, such materials offer no advantage because the modulus of elasticity for all steels approximately is the same. For these materials, tests are in good agreement and the small discrepancies in the immediate region of the intersection of the elastic buckling and plastic yield curves, point B Fig. 8.5*a*, can be taken care of by the factor of safety. Greater accuracy in calculating the critical stress in this region or with materials that do not have a definite yield point can be accomplished by employing the tangent modulus, E_t, instead of E so

$$\sigma_c = \frac{E_t}{4(1-\mu^2)}\left(\frac{h}{r}\right)^2 = \frac{E_t}{(1-\mu^2)}\left(\frac{h}{d}\right)^2 \tag{8.2.39}$$

This is illustrated in Fig. 8.6 for a steel which has the compression test stress-strain diagram and tangent modulus values shown in Fig. 8.7. This material has a proportional limit of 30,000 psi and up to which limit the elastic modulus remains constant at $E = 30{,}000{,}000$ psi and $\mu = 0.3$. The material does not have a definite yield point; and accordingly, the yield strength is taken at the 0.2 percent offset strain value of 40,000 psi, Par. 1.1. In Fig. 8.6 the elastic portion of the buckling curve AB is obtained by plotting σ_c against d/h using Eq. 8.2.38. The curve gives the critical stress only if its magnitude does not exceed the proportional limit. Therefore, for a proportional limit of 30,000 psi, Eq. 8.2.38 gives a value of $d/h \cong 33$, point B, as the smallest value for which this equation applies. These are relatively thin tubes. For thick walled tubes buckling, per se, is not restrictive; rather, the plastic instability of the material itself at yield strength

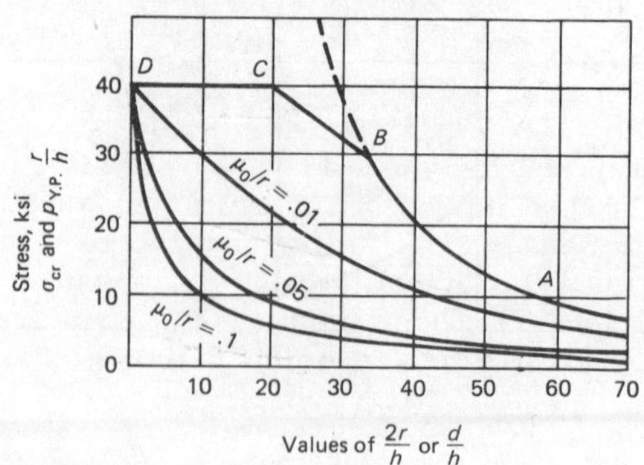

Fig. 8.6. Cylindrical Vessel, Critical Buckling and Yield Strength Pressure

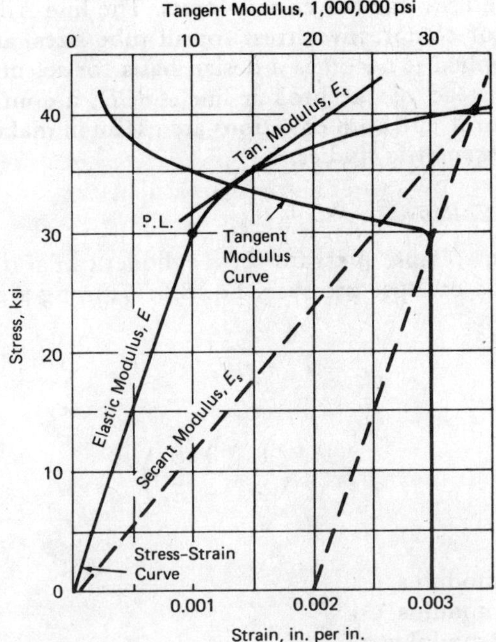

Fig. 8.7. Tangent Modulus for a Steel Without a Definite Yield Strength. (Proportional Limit = 30,000 psi, Yield Strength at 0.2 Percent Offset Strain = 40,000 psi)

governs. Here the yield strength line CD must be taken as the critical stress for thick walled tubes (low d/h ratios) where

$$\sigma_c = \sigma_{y.p.} = \frac{pr}{h} \tag{8.2.40}$$

Between the proportional limit and the yield strength, the modulus, E_t, is continually changing in relation with its position on the stress-strain curve. Hence, beyond the proportional limit the tangent modulus, E_t, must be used instead of the elastic modulus, E. When the compression stress-strain curve for the material is known, E_t can be calculated for values of σ_c, and then the corresponding value of d/h found from Eq. 8.2.39. The portion beyond the proportional limit can be approximated by the straight line BC when it is noted that E_t appears linearly in Eq. 8.2.39, and in Fig. 8.7 this material property exhibits a substantially linear behavior over a wide range. Thus, for intermediate tube thicknesses of approximately $20 < d/h < 33$, the line BC can be used for the critical stress. The line $ABCD$ gives the critical values of compressive stress for all tube sizes, and if a factor of safety is applied it becomes a design basis for selecting safe tube thicknesses. Instead of the broken line $ABCD$, a continuous curve can be fitted to it and such equations are useful in making computer analyses and parametric studies.

4. *Inelastic Collapse*

The inelastic collapse pressure for a cylinder can also be predicted from the elastic collapse pressure, Eq. 8.2.37, using a plastic reduction factor, η,

$$P_p = \eta P_c \tag{8.2.41}$$

where

$$\eta = \frac{E_s}{E}\left(\frac{1}{4} + \frac{3}{4}\frac{E_t}{E_s}\right) \tag{8.2.42}$$

and

E = elastic modulus, psi
E_s = secant modulus, psi
E_t = tangent modulus, psi
P_c = elastic collapse pressure, psi
P_p = inelastic collapse pressure, psi

This gives results in good agreement with experiment.[32] A similar correlation of the elastic and inelastic collapse pressures for spheres is given in Par. 8.6.2. The inelastic collapse pressure is always less than the elastic collapse pressure and converges on this value as the stress approaches the proportional limit, where $E = E_s = E_t$. Combinations of these three, or other reduced, moduli are frequently used to reconcile theory with experiment. In general, the tangent modulus is conservative in the case of fully plastic buckling in stiff vessels, and gives reasonable results in less stiff or flexible ones.[52, 53]

In the inelastic range the value of Poisson's Ratio in Eq. 8.2.39 is also changing and varies from a value for steel of approximately $\mu = 0.3$ for elastic strain to $\mu = 0.5$ for inelastic strain. It is observed in Eq. 8.2.39 that the effect of this variable is not as pronounced as that of E_t and it is usual practice to simplify the use of this equation by retaining a constant value of Poisson's Ratio; say, $\mu = 0.3$ or an average value $\mu = 0.4$, since strains in the tube wall when bending and buckling occur will be partly elastic and partly inelastic.

8.3 Cylinders and Tubes With Initial Noncircularity

The failure of cylinders such as boiler tubes, submersibles, etc. under external pressure are dependent in great measure upon the various kinds of imperfections in them. The most important one is that of noncircularity, which is usually of an elliptical or oval shape; for this reason, purchasing specifications for these vessels contain limiting acceptance standards.[2,6] Deviations from true circular cross sections are readily obtained from inspection measurements. The solid line in Fig. 8.4*b* shows such a tube and its deviation from the shape of a perfect circular form, shown by the broken line. This deviation is given by the equation

$$u_1 = u_0 \cos 2\phi \tag{8.3.1}$$

in which u_0 is the maximum initial radial deviation, which is considered small in comparison with the radius, r. When external pressure, p, is applied, a further flattening of the tube occurs. Considering an elemental ring of unit width and denoting the radial displacements corresponding to the latter flattening by u_2, Eq. 8.2.17 gives, noting $ds^2 = r^2\,d\phi^2$,

$$\frac{d^2u_2}{d\phi^2} + u_2 = -\frac{r^2M}{D} \tag{8.3.2}$$

In the portions *ab* and *cd* of the elemental ring, the bending moment, M, due to the pressure, p, causes a decrease in curvature to occur; hence, M is positive in these regions and negative in the portions *ad* and *bc*. At points *a*, *b*, *c*, and *d*, the bending moment is zero and the interaction between these parts of the elemental ring is given by the compressive force F tangential to the broken line representing the perfect circular shape of the tube. The compressive force along this curve is constant and equal to $F = pr$. The bending moment curve is represented by that area between this curve and that for the ellipse-like cross section. The bending moment at any cross section is equal to this compressive force multiplied by the total radial deviation $u_1 + u_2$ at this cross section, or

$$M = pr(u_2 + u_0 \cos 2\phi) \tag{8.3.3}$$

Substituting Eq. 8.3.3 in Eq. 8.3.2 gives

$$\frac{d^2u_2}{d\phi^2} + u_2 = -\frac{pr^3}{D}(u_2 + u_0 \cos 2\phi) \tag{8.3.4}$$

or

$$\frac{d^2u_2}{d\phi^2} + u_2\left(1 + \frac{pr^3}{D}\right) = -\frac{pr^3 u_0 \cos 2\phi}{D} \tag{8.3.5}$$

The solution of this equation, satisfying continuity conditions at a, b, c, and d is

$$u_2 = \frac{u_0 p \cos 2\phi}{p_c - p} \tag{8.3.6}$$

in which p_c is that given by Eq. 8.2.37. At the points a, b, c, and d, the displacement, u_2, and the second derivative, $d^2u/d\phi^2$, are zero; hence, the bending moments at these locations are zero as assumed. The maximum moment occurs at $\phi = 0$ and at $\phi = \pi$, where

$$M_{\max} = pr\left(u_0 + \frac{u_0 p}{p_c - p}\right) = \frac{pu_0 r}{1 - (p/p_c)} \tag{8.3.7}$$

Equation 8.3.7 shows that for small values of the ratio p/p_c the change in the ellipticity of the tube due to the external pressure, p, can be neglected and the maximum bending moment is obtained by multiplying the compressive force, $F = pr$, by the initial deviation, u_0. If the ratio p/p_c is not small, the change in the initial ellipticity of the tube must be considered and Eq. 8.3.7 used to calculate $M_{\max}$.

The total maximum compressive stress is obtained by adding to the stress produced by the compressive force, pr, the maximum compressive bending stress due to $M_{\max}$, Eq. 8.3.7, to give

$$\sigma_{\max} = \frac{pr}{h} + \frac{6}{h^2} \cdot \frac{pu_0 r}{1 - (p/p_c)} \tag{8.3.8}$$

Assuming that this equation can be used with sufficient accuracy up to the yield strength of the material, it follows that the limiting value of the pressure, p, is that value at which yielding of the material begins. Equation 8.3.8 can then be written

$$\sigma_{Y.P.} = \frac{p_{Y.P.} r}{h} + 6p_{Y.P.} \frac{r^2}{h^2} \frac{u_0}{r} \frac{1}{1 - (p_{Y.P.}/p_c)} \tag{8.3.9}$$

or

$$p_{Y.P.}^2 - \left\{\sigma_{Y.P.}\left(\frac{h}{r}\right) + \left[1 + 6\left(\frac{r}{h}\right)\left(\frac{u_0}{r}\right)\right]p_c\right\}p_{Y.P.} + \sigma_{Y.P.} p_c\left(\frac{h}{r}\right) = 0 \tag{8.3.10}$$

The pressure $p_{Y.P.}$ determined in this manner is always smaller than the pressure at which collapse of the tube occurs. It becomes equal to the latter in the case of a perfectly round tube as is seen when the value $u_0/r = 0$ is replaced in Eq. 8.3.10. Fig. 8.6 shows several curves giving the value of the average tangential compressive stress $p_{Y.P.}r/h$ at which yielding starts as calculated from Eq. 8.3.10 for several values of u_0/r, and $\sigma_{Y.P.} = 40{,}000$ psi. Curves of this type are used to calculate safe pressure when the noncircularity of the tube is known and a suitable factor of safety is employed.

The elliptical shape imperfection considered here is very characteristic of that found in tubes, pipes and fabricated cylinders. If the exact shape of the imperfection is known, it can be described by a Fourier series and the analysis performed. However, such a task is formidable and is of no help in the design stage when the actual shape is unknown. Accordingly, simple expressions describing the noncircularity are adequate and are used by codes to assure design safety. The ASME Code for Pressure Vessels is an example, Par. 8.8. It specifies the maximum permissible plus-or-minus deviation, e, from the true circular cross section and provides for the compatibility of these deviations in the associated design formulas and factor of safety employed by this code to assure design adequacy.

1. *Propagation of Buckles in Long Cylinders Under External Pressure*

Buckling in cylinders starts at the critical pressure, p_c, Eq. 8.2.35, and ovalization proceeds under a very slight rise in the regions of the highest curvature change. Once yielding starts, the pressure carrying capacity falls precipitously. As ovalization progresses, the deformation becomes confined to four "hinges" at the quarter points of the ring and the pressure falls more slowly. When the flow area reduction, ΔA, reaches approximately 75 percent of the original cross-sectional area, $A = \pi r^2$, touching of the two opposite quarter points occurs. This immediately results in a substantial stiffening so that the pressure increases very rapidly, Fig. 8.1*a*, thereby halting an unhampered progressive longitudinal collapse of the entire length of the cylinder. The transition between the collapsed and uncollapsed sections is gradual and occurs over approximately 10 cylinder diameters. The pressure to cause collapse propagation of the full length of the type p^* varies from a value of 0.1–0.3 of the classical elastic buckling pressure, Eq. 8.2.35, for low r/h in the range of 15–20, to a value of 0.3–0.5 for r/h ratios in the 45–50 range. In Fig. 8.1*b* the applied collapse pressure was maintained well above that to cause collapse and flattening of the full length of the tube.

A pipe or tube meeting normal commercial tolerances should have no difficulty supporting pressures that are several times p^* as long as they are below p_c. However, if substantial dents of blows are possible, the pressure must be below p^* to assure that full length progressive collapse will not occur.[47]

8.4 Collapse of Thick-Walled Cylinders or Tubes under External Pressure

Thick-walled cylinders and tubes are finding increasing use in undersea deep drilling operations, and in the construction of ultra-high pressure components for the petrochemical and other industries. The collapse behavior of thick-walled cylinders under external pressure is that of first yielding at the inside diameter where the stress is a maximum, Eq. 2.8.23. This yielded region continues to expand toward the outside diameter until the entire wall is plastically deformed, Par. 2.10. At this time, increased pressure will at first collapse the tube uniformly inward, followed by ovalization and flattening of the cylinder as initial imperfections of noncircularity, thickness variations and material property variations accelerate wall bending. The thicker the wall relative to the diameter (low d/h ratio), and the higher the rate of strain hardening (increase in yield strength) of the material, Par. 5.7, the greater the resistance of the cylinder to ovalization.

1. *Fully Plastic Wall*

This fully plastic condition must first be present for the second, or buckling mode, to occur.

Equations for the average stress through the cylinder wall are:

hoop stress, $$\sigma_2 = -\frac{pr}{h} \tag{8.4.1}$$

radial stress, $$\sigma_r = -\frac{p}{2} \tag{8.4.2}$$

longitudinal stress, $$\sigma_1 = -\frac{pr^2}{r^2 - r_i^2} \tag{8.4.3}$$

Full wall yielding occurs when the pressure is sufficient to increase the effective stress to the material yield strength. This gives for the

Tresca failure theory, Eq. 5.15.8,

$$-\sigma_{Y.P.} = \sigma_2 - \sigma_r \tag{8.4.4}$$

and for the von Mises theory, Eq. 5.15.23,

$$2\sigma_{Y.P.}^2 = (\sigma_r - \sigma_2)^2 + (\sigma_r - \sigma_1)^2 + (\sigma_2 - \sigma_1)^2 \tag{8.4.5}$$

Experimental data for a triaxial stress state is very limited and not always in agreement with the generally accepted theories of ductile failure; namely, maximum shear stress, Tresca, Para. 5.15.1(b) and distortion energy, von Mises, Para. 5.15.1(c). These theories assume that the hydrostatic component of stress does not contribute to plastic deformation but it is governed by the difference of the maximum and minimum principal stress in the case of Tresca theory, or the sum of the difference of each of the principal stresses in the case of the von Mises theory. Tests of thick walled cylinders show much lower results than predicted by these theories, Fig. 8.8, when the material yield strength employed is taken as the maximum of the actual measured triaxial values.[7] However, when the lowest of the measured triaxial yield strength values is used, good agreement with experimental results is obtained. This minimum material property is usually in the radial direction, Table 8.1, the same direction as the minimum principal stress, Eq. 8.4.2. This indicates that a slip system

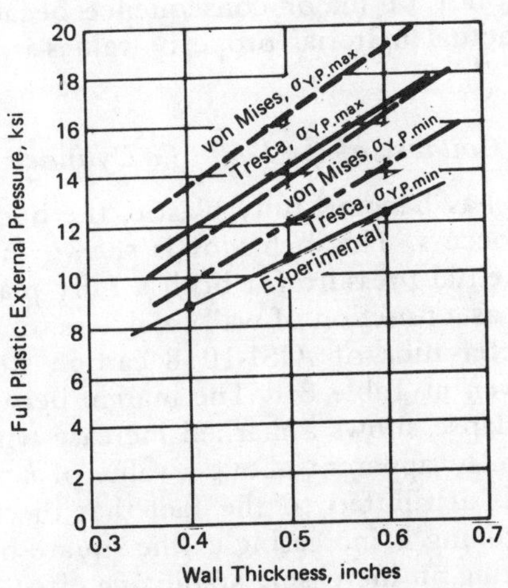

Fig. 8.8. Predicted vs. Experimental Yield Pressures for 2.375 in. Inside Diameter Thick Walled Cylinders of AISI-1018 Steel, Showing the Effect of the Triaxial Properties of the Material[7]

TABLE 8.1 TRIAXIAL YIELD STRENGTH VALUES[7] OF AISI-1018 CARBON STEEL THICK-WALLED CYLINDERS WHOSE RESULTS ARE SHOWN IN FIGS. 8.8, 8.9 AND 8.10

	Yield Strength, psi		
	Longitudinal	*Circumferential*	*Radial*
Tension Test	40,400	—	—
Compression Test	37,800	39,400	33,900

related to the lowest material yield strength in the triaxial principal stress directions is controlling for the ductile failure of thick walled cylinders. This effect is shown in Fig. 8.8 for a test of AISI-1018 carbon steel cylinders[7] for which the triaxial yield strength is shown in Table 8.1. Here the maximum shear stress theory, Tresca, employing the minimum radial direction material yield strength is in good agreement with experimental results. Is is reasonable that material strength property should be the minimum value in the direction of the imposed principal stresses represented in the theory. The maximum difference in failure predictions by the Tresca and von Mises theories is only in the order of 15 percent, Par. 5.15. Hence, a choice between them is not of major consequence because the accuracy with which the actual material property values is known is usually much less.

2. *Buckling or Collapse of the Wall of a Cylinder*

After the wall has become fully plastic, the buckling or collapse mode condition occurs. This behavior is shown in Fig. 8.9, which gives the experimental pressure for both a fully plastic wall and for the final collapse as a function of wall thickness for a constant 2.375 inch inside diameter tube of AISI-1018 carbon steel material with yield strengths given in Table 8.1. The margin between the full-wall plasticity and collapse shows a marked increase with wall thickness which is significantly apparent above a value of $h = 0.3$ ($2r/h$ value below 10). This is attributed to the fact that the internal moment which resists flattening is increasing as the square of the wall thickness thereby creating an increasing stabilizing effect. This increase in buckling pressure above the full plastic thick-walled yield stress should not be used in design because it can only be obtained if special conditions are present to restrict ovalization, such as very thick walls. A

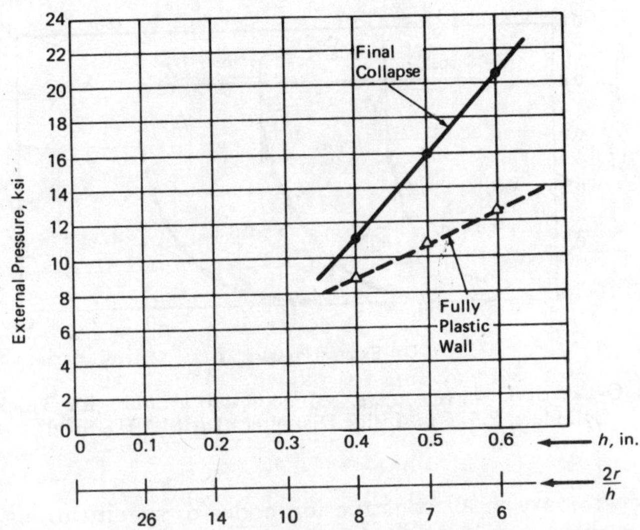

Fig. 8.9. Experimental Yield and Collapse Pressure for Thick Walled Cylinders, 2.375 in. Inside Diameter, of AISI 1018 Steel[7]

similar effect also occurs with thick, low slenderness ratio columns, where buckling stresses above the yield strength can be obtained when special precautions are taken to forestall column buckling at the yield stress.

The mode of final failure is still a buckling two-lobe one, and tube noncircularity is the prime parameter establishing the collapse pressure. This is shown in Fig. 8.10, where the external pressure is plotted against cross-sectional noncircularity as measured by the difference between the maximum and minimum diameter for the same series of tests mentioned above. The final collapse is controlled by the amount of this noncircularity, $d_{max} - d_{min}$, Fig. 8.10. For these tube test results, this was in the order of $d_{max} - d_{min} = 0.3\text{–}0.4$ inches, or, expressed as a percentage of the nominal or average diameter

$$\text{Noncircularity} = 2\,\frac{(d_{max} - d_{min})}{(d_{max} + d_{min})} \times 100 \tag{8.4.6}$$

was in the order of 10–15 percent. The importance of noncircularity is emphasized by Fig. 8.10, which shows that all these thick cylinders reached final collapse at essentially the same amount. The greater the initial noncircularity, the less remaining resistance to collapse.

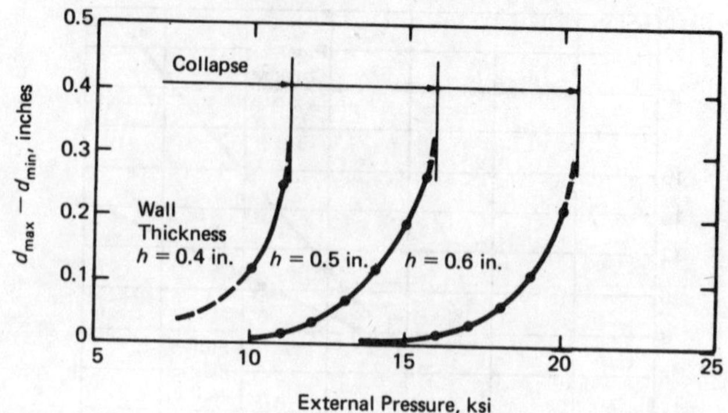

Fig. 8.10. Ovalization, $d_{max.} - d_{min.}$ vs. External Pressure for Thick Walled Cylinders, 2.375 in. Inside Diameter of AISI 1018 Steel[7]

It is customary in all construction codes or specifications to prescribe a maximum initial noncircularity tolerance because this is the critical fabrication or quality control parameter for vessels subject to external pressure. It is frequently specified as 1 percent, Par. 8.8; however, for deep diving submersibles or other critical lightweight components, even more severe restrictions are stipulated, thereby requiring machining and grinding to control this prime buckling parameter.

8.5 Effect of Supports on Elastic Buckling of Cylinders

In Pars. 8.2.2 and 8.2.3 the buckling strength of an externally loaded ring was developed and shown that this was also applicable to the case of a long cylinder with free ends subjected to a uniform external pressure. These same formulas also apply to cylinders with end supports if the length is large; consequently, their stiffening effect can be neglected, with the result that collapse will occur in the central portion of the cylinder in the form of two lobes, $n = 2$, which is the form associated with the minimum critical pressure. Figure 8.1*b* shows this condition. Here, a long cylinder is supported at the left by a diaphragm, while at the right end, the support that resists buckling is obtained by the double curvature of the bend. If the length of the cylinder is not large relative to its diameter, the end supports must be considered in calculating the buckling collapse pressure because they will prevent the cylinder from collapsing in two lobes and force it into higher modes, thereby increasing the minimum critical pressure before elastic buckling occurs. Equation 8.5.1, developed by R. von Mises, can be used to appraise this effect:[8,48,49]

$$p_c = \frac{Eh}{r(1-\mu^2)}\left[\frac{1-\mu^2}{(n^2-1)\left(1+\dfrac{n^2\ell^2}{\pi^2 r^2}\right)} + \frac{h^2}{12r^2}\left(n^2-1+\frac{2n^2-1-\mu}{1+\dfrac{n^2\ell^2}{\pi^2 r^2}}\right)\right] \tag{8.5.1}$$

The condition of a cylinder subject to axial pressure in addition to radial pressure, such as a submarine, has also been solved.[8,9,50,51]

In Eq. 8.5.1 ℓ is the length between supports; i.e., distance between end supports, stiffener rings, or bend tangents where the initial double curvature provides support, Fig. 8.1. When the cylinder is long, ℓ/r is large, and the terms containing the square of this ratio in the denominator of Eq. 8.5.1 can be neglected to give

$$p_c = \frac{Eh^3(n^2-1)}{12r^3(1-\mu^2)} \tag{8.5.2}$$

This becomes Eq. 8.2.37 for the buckling of a long cylinder when $n = 2$. Equation 8.5.1 can be written in the form

$$p_c = KE\left(\frac{h}{d}\right)^3 \tag{8.5.3}$$

where the effect of the number of lobes and support spacing obtained from Eq. 8.5.1 is shown in Fig. 8.11.[10] The value of K corresponding to a given value of ℓ/r (or ℓ/d) and d/h can be found from Fig. 8.11. For instance, Fig. 8.11 shows that a cylinder with a value of $\ell/r = 5$ and $d/h = 50$ will collapse in three lobes and that the value of K for use in Eq. 8.5.3 is approximately 8.0. It is again noted that as the cylinder becomes long so that its collapse is in the form of two lobes, K becomes constant at $2/(1-\mu^2)$ in region AB of Fig. 8.11 and Eq. 8.5.3 reduces to Eq. 8.2.37. The length at which this condition is first reached is called the "critical length" and for $\mu = 0.3$ is given by

$$L_{cr} = 1.11d\sqrt{\frac{d}{h}} \tag{8.5.4}$$

If a lower bound envelope is fitted to each of the d/h curves in Fig. 8.11, the engineering design curves in terms of the parameters d/h and ℓ/d shown in Fig. 8.12 result.[11] The use of these curves with

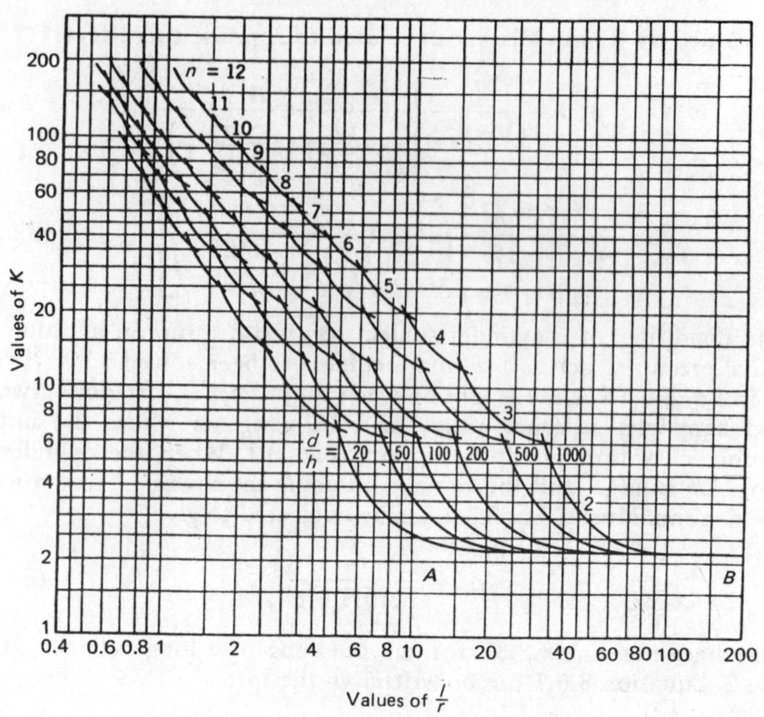

Fig. 8.11. Buckling Coefficients for Cylinders Subject to External Pressure[10]

Eq. 8.5.3 requires a trial-and-error process when endeavoring to establish a thickness for a given pressure because of the interdependence of the variables.

The degree of fixity at the ends creates only a very local effect. The deflection curve for a longitudinal generator of the cylinder is given by Eq. 4.4.5 and has a rapidly dampened full wave length of $4.9\sqrt{rh}$, Par. 4.6. In thin cylinders, h is small in comparison with r and the wavelength is much shorter than the radius. So for a cylinder of length several times its radius, bending at the ends is considered local and of little effect on the buckling load. Accordingly, Fig. 8.12 can be used for either simply supported or clamped ends because the manner of support does not have a significant influence on the critical buckling pressure and this is customary engineering design practice. This also applies to long cylinders stiffened by structural rings provided their flexural rigidity is sufficient to support the

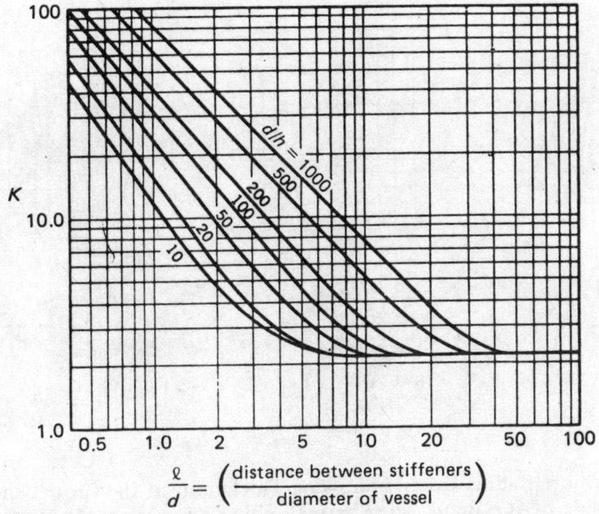

Fig. 8.12. Collapse Coefficients for Cylinders Subject to External Pressure[11]

hoop or lateral load without buckling. In such a case ℓ is the distance between ring supports, Fig. 8.13. Initial double curvatures also provide elastic buckling resisting support locations, such as a dished cover where the support is estimated at one-third the crown depth

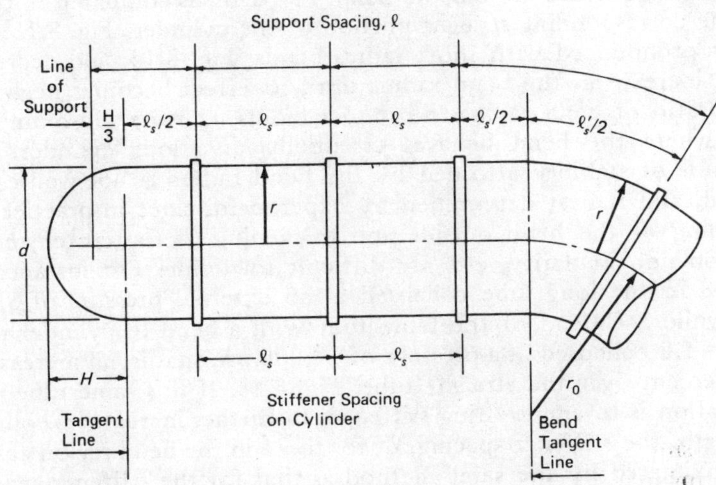

Fig. 8.13. Support Locations for Cylinder with Closure Head and Bend Effects

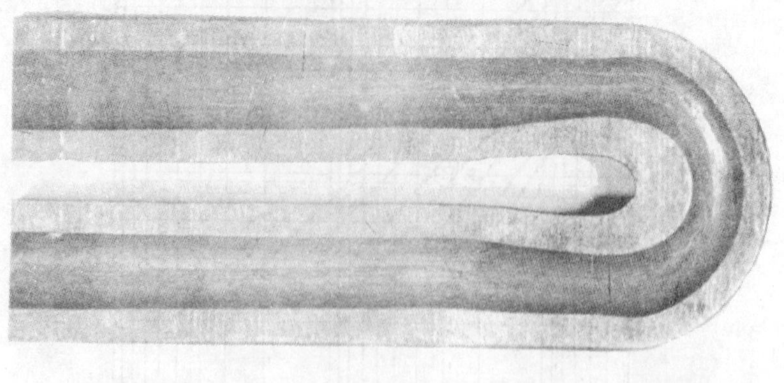

Fig. 8.14. Short Radius U-Bend Showing Thickening at the Crotch and Thinning at the Outside of the Bend. The Sharp Double Curvature at the Bend Region Increases its Buckling Resistance over That of the Connecting Straight Cylinder

from the tangent line, Fig. 8.13. The double curvature offered by bends (elbows, U-bends, or portions of a torus) also offer an additional stabilizing stiffness against buckling to external pressure. This effect is characterized by the shorter distance between the longitudinal buckling nodes around the bend, Fig. 8.1*c*, as compared to those in the corresponding straight portion of the cylinder, Fig. 8.1*b*. It is most pronounced with short radius bends, Fig. 8.14, and decreases with increase in the bend radius until its effect becomes negligible at a ratio of bend radius to cylinder radius, r_0/r, of approximately 10, where the bend behaves essentially as a long cylinder. The amount of stability afforded by the bend radius is not well established, and is best determined by experiment, since in practice the thinning of the bend outside and thickening of the crotch, cross-section noncircularity, etc. are difficult to define. For instance, in Fig. 8.1*b* the long tube collapsed at an external pressure of 6,800 psi while a U-bend of the same tube with a bend to cylinder ratio $r_0/r = 4.2$ collapsed at a pressure of 12,200 psi, that is, an increase of 93 percent over the straight tube, Fig. 8.1*c*. If this same tube configuration is to employ ring stiffeners to further increase its collapse strength, the stiffener spacing, ℓ', for those in the bend region can be approximated by the same method as that for the stiffener spacing in a straight cylinder by using a virtual pressure $p' = p/1.93$.

Example: A long cylinder (20 inch outsider diameter and 0.2 inches thick) has ring stiffeners spaced at $\ell = 40$ inches and is subjected to external pressure. The material is steel with a yield strength of 40,000 psi, $E = 30{,}000{,}000$, and $\mu = 0.3$. Will the cylinder fail by elastic buckling before the wall material reaches yield strength due to the tangential hoop stress?

From Fig. 8.11 for a value of

$$\frac{\ell}{r} = \frac{40}{10} = 4 \qquad \text{and} \qquad \frac{d}{h} = \frac{20}{0.2} = 100$$

K is 15, so the elastic buckling pressure from Eq. 8.5.3 is

$$p_c = KE\left(\frac{h}{d}\right)^3 = 15 \times 30{,}000{,}000 \times \left(\frac{0.2}{20}\right)^3 = 450 \text{ psi}$$

The hoop stress resulting from a pressure of 450 psi, Eq. 2.2.4, is

$$\sigma = \frac{pd}{2h} = \frac{450 \times 20}{2 \times 0.2} = 22{,}500 \text{ psi}$$

Since this value is less than the yield strength of the material, failure will occur by elastic buckling at a pressure of 450 psi.

1. *Cylinders with Ring Stiffeners*

When external pressures are high and/or diameters large, plain cylinders become extremely thick and expensive. Savings in weight and material can be made by the use of ring stiffeners attached to these cylinders at intervals along their length. Three primary modes of failure may be encountered in the stiffened cylinder and are characterized by their visual appearance as follows:[12]

(a) Buckling of the cylinder between rings, manifested by a number of lobes girding the circumference, Fig. 8.15. This type of failure occurs when the ring stiffeners are too widely spaced.

(b) Yielding of the cylinder between rings, which is characterized by axisymmetrical accordion-shaped pleats, Fig. 8.16. This occurs when the rings are heavy and closely spaced. The cylinder yields circumferentially in compression and the radius decreases until longitudinal collapse occurs.

(c) General instability collapse of the cylinder and rings together. This embodies many rings and occurs over a considerable length of the stiffened cylinder, Fig. 8.17.

Ring stiffeners may be attached to either the inside or outside of the cylinder.[13] Naval submarines utilize outside ring stiffeners which are less susceptible to lateral instability than those placed on the inside. Likewise, they are generally placed on the outside of industrial

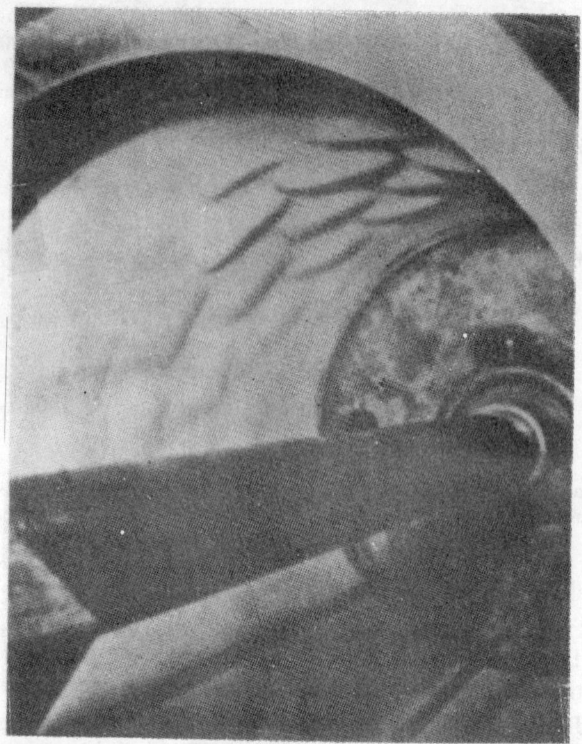

Fig. 8.15. Buckling Collapse of Cylinder Between Ring Stiffeners Located on the Outside. The Multitude of Lobes Girdling the Circumference Occur Alternately Spaced on Each Side of the Stiffeners[12]

vessels so they do not obstruct flow and facilitate cleaning. Under external pressure, the web of an external stiffener is in radial tension and circumferential compression, whereas the web of an internal stiffener is in compression in both directions which can cause lateral instability.

Each stiffener is considered to resist the external load for a distance $\ell/2$ on each side of the stiffener, Fig. 8.13. The load per unit length of circumference of the combined cylinder and stiffener ring at collapse is equal to $p\ell$. Substituting $P_c = p\ell$ in Eq. 8.2.35 gives

$$p\ell = \frac{24EI}{d^3} \tag{8.5.5}$$

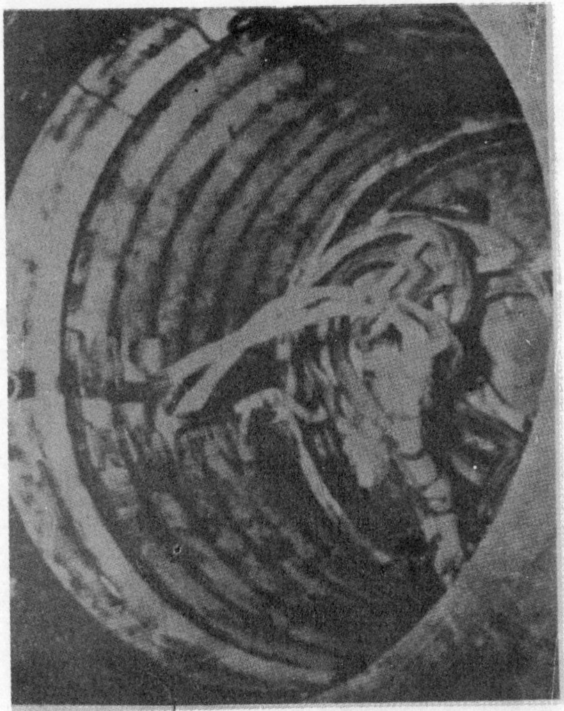

Fig. 8.16. Yield Collapse of Cylinder Between Ring Stiffeners Which Are Located on the Outside of the Cylinder[12]

Replacing p in terms of the maximum hoop stress, Eq. 2.2.4, gives

$$I = \frac{h\ell d^2}{12}\left(\frac{\sigma}{E}\right) \tag{8.5.6}$$

and further noting $\sigma = Ee$, Eq. 1.1.3, this can be written

$$I = \frac{h\ell d^2 e}{12} \tag{8.5.7}$$

The combined moment of inertia of the cylinder and stiffener may be considered as equivalent to a single thicker cylinder

$$h_e = h + \frac{A}{\ell} \tag{8.5.8}$$

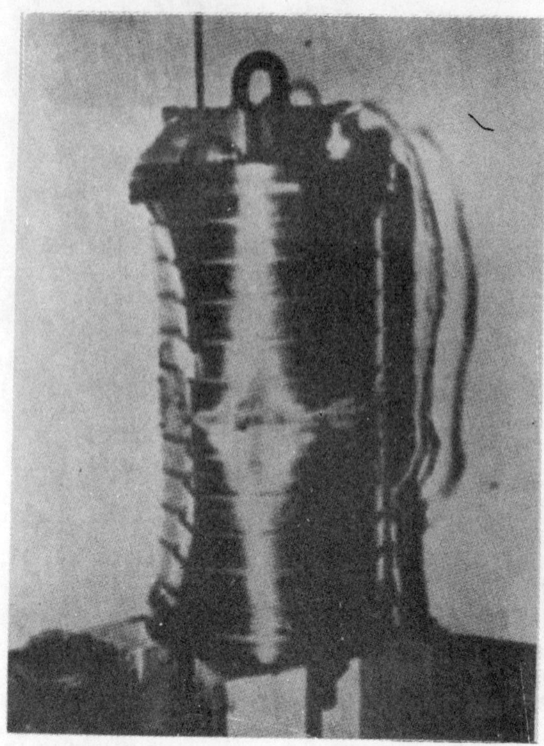

Fig. 8.17. General Instability Collapse of Cylinder and Ring Stiffeners Which Are Located on the Outside of the Cylinder[12]

where h_e is the equivalent thickness of the cylinder, and A is the cross-sectional area of a stiffener. This is a good approximation based on the premise that a longitudinal generator of the cylinder remains straight throughout and not distorted by inter-stiffener deformation. The required moment of inertia of the ring stiffener is found by substituting Eq. 8.5.8 into Eq. 8.5.6 or Eq. 8.5.7 to give

$$I = \frac{d^2 \ell}{12} \left(h + \frac{A}{\ell} \right) \frac{\sigma}{E} \tag{8.5.9}$$

or

$$I = \frac{d^2 \ell}{12} \left(h + \frac{A}{\ell} \right) e \tag{8.5.10}$$

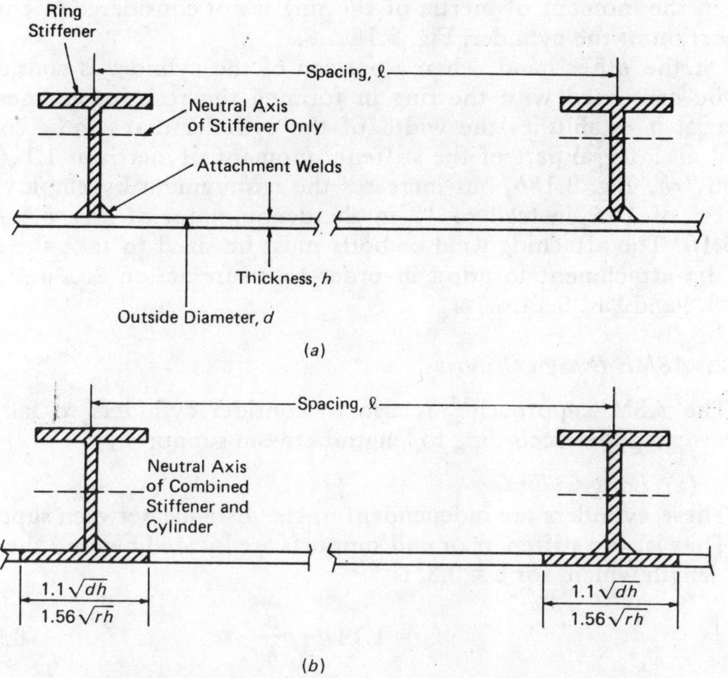

Fig. 8.18. Structural Tee Ring Stiffener. Stiffener Assumed to Act (*a*) Independently of Cylinder, (*b*) in Combination with Portion of Cylinder

If the stiffener is attached to the cylinder so that continuity prevails, it is reasonable to assume that a portion of the cylinder also acts as an integral part of the stiffener moment of inertia geometry. As in Pars. 4.6 and 6.14 this may be taken as $1/\beta$ on each side of the stiffener or $2/\beta = 1.56\sqrt{rh}$, Fig. 8.18. The most commonly used stiffeners are rectangular bars or structural shapes such as tees, channels, and I-beams.

The formulas for ring-stiffened cylinders are basic ones to which modifiers can be applied to reflect experience and refine collapse prediction. Codes, regulations and standards use this approach to focus on their pertinent experiences and concerns for the safety of their components. For instance, the ASME Code Section VIII, Division 1, takes the collapsing strength of the ring[14] somewhat higher (in the order of 10 percent) than the cylinder by using a constant 14 instead of 12 in the denominator of Eqs. 8.5.9 and 8.5.10

when the moment of inertia of the ring is not considered to contain a portion of the cylinder, Fig. 8.18*a*.

On the other hand, when a portion of the cylinder is considered to be combined with the ring in forming the resisting moment of inertia, it establishes the width of the cylinder that can be considered an integral part of the stiffener moment of inertia as $1.1\sqrt{dh} = 1.56\sqrt{rh}$, Fig. 8.18*b*, but increases the requirement by employing a factor of 10.9 instead of 12 in the denominator of Eqs. 8.5.9 and 8.5.10. The attaching weld or bolts must be sized to take the shear at the attachment location in order to insure action as a unit, Par. 3.13.2 and Par. 6.14.4.

2. *ASME Design Criteria*

The ASME approach[15] is also to consider cylinders to fall into three categories according to length between supports.

(a) *Long Cylinders*

These cylinders are independent of the distance between supports, ℓ. That is, the stiffeners or end supports are located beyond the critical length which, for $\mu = 0.3$, is

$$\ell_c = 1.11d\sqrt{\frac{d}{h}} \tag{8.5.11}$$

Here, the collapse pressure is given by Eq. 8.2.37.

(b) *Short Cylinders*

In these cylinders failure is by plastic yielding at the yield strength of the material. This type of failure is characteristic of thick walled cylinders in which the influence of ℓ is negligible. Here, failure is predicted by the membrane stress as given by Eq. 8.2.40.

(c) *Intermediate Length Cylinder*

The theoretical elastic equations for the collapse pressure at which intermediate length cylinders, $\ell < \ell_c$, under uniform external radial pressure, or uniform external radial and axial pressure, depend on n, the number of lobes at collapse, Eq. 8.5.1. These are cumbersome for design purposes and it is desirable to eliminate n from them. The U.S. Experimental Model Basin developed such an approximation of the von Mises equation for a cylinder subject to uniform external pressure in both the radial and axial direction.

It is independent of the number of lobes, and the average deviation is about one percent. This equation, for $\mu = 0.30$, s

$$p_c = \frac{2.6E(h/d)^{2.5}}{\frac{\ell}{d} + 0.45\sqrt{\frac{h}{d}}} \tag{8.5.12}$$

The second term of the denominator has no discernible effect, except for very small values of ℓ/d, and can be nelgected, giving

$$p_c = \frac{2.6E(h/d)^{2.5}}{\ell/d} \tag{8.5.13}$$

Assuming the stress-strain diagram of the material is an elastic-plastic one described by Fig. 1.5*c* or Fig. 8.5*b*, a series of curves representing constant values of h/d can be plotted on a log-log scale against values of failure or collapse as ordinates and ℓ/d as abscissas. Such an h/d curve consists of three straight line segments, each of which is determined by a different equation dependent upon the cylinder length to diameter ratio, Fig. 8.19. This basic curve permits modifiers to be applied to each segment as experiment and experi-

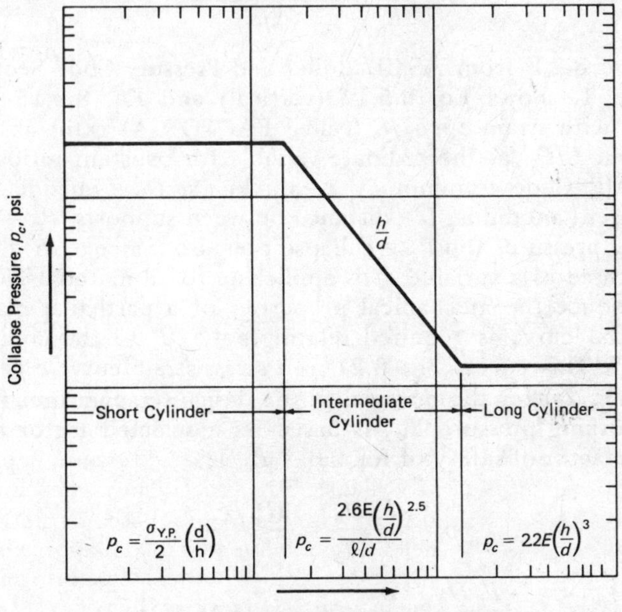

Fig. 8.19. Failure Curve for a Cylinder under External Pressure

ence dictates. For instance, the effect of noncircularity tolerances on long cylinders as compared to short ones, etc. can be accounted for. When a stress-strain curve more representative of the actual material is employed and the tangent modulus is used for buckling beyond the proportional limit a transition between the segments results. Further, when the instability equations Eq. 8.2.37 and Eq. 8.5.13 are written so that the material mechanical properties are on one side of the equation and the geometry dimensional variables are on the other, they form a basis for the ASME charts for external pressure design, Figs. 8.20 and 8.21. Substituting the value of the pressure from Eq. 8.2.37 and Eq. 8.5.13 in the average tangential stress equation, Eq. 2.2.4, gives the stress at collapse for long cylinders with $\ell > \ell_c$ as

$$\frac{\sigma_c}{E} = 1.1\left(\frac{h}{d}\right)^2 \tag{8.5.14}$$

and for intermediate length cylinders with $\ell < \ell_c$ as

$$\frac{\sigma_c}{E} = \frac{1.30(h/d)^{1.5}}{\ell/d} \tag{8.5.15}$$

Figure 8.20, from ASME Boiler and Pressure Code Section VIII, Division 1, shows Eq. 8.5.14 (vertical) and Eq. 8.5.15 (slanting) plotted with strain $e = \sigma_c/E$ (called FACTOR A) as the abscissa variable, and L/D_o as the ordinate variable for constant ratios of D_o/t. In ASME Code terminology t = thickness, D_o = outside diameter, E = tangent modulus, L = distance between supports, P_a = allowable working pressure, and P_c = collapse pressure. Since σ_c/E = FACTOR A is treated as a variable, it is applicable to all materials. Therefore, to introduce the mechanical properties of a particular material, an additional curve is required relating $e = \sigma_c/E$ to the collapse pressure, P_c. This curve, Fig. 8.21, is a stress-strain curve e $(=\sigma_c/E)$ vs. $\sigma_c(=P_cD_o/2t)$ for the material at the design temperature. The allowable working pressure, P_a, is based on a selected factor of safety. Using a factor of safety of four, $P_a = P_c/4$,

$$\sigma_c = \frac{P_c D_o}{2t} = \frac{2P_a D_o}{t} \tag{8.5.16}$$

$$\frac{P_a D_o}{t} = \frac{\sigma_c}{2} = B \qquad \text{(FACTOR B)} \tag{8.5.17}$$

where $B = \sigma_c/2$.

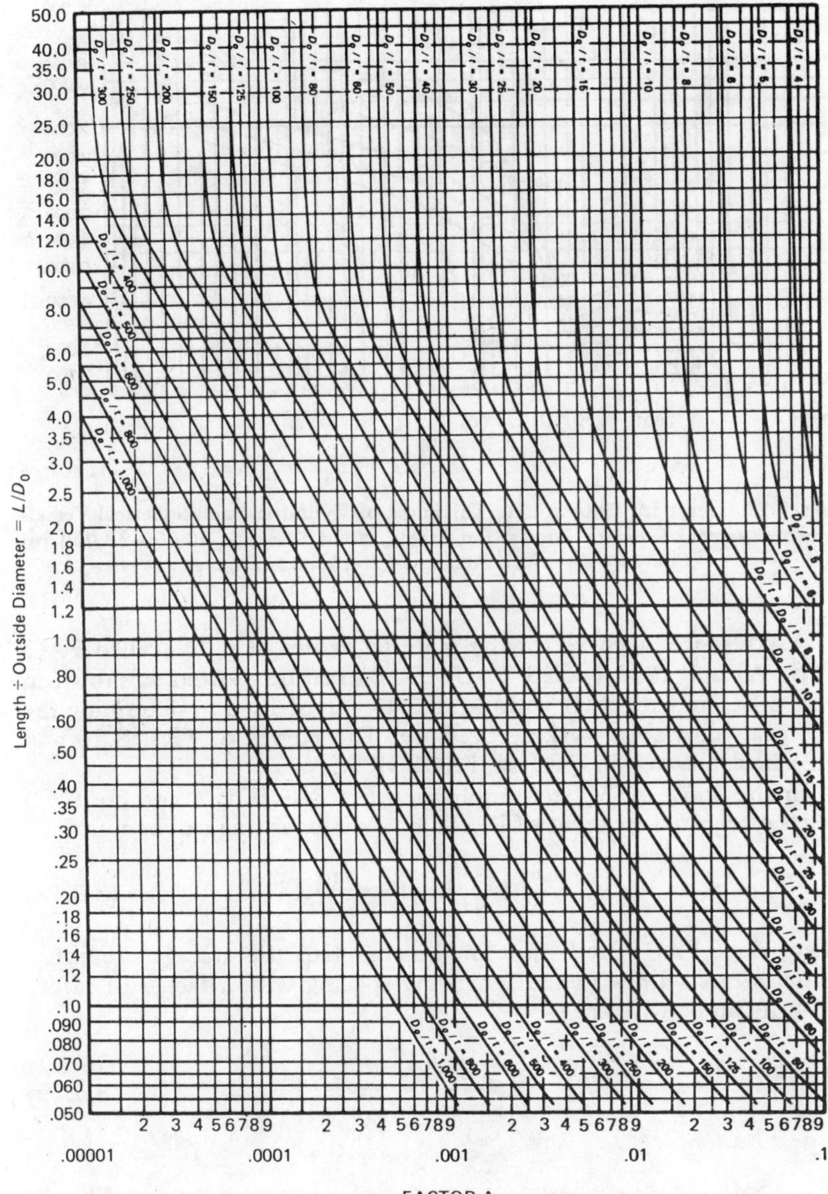

Fig. 8.20. Geometric Chart for Cylindrical Vessels under External Pressure (*Courtesy American Society of Mechanical Engineers*)

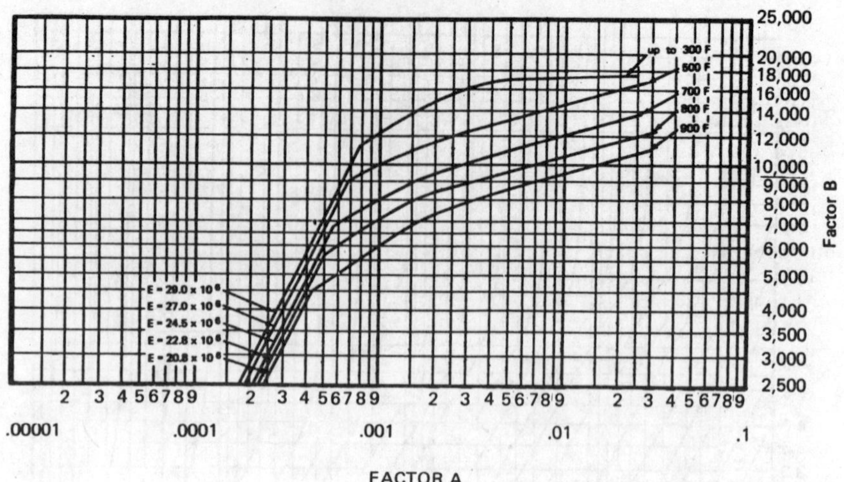

Fig. 8.21. Chart for Determining Thickness of Cylindrical and Spherical Vessels under External Pressure for Carbon Steels, Yield Strength 30,000–38,000 psi (*Courtesy American Society of Mechanical Engineers*)

A stress-strain curve for the material values of σ_c/E (called FACTOR A) and $\sigma_c/2$ (called FACTOR B) can be plotted as shown in Fig. 8.21, and together with Eq. 8.5.17 can be used to determine the external allowable working pressure. The curves of Figs. 8.20 and 8.21 have a common abscissa, FACTOR A.

In the above ASME terminology, σ/E = FACTOR A and Eq. 8.5.9 for the required moment of inertia of a stiffener-ring can be written

$$I_{\text{reqd}} = \frac{D_o^2 L}{10.9}\left(t + \frac{A_s}{L}\right) A \tag{8.5.18}$$

Also, since FACTOR B is defined as one-half the average circumferential stress within the stiffener-ring spacing with a factor of safety of 4, it can be written

$$B = \frac{\sigma_c}{2} = \frac{PD_o}{4t} = \frac{P_a D_o}{\left(t + \dfrac{A_s}{L}\right)}. \tag{8.5.19}$$

The ASME charts are based on a factor of safety of 4, so for a SF = 3, Eq. 8.5.19 is modified to

$$B = \frac{3}{4} \frac{P_a D_o}{\left(t + \frac{A_s}{L}\right)} \tag{8.5.20}$$

This method of adjustment allows the ASME charts per se, which are based on a factor of safety of 4, to remain unaltered and only change Eq. 8.5.19 for the desired factor of safety.

Example: A cylindrical vessel 14 ft ID × $\frac{3}{8}$ in. thick has ring stiffeners located on 40 in. spacing, and is subject to an external pressure of 15 psi at a temperature of 700°F. The material of both the vessel and stiffener rings is carbon steel with a yield strength of 30,000–38,000 psi.

(a) Is the $\frac{3}{8}$ in. thickness adequate for a design factor of safety of 4?

$$\frac{L}{D_o} = \frac{40}{168 + 2 \times .375} = 0.237$$

$$\frac{D_o}{t} = \frac{168.75}{0.375} = 450$$

Enter Fig. 8.20 at a value of $L/D_o = 0.237$ and $D_o/t = 450$ to read a value of FACTOR A = .0007. With this value of FACTOR A, enter Fig. 8.21 and for a temperature of 700°F read a value of FACTOR B = 7200. Then using Eq. 8.5.17.

$$\frac{P_a D_o}{t} = B = 7200$$

$$P_a = 16 \text{ psi}$$

Accordingly, the $\frac{3}{8}$ in. thickness is satisfactory for a factor of safety of 4.

(b) What is the allowable external pressure for a factor of safety of 3?

Since the development of Eq. 8.5.17 incorporated a factor of safety of 4, it is only necessary in introducing the fraction $\frac{4}{3}$ in this equation to obtain

$$P_a = \frac{4Bt}{3D_o} = 21.3 \text{ psi}$$

(c) What is the required thickness for the same ID vessel based on a factor of safety of 3?

This must be determined on a trial-and-error basis. Assume a thickness of $\frac{5}{16}$ in.

$$\frac{L}{D_o} = \frac{40}{168 + 2 \times .3125} = 0.237$$

$$\frac{D_o}{t} = \frac{168.625}{.3125} = 540$$

With these values, enter Fig. 8.20 to read FACTOR A = 0.0005. With this value of FACTOR A, enter Fig. 8.21 and, for a temperature of 700°F, read a value of FACTOR B = 6,100. Then using Eq. 8.5.17 and a modifier of $\frac{4}{3}$ to compensate for the factor of safety = 3,

$$P_a = \frac{4Bt}{3D_o} = \frac{4 \times 6{,}100 \times .3125}{3 \times 168.625} = 15.1 \text{ psi}$$

Therefore, the selected thickness of $\frac{5}{16}$ in. is satisfactory.

(d) Determine the stiffener ring requirements for the vessel in (c) above (D_o = 168.625 in., L = 40 in., t = .3125 in., P_a = 15 psi, T = 700°F, factor of safety = 3) if this stiffener ring is to be made from a structural tee shape.

As a first trial select a 4 inch wide flange tee at 12 lb per ft, A_s = 3.53 sq. in. and I = 3.53 in.^{4}. The combined ring-cylinder moment of inertia, Fig. 8.18*b* is approximately 20.6 in.^{4}. FACTOR B is calculated from Eq. 8.5.20:

$$B = \frac{3}{4}\left(\frac{15 \times 168.625}{.3125 + \frac{3.53}{40}}\right) = 4{,}733 \text{ psi}$$

and entering Fig. 8.21 with this value of FACTOR B gives FACTOR A = 0.0004. Then substituting the respective values in Eq. 8.5.18 gives

$$I_{\text{reqd}} = \frac{168.625^2 \times 40}{10.9}\left(.3125 + \frac{3.53}{40}\right) 0.0004 = 16.7 \text{ in.}^4$$

The required moment of inertia of 16.7 in.^{4} is less than that provided by the combined ring-cylinder moment of inertia of 20.6 in.^{4}; therefore, the chosen size tee is satisfactory. Normally, several trials are required to select a satisfactory stiffener-ring size, and one giving the most economical weight.

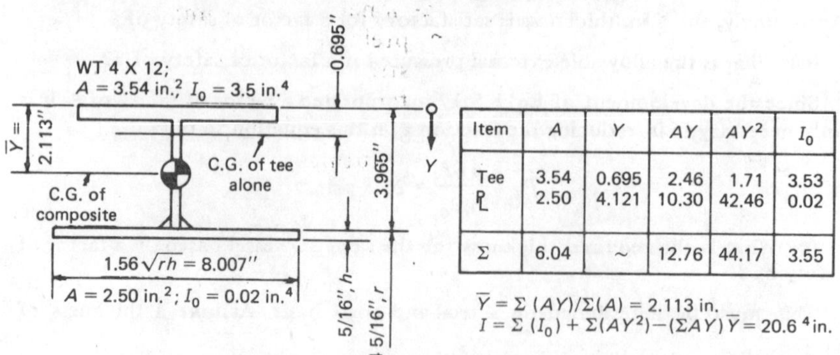

Item	A	Y	AY	AY^2	I_o
Tee	3.54	0.695	2.46	1.71	3.53
℄	2.50	4.121	10.30	42.46	0.02
Σ	6.04	~	12.76	44.17	3.55

$\bar{Y} = \Sigma(AY)/\Sigma(A) = 2.113$ in.

$I = \Sigma(I_o) + \Sigma(AY^2) - (\Sigma AY)\bar{Y} = 20.6$ ^{4}in.

8.6 Buckling of Spheres

Spheres, or portions thereof, are optimum shapes for externally pressurized vessels, tanks and closures. Deep diving submersibles with pressure hulls of single or multiple spheres are an example, Fig. 7.34. A spherical shell subjected to a uniform external pressure may retain its spherical shape and undergo only a compressive stress, Eq. 2.2.7, of

$$\sigma = \frac{pr}{2h} \tag{8.6.1}$$

or one-half of that for a cylinder. When the pressure increases beyond a critical limit the spherical form of equilibrium becomes unstable and buckling occurs at a critical buckling pressure[16,17] of

$$p_c = \frac{2E}{[3(1-\mu^2)]^{1/2}} \left(\frac{h}{r}\right)^2 \tag{8.6.2}$$

or for $\mu = 0.3$

$$p_c = 1.2E\left(\frac{h}{r}\right)^2 \tag{8.6.3}$$

Equating Eqs. 8.6.1 and 8.6.2 gives the critical buckling stress as

$$\sigma_c = \frac{Eh}{r\sqrt{3(1-\mu^2)}} \tag{8.6.4}$$

or for $\mu = 0.3$

$$\sigma_c = 0.605E\frac{h}{r} \tag{8.6.5}$$

These equations are based on perfect sphericity, no residual stress, and a linear stress-strain relationship. Of these, perfect sphericity and no residual stress are essentially impossible to obtain; hence, the actual buckling strength of practical spherical vessels is appreciably less than that given by Eq. 8.6.3. The basic design approach is to modify existing theoretical solutions by experimental results to compensate for the effects of out-of-roundness and residual stress.

1. *Nonsphericity*

Experiments have shown that the collapse pressure occurs at values much smaller than those predicted by elastic theory, Eq. 8.6.3, for a perfect spherical shape. This is due to the existence of forms of equilibrium in the vicinity of the spherical shape slightly deviated from the spherical shape[18] which requires collapse pressures smaller than that given by Eq. 8.6.3. This sensitivity to general nonsphericity

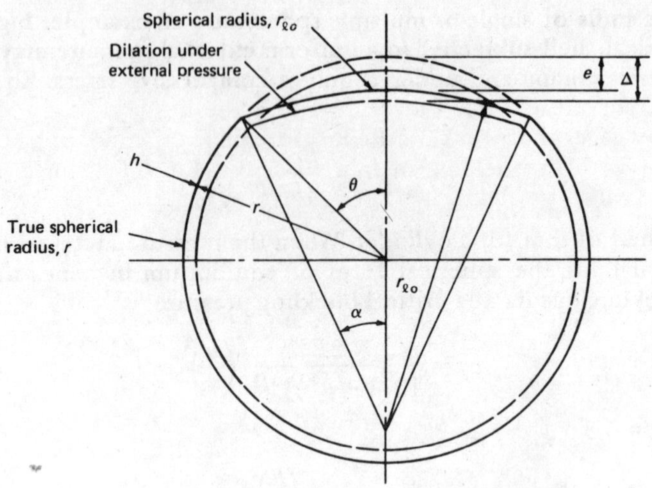

Fig. 8.22. Sphere with Flat Spot Imperfection

and local flat spots on the elastic buckling pressure has been correlated with experimental results to give[19]

$$p_c = 0.84E\left(\frac{h}{r_{\ell o}}\right)^2 \tag{8.6.6}$$

where

p_c = elastic buckling pressure
$r_{\ell o}$ = local radius of curvature, Fig. 8.22

from which it is noted that the constant in Eq. 8.6.6 is 70 percent of that in Eq. 8.6.3, and the local radius is used instead of the nominal radius, Fig. 8.22. These experiments have also defined the critical arc length of local defects, which include flat spots, thin spots and mismatch, as

$$L_c = \frac{2.2\sqrt{r_{\ell o}h}}{1 - \mu^2} \tag{8.6.7}$$

or for $\mu = 0.3$

$$L_c = 2.4\sqrt{r_{\ell o}h} \tag{8.6.8}$$

This provides a definition of the local region over which such a defect is assumed critical from a buckling viewpoint; and also provides a uniform basis for correlating buckling test data.[20] If the arc length of the local defect is considered to be critical, the local radius must be used in evaluating both the membrane and buckling stress. Local radii of curvature* may be measured on the inside or outside of the vessel by gages that work on the measurement principle of Fig. 3.7.

Since both prebuckling membrane stress, Eq. 8.6.1, and buckling pressure must be reconciled, this equation is modified by using the value of the local radius to give

$$\sigma = \frac{pr_{\ell o}}{2h} \tag{8.6.9}$$

In addition to the initial nonsphericity, e, the larger local radius, $r_{\ell o}$, produces a greater membrane stress and dilation; hence, increases this total offset, Δ, (that produces the local bending) per Eq. 2.4.4, to approximately

$$\Delta = e + \frac{p(1-\mu)}{2hE}(r_{\ell o}^2 - r^2) \tag{8.6.10}$$

The consequence is that the deflection curve is nonlinear. The buckling collapse of shells under external pressure is sudden and explosivelike because small deformations increase the bending moments greatly due to the large compressive forces present.

2. *Inelastic Buckling*

Elastic buckling, Eqs. 8.6.3 and 8.6.6, prevails up to the proportional limit and is controlled by the elastic modulus, E. When the prebuckling stresses are between the proportional limit and the yield stress, the concept of using the geometric mean of the secant E_s, and tangent E_t, modulus, in lieu of the tangent modulus only, seems to best predict buckling of spheres in this region,[21,22] Fig. 8.7. Substituting

$$E = \sqrt{E_s E_t} \tag{8.6.11}$$

in Eq. 8.6.6 gives

$$p_p = 0.84\sqrt{E_s E_t}\left(\frac{h}{r_{\ell o}}\right)^2 \tag{8.6.12}$$

*Elastic buckling theory is based on mid-thickness radius; however, the basic premise in thin vessel analysis is that the difference between the inside radius and outside radius is nil. It is customary practice in pressure vessel design to use the radius of the pressure boundary surface for establishing both static equilibrium and buckling stability. Accordingly, when the applied pressure is external, the outside radius is used. Correspondingly, when the applied pressure is internal, Chapter 2, the pressure boundary is established by the inside surface dimensions and these are used for stress and stability evaluation.

where

E_s = secant modulus
E_t = tangent modulus
p_p = inelastic buckling pressure, psi.

Equation 8.6.12 reduces to Eq. 8.6.6 for prebuckling stresses at the proportional limit where $E = E_s = E_t$.

The basic control that construction codes and regulations utilize to insure safety against buckling is to restrict the initial maximum plus-or-minus radial deviation from the theoretical true circular form as a function of thickness and radius. The ASME Code Section VIII, Rules for Construction of Pressure Vessels, uses a similar variable arc length for establishing this allowable variation from the circular form, Par. 8.8.

3. *Residual Stress*

The fabricating processes of forming and welding create residual stresses.[23] Residual compressive stresses are important in establishing collapse pressures, hence they must also be evaluated. Since residual stresses are difficult to measure, it is seldom done, but their effect has been correlated with experimental collapse pressure by introducing a factor C_m, Table 8.2, into Eq. 8.6.6 to account for the residual compressive stresses as introduced by the particular method of fabrication:

$$p_c = 0.84EC_m \left(\frac{h}{r_{\ell o}}\right)^2 \tag{8.6.14}$$

Stress relieving does not completely remove the residual stress, and it can have the associated effect of lowering the yield strength value a

TABLE 8.2. APPROXIMATE EFFECT OF RESIDUAL COMPRESSIVE STRESS ON THE ELASTIC COLLAPSE PRESSURE OF SPHERICAL VESSELS

Method of Fabrication	*Approximate Residual Compressive Stress Reduction Factor, C_m*
Machined and Stress Relieved	0.95
Welded and Stress Relieved	0.85
As Welded	0.70

small amount. This is reflected in Table 8.2, which shows a 0.95 factor for a machined and stress relieved sphere.

4. *ASME Evaluation of Spheres and Formed Heads*

The ASME Pressure Vessel Code approach to the design of formed heads, such as hemispheres, under external pressure is similar to that for cylinders, Par. 8.5.2. It considers relatively thin heads in which failure is a consequence of buckling instability, and also relatively thick ones where failure starts with plastic deformation. The instability formula employed is based on that developed by Kármán and Tsien,[24] which appraises the sensitivity of small local disturbances and lower associated critical buckling forms, and for the ASME Code out-of-roundness and factor of safety is

$$P'_a = 0.0625E\left(\frac{t}{r}\right)^2 \tag{8.6.15}$$

This equation is similar to the classical solution Eq. 8.6.3, to which a suitable factor of safety must be applied.[25]

The material chart, Fig. 8.21, for determining the allowable external pressure, P_a, for cylinders can also be used for spheres or portions thereof. The material chart is plotted with $\sigma_c/2 = P_a D_o/t = B$ as the ordinate and with $\sigma_c/E = A$ as the abscissa, where σ_c is the critical stress in a cylinder at collapse and P_a is the maximum allowable external pressure for a cylinder. The abscissa for a sphere, noting that in Par. 8.5.2.3 a factor of safety of four was used, so $P'_c = 4P'_a$, is

$$\sigma'_c = \frac{P'_c r}{2t} = \frac{4P'_a r}{2t} \tag{8.6.16}$$

and substituting the value of P'_a from Eq. 8.6.15 in Eq. 8.6.16 gives

$$\frac{\sigma'_c}{E} = 0.125\left(\frac{t}{r}\right) = \text{FACTOR A} \tag{8.6.17}$$

in which σ'_c is the critical stress in a sphere at collapse and P'_a is the maximum allowable external pressure on the sphere. Further, since $\sigma'_c/E = 2\sigma_c/E$, FACTOR B must be adjusted by multiplying it by 2, so that

$$2B = P'_a D_o/t \tag{8.6.18}$$

$$P'_a = B\left(\frac{t}{r}\right) \tag{8.6.19}$$

The buckling of an ellipsoidal or torispherical head under external pressure is evaluated by considering it to be analogous to a sphere with radius equivalent to the maximum crown radius, Fig. 2.15.

Example: Determine the minimum thickness for a spherical vessel with a 92 inch inside radius subjected to an external pressure of 15 psi, operating at a temperature of 700°F, and fabricated from a carbon steel material with a minimum yield strength of 30,000–38,000 psi.

This involves a trial-and-error procedure. Select $t = 0.5$ in., and from Eq. 8.6.17:

$$A = 0.125\left(\frac{t}{r}\right) = \frac{0.125 \times 0.5}{92 + 0.5} = 0.00068$$

Enter Fig. 8.21 with a value of $A = 0.00068$ and a temperature of 700°F to read a value of $B = 7{,}000$. Substituting this value of B in Eq. 8.6.19 gives

$$P_a' = B\left(\frac{t}{r}\right) = \frac{7{,}000 \times 0.5}{92 + 0.5} = 37 \text{ psi}$$

This exceeds the pressure requirement of 15 psi and is satisfactory; however, if it is desirable to reduce this thickness, while maintaining the required pressure, this can be done by selecting a new, smaller thickness and repeating the above process. For instance, select a thickness of $t = \frac{5}{16}$ in. Then:

$$A = \frac{0.125 \times 0.3125}{92 + 0.3125} = 0.00042$$

$$B = 5{,}000.$$

and from Eq. 8.6.19, $P_a' = 17$ psi. This exceeds the required pressure; hence, the thickness of $\frac{5}{16}$ in. is satisfactory.

5. *Compressive Buckling Effect of Ring Stiffeners on the Stability of Ellipsoidal or Torospherical Heads Under Internal Pressure*

Ellipsoidal heads, and those closely approximating torospherical ones, Para. 2.6.4, are widely used in vessel construction because they offer a good compromise between a full hemisphere with its low stress but unused internal space and external site space requirements, and a flat head with its high bending and discontinuity meridional stresses. When this type of vessel is subject to internal pressure, however, high circumferential compressive (hoop) stresses develop in the knuckle region which can cause instability in the form of local meridional wrinkles, Fig. 2.13. In ellipsoidal heads under internal pressure a compressive stress occurs only when the major to minor axis ratio exceeds 1.42 (ratio of equator radius to crown height) Para. 2.6.4, and increases rapidly in the knuckle region as the head becomes shallower. It is relieved to a considerable extent when wrinkling occurs. Hence, the instability remains confined to the local knuckle region and is not catastrophic, as in the case of a spherical or cylindrical vessel under external pressure.

If a ring stiffener, Fig. 8.23*a*, is used to stabilize the region of high compressive stress, its effect on the vessel can be approximated by the deflection curve of a generator of an equivalent cylinder, Para. 4.7, as

$$y = \delta_0 e^{-\beta x} (\cos \beta x + \sin \beta x) \tag{8.6.20}$$

where δ_0 is the displacement of the vessel at the stiffener location, Para. 4.3, Fig. 8.23*b*. The bending moment variation will be of the

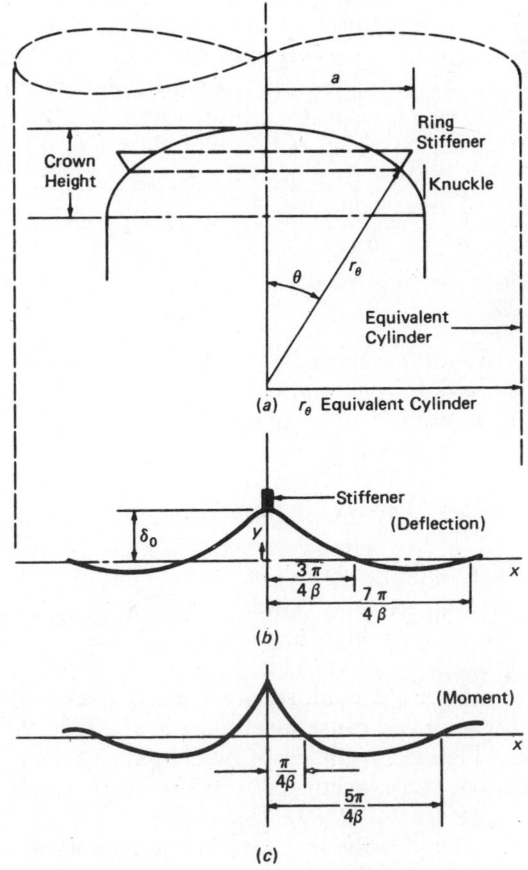

Fig. 8.23. Ellipsoidal Head with Ring Stiffener

type shown in Fig. 8.23*c*. The damping factor β is given by Eq. 4.5.5 when the radius r_θ of the equivalent cylinder is used, and gives a rapidly damped waveform of length, Para. 4.6.1,

$$L = \frac{2\pi}{\beta} = 4.9\sqrt{r_\theta h} \tag{8.6.21}$$

In the case of thin, large diameter vessels, such as liquid storage tanks, h is small in comparison to the radius r_θ and the bending produced at the ring stiffener is very local and has little effect on the critical load, Fig. 4.1 and Fig. 8.23*c*. Hence, for thin vessels a single discrete stiffener cannot reduce the circumferential compressive stress over a wide enough area to affect the buckling phenomenon. Two or more discrete rings would improve the situation slightly more, and so on with additional rings.[26] A substantial improvement requires that either the compressive stress be reduced, or the bending stiffness must be increased over the entire knuckle region. The latter can be accomplished in two ways:

A. Use sandwich wall construction when structural weight is critical, such as in aircraft.
B. Increase the solid wall thickness when structural weight is not critical.

The latter is usually employed in pressure vessel construction because ring stiffeners would have to be so closely spaced to be effective that it is more economical to use an increased wall thickness, Para. 2.6.4. The critical pressure will increase with the square of the wall thickness, Eq. 8.6.15; hence, substantial increases in wrinkling pressure occur with modest increases in knuckle wall thickness.

8.7 Local Buckling

Local elastic buckling is that involving only a small part of the member or component. An initial compressive residual stress remaining from the fabricating process, or the high stress concentration at the edge of an unreinforced opening in conjunction with the applied loading can result in such local buckling.

1. *Initial Local Buckling*

Initial local buckling is not always equivalent to total structural collapse. For instance, the basic premise in the design of ring stiffeners for cylinders is to prevent deflection normal to the cylinder

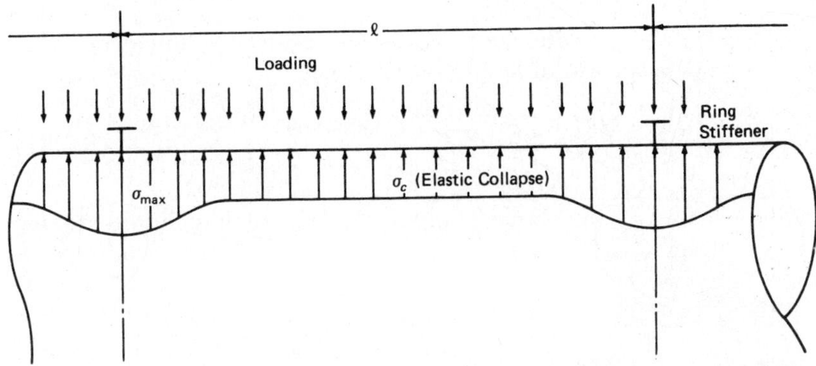

Fig. 8.24. Local Elastic Buckling of a Thin Cylinder Under External Pressure

surface and to first force buckling of the thin cylinder, Par. 8.5.1, Fig. 8.15. The load or stress at which this first occurs is not the maximum. As the load is increased above the critical value, the stress close to the more rigid stiffeners increases, Fig. 8.24. The critical stress in the mid-region where the bulges first form remains approximately constant as their deflection continues to increase. The result is an increase in the maximum load associated with buckling failure of the cylinder and is dependent upon how much deflection of these bulges may be permitted to develop. The maximum is assumed to be reached when the yield strength is reached in these more rigid regions. The margin between this maximum buckling strength and the normal buckling strength increases with decreasing critical stress in the mid-region, σ_c, and becomes considerable when σ_c lies below the proportional limit. On the other hand, this margin approaches zero when σ_c nears the yield strength.

2. *Effect of Openings, Unreinforced and Reinforced*

Openings are frequently used in vessels and these can cause local buckling.[33] When they are reinforced per Par. 6.6, there is no reduction in the buckling pressure over that in a similar unpierced vessel, and the same formulas are applicable. If this is not done, the effect of these openings must be considered in establishing the buckling collapse pressure, Fig. 8.25*a*. This is due to the free unstabilized edge of the hole being subjected to lateral deflection under the high local compressive stress concentration tangent to the free edge of the hole, Eqs. 5.5.2, 6.6.1, and 6.6.3. Figures 8.25*a* and *b* show the local na-

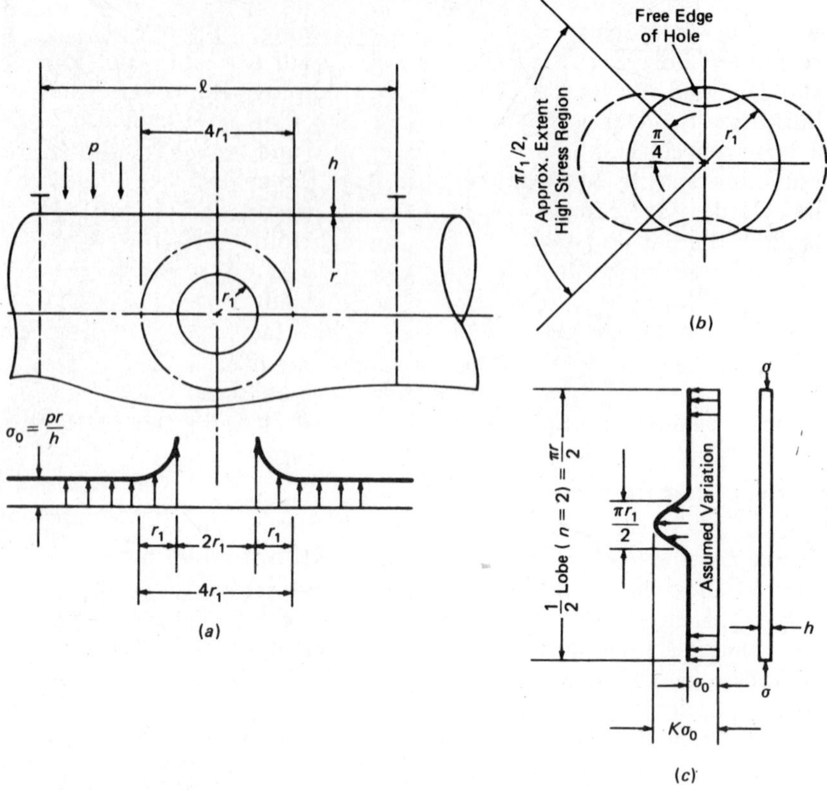

Fig. 8.25. Representation of Local Buckling in the Region of a Circular Unreinforced Opening in a Cylinder under External Pressure

ture of this high compressive stress which varies from a peak maximum at the longitudinal centerline to a nominal value at a distance $\pi r_1/4$ around the edge of the hole, Fig. 8.25*b*. For practical purposes, the nature of this high compressive stress concentration may be considered to extend a radial distance of one hole radius, r_1, from the edge of the hole much as was considered in the case of a tension field, Fig. 6.17. Hence the effect of a small hole in a large cylinder is nil, and for this reason many pressure vessel construction codes, such as those published by ASME, permit a small hole of radius up to a value[27]

$$r_1 = 0.1\sqrt{rh} \tag{8.7.1}$$

without added reinforcement for the opening. This permissible unreinforced hole size may vary somewhat with the applicable code or standard and the factors of safety they employ. However, Eq. 8.7.1 indicates the relative inconsequential effect of a very small hole in a large vessel. This is a conservative value and tests have shown no reduction in the buckling collapse load for unreinforced openings in cylinders subjected to axial, external pressure or a combination loading thereof up to $r_1 = .85\sqrt{rh}$, an eightfold margin of safety[44,45] over that permitted by ASME, Eq. 8.7.1. As the opening becomes larger, the length $\pi r_1/2$ of the free unstablized edge of the hole subjected to a high stress concentration (above a factor of 1) increases to a significant portion of the primary buckling half wavelength, $\pi r/2$, Fig. 8.25*c*, and so does the susceptibility to buckling. This condition becomes more pronounced when a series of closely spaced openings are present, in which case the acting stress must be adjusted by the ligament efficiency.

(a) *Thick Walls*

In heavy walled cylinders, say, with a ratio of $2r/h$ less than 20, Fig. 8.6, failure occurs in the weakest cross section when the yield strength of the material is reached. The effect of the hole is primarily one of loss of cross-sectional area as compared to local hole edge buckling. The critical elastic buckling load has no physical significance unless that critical value is reached before the yield strength of the material is developed.

(b) *Intermediate Walls*

In the intermediate thickness range of ring stiffened cylinders, the equivalent thickness, Eq. 8.5.8, may be approximated by introducing the ligament efficiency effect to compensate for the loss of cross-sectional area, Fig. 8.25(a), to give an effective thickness

$$h_e = \left(h + \frac{A}{\ell}\right) \text{lig. eff.} \tag{8.7.2}$$

For design purposes, the local hole effect may be compensated for by subtracting an additional distance of one radius from the hole edge to which region the high stress concentration is confined, Fig. 8.25*a*, and employing a pseudo-ligament efficiency $(\ell - 4r_1)/\ell$ in Eq. 8.7.2 to give

$$h_e = \left(h + \frac{A}{\ell}\right)\left(\frac{\ell - 4r_1}{\ell}\right) \tag{8.7.3}$$

Edge buckling at small rivet and bolt holes is not a practical concern because the clamping action of these fastener heads prevents lateral edge buckling. Hence, collapse is a function of the remaining ligament.

(c) Thin Walls

Port holes in the hulls of ships, openings in thin vessels, and perforated webs or batten plates in the compression members of lightweight aircraft structures must be reinforced by additional material or edge crimping or flanging to prevent this locally induced buckling, Fig. 8.25.

8.8 Imperfections and Their Effect on the Buckling Strength of Cylinders and Spheres

The design of vessels subjected to external pressure differs from those internally pressurized due to the buckling instability it produces. Further, in order to increase structural efficiency and reduce weight by raising these instability stresses, stiffeners are usually employed which add to the complexity of the stress pattern. Correlations of actual buckling tests of cylinders and spheres with theory have not been good. This is due in great measure to the pronounced effect of imperfections, such as out-of-roundness, because the departure from perfect circularity grows from the initial "as manufactured" value when the external pressure is increased from zero. Collapse will finally occur at the point of maximum initial deviation from circularity, hence the salient part fabrication precision plays in establishing the buckling pressure.[28,29] This effect increases under external pressure, while it decreases under internal pressure. The effect of small initial local random geometric imperfections, such as wrinkles or bulges, is influenced by the stiffness of the structure. In stiff structures, thick walled vessels, these imperfections have little influence and it is not necessary to take them into account. However, in flexible structures, thin walled vessels, imperfections can dramatically reduce the buckling load. For instance, small random imperfections, less than 20 percent of the wall thickness, in cylinders do not influence the buckling load; but imperfections much above this value reduce it 50 percent.[54,55,56]

The overall collapse mode is to be avoided in favor of the inter-stiffener buckling mode as a design objective. The overall collapse mode is very sensitive to shape imperfections; accordingly, it is not possible to design the stiffening to achieve a given minimum overall collapse pressure objective without a knowledge of the degree of out-of-roundness. Hence, it is mandatory that an applicable tolerance or standard be first established for design and then rigidly adhered to throughout construction. The ASME Boiler and Pressure Vessel Code requires that the difference between the maximum and minimum diameter of a section shall not exceed one percent of the nominal diameter for that section and is for the purpose of eliminating general out-of-roundness. A second requirement is that the maximum plus-or-minus radial deviation from the true circular form of a cylinder does not exceed that given in Fig. 8.26a, when measured over twice the arc length given in Fig. 8.26*b*. This is to detect bulges and flat spots that may have occurred during fabrication. In using Fig. 8.26a and *b*, the value of the design length is the distance between lines of support for cylinders, Fig. 8.13, and for spheres it is one-half the outside diameter of the sphere. The permissible deviation, *e*, is chosen so that the resulting collapse pressure will not be less than 80 percent of that of the corresponding perfect cylinder. The curves given in Fig. 8.26*a* represent an attempt to meet this requirement. They are based on the empirical formula:[30,31]

$$\frac{e}{t} = \frac{1.8}{n(100t/D_o)} + 0.15n \tag{8.8.1}$$

where e is the maximum permissible deviation from a true circle, D_o is the outside diameter, t is the thickness, L is the distance between support lines, and n is the number of lobes or waves in a complete circumference at collapse, Fig. 8.27*a*. While Eq. 8.8.1 is empirical and embodies experience factors, it is based on an analogy between a cylindrical pressure vessel and a column by considering the cylinder to be made up of a series of columns, one-half lobe length long, extending end-to-end circumferentially around the vessel, Fig. 8.27*b*. The pressure vessel analog of the slenderness ratio of a column is

$$\frac{\ell}{r_g} \,\hat{=}\, \frac{\text{constant}}{n(t/D_o)} \tag{8.8.2}$$

where r_g is the radius of gyration. The eccentricity of a column corresponds to the out-of-roundness of a pressure vessel and the eccentricity ratio e/r_g of a column has the pressure vessel analog

$$\frac{e}{r_g} \,\hat{=}\, \text{constant}\left(\frac{e}{t}\right) \tag{8.8.3}$$

since the radius of gyration of the cylinder thickness is $t/\sqrt{12}$. If the permissible eccentricity, e, is taken as that which will reduce the strength of a column or pressure vessel by a given percentage, 1%,

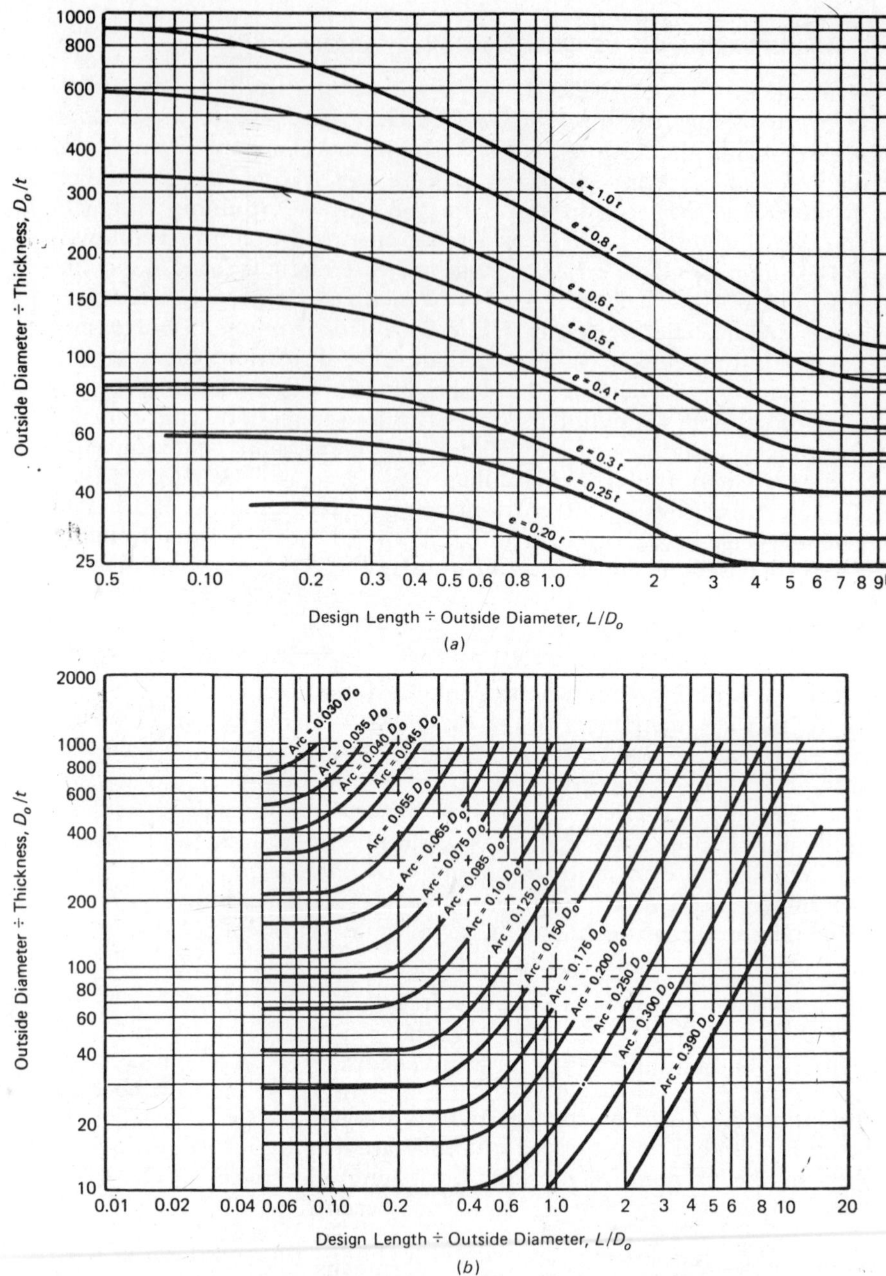

Fig. 8.26. (*a*) Maximum Permissible Deviation *e* from a Circular Form for Vessels under External Pressure. (*b*) Maximum Arc Length for Determining Deviation, *e*. (*Courtesy American Society of Mechanical Engineers*)

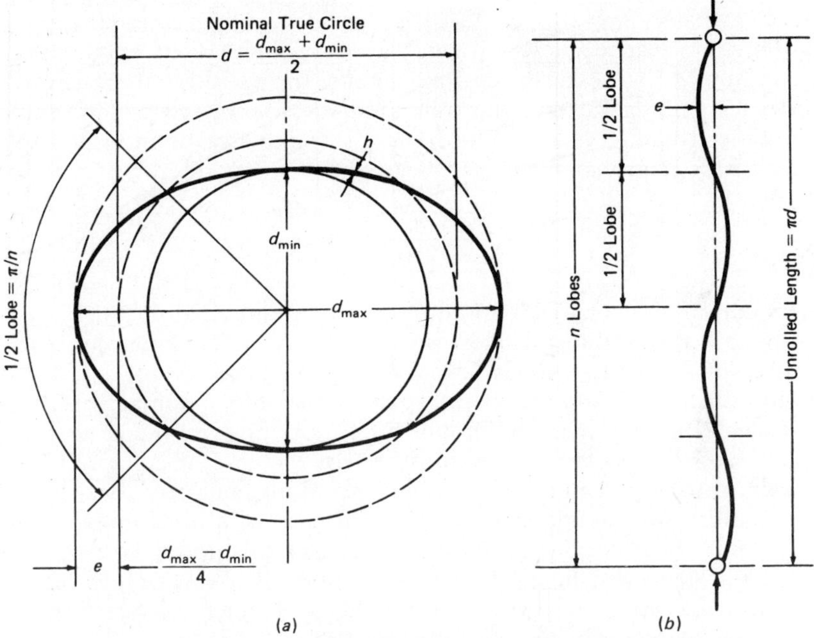

Fig. 8.27. Column Analog of Pressure Vessel Noncircularity

then for columns

$$\frac{e}{r_g} \propto \frac{\ell}{r_g} \tag{8.8.4}$$

and for pressure vessels

$$\frac{e}{t} = \frac{\text{constant}}{n(100t/D_o)} \tag{8.8.5}$$

Equation 8.8.5 can be represented on a logarithmic L/D_o-t/D_o plot, as in Fig. 8.26*a*, for vessels of intermediate length, but must be modified in the extreme regions for both long and short vessels. It has been found by experiment that a given eccentricity has less effect on the collapse pressure of a long cylinder than on a short one of the same thickness and diameter.[30] This means the allowable out-of-roundness decreases with decreasing length L between supports until a certain limit is reached; i.e., the e/t curve flattens and continues parallel to the L/D_o axis for small values of L/D_o. Likewise, on the other extreme when the length of the cylinder L is increased, a critical length is reached beyond which a further increase in length does not influence the collapse pressure; hence, the e/t curve flattens and continues parallel to the L/D_o axis as L/D_o is increased beyond this critical length, Fig. 8.26*a*. The basic Eq. 8.8.5 can be modified to reflect these requirements as

$$\frac{e}{t} = \frac{a}{n(100t/D_o)} + bn \tag{8.8.6}$$

The value of the constants a and b were then obtained to satisfy experimental test results, and provide a margin on buckling collapse pressure of 20 percent, to give Eq. 8.8.1, Fig. 8.26*a*.

The control of imperfections, such as, out-of-roundness, bulges, dents and flat spots, is paramount to forestalling the buckling collapse of vessels under external pressure. In the case of collapse of a cylinder into a high number of lobe formations, an initial local dent or bulge can approximate the size of one part of a lobe and be the equivalent of an eccentric column loading at this point. Hence, for short cylinders, which means a high number of lobes, simple variations in diameter are a less important factor than a local variation in the region of one lobe. In fact, the point of initial failure can be predicted from an inspection of the deformation diagrams depicting the exact initial radial profile form. Collapse will finally occur at the location where the initial variation in radial deviation is the greatest. Consequently, safety codes and regulations establish permissible variations in over-all roundness and local deviations from true circularity and require that these be verified by radial measurements taken around the circumference and at intervals along the length of the vessel, Fig. 8.26. These tolerances and inspections also serve as workmanship and quality control guides.[34,57]

8.9 Buckling Under Combined External Pressure and Axial Loading

The buckling of cylindrical vessels under the combined action of external pressure and compressive axial loads has received increased attention for floating platforms in the offshore oil industry and containment vessels in the nuclear industry.[35] Buckling under combined axial and radial loading is difficult to predict because failure is dependent on the way the loads are combined. Hence it is convenient for design purposes to estimate failure by the interaction method.[36,46]

1. *Interaction Method*

The interaction method is based on the following principles:

a. The strength under each individual loading condition (compression, bending, torsion, etc.) is first determined by analysis or tests.
b. The combined stress, or loading, condition is represented by the sum of ratios R_i, in which

$$R_i = \frac{\text{applied stress of type } i}{\text{failure stress of type } i} \qquad (8.9.1)$$

Failure stress refers to that at buckling, rupture, etc. As an example, for an axial compressive stress

$$R_a = \frac{\sigma_a}{\sigma_{k,a}} \qquad (8.9.2)$$

where:

R_a = stress ratio for axial loading
σ_a = applied axial stress
$\sigma_{k,a}$ = axial failure stress.

Failure under only an axial compressive stress condition is then

$$R_a = \frac{\sigma_a}{\sigma_{k,a}} = 1 \qquad (8.9.3)$$

c. The effect of one loading, represented by R_1, on the allowable value of another simultaneous loading, R_2, is then assumed to be given by

$$\sum_1^2 R_i^{\beta_i} = 1 \qquad (8.9.4)$$

The exponents β_i reflect the mathematical interaction between the combined stresses, or loads. Their values used in Eq. 8.9.4 should be verified by experiment or proven established engineering practice. The interaction equation for a cylindrical vessel under external pressure, and an additional axial load would be represented by

$$R_p^{\beta_p} + R_a^{\beta_a} = 1 \qquad (8.9.5)$$

where

$$R_p = \frac{\sigma_\theta}{\sigma_{k,\theta}} \tag{8.9.6}$$

$$R_a = \frac{\sigma_a}{\sigma_{k,a}} \tag{8.9.7}$$

and

σ_a = axial compressive stress in the cylinder including that due to external pressure
σ_θ = hoop stress in the cylinder due to external pressure
$\sigma_{k,a}$ = minimum predicted buckling stress in a welded fabricated cylinder in the axial direction a
$\sigma_{k,\theta}$ = minimum predicted buckling stress in a welded fabricated cylinder in the hoop direction due to external pressure.

Some simple interaction equations which follow from Eq. 8.9.5, and are shown in Fig. 8.28, are:

$$\beta_p = \beta_a = 1: \quad R_p + R_a = 1 \quad \text{(straight line)} \tag{8.9.8}$$

$$\beta_p = \beta_a = 2: \quad R_p^2 + R_a^2 = 1 \quad \text{(circle)} \tag{8.9.9}$$

$$\beta_p = 2, \beta_a = 1: \quad R_p^2 + R_a = 1 \quad \text{(parabola)} \tag{8.9.10}$$

These interaction curves terminate at the points (1, 0) and (0, 1).

2. *Interaction Equations in Design*

The minimum predicted buckling quantities $\sigma_{k,\theta}$ and $\sigma_{k,a}$ are prescribed in applicable codes and standards to which the structure is to be built and are based on failure test results; hence, they incorporate failure data consistent with the welding imperfections, residual stress, and fabrication tolerances of the final vessel.[37,38] The failure stress values used in the denominators of the stress ratios, Eqs. 8.9.6 and 8.9.7 are not the theoretical critical stresses for failure calculated for perfect elastic cylinders. Instead, the lower-bound curves of buckling tests are used for the particular loading condition being considered.[39] Safety factors are also introduced for design applications.

The prediction of buckling stresses for imperfect cylinders raises some problems, and different countries have different rules for determining them. A comparison of the buckling tests on welded steel

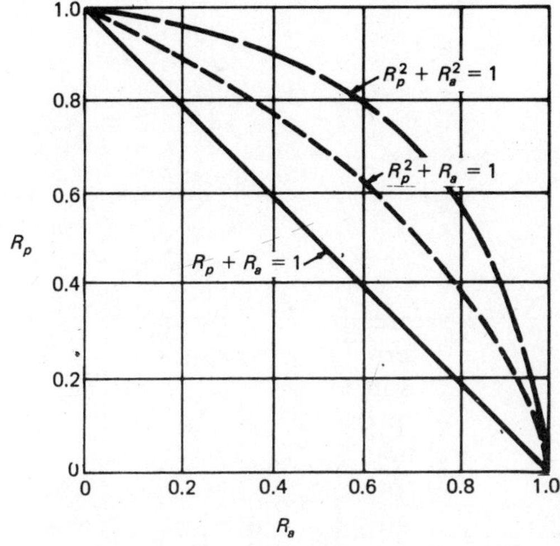

Fig. 8.28. Interaction Curves for Combined External Pressure and Axial Loading

cylindrical vessels under external pressure and axial loading in accordance with the following four codes is shown in Fig. 8.29:

1. DASt (Deutscher Ausschuss für Stahlbau) Richtlinie 013, "Beulsicherheitsnachweise für Schalen," Cologne, Germany.
2. DnV (Det norske Veritas) "Rules for the Design, Construction and Inspection of Offshore Structures", Appendix C, Steel Structures, Oslo, Norway.
3. ECCS (European Convention for Construction Steelwork) "European recommendations for steel construction, Section 4.6, Buckling of shells," The Construction Press, London.
4. ASME (American Society of Mechanical Engineers) "Boiler and Pressure Vessel Code Case N-284, Metal containment shell buckling design methods," Section III, Div. 1, Class MC, New York, New York.

These codes have interaction relations which are similar to Eqs. 8.9.8, 8.9.9, and 8.9.10. For instance, Eq. 8.9.8 is similar to that suggested by the ECCS recommendations. Equation 8.9.9, multiplied by a factor, is used by DnV. The DASt Code is similar to Eq. 8.9.8 but has exponents of 1.1 instead of 1.0 for β_p and β_a. The ASME Code

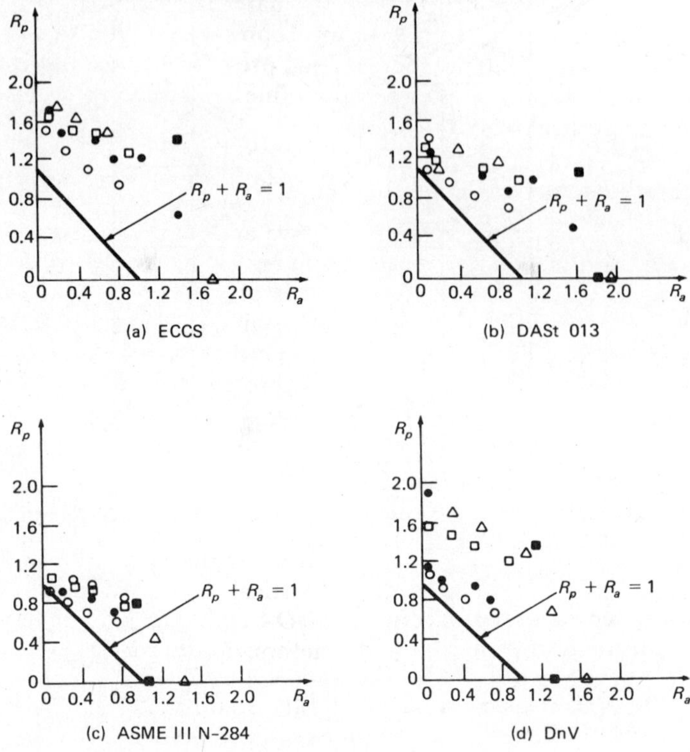

Fig. 8.29. Elastic Buckling Tests of Welded Cylinders[40] Plotted in Terms of R_p and R_a. According to Four Different Codes[41]

Case N-284 interaction curve is a combination of straight lines and parabolas that basically conforms to Eq. 8.9.10.

The difference in the results from the codes is sometimes considerable, as shown in Fig. 8.29. The linear interaction relation, Eq. 8.9.8, used in conjunction with the buckling stresses of imperfect cylinders (under external radial pressure alone, and compressive axial load alone) gives a conservative approximation of the lower bound of these buckling tests, and this case is shown on all the experimental data plots of Fig. 8.29 as a reference conservative lower bound. In contrast, the parabolic relation, Eq. 8.9.10, seems to offer the best fit of the available data and is close to the ASME recommendation.[39] The present codes (ASME III N-284, DASt 013, DnV, and ECCS) give conservative predictions. Refinements of these basic interaction equations to account for differences in material type, yield strength, and methods of stiffening have been proposed by Miller[42] and Odlund,[43] and tests of combined external pressure and axially loaded cylinders are continuing in order to refine construction and safety codes for these structures.[44]

3. *Safety Factors in Design*

Figure 8.30 shows an interaction curve and represents all the conditions R_p and R_a that will cause failure. Thus, if a value of R_p is known to be acting, the maximum value of R_a that can be simultaneously applied without failure can be obtained from this interaction curve or equation. Each end point of the interaction curve represents a single loading condition. If an applied loading condition is below the interaction curve, point *A*, Fig. 8.30, the factor of safety can be found by projecting a line from the origin through this applied stress point to intersect the interaction curve at the critical stress point. If the stress ratios R_p and R_a are known to be proportional, the factor of safety is then represented by OB/OA. However, if one loading condition is known to remain constant while the other increases, the projection line will be horizontal or vertical, as shown in Fig. 8.30, to give a factor of safety of $(OA + AC)/OA$ or $(OA + AD)/OA$.

Safety factors, or knock-down factors, are customarily used in design to take care of uncertainties in load, material properties, size

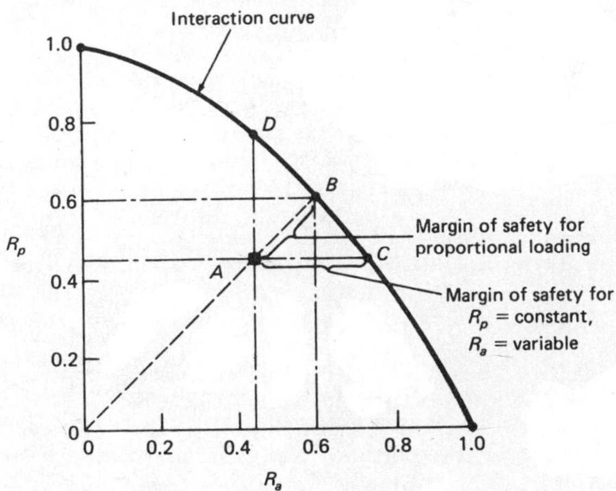

Fig. 8.30. Schematic Representation of the Interaction Method of Establishing Safety Factors

effects, etc., and are usually introduced into the interaction equation. Denoting the safety factor by SF, and assuming the safety factor is the same in the axial and hoop directions, the linear interaction equation for design becomes

$$R_p + R_a = \frac{1}{\text{SF}} \tag{8.9.11}$$

or

$$\frac{\sigma_\theta}{\left(\dfrac{\sigma_{k,\theta}}{\text{SF}}\right)} + \frac{\sigma_a}{\left(\dfrac{\sigma_{k,a}}{\text{SF}}\right)} = 1 \tag{8.9.12}$$

References

1. F. B. Seeley and J. O. Smith, *Advanced Mechanics of Materials*, John Wiley & Sons, Inc., New York, 1960.

2. S. P. Timoshenko, *Strength of Materials: Part II. Advanced Theory and Problems*, D. Van Nostrand Co., Inc., New York, 1956.

3. J. P. Den Hartog, *Advanced Strength of Materials*, McGraw-Hill Book Co., Inc., New York, 1952.

4. C. G. Fostor, "Interaction of Buckling Modes in Thin-Walled Cylinders," *Experimental Mechanics*, March 1981.

5. H. E. Saunders and D. F. Windenburg, "Strength of Thin Cylindrical Shells Under External Pressure," ASME Pressure Vessel and Piping Design, Collected Papers, 1960.

6. B. G. Johnston, *Guide to Stability Design Criteria for Metal Structures*, John Wiley & Sons, New York, NY 1976.

7. F. A. Simonen and R. J. Shuppell, Jr., "Collapse of Thick-Walled Cylinders Under External Pressure," *Experimental Mechanics*, February 1982.

8. S. P. Timoshenko and J. M. Gere, *Theory of Elastic Stability*, McGraw-Hill Book Co., Inc., New York, 1961.

9. D. F. Windenburg and C. Trilling, "Collapse by Instability of Thin Cylindrical Shells Under External Pressure," ASME Pressure Vessel and Piping Design, Collected Papers, 1960.

10. R. G. Sturm, "A Study of the Collapsing Pressure of Thin-Walled Cylinders," Bulletin 329, Engineering Experiment Station, University of Illinois, 1941,

11. L. E. Brownell and E. H. Young, *Process Equipment Design: Vessel Design*, John Wiley & Sons, New York, 1959.

12. E. Wenk, Jr., "Feasibility of Pressure Hulls for Ultradeep Running Submarines", ASME Publication Paper No. 61-WA-187, 1961.

13. C. Garland, "Design and Fabrication of Deep-Diving Submersible Pressure Hulls", *Transactions, The Society of Naval Architects and Marine Engineers*, Vol. 76, 1968.

14. "Rules for Construction of Pressure Vessels," Section VIII, ASME Boiler and Pressure Vessel Code.

15. E. O. Bergman, "The New Type Code Chart for the Design of Vessels Under External Pressure", ASME Pressure Vessel and Piping Design, Collected Papers, 1960.

16. S. S. Gill, *The Stress Analysis of Pressure Vessels and Pressure Vessel Components*, Pergamon Press, Oxford, 1970.

17. J. Morton, P. R. Murray and C. Ruiz, "On the Buckling Design of Spherical Shells," ASME Publication Paper No. 81-PVP-32, 1981.

18. R. G. Sturm, L. W. Smith and H. L. O'Brien, "Allowable Eccentricity of Spherical Heads Convex to Pressure," ASME Pressure Vessel and Piping Design, Collected Papers, 1960.

19. T. J. Kiernan and K. Nishida, "The Buckling Strength of Fabricated HY-80 Steel Spherical Shells," David Taylor Model Basin Report 1721, July 1966.

20. A. Blake, "A Practical Approach to Spherical Shell Design," *Machine Design*, February 26, 1976.

21. M. Esslinger and B. Geir, *Postbuckling Behavior of Structures*, Springer-Verlag, New York, 1975.

22. M. A. Save and C. E. Massonnet, *Plastic Analysis and Design of Plates, Shells, and Disks*, North-Holland Publishing Co., Amsterdam, 1972.

23. S. R. Heller, "Structural Design of Spherical Shells Subject to Uniform External Pressure," *Naval Engineers Journal*, Washington D.C., December 1974.

24. Th. Von Kármán and Hsue-Shen Tsien, "The Buckling of Spherical Shells by External Pressure," ASME Pressure Vessel and Piping Design, Collected Papers, 1960.

25. C. E. Washington, R. J. Clifton, and B. W. Costerus, "Tests of Torispherical Pressure Vessel Heads Convex to Pressure," Welding Research Council Bulletin 227, June 1977.

26. C. R. Steele and G. V. Ranjan, "Effect of Stiffening Rings on the Stability of Torospherical Heads," ASME Publication, PVP-Vol. 57, 1982.

27. E. C. Rodabaugh and R. L. Cloud, "Proposed Reinforcement Design Procedure for Radial Nozzles in Cylindrical Shells with Internal Pressure," Welding Research Council Bulletin 133, September 1968.

28. G. D. Gallently, "Buckling of Fabricated Cylinders Subjected to Compressive Axial Loads and/or External Pressure—A Comparison of Several Codes," ASME Publication PVP-Vol. 57, 1982.

29. R. L. Citerley, "The Use of Manufacturing Tolerance in Imperfection Analysis of Shells," ASME Publication PVP-Vol. 57, 1982

30. D. F. Windenburg, "Vessels Under External Pressure," ASME Pressure Vessel and Piping Design, Collected Papers, 1960.

31. M. Holt, "A Procedure for Determining the Allowable Out-of-Roundness for Vessels Under External Pressure," ASME Pressure Vessel and Piping Design, Collected Papers, 1960.

32. E. Tschoepe and J. R. Maison, "The External Pressure Collapse Tests of Tubes," Welding Research Council Bulletin 284, April 1983.

33. D. Bushnell and E. Meller, "Elastic-Plastic Collapse of Nonuniformly Axial Compressed Ring-Stiffened Cylindrical Shells With Reinforced Openings," ASME Publication #H00262, 1983.

34. J. R. Fowler, E. F. Klementich, and J. F. Chappell, "Analysis and Testing of Factors Affecting Collapse Performance of Casing," ASME Publication Paper 83-PET-12, 1983.

35. C. D. Miller, P. A. Frieze, R. A. Zimmer, and H. Y. Jan, "Collapse Test of

Fabricated Stiffened Steel Cylinders Under Combined Loads," ASME Publication 83-PVP-7, 1983.

36. F. R. Shanley, *Strength of Materials*, McGraw-Hill Book Company, Inc., 1957.

37. S. Kendrick, "The Technical Basis of the External Pressure Section of BS-5500," ASME Publication Paper 83-PVP-6, 1983.

38. S. K. Tsang, J. E. Harding, A. C. Walker, and L. Andronicou, "Buckling of Ring Stiffened Cylinders Subjected to Combined Pressure and Axial Compressive Loading," ASME Publication Paper 83-PVP-4, 1983.

39. D. Faulkner, Y. N. Chen and J. G. DeOliveira, "Limit State Design Criteria for Stiffened Cylinders of Offshore Structures," ASME Publication Paper 83-PVP-8, 1983.

40. C. D. Miller, "Summary of Buckling Tests on Fabricated Steel Cylindrical Shells in USA," Paper No. 16, Buckling of Shells in Offshore Structures, (eds.) J. E. Harding, P. J. Dowling, and N. Agelidis, Granada, London, 1982.

41. G. D. Galletly and K. Pemsing, "On Design Procedures for the Buckling of Cylinders Under Combined Axial Compression and External Pressure," ASME Publication Paper 83-PVP-5, 1983.

42. C. D. Miller, "Buckling Design Methods for Steel Structures—A State of the Art," Paper 24, Proc. 2nd Inter. Symp. on Integrity of Offshore Structures, Glasgow University, Applied Science Publishers, London and New Jersey, 1981.

43. J. Odland, "On the Strength of Welded Ring-Stiffened Cylindrical Shells Primarily Subjected to Axial Compression," University of Trondheim, Norway, 1981.

44. C. D. Miller, "Experimental Study of the Buckling of Cylindrical Shells with Reinforced Openings," *Advances in Containment Design and Analysis*, ASME/ANS Joint Conference, Portland, Oregon, July 26–28, 1982.

45. S. Toda, "Buckling of Cylinders With Cutouts Under Axial Compression," SESA-JASME Joint Conference held in Oahu and Maui, HI, May 23–30, 1982.

46. C. G. Foster, "Interaction of Buckling Modes in Thin-Walled Cylinders", *Experimental Mechanics*, March 1981.

47. E. Chater and J. W. Hutchison, "On the Propagation of Bulges and Buckles," ASME Publication Paper 84-APM-18, 1984.

48. D. Bushnell, "Survey of Collapse Analysis Methods," ASME Publication Volume H00298, 1984.

49. E. Riks, "Progress in Collapse Analysis," ASME Publication Volume H00298, 1984.

50. E. G. Carnoy, I. Duvernoy and G. Panosyan, "Elasto-Plastic Buckling Analysis of Cylindrical Shells Under Axial Compression," ASME Publication Volume PVP-Vol. 89, 1984.

51. A. Combescure and R. Bastien, "Buckling Analysis of Imperfect Cylinders Under Axial Compression," ASME Publication Volume PVP-Vol. 89, 1984.

52. A. Combescure, A. Hoffman, J. Devos, and G. Baylac, "A Review of Ten Years of Theoretical and Experimental Work on Buckling," ASME Publication Volume PVP-Vol. 89, 1984.

53. A. Combescure, "Design Against Elasto-Plastic Buckling of Shells—Proposition of a Methodology," ASME Publication Volume PVP-Vol. 89, 1984.

54. A. Morel, R. Del Beccaro, and G. Baylac, "Design and Fabrication Studies for Large French LMFBR Components," ASME Publication Volume PVP-Vol. 89, 1984.

55. N. Wackel, J. F. Jullien, and P. Ledermann, "Experimental Studies on the Instability of Cylindrical Shells with Initial Geometric Imperfections," ASME Publication Volume PVP-Vol. 89, 1984.

56. N. Waeckel, M. Cousin, A. Combescure, and E. Carnoy, "Buckling of Imperfect Cylinders—A Synthesis," ASME Publication Volume PVP-Vol. 89, 1984.

57. D. Moulin, "Appraisal of Allowable Loads by Simplified Rules," ASME Publication Volume PVP-Vol. 89, 1984.

INDEX

Numbers Refer to Articles